Tedd R Coffman

Methods of Geological Engineering

in Discontinuous Rocks

RICHARD E. GOODMAN

Professor of Geological Engineering
University of California, Berkeley

WEST PUBLISHING COMPANY

ST. PAUL • NEW YORK • BOSTON • LOS ANGELES • SAN FRANCISCO

Printed in the United States of America
Library of Congress Cataloging in Publication Data

Goodman, Richard E.
Methods of geological engineering in discontinuous rocks.
Includes index.
1. Rock mechanics. I. Title.

TA706.G66 624'.1513 75–42152

ISBN 0–8299–0066–7

preface

This work contains material from my courses at Berkeley in Engineering Geology and Applied Rock Mechanics. It could serve as a supplementary source for general courses in these fields, as well as a text in a more specialized course in geological engineering analysis. I have not tried to separate the disciplines of geology and civil engineering as I think it unnatural to do so. The geologist can decide on the relevancy of the features he maps and describes only if he understands how they relate to the analytical process. The engineer can not handle the geological data correctly without the respect for its "fuzziness" that comes from a personal acquaintance with geology.

Interest in geological engineering methods has occupied me since I visited the sites of the Malpasset and Vajont failures in the early 60's. I wanted to know if a person with reasonable education and experience in engineering geology could have foreseen the difficulties. It seemed to me that objective, rational procedures for evaluating such sites were inadequate and that we were, perhaps, relying too heavily on intuition. Wisdom derived from real experiences will always be an important and necessary ingredient for predicting rock behavior; but we must also search for objective tools. Since 1960, a number of such tools have appeared and it is timely and useful to set them forth for students, and for practicing technicians who can apply them in their work.

Most of this Monograph was written at Imperial College, London, during the term of a Guggenheim Fellowship, while I was on a sabbatical leave from Berkeley. I wish to thank the John Simon Guggenheim Foundation and the Regents of the University of California for this opportunity to reflect on the whole of the field of geological engineering methodology. In residency at the Royal School of Mines during 1973, I was privileged to frequent discussions with Professor Evert Hoek, Dr. John Bray, John Boyd, and other faculty members. And I profited from the work of a number of their students, past and contemporary, including Peter Cundall, Christopher St. John, Nick Barton, Tidu Maini, John Franklin, John Sharp, Laurie Richards, Dermot Ross-Brown, Peter Riley, Ross Hammett, Peter Kelsall, Don Moy, Tim Harper, and Graeme Major. I am particularly indebted to Dr. St. John who prepared the special finite element program listed in Appendix 1. We were anxious to have a small finite element program written expressly for a student trying to bridge between theory and application.

In the text, I have indicated the sources for ideas by references to the works of numerous authors. I have profited from personal communications over the years with a number of these authors - - Walter Wittke, Klaus John, Pierre Londe, and "Skip" Hendron with respect to limit equilibrium analyses - - Dr. Leopold Müller and Karel Drozd with respect to physical models - - E. J. Polak, Tor Brekke, and Dan Moye with respect to geological and geophysical exploration - - and Robert Taylor, Edward Wilson, Hugh Trollope, and Ann Bornstein with respect to computer methods. I have also benefitted from the work of past and present students at Berkeley including Yuzo Ohnishi, P. N. Sundaram, Ashraf Mahtab, Rudolfo de la Cruz, John Cadman, Jacques Dubois, Alain de Rouvray, and Francois Heuze, and of post Doctoral scholars Kemal Erguvanli, Jean Luc Dessenne, and Karel Drozd. The critical comments and suggestions of Prof. Arvid Johnson of Stanford University, and Prof. Hendron and his colleagues at the University of Illinois were quite helpful.

A number of persons kindly loaned materials or gave me permission to refer to their work. These include Nick Barton, Z. T. Bieniawski, Tor Brekke, W. Chinn, Lloyd Cluff, Phillip Cole, James Coulson, Walter Day, William Dearman, G. Everling, Irving Fatt, Alena Gralewska-Vickery, Richard Hay, Francois Heuze, Larry James, Dennis Lachel, Branko Ladanyi, Thomas Lang, Pierre Londe, Ken Matthews, J. Myung, Carlos Ospina, Marc Panet, M. Popovic, Howard Pratt, Hernando Quijano, Dorothy Radbruch, Nick Rengers, Manuel Rocha, Fritz Rummel, F. Sabarly, Ed. Slebir, Gerardo Tama, Jose Tejada, Ruth Terzaghi, Cl. Tourenq, Lloyd Underwood, and Joel Verdier.

The following organizations generously permitted me to refer to or borrow their materials: Atlas Copco ABEM (Stockholm), Bergbau Forschung (Essen), Birdwell Division of Seismograph Service Corp. (Tulsa), California Department of Water Resources (Sacramento), Christensen Diamond Products Co. (Salt Lake City), Coyne and Bellier (Paris), Engineering Laboratory Equipment Ltd. (Hemel Hempstead, U. K.),

Golder and Brawner and Assoc. (Vancouver), Ingetec Ltda. (Bogota), Integral Ltda. (Medellin), Laboratoire des Ponts et Chaussees (Paris), Laboratorio Nacional de Engenharia Civil (Lisbon), Longyear Co. (Minneapolis), Joy Manufacturing Co. (Montgomeryville, Pa.), Mindrill Ltd. (Melbourne), Norwegian Geotechnical Institute (Oslo), Soil Mechanics Equipment Co. (Glen Ellyn, Ill.), Sprague and Henwood, Inc. (Scranton, Pa.), Tacoma City Light, TerraTek (Salt Lake City), U. S. Army Corps of Engineers (Libby Resident Office; Missouri River Division, Omaha District, and Explosives Excavation Research Lab.), U. S. Bureau of Mines, (Denver), U. S. Bureau of Reclamation (Denver), U. S. Geological Survey (Menlo Park), Woodward Clyde and Associates (Oakland), and Zavod Za Geotehniku I Fundiranje (Sarajevo).

Finally, I wish to thank the persons who helped me with the work of producing the manuscript: Fran Riley, Laurie Wilson, Gloria Pelatowski, and Lillian Goodman.

to the memory of

Parker D. Trask

contents

Methods of Geological Engineering in Discontinuous Rocks

1 introduction

This book discusses methods and procedures available to assess the influence of discontinuities on the behavior of rocks in engineering applications. Most rock masses in the region of influence of works such as quarries, road cuts, foundations, dam abutments, tunnels, and underground chambers contain planar surfaces of potential or real weakness. These weakness planes come in all lengths and spacings and have varying degrees of influence on the overall mass properties. We rarely can afford to close our eyes to their presence in attempting to calculate rock performance.

Use of the phrase "discontinuous rocks" in the title implies that there are other rocks which are truly continuous. This is not strictly correct for even the mightiest wall of granite has exfoliation surfaces and other widely spaced joints and faults in various orientations. There are many rock masses, however, in which the discontinuities though present are not the weakest link in the list of components which collectively give the rock its strength and other physical attributes. In friable sandstones of Tertiary age, for example, the sand grains may be so poorly bonded that failure through the rock material itself is more likely than failure by sliding on bedding planes or joints. This may also be the case in shales which though disrupted and loosened by anastamosing cracks due to slaking near exposed surfaces, will tend to fracture through the body of material rather than on structurally controlled surfaces. The younger,

Figure 1-1. A discontinuous rock mass. Columnar joints and flow banding in a basaltic flow—Iceland; (courtesy of Dr. Tor Brekke).

weaker rocks as a rule tend more closely to fit a "continuous model", while hard rocks invariably are controlled in their failure modes by their pattern of discontinuities. How else could a rock like quartzite fail? It has a compressive strength some ten times that of mass concrete.

Soils are not continuous materials; they have grains and pores. But they have been successfully analyzed using a continuum model -- heterogeneous, if necessary, but continuous. The discontinuous rocks with which we are concerned here might at first thought be likened to soils and treated using soil mechanics theory and techniques. In fact, some early attempts were made in this direction. But there are fundamental differences. The discontinuous rocks have essentially no pore space, except that of the rock material itself (pores in the rock material are analogous to pores in the grains of the soil). Thus the discontinuous rock is locked together into a perfectly fitted pattern. To create failure, pore space must be created and this implies dilatancy, or bulking in the construction man's parlance. Not only normal and shear forces act inside such rock masses, but moments as well. Soil grains may be free to turn in place; rock blocks are not.

It might seem a hopeless quest to rationalize the design process when dealing with such a material as discontinuous rock. Sometimes it is hopeless, and only previous experience, or trial and error, can be used. Other times, fortunately more frequently as we gain experience, the network of discontinuities can be accurately described and mapped and its influence on the mass behavior can be adequately evaluated. The elaboration of these methods is the subject of this book. First we must measure the orientations of the various sets of planes which penetrate the rock in question. This can be done by geological observations on outcrops, by inspection of natural and artificial cuttings, by study of aerial photographs, by measurements on drill cores and the walls of exploratory borings, and by geophysical traverses using a number of available techniques. Chapter four reviews some of the relevant methods of exploration aimed at providing a description of the orientation and spacing of the discontinuity network in a rock body. Chapter five then introduces the mechanical properties of surfaces of discontinuity and considers their measurement and numerical values. When we are dealing with single, very important weakness surfaces, whose orientation and position with respect to a project are known with precision, it is possible to make explicit analyses of the resulting stresses and deformations; this can be done by kinematics and statics, using stereographic projection to handle the three dimensional aspect of the problem, as discussed in chapter 6. In chapter 7, physical model methods are introduced. An emphasis is placed on kinematical models which examine the various possible modes of failure of a discontinuous rock mass in an engineering context. Analyses can also be performed by numerical methods; the finite element method is introduced in chapter 8 and a digital computer program, designed so that it can be read along with the theoretical discussion of chapter 8, is presented in the Appendix. Because stereographic projection procedures are used frequently throughout the book as a means of solving spatial problems, such as orienting planes in drill core, measuring angles on terrestrial photographs, resolving stresses on planes of given orientation, and operating with vectors, a chapter has been addressed specifically to techniques of stereographic projection (chapter 3). In dealing with vector quantities, we must use the whole sphere so the subject is treated somewhat differently than in works on structural geology. Chapter 2, on classification of rock, has been written to relate the discontinuous rocks to other categories of rock, i.e. to set this work in its proper context.

Geological engineering is concerned with a broad spectrum of natural processes. At one end of the spectrum are those geologic hazards, such as large landslides, active faults, and cavernous terrain, which dwarf an intended project in terms of size, potential energy, or the cost of neutralizing the hazard; with such hazards, the geological engineer can do little more than recognize, describe and be responsive to eventualities. He uses various methods to study their potential and to observe their activity, but he has little effect on the phenomena themselves. At the other end of the spectrum of geological engineering applications are mining and quarrying activities where the geology is not only studied and evaluated, but

Figure 1-2. A concrete arch dam. Mossyrock Dam, Cowlitz River, Washington: a doubly curved, thin arch dam 365 feet above riverbed, 606 feet above the basalt bedrock; (courtesy of Tacoma City Light).

Figure 1-3. Malpasset Dam site, looking at the left abutment and into the reservoir area.

wherein the rock is removed, crushed, stockpiled and perhaps even emplaced in a hostile and caustic environment, for example, as aggregate in cement. In between are those constructions and excavations, such as dams, underground openings and open cuts, which apply static or dynamic loads or unloads at the surface or subsurface. It is with these that the methods discussed in this book are primarily concerned.

Large dams, especially concrete arch dams as in figure 1-2 combine large loads with the hydraulic and chemical effects of water and therefore place challenging demands on geological engineering investigations. Much of the recent interest in geological engineering and rock mechanics has in fact been motivated by concern about the safety of dams and reservoirs following the catastrophes at Malpasset dam in France and Vajont reservoir in Italy. At Malpasset dam, figure 1-3, a complicated set of circumstances deriving from the behavior of the schistose gneiss bedrock caused a rupture of an arch dam. French investigators determined that a wedge of rock in the abutment, bounded by intersecting weakness surfaces, moved due to the thrust of the dam and high water pressure within the abutment (Bernaix, 1966)*. The high water pressure was generated by the development of

---

* References will be found in the Bibliography, on page 419

Figure 1-4. Kukuan Dam, during construction; (courtesy of Coyne and Bellier).

a natural flow barrier under the line of action of the dam as fissures within the rock mass closed in response to applied load. The Vajont failure (Müller, 1964 and 1968) occurred when a massive landslide moved on bedding surfaces into a relatively small reservoir, causing overtopping and flooding. The landslide was triggered by uplift forces associated with reservoir filling.

The large influence of discontinuities on construction operations in rock is well illustrated by the Kukuan arch dam, designed by Coyne and Bellier for Taiwan Power Company. This dam, 86 meters high, was constructed in a valley cut 500 meters deep into alternating layers of slate and quartzite. Thin clay seams containing graphite compromise the stability of unfavorably oriented layers of the site. The right bank (figure 1-4) is a 60 - 70 degree dip slope. To found the dam in solid rock, it was necessary to excavate through 20 to 40 meters of loosened slabs, but conventional excavation was undesirable because of the slide potential. Grouting and "dental work" (localized replacement of weak rock with concrete) were unsuccessful. A solution was obtained in which tunnels up to 10.7

meters wide were driven well into the abutments and backfilled with concrete. Since the tunnels cut across the bedding, they were stable. After driving a tunnel to the full depth and width, it was concreted to within several meters of the crown. Then, after two to three weeks, a stone protection was laid on the concrete fill and a tunnel was excavated above. The process was repeated until eight tunnels had been constructed, producing a stable concrete structural abutment.

Activities in advancing the construction and utilization of tunnels and underground chambers have also created interest in methods of geological engineering. Investigations of tunnel sites remain fairly primitive because the sites are long, and remain inaccessible until construction. Some attention has been focused on assessing the excavatability of the rock from tests on samples, but geological and geophysical prediction techniques, and analytical methods to forecast formation conditions are not yet generally available.

Techniques for investigating and analyzing rock behavior for underground works such as subsurface power plant chambers (figure 1-5),

(a)

(b)

Figure 1-5. (a) Oroville Dam project. The dam has shells of gravel while the core is derived from a vast alluvial fan; (courtesy Calif. Dept. of Water Resources).
(b) Oroville underground power station machine hall during construction. The man standing in the lower left gives the scale; (courtesy Calif. Dept. of Water Resources).

Figure 1-6. Spillway excavation on left abutment of Chivor Dam, Colombia. Notice the truck and shovel for scale. The smooth surface of discontinuity in the middle left was exposed during construction and caused a design change. The benches are 5 meters wide and spaced every 10 meters; (courtesy of Ingetec Ltda., Bogota).

subterranean factories, defense installations, storage chambers, and mine shafts, on the other hand, are better developed. It is usually feasible in such projects to make detailed investigations including determination of rock properties, analysis, and instrumentation. An additional aspect of investigations for underground structures not addressed in investigations for dams, is the role of in-situ stresses. At great depth, such as in some mines in South Africa and Canada, one occasionally reaches the natural strength of the rock.

Surface excavations for spillways (figure 1-6), mine pits (figure 1-7), transportation routes, power plants, and for access to the underground, are other important areas of rock engineering. In mines, important savings in excavation volumes can be achieved by application of simple theory supported by field observations of geological details, back calculations of failures, piezometric measurements, and analysis of the response of instruments (Hoek and Bray, 1974). Careful blasting practise and instrumentation can insure safe operation of engineering works immediately adjacent to rock slopes, which themselves can be regarded as engineering structures (figure 1-8).

Though the specific choices of methodology will differ among all these types of projects, basic similarities of purpose prevail.

Figure 1-7. Chambishi Mine, Zambia (courtesy R.S.T. Ltd. and Prof. E. Hoek).

First, the geology of the site must be defined; this entails mapping of field exposures, study of aerial photographs, and specific exploration with excavations or drill holes. Then, the properties of the rocks must be assessed. Here there can be different choices of

Figure 1-8. Pre-split rock excavation for Stockton Dam; (courtesy Mr. Lloyd Underwood, Corps of Engineers, Missouri River Division).

methods since the relevant properties to be evaluated vary greatly according to the purpose of the project. The behavior of a complex of underground openings reflects the initial state of stress; in some analytically based design processes the in-situ stress will need to be measured, or otherwise determined. In shallow rock excavations, on the other hand, the shear strength and water pressure levels are more critical, while for foundations the deformability of the rock is foremost. Thirdly, through model studies, computer analysis, or reference to appropriate similar experiences, the response of the work at the specific site with the assigned properties is evaluated. If unsatisfactory, the structure may be relocated or the properties may be changed in some measure by excavation, grouting, drainage, bolting, or other means. In this case new explorations, tests, and studies will be inaugurated. The designer will have the most economical solution if he is able to adapt the style of structure to the particular attributes of the site, most of which have been provided naturally. The methods and work of geological engineering are therefore mainly devoted to discerning just what is already there.

The nature of rock is vastly different from other types of engineering materials. Therefore it is natural that the methodology employed for its characterization should be peculiar to the field of geological engineering. Nevertheless, each of the methods employed and discussed here has its cousins in other disciplines, and a book such as this must cross the borders of many fields. These include mining; petroleum; geophysics; cartography; planning; soil mechanics; hydraulics; mechanics of materials; concrete technology; structural engineering; statistics; aeronautics; and computer science. The obvious consequence is that sources of literature of interest for further reading are scattered among numerous journals, and reference books. However, a number of basic references and journals can be singled out as especially relevant. These are listed in Table 1-1.

TABLE 1-1

Some Sources of Information

Bibliographies and Abstracts

KWIC Index of Rock Mechanics literature published before 1969 - 2 volumes. Produced by Rock Mechanics Information Service, Imperial College, London.
Published by AIME, 345 East 47th St., New York, N.Y. 10017

Geomechanics Abstracts - Part II of the Inter. Jour. Rock Mechanics and Mining Science
Published by Pergamon Press from volume 4 (1973) onward (Originally called Rock Mechanics Abstracts; produced by Imperial College).

Geotechnical Abstracts - Monthly with annual indexes
Deutsche Gesellschaft fur Erd-und Grundbau (for Inter. Soc. for Soil Mechanics and Foundation Engineering).
(Published also in a card format called "Geodex Retrieval System").

Bibliography and Index of Geology - Monthly
Geological Society of America.

National Technical Information Service, Springfield, Va. 22151 (Bibliography and source for U.S. Government documents).

Geoscience Abstracts - Monthly.
American Geological Institute, Washington 25 D.C.
(A special supplement is devoted to a "Bibliography of bibliographies of the States").

Chronique des Mines et de la Recherche Miniere
published 10 times per year by Centre d'etudes geologiques et Minieres

Annotated Bibliography of Economic Geology - semi-annual.
Economic Geology Publishing Co.

Journals and Serials

Rock Mechanics (Inter. Soc. for Rock Mechanics) Formerly "Rock Mechanics and Engineering Geology".

International Journal of Rock Mechanics and Mining Science (Pergamon Press).

Engineering Geology (Elsevier).

Quarterly Journal of Engineering Geology (Geological Soc. of London).

Bulletin of the Association of Engineering Geologists.

U.S. Bureau of Mines, Reports of Investigations and other publications.

Canadian Geotechnical Journal.

Geotechnique.

Bulletin of the Inter. Association of Engineering Geology.

Proceedings of Congresses and Symposia of the International Society for Rock Mechanics*

First Congress - Lisbon 1966 - 3 volumes.

Second Congress - Belgrade 1970 - 4 volumes

Third Congress - Denver 1974 - 5 volumes.

Symposium on Rock Mechanics - Madrid 1968 - 1 volume.

Symposium on Stress Measurement - Lisbon 1970 - 1 volume.

Symposium on Large Permanent Underground Openings - Oslo 1969 - 1 volume

Symposium on Rock Fractures - Nancy 1971.

Symposium on Percolation through Fractured Rock - Stuttgart 1972.

Proceedings of Symposia on Rock Mechanics—U.S.A.

8th to 12th, 1966 - 1970 (AIME).

13th to 15th, 1971 - 1973 (ASCE).

Previous Symposia are listed in preface material for above Symposia.

Other proceedings of interest are listed in volume 2 of KWIC Index, (see "Bibliographies and Abstracts" above).

Textbooks

Coates, D.F., (1967) "Rock Mechanics Principles", Canadian Dept. Energy, Mines and Resources, Monograph 874.

Hoek, E., and Bray, J. (1974) "Rock Slope Engineering", (Inst. of Min and Metal, London).

*Can be ordered through ISRM, Laboratorio Nacional de Engenharia Civil, Avenida de Brazil, Lisbon, Portugal.

Jaeger, J.C., and Cook, N.G.W. (1969) "Fundamentals of Rock Mechanics", (Methuen).

Krynine, D., and Judd, W. (1959) "Principles of Engineering Geology and Geotechnics", (McGraw Hill).

Obert, L., and Duvall, W., (1967) "Rock Mechanics and the Design of Structures in Rock", (Wiley).

Scott, R.F., (1963) "Principles of Soil Mechanics", (Addison Wesley).

2
rock classification

While this book primarily concerns the discontinuous rocks, it is necessary to see this rock class in context and accordingly the question of rock classification in general will be explored. The object of rock investigations and measurements is to make judgments about the rock as a prelude to some action. The properties used to classify the rock will vary according to the designer's purposes and may include various subsets of: shear strength; flexural strength; tensile strength; elasticity; permanent deformability; creep-rate; water flow and water storage properties; in-situ stress; drillability; fragmentation characteristics; and sometimes density, thermal expansion, mineralogy, and color.

THE NATURE OF ROCK

One can not assign rock properties to a design calculation with the same degree of certainty as with some other types of engineering materials. The reason is that there is rarely a wholly dependable large sample of the total population available from which test results can be extracted. The application of principles of structural geology makes the sampling problem solvable. But we must realize that most of the volume of rock of immediate concern is hidden and inaccessible and, unfortunately, what we do see is rarely representative of what we don't. It is almost a law of geological engineering that the hidden, mantled material is the weakest and potentially most troublesome; only the sandstone layers will cropout in a formation composed of sandstone and shale; only the flow rocks will form ledges in a volcanic series of basalts and pyroclastics. The granite will form a hill, but the fault zone through it will form a valley.

Nor can the designer of a work in rock make use of rock properties with the same rigor as he might for other types of structural and hydraulic computations, because rocks seldom lend themselves to the usual sort of idealizing assumptions. First, most rock formations have directionality, such as bedding in sedimentary rocks, flow banding in volcanic rocks, and foliation in metamorphic rocks, and are consequently moderately to highly anisotropic. Then we find rock responding differently to excavation according to the initial state of stress, particularly in underground applications, and this is heavily dependent on the stress history which will be known only occasionally. Many rocks are semi-discontinuous on the hand specimen scale owing to a network of fissures and flaws, and almost all rocks on the formation scale are penetrated by surfaces of potential or real discontinuity. At the depths reachable in deep mines, deep drill holes, and some tunnels, some rocks are ductile, and very few rocks behave entirely elastically even at low pressures. Some rocks are chemically changeable within the lifetime of an engineering work and even more show great variability vertically and horizontally, due to different degrees of weathering. In the face of these difficulties, results of computations are to be utilized with restraint, and controlled by observations during construction. Fortunately, it is often sufficient for engineering purposes to produce only a reasonable estimate of the final behavior - an estimate that can be arrived at satisfactorily by rock classification.

ROCK SPECIMEN VERSUS ROCK-MASS

In a discussion of rock classification, we must carefully distinguish characteristics of a specimen of rock from properties of a body of rock in situ which, in the language of rock mechanics, we call the rock mass. The mass is comprised of the rock, its network of discontinuities and its weathering profile. The behavior of the rock mass reflects all of these components as well as water and stress regimes, strength, deformability, and permeability, which may be largely unrelated to material properties.

Classification of the entire realm of rock masses for the totality of applications would demand an unwieldy number of independent factors because different pursuits require different parameters. In assessing the suitability of facing stone, aggregates, embankment materials, and other rock products, we need rock specimen attributes describing durability, strength, thermal expansion, shrinkage, swell, absorption, and specific gravity. Rock mass characteristics affect items related to the cost of production. In regard to excavations, both specimen and mass characteristics are essential, the former affecting drillability and durability and the latter being basic to stability while also influencing excavatability. The essential factors for foundations, particularly for hydraulic structures, are those descriptive of rock mass deformability, stability, and permeability which derive principally from the discontinuities, (although rock specimen characteristics may sometimes control the design, as for example in non-durable, fissured, weathered, or permeable rocks).

First, we will examine classification of rock specimens then the weathering profiles and systems of discontinuities and finally the classification problem for rock masses.

PETROLOGIC CLASSIFICATION OF ROCK SPECIMENS

Geological methods of classifying rock specimens are based on a number of different criteria, which can be studied in Williams, Turner and Gilbert (1958). We will explore the wisdom of using geological rock names and petrological descriptions for engineering purposes.

A description of a rock's texture and fabric affords a basis for understanding its mechanical properties, which are closely related to interparticle bonding, interlocking and imperfections. The crystalline rocks (figure 2-1a) have tightly interlocked particle arrangements sometimes impaired by micro fissures within and between crystals. Coarse grained crystalline rocks tend to be weaker and less stiff than fine grained or aphanitic crystalline rocks. Foliation, the

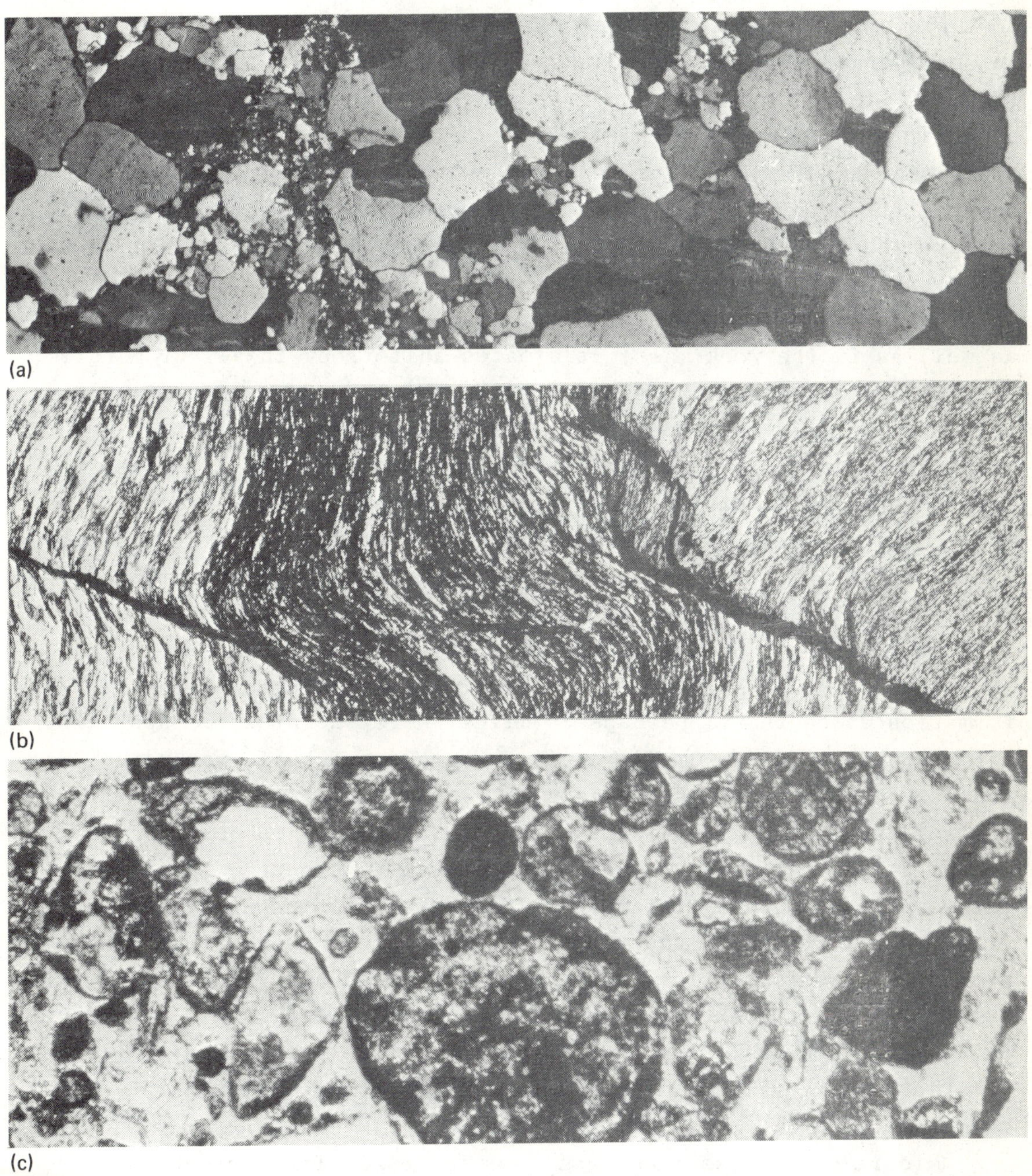

(a)

(b)

(c)

Figure 2-1. (a) Highly interlocked, crystalline texture of a Mesozoic quartzite; (courtesy of Prof. R. Wenk). 25.5X. (b) Highly anisotropic crystalline texture with oriented fissures (fracture cleavage) in chlorite schist—Homestake Mine, S.D.; (courtesy of Dr. W. Chinn). 40X. (c) Porous, clastic texture; eolian sandstone from Olduvai gorge, Tanzania, consisting of poorly sorted rock fragments and grains, some coated with clay. Cavities occur in altered nepheline grains (N); (courtesy of Prof. R. Hay). 136X.

most predominant fabric element of metamorphic rocks, causes strong anisotropy and surfaces of weakness within the scale of the specimen (figure 2-1b). Foliation is particularly pronounced when formed by coplanar platy minerals like mica. In clastic rocks (figure 2-1c), grain size has far less influence on mechanical properties than the nature, strength and durability of the binder or cement. Properties of cemented and compacted varieties of shale, for example, can be as different mechanically as soil and rock. Bedding is the most important structural feature of sedimentary rock on the specimen scale, as well as in the rock mass; it creates anisotropy in all properties.

Since geological names for rocks are intended to classify rocks according to differing modes of origin, one may wonder if they are meaningful for geological engineering practice. In the igneous rock group, the genetic division between intrusive and extrusive rocks <u>is</u> meaningful in terms of engineering attributes since it concerns the depth of formation. Features derived from the surface environment -- vugs, amygdules, and flow structures -- partly determine the mechanical properties of the volcanic flow rocks. The plutonic rocks, on the other hand, present quite different aspects linked to their formation at depths of perhaps 30 miles where the pressure approximates 150,000 psi (1000 MN/m^2). For example, plutonic rocks such as granite may possess large horizontal stress and fissures from unloading and a strong inclination towards chemical weathering. Dynamically metamorphosed rocks (as opposed to products of thermal metamorphism) contain miniature fold and fault structures and minerals oriented during growth under deviatoric stress. The various genetic processes responsible for the sedimentary rocks also produce distinct assemblages of properties linked to the mode of origin, -- although on the specimen scale the mechanical properties are more directly related to textural and mineralogic considerations independent of origin.

Mineralogic classifications form the basis for the actual rock names in the igneous and metamorphic rocks and to some extent in the sedimentary rocks. The mineral composition of crystalline rocks is not vital to a classification of mechanical properties and consequently many of the rock distinctions important to petrologists are useless for engineering purposes; for example we usually don't care whether a rock is classified as a granodiorite or diorite or tonalite. However, the accessory minerals may vary from one species to another and these, more than the proportions of quartz and feldspar may affect engineering response. Pyrrhotite (Martna, 1970), possibly pyrite, iron-rich micas, nepheline, leucite and nontronite have been identified as instrumental in deterioration of originally solid rocks quarried for aggregate and building stone. Minerals containing vugs filled with carbon dioxide can lower the pH of the groundwater, contributing to rapid weathering as at Bergeforsen Dam (Aastrop and Sallstrom, 1964). Any of the sheet silicate minerals, e.g. mica, chlorite, talc, and serpentine, introduce low shear strength, especially if in coplanar orientation; mica schist, serpentine and talc schist can be hazardous rocks in foundations and excavations. Glass, and secondary minerals zeolite and opal, can promote chemical reactions with cement even when present in small quantities in rock aggregate. Crystalline sedimentary rocks include some varieties largely or partially composed of weak, soluble, or non-durable grains, e.g. clays of the montmorillonite group, gypsum, halite, sylvite, weak shales, coal, chalk and chert.

In summary, though the science of petrology has evolved according to the needs of classical geology, its refined terminology and class distinctions are frequently meaningful for engineering work. Moreover, as geologists are familiar with it, and its rock classes are generally mappable, the geological nomenclature, especially when accompanied by textural descriptions and mineralogic details, is the most appropriate rock material classification for engineering purposes. The complete classification of the rock material must of course describe the state of weathering, the durability, and the degree of fissuring.

ROCK VERSUS SOIL AND WEATHERED ROCK

The most vital distinction to be recognized is between rock, weathered rock and inherently soil-like rock. The distinction is essential for all engineering work in rock, and yet it is not an elementary proposition. The fundamental precept is that to be rock, the material must be strong and durable. It is solid when first encountered and can not be softened, disaggregated, or easily weakened by accelerated weathering. Furthermore, it does not swell or shrink appreciably upon soaking. These requirements are pragmatic but do not coincide with geological nomenclature, in which a rock is defined as "any consolidated or coherent and relatively hard, naturally formed mass of mineral matter".*

Table 2-1 was based upon one devised by Karl Terzaghi for students in his engineering geology class** to distinguish between rock, weathered rock, and soil-like rocks. One may apply the term "solid rock", according to Terzaghi only if a rock is solid with a ringing sound when struck by a hammer and remains solid throughout weathering tests and soaking. Moderately soluble varieties, such as limestone and dolomite, will still be classified as solid rock, but greatly soluble rocks such as salt and gypsum will not survive a reasonable weathering test intact. Rocks which are originally solid but break up into small, hard pieces with a clean surface on weathering are termed fissured or crushed unaltered rocks, whereas if the rock disaggregates or yields greasy surfaces, it is an unstable or slightly decomposed rock. If such a rock exhibits perceptible volume change upon soaking, Terzaghi thought "rock" would be a dangerous misnomer; he preferred to designate swelling materials as "intermediate between rock and clay, rock characteristics predominat-

---

* Dictionary of Geological Terms" Dolphin Reference Book C36D. The above is the ordinary usage but this dictionary gives as a strict definition "any naturally formed aggregate or mass of mineral matter whether or not coherent, constituting an essential and appreciable part of the earth's crust". The word consolidated in the first definition is troublesome to engineers familiar with soil consolidation theory which refers to the expulsion of water from the voids of a soil under pressure. The geological usage means firm.

** Table 2-1 is based upon one devised by Karl Terzaghi and distributed to students in his course on engineering geology at Harvard University in the 1950's. A copy revised shortly before his death was generously supplied by Dr. Ruth Terzaghi. A somewhat similar approach is used by the National Institute for Road Research, South Africa, as published by Weinert (1964); see Fookes, Dearman, and Franklin (1971).

TABLE 2-1
EFFECTS OF SATURATION ON ROCKS AND ROCK-LIKE MATERIALS

Terzaghi's Guides for Distinguishing Rock, Weathered Rock, and Soil*

In original state	After repeated drying, immersing, and shaking, or upon prolonged exposure to the atmosphere	Volume change produced by saturating dried fragments with water	Group
Solid with ringing sound when struck with a hammer	unchanged	imperceptible	a) solid rock
	breaks up into small hard pieces with clean surfaces		b) finely fissured or crushed unaltered rock
	breaks up into small fragments with "greasy" surfaces owing to the presence of fine-grained weathering products		c) slightly decomposed fissured rock
	breaks up into individual sand or silt particles		d) sandstone or mudstone with unstable cement
	breaks up into small angular fragments without any indication of chemical alteration	measurable	e) intermediate between rock and clay, rock characteristics dominant
Solid with dull sound when struck with a hammer	gradually transformed into a suspension of soil particles		f) intermediate between rock and clay, clay characteristics dominant
	gradually transformed into a suspension of clay particles and a sediment consisting of angular rock fragments		g) thoroughly decomposed rock
	completely transformed into a suspension and/or a loose sediment	imperceptible to important	h) clay, silt, and very fine sand in dry or a very compacted condition

* From Professor Karl Terzaghi's course notes for Engineering Geology at Harvard University; included with kind permission of Dr. Ruth Terzaghi (with minor editorial changes) and including revisions made by Karl Terzaghi shortly before his death.

ing". Materials that are not solid with a ringing sound when struck by a hammer when first encountered should not be referred to as rock at all, according to this scheme. Many sedimentary rocks would accordingly be termed "soil-like rocks" in maps and reports, and the resulting impression would be correct for the engineer.

Geological investigations must correctly diagnose a specific soil-like condition as either inherent or localized. Weathering, and hydrothermal alteration -- the first usually intensifying towards the surface and the other, with depth or laterally -- may produce spotty and variable degrees of localized softness. In contrast, some sedimentary rocks are inherently soft either through incomplete cementation, intense fissuring, or regional alterations; neither "dental work" nor outright "extraction" can improve the rock conditions in this case.

The distinction between rock and soil is especially important as regards specifications for excavation contracts. So many legal controversies have revolved about this point that agencies such as the U.S. Bureau of Reclamation have been forced to adopt almost comically detailed wording for contracts, as in Table 2-2. The main ideas are that the material to be excavated is rock only if it is both in place, (or of large mass) and solid. If it is too risky to attempt a classification, the excavation receives one name--unclassified excavation -- and one price throughout. This can happen in deeply weathered materials, with their extreme variability and gradational qualities, in soil-like soft rocks, in bedded rocks alternating in hardness, and in very dense or cemented soils.

WEATHERING

Closely related to the question of differentiating soil from rock, is evaluation of the degree of weathering of the rock material. The importance of the subject is suggested by a voluminous literature, a selection of which is included in the list of references. Rocks respond to prolonged weathering in many ways. The granitic rocks become cracked and then decayed by the carbonic acid developed as rain water filters through the soil; this reagent attacks the feldspars and dark minerals releasing soluble salts of K, Mg, Fe, and Na,

TABLE 2-2

Classification of Excavation According to U.S. Bureau of Reclamation Contract Specifications

"Except as otherwise provided in these specifications, material excavated will be measured and classified in excavation, to the lines shown on the drawings or as provided in these specifications, and will be classified for payment as follows:

Rock Excavation. For purposes of classification of excavation, rock is defined as sound and solid masses, layers, or ledges of mineral matter in place and of such hardness and texture that it:

(1) Cannot be effectively loosened or broken down by ripping in a single pass with a late model tractor-mounted hydraulic ripper equipped with one digging point of standard manufacturer's design adequately sized for use with and propelled by a crawler-type tractor rated between 210- and 240-net flywheel horsepower, operating in low gear, or

(2) In areas where it is impracticable to classify by use of the ripper described above, rock excavation is defined as sound material of such hardness and texture that it cannot be loosened or broken down by a 6-pound drifting pick. The drifting pick shall be Class D, Federal Specification GGG-H-506d, with handle not less than 34 inches in length.

All boulders or detached pieces of solid rock more than 1 cubic yard in volume will be classified as rock excavation.

Common Excavation. Common excavation includes all material other than rock excavation. All boulders or detached pieces of solid rock less than 1 cubic yard in volume will be classified as common excavation."

as well as free silica which may be transported out of the weathering environment, and detrital clay and resistant quartz grains which usually remain. The rock is gradually transformed into a "saprolyte", figure 2-2a, which resembles rock but has the strength of a dense soil. Vargas (1953), Ruxton and Berry (1957), Lumb (1962), Deere and Patton (1971), and others have described the transitional states

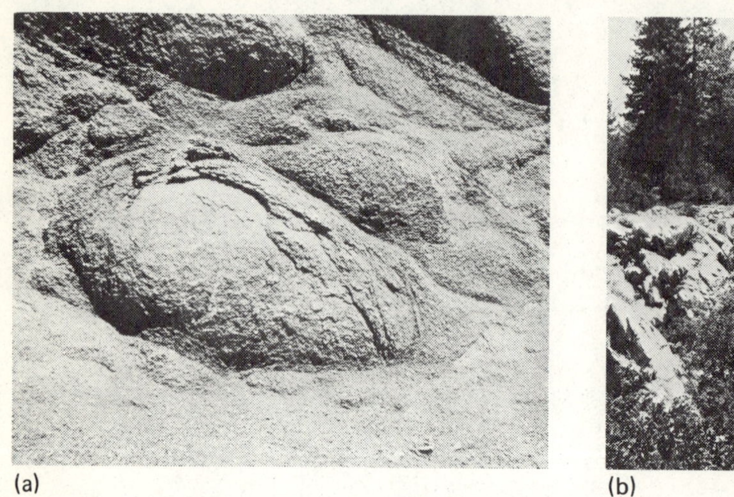

(a)

(b)

Figure 2-2. (a) Decomposed granite. Former joint blocks contain hard "core stones" in their centers while relict joints are now sandy clay seams or partings in the thoroughly weathered rock (saprolyte). (b) The top-of-rock surface in soluble marble; Columbia, California. The soil was removed by hydraulic monitors to obtain placer gold.

of granitic rocks and their properties. Basic igneous rocks follow a similar sequence but tend to produce a residuum richer in clay. The soluble rocks become enriched in impurities, often clayey, and develop stable or unstable vugs according to their strength. Limestones are often karstic whereas gypsum and halite are simply thinned or removed altogether, the karstification inducing almost immediate collapse, (Brune, 1965), (figure 2-2b). Intermediate weathering states consisting of vuggy rock are less common in gypsum than in limestone and dolomite rock. Anhydrite expands, relative to the initial solid volume, as it is converted to gypsum by hydration (but relative to the total volume of reactants it contracts).* Compacted shales and

---

* There is a difference of opinion on hazards of anhydrite. A thesis by Sahores (1962) considered the engineering problems implied by volume expansion to be overstated. Brune (1965) on the other hand, described uplifts and explosions occurring naturally in an area of West Texas underlain by anhydrite at depth; moreover the anhydrite grades into a thinner, folded gypsum layer updip and the uplifts occur directly over the locus of anhydrite - gypsum interfingering leaving no doubt that conversion of anhydrite to gypsum is responsible for these violent phenomena.

poorly-cemented sandstones -- the soil-like rocks -- disaggregate and return to sediment in response to weathering, and montmorillonitic varieties swell. In general, as the non-soluble rocks pass through intermediate stages of weathering they gain porosity and deformability, lose strength and elasticity, and become first more and then less permeable; (see for example Iliev's (1966) discussion of property changes in weathered monzonite). To classify the materials and attributes of the weathered zones, one must consider two independent criteria: first it is necessary to distinguish differing degrees of weathering of the rock itself; then this distinction must be superimposed on a classification of differing styles and arrangements of the weathering products.

The Degree of Weathering

Appraisal of the degree of weathering actually attained by a particular specimen of rock material is basic to any meaningful classification of rock masses within the weathered zone, which in the tropics and in particularly susceptible rocks such as granite, may extend more than 300 feet below the ground surface. No single index derived from simple field observations or laboratory tests can be expected to apply appropriately for all materials in the vast range of weathering products derivable from intermediate stages of decomposition of rock. Several approaches useful in particular rock types are offered as examples to be emulated in principle or detail as the case warrants.

Lumb (1962, 1965) discussed correlation between soil and rock properties in granites of Hong Kong ordered by a mineralogic weathering index X_d. Lumb's index, appropriate for quartz bearing granitic rocks in which the feldspars are attacked during the decomposition process, is assessed by hand lens examination of weathered and fresh rock to determine the percentages of feldspar and quartz as follows:

$$X_d = (N_q - N_{qo}) / (1 - N_{qo})$$

N_q is the weight ratio of quartz to quartz + feldspar in the weathered specimen, and N_{qo} is the corresponding ratio in the intact, unweathered specimen. N_{qo} is of the order of 1/3 for a fresh granite and increases toward 1 as the weathering progresses. Thus the index varies over the range 0 to 1 with an increasing degree of weathering.

Ege (1968) used a similar approach as one of four classification indices for granitic rocks at the Nevada Test Site. The degree of weathering is expressed by estimating the percent of altered minerals in the rock, without reference to an unweathered standard. The rock is classed as unweathered, slightly weathered, moderately weathered, or severely weathered respectively as the percent of altered minerals falls within the classes 0-10%, 10-25%, 25-75%, and 75-100%. The degree of weathering can also be classified on less formal divisions as in the example by Kiersch and Treasher (1955) for granodiorite at Folsom dam, California where: fresh rock was totally unaltered; slightly weathered rock showed slight fissuring in the feldspars and bleaching of their original color; moderately weathered rock showed more intense bleaching and fissuring in the feldspars, bleaching of the biotite, limonite appearing as specks and coatings of other minerals and slight rounding of quartz grains; and highly weathered rock showed strongly bleached biotite, the feldspars highly fractured and bleached, the quartz grains highly rounded, and limonite common as an accessory; further, the highly weathered rock could be scratched readily with a steel nail. This simple classification could be mapped and was successfully correlated with variations in resitivities, seismic velocities, drilling rates with diamond and percussion drilling, blasting patterns and powder factors, rippability, grout takes, and suitability of stone for rock fill and rip-rap.

Iliev (1966) introduced an index K based upon the reduction of longitudinal wave velocity with weathering.*

$$K = \frac{V_o - V_w}{V_o} \qquad (1)$$

The subscripts o and w identify the unweathered and weathered states. Like Lumb's index, this one goes from 0 to 1 as weathering progresses.

---

* Such an index can be applied in the laboratory or in the field; in the latter case characteristics other than specimen properties are involved and classification by the application of this simple parameter can be wrong.

Hamrol (1961) proposed a simple measurement of apparent porosity by the water content of a rock (dry weight basis) as an index of degree of weathering after quick immersion. The water content is determined after oven drying at 105 degrees centigrade. Lumb (1962), Pender (1971) and others have shown that porosity increases with weathering (see figure 4-16); since some engineering properties of rocks are directly associated with porosity or indirectly sensitive to its changes (Griffith, 1937), it is not surprising that Hamrol's index has met with success (Serafim, 1964) in recognizing rock grade boundaries within a single rock type at a single engineering site and in extrapolating results of field tests from one part of a foundation to another. There has been little quantitative work on the changes in properties of joints resulting from weathering.

The Profile of Weathering

Most engineering projects involve rock work in various levels within the weathered zone, which may extend as deeply as 100 meters below the surface. The outstanding feature of the weathering zone is extreme variability of rock quality, both laterally and vertically, (figure 2-2a); rocks of various degrees of weathering grade into one another insensibly. Classification of the weathered rock mass can be meaningful if described in terms of percentages of various weathering products at any given level (weathering horizon). Deere and Patton (1971) reviewed the weathering profiles of different rock types and suggested standard terminology based upon the approach used by Ruxton and Berry (1957) for granite soils of Hong Kong. These papers, as well as the work of Fookes and Horswill (1970) Spears and Taylor (1972), and others listed in the references should be consulted.

Durability

The discussion of weathered rock has considered only observed or measured attributes of a present sample. What will the properties be some years later, in response to construction and service? The question of durability and its inverse, weatherability, is only beginning to be answered by testing techniques and comparative data -- meager data in view of the variety of engineering requirements. Some of the minerals suspected of contributing to weatherability in rocks were listed earlier; now, we will consider a simple index test.

Franklin's slake durability test. Fookes, Dearman and Franklin (1971), and Franklin (1970) developed a durability test consisting of a standardized measurement of the weight loss of rock lumps when repeatedly rotated through a water-air interface. Ten lumps of 40 to 60 grams each are oven dried and weighed, and then placed in a standard test drum (figure 2-3a) whose circumferential wall is constructed of sieve mesh (2 mm opening). The drum is rotated at

(a)

(b)

Figure 2-3. (a) Slake durability apparatus (courtesy Soil Mechanics Equipment Co., Glen Ellyn, Illinois). (b) Franklin Point Load Testing Device (courtesy Soil Mechanics Equipment Co., Glen Ellyn, Illinois).

20 revolutions per minute for ten minutes. The slow speed reduces mechanical wear effect in the agitating process. The dry weight retained after the weathering cycle, expressed as a percentage of the original weight, is reported as the Slake Durability Index (I_d). Gamble (1971), who evaluated this index in relation to other durability and abrasion tests used for aggregates, found the slake durability to be far gentler and better able to cope with the large range of durability response offered by rocks (he preferred a modified durability index based on 2 cycles of rotation and drying). Tests such as the Los Angeles abrasion test* are more sensitive to slight variations in durability among rocks to be considered for aggregate.

To assess the weatherability of a rock, it is meaningful to attempt to simulate a project's anticipated weathering environment at an accelerated rate. The pitfall is that unless one is able to incorporate all pertinent factors in the laboratory simulation, the results will be difficult to interpret. The advantage of a standardized test, such as the one described, is that experience gained in assorted projects will eventually be grouped in a useful format for future reference.

Slaking of claystones and shales can be caused by swelling of clays. The slake durability test is not suitable for swelling materials as the lumps tend to build protective clay coatings. Durability problems associated with expansive clay minerals can be predicted by standard methods of testing for the presence of swelling clays and measurement of swell pressure. A consolidometer especially suited to this purpose is the GeoNor Swelling apparatus** in which disc-shaped rock samples, or pulverized and elutriated samples, are preconsolidated and then allowed to swell under imposed displacement constraints (Bjerrum, Brekke, et al, 1963). Free swell of altered, hard rock samples was measured with sufficient precision very simply by Nascimento (1970) with a jewelled dial gauge. The rock core specimen stands in a beaker of water on a point con-

---

* ASTM Standard Methods of Test C535-69 and C131-69.

** Sold by the Norwegian Geotechnical Institute, Oslo.

tact. In several altered granites and gneisses Nascimento monitored, swelling began almost immediately after the water was added and essentially terminated after two to five hours.

A meaningful and potentially rewarding area of inquiry into the weatherability of rocks considers the changing content of dissolved solids in water percolating steadily through rock specimens. The Bernaix radial permeameter, discussed later in connection with fissuring, is suited to this approach.

INDEX TESTS FOR THE QUALITY OF THE ROCK MATERIAL

Other simple laboratory tests or quick field measurements can serve as quantitative indices of rock quality and degree of weathering and as basic components of applied classifications. Deere and Miller (1966) studied the use of the Schmidt hammer which can be carried in the field. Defects on the surface against which the hammer is activated can give low readings unrelated to the rock material quality, a problem which can be avoided by exercising care in preparing the test surface. A more revealing measurement is provided by any strength test, especially one which demands a small sample and which can be done routinely on a large number of specimens. The point load test (figure 2-3b) introduced by Franklin (1970) and Broch and Franklin (1972) is one such method. Tests are conducted by squeezing pieces of rock drill core diametrically between standard steel cones until rupture. The point load index I_s is P/D^2 where P is the load at rupture and D is the diameter. The results are affected by the value of D, but size correction charts given by Broch and Franklin allow all results to be expressed in terms of a standard size (50 mm is recommended). The point load index $I_{s,50}$ correlates fairly well with the uniaxial compressive strength divided by 24. The test can also be applied to irregular chunks, approximating 50 mm in size.

FISSURED ROCKS

Small cracks and fissures may be contained in apparently intact rock specimens. As opposed to pores, which are three dimensional

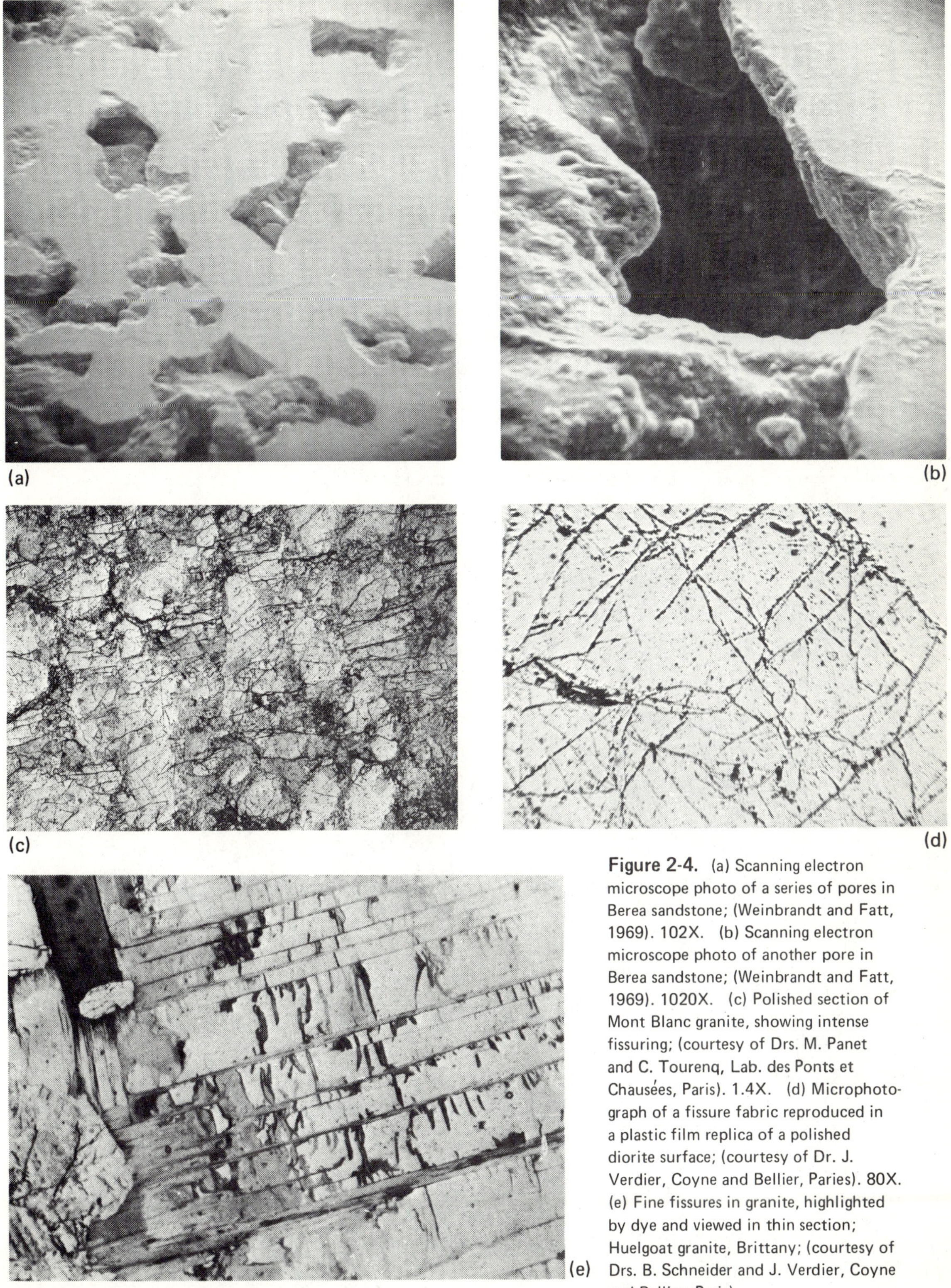

Figure 2-4. (a) Scanning electron microscope photo of a series of pores in Berea sandstone; (Weinbrandt and Fatt, 1969). 102X. (b) Scanning electron microscope photo of another pore in Berea sandstone; (Weinbrandt and Fatt, 1969). 1020X. (c) Polished section of Mont Blanc granite, showing intense fissuring; (courtesy of Drs. M. Panet and C. Tourenq, Lab. des Ponts et Chausées, Paris). 1.4X. (d) Microphotograph of a fissure fabric reproduced in a plastic film replica of a polished diorite surface; (courtesy of Dr. J. Verdier, Coyne and Bellier, Paries). 80X. (e) Fine fissures in granite, highlighted by dye and viewed in thin section; Huelgoat granite, Brittany; (courtesy of Drs. B. Schneider and J. Verdier, Coyne and Bellier, Paris).

(figures 2-4ab), fissures are short planar cracks of microscopic or macroscopic size (figures 2-4 c,d,e). They occur as intercrystalline cracks less than 1 micron to macroscopic (> 1mm) in size, as inter-grain cracks, and as multi-grain fractures. The presence of such cracks as well as their significance in reducing the tensile strength of brittle materials were appreciated by Hoek, Brace, McClintock and Walsh, and others with regard to the Griffith theory of failure which is based upon stress concentrations around such fissures. Habib and Bernaix (1966) linked the degree of fissuring also with scale effects in strength and deformation measurements, dispersion of results in repeated measurements, and stress dependency of specimen permeability. All of these effects were shown to be large in highly fissured rock at low pressure and to disappear in non-fissured rocks, and in fissured rocks at elevated pressure, within which the fissures have closed. With respect to mechanical properties, it is the presence of fissures more than any other aspect, wrote Habib and Bernaix, that distinguishes rock from other solids. The French have held two colloquia and an international symposium on rock fissures* and correlations have emerged reinforcing Habib's belief that in fissured rocks, mechanical properties are more closely dependent upon fissure fabric than on mineral composition or texture.** It appears that fissuring has a primary influence on static elastic modulus values, hysteresis in load cycling, sound wave velocity, direct tensile strength, resistivity and thermal conductivity of rock specimens.

The degree of fissuring in a rock reflects its history. Fissures

---

* The 1st and 2nd Colloquia on Fissuring of Rocks were published in special numbers of "Revue de l'Industrie Minerale" respectively 15 May, 1968 and 15 July, 1969. The Nancy Symposium held October 1971 was published by ISRM.

** In introducing the 2nd Colloquium on fissuring in rocks, Pierre Habib wrote: "One can now say that the properties of rocks are essentially those of their fissures. The mineral matrix has only a discreet role in the sense that if the rock is continuous it is always over endowed either in rigidity or in strength. To describe the fissuring of a rock is thus to define its present state and the study of its mechanical behavior is first of all the examination of the arrangement and development of fissures up to destruction."

can be generated by chemical weathering, unloading, heating and cooling, and most importantly by localized cracking accompanying deformation. Rocks likely to be found in a fissured state are: volcanic flow rocks; foliated metamorphics, especially schistose varieties; marble; pegmatites and porphyritic or hypidiomorphic granitic rocks; granites exhibiting cleavage; serpentine; chert and siliceous shales; and quartzite.

The degree of fissuring should be a basic component of any rock classification scheme. It can be characterized through direct observation, or more simply through index tests.

A polished surface will often enable prominent fissures to be observed with the naked eye. A hand lens, binocular microscope, or best of all, an ore microscope allows study of fissure distribution in the polished section. Tourenq (1969) displayed fissures in polished surfaces by preparing replicas of the surface adapting techniques for electron microscope specimens described by Bradley (1954) and Jacquet and Mencarelli (1959), (figure 2-4d). Schneider (1967) used dyes: Fuschine ASA (basic), Victoria Blue (basic) and Auramine J (acid) to show fine fissure detail in thin sections viewed in transmitted light with a petrographic microscope, (figure 2-4e). Study of fluid-filled epigenetic inclusions, e.g. in quartz, can allow relative evaluation of different directions of fissuring, (Verdier and Deicha, 1971). These processes can be tedious and for practical engineering work it may be more appropriate to characterize the degree of fissuring implicitly.

Parameters of fissuring can be derived from pressure - volume change curves, shear and longitudinal wave velocity measurements, comparison of direct and indirect tensile strengths, and the ratio of permeabilities in tension and compression. Morlier (1968), following work of Walsh (1965), suggested computation of the volume of fissures --fissure porosity -- from the shape of the pressure - volume change curve (p versus $\frac{\Delta V}{V}$). As shown in figure 2-5, this curve is concave upward, becoming asymptotic to a line whose slope is defined by the compressibility of the rock $(k = \frac{E}{3(1-2\nu)})$. The initial fissure porosity $n_f(o)$ is estimated by the value of $\frac{\Delta V}{V}$ at intercept of the asymptote. The concavity of the curve can also be interpreted to

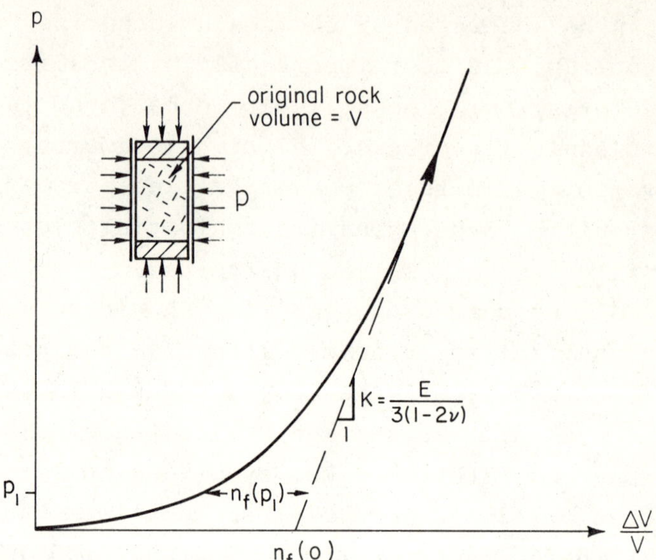

Figure 2-5. Compressibility of fissured rock.

yield a fissure shape distribution function ("fissure spectrum").

Tourenq, Fourmaintraux, and Denis (1971) offered a second approach based upon a comparison of actual and theoretical wave propagation velocities. A crystalline rock composed of given percentages of stated minerals has theoretical elastic properties close to the weighted average of the elastic properties of the components. Table 2-3 gives values for Young's Modulus, Poisson's ratio, and longitudinal and transverse wave velocities for the common rock forming minerals. If a rock is fissured, measured properties will be lower than the theoretical values calculated from Table 2-3. The degree of fissuring is expressed in terms of a quality index, IQ, defined as the ratio of measured to calculated longitudinal wave velocities.

$$IQ = \frac{V_{\ell \text{ measured}}}{V_{\ell \text{ calculated}}} \times 100\% \qquad (2)$$

Fissure porosity (n_f) drives the quality index downward linearly approximately 15 times as fast as normal porosity (n_p) (spherical pores). If one measures the total porosity n% (= $n_p + n_f$) as well as IQ, figure 2-6 can be used to determine the value of n_f. This

TABLE 2-3

Average Elastic Modulus and Velocity of Longitudinal Waves for Common Rock Forming Minerals*

	Young's Modulus E (10^5 bar)	Poisson's Ratio ν	Longitudinal Velocity V_ℓ km/sec	Transverse Velocity V_t km/sec
quartz	9.6	0.08	6.0	4.1
orthoclase	6.7	0.27	5.7	3.3
plagioclase	8.1	0.28	6.3	3.5
biotite	7.0	0.25	5.1	3.0
calcite	8.1	0.30	6.7	3.4
muscovite	7.9	0.25	5.8	3.4
amphibole	12.9	0.28	7.2	4.0
pyroxene	14.4	0.24	7.2	4.2
olivine	20.0	0.24	8.4	5.2
magnetite	23.1	0.26	7.4	4.2

* From data of Alexsandrov, Belikov and Ryzova, a reference cited by Fourmaintraux and Tourenq (1970).

figure also shows the relative effects of pores and fissures on the ratio of measured to calculated elastic modulus values. A value of n_f = 2% reduces the ideal elastic modulus almost by half, whereas a value of n_p = 15% would be required to achieve this effect.

If both the transverse and longitudinal wave velocities are measured, the degree of fissuring can be derived from their ratio. Fissured rocks are not "ideal" materials and one should not automatically try to report the ratio V_t/V_ℓ in terms of a "dynamic Poisson's ratio" value. Instead, Tourenq et al suggest Table 2-4.

A third method of evaluating the degree of fissuring is based upon the ratio of strengths in direct and indirect tension tests, (Tourenq and Denis, 1970). Direct tension tests can be performed by bonding moment-free end pieces to cylindrical rock specimens. The

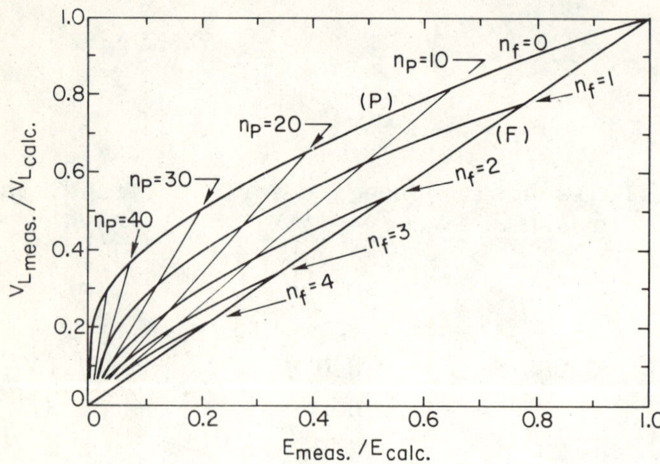

Figure 2-6. Relative effect of fissures and pores on the longitudinal wave velocity and the modulus of elasticity; (Tourenq, Fourmaintreau, and Denis, 1971).

direct tensile strength, σ_t, i.e. the average tensile stress at failure, is greatly reduced by fissuring. An indirect tension test (Brazilian test) can be obtained by compressing the opposite diameters of a rock disc* causing a uniform state of tension across the vertical diameter; the Brazilian tensile strength $\sigma_{t,B}$ i.e., the tensile stress at failure in a Brazilian test, is only slightly affected by fissuring. Therefore, the ratio $\sigma_t / \sigma_{t,B}$ is descriptive of the degree of fissuring, as shown by Table 2-5 summarizing some data presented by Tourenq and Denis. They recommend that the rock be classed as: essentially non-fissured if $\sigma_t / \sigma_{t,B} > 0.8$; very fissured if $\sigma_t / \sigma_{t,B} < 0.2$.

Bernaix (1969) developed an index of fissuring intensity based upon a radial permeability test. Water introduced under pressure in the center of a thick walled cylinder of rock, figure 2-7a, produces tangential tension stress as it flows divergently towards the outer circumference. Conversely, convergent flow produced by directing water from the outer circumference to the inner produces a tangential compression. Assuming that the flow net is not altered by stress

* In Tourenq and Denis' tests, the length to diameter ratio of the discs was unity.

TABLE 2-4

Index to Degree of Fissuring According to the Ratio of Transverse to Longitudinal Wave Volocities

V_t / V_ℓ	Description
< 0.6	non fissured
0.6 to 0.7	fissured
> 0.7	very fissured

TABLE 2-5

Tension Test Index to Fissuring

Data from Tourenq and Denis (1970).

Rock	Fissure length (mm)	$\frac{\sigma_t}{\sigma_{t,B}}$ *
Limestone	0.2	1.0
Limestone	1.5	0.45
Granite	0.1	0.93
Granite	0.3	0.7
Granite	1.3	0.50
Granite	2.5	0.34
Granite	1 to 10	0.14
Granite	3 to 20	0.07
Basalt	0.1	0.9
Basalt	2 to 10	0.15

* σ_t = direct tension strength; $\sigma_{t,B}$ = Brazilian tensile strength.

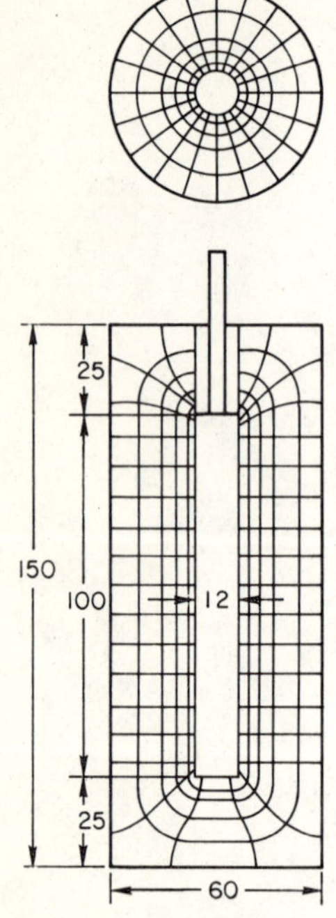

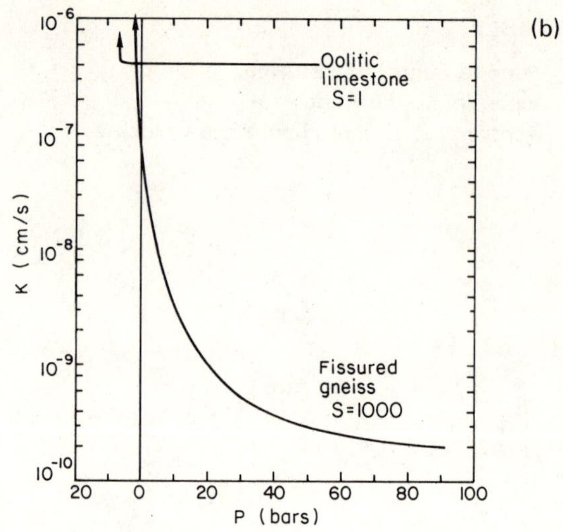

Figure 2-7. Radial permeability test; redrawn from Habib and Bernaix (1966). (a) Radial permeameter (dimensions in mm). (b) Results for a porous and a fissured rock; $S = K_{-1}/K_{50}$

dependency of permeability*, Bernaix derived expressions for the water pressure gradient dp/dr as a function of r, added this as a body force to the equilibrium equations and solved for σ_r and σ_θ ; the value of σ_θ on the inner wall of the cylinder ($r = R_1$) is

$$\sigma_\theta = \frac{\pm P}{2(1-\nu)} \left(\frac{2R_2^{\,2}}{R_2^{\,2} - R_1^{\,2}} + \frac{1-2\nu}{\ell n\ (R_2/R_1)} \right) \qquad (3)$$

σ_θ is tension for divergent flow where P is the water pressure on the inner radius $r = R_1$; $(P(R_2) = 0)$

and

σ_θ is compression for convergent flow, where P is the water pressure on the outer radius $r = R_2$; $(P(R_1) = 0)$.

---

* In actual fact the permeability K is a function of σ and since σ is a function of r, K depends on r and dp/dr can not be solved as readily as in the paper. The problem is one of "coupled flow"; see Noorishad et al (1972).

For the conditions of Bernaix's tests, R_1 = 0.6 cm, R_2 = 3 cm and Poisson's ratio ν = 0.2 giving σ_θ = 1.53 P. Figure 2-7b shows the variation of permeability K for an oolitic limestone and for Malpasset gneiss as P was varied between 100 bars outside to 1 bar inside. The permeability K was computed according to the relationship

$$K = \frac{Q}{2\pi\ LP} \ln \frac{R_2}{R_1} \tag{4}$$

wherein Q is the steady state flow rate (1^3 / t) and L is the length of the cylinder. The permeability varied continuously over more than 3 orders of magnitude for the fissured gneiss whereas it remained constant for the limestone. Bernaix recommends as an index of fissuring to report the ratio (S) of permeability at 1 bar in divergent flow (K_{-1}) to permeability at 50 bars in convergent flow (K_{50}). A sampling of results with a number of rocks are summarized in Table 2-6.

TABLE 2-6

Radial Permeability Test Index to Fissuring

Rock	Description of fissuring	$S = K_{-1}/K_{50}$
Limestone	porous, non fissured	1
Limestone	porous, some fissuring	1.3
Limestone	fissured	2.6
Granite	slightly microfissured	1.2
Quartzite	microfissured	1.8
Mica schist	fissured	4.8
Schist	highly fissured	10 to 100
Malpasset gneiss	highly fissured	7 to 200 right bank 1 to 50,000 left bank*

* The failure originated on the left bank.

DISCONTINUITIES

Rock masses invariably include numerous surfaces of real or potential discontinuity. Though somewhat artificial, we will distinguish between discontinuities and fissures. Obviously, there is a continuous distribution of discontinuity surfaces according to length. However, fissures within a specimen are included in a sample of the specimen, thus subject to meaningful inquiry in the laboratory. Laboratory techniques for samples of larger surfaces of weakness are developing but the results are seldom exportable to the field without additional field observations and tests. Thus fissures can be considered as rock specimen features whereas discontinuities cannot.

A single discontinuity includes two mating surfaces and a space, or filling. The term "joint" which has come to be used in engineering contexts for all or part of the family of discontinuities in rock masses, is unfortunately potentially confusing for structural engineers, who use the term joints to describe points of connection in steel structures; in geological usage the term is applied only to penetrative, repetitive discontinuities without appreciable shear displacement. However, as the term joint is entrenched in its engineering geology context, it will be retained here.

On a geological basis, we can distinguish extension and shear joints, bedding, banding, contacts, cleavage, schistocity, foliation, sheared zones and faults, as discussed in standard works in geology, for example Leet and Judson (1971) and Price (1966). With reference to mechanical and flow properties of a discontinuous rock mass, we require considerably more information than the geological identification. In particular, load-deformation and strength properties of discontinuities (see Chapter 5) make specific reference to a number of controlling quantities, including the parameters of the peak and residual shear strength-variation with normal stress, the initial angle of dilatancy, the normal pressure required to prevent all dilatancy, the maximum amount a joint can close, the peak and residual shear displacement, the tensile strength, and the normal and shear stiffness. Though no rational formula exists for extracting the explicit joint parameters required for an analysis from field observations, careful and detailed descriptions of the many encounters in outcrops, excavations, and in the core box allow the whole system of discontinuities in any project to be divided into a relatively small number of types. Usually the field description will permit reasonable estimates for certain of the quantities mentioned and their contribution to the rock mass characteristics. The joint system properties are derived from observable features of: 1) the discontinuity surfaces; 2) the blocks they define through their repetition and intersection; 3) and the properties of the space between the blocks. Most of these features can be described deterministically, but are better expressed statistically through distribution curves and numerical indices.

Properties of the joint surfaces themselves include orientation, extent, planarity, roughness and waviness, and the strength of wall rock asperities. Joint systems usually display several preferred orientations and this aspect alone results in wholly different classes of rock masses. This subject lends itself to statistical and graphical treatment by means of stereographic projection, which will be elaborated in Chapter 3. Joint "extent", meaning the total area or length, usually cannot be measured directly in the field; however it can be estimated occasionally from aerial photos (see Chapter 4). Roughness and waviness, which influence the friction angles, dilatancy, and peak shear (Patton 1966, Goodman and Dubois, 1971) refer to the local departures from planarity at small and large scales respectively (figure 2-8). The most convenient roughness measure for rock mechanics purposes is in terms of the local angles with respect to the mean plane through all the hills and valleys of a joint surface. Most joints can be represented quite well by planes. Patton (1966) measured roughness angles from edge views of surfaces in outcrops and cliff faces; the required data can be obtained from photographs as discussed in Chapter 4. When the joint surface itself is exposed in outcrop, repeated measurements of dip will generate a scattered distribution of values, whose standard deviation or mean departure may be an estimate of the mean roughness angle, as discussed in Chapter 5. The roughness angles increase joint shear strength at low normal pressure, but at higher normal pressure, the strength of the wall rock asperities controls the shear strength of the joints. A good

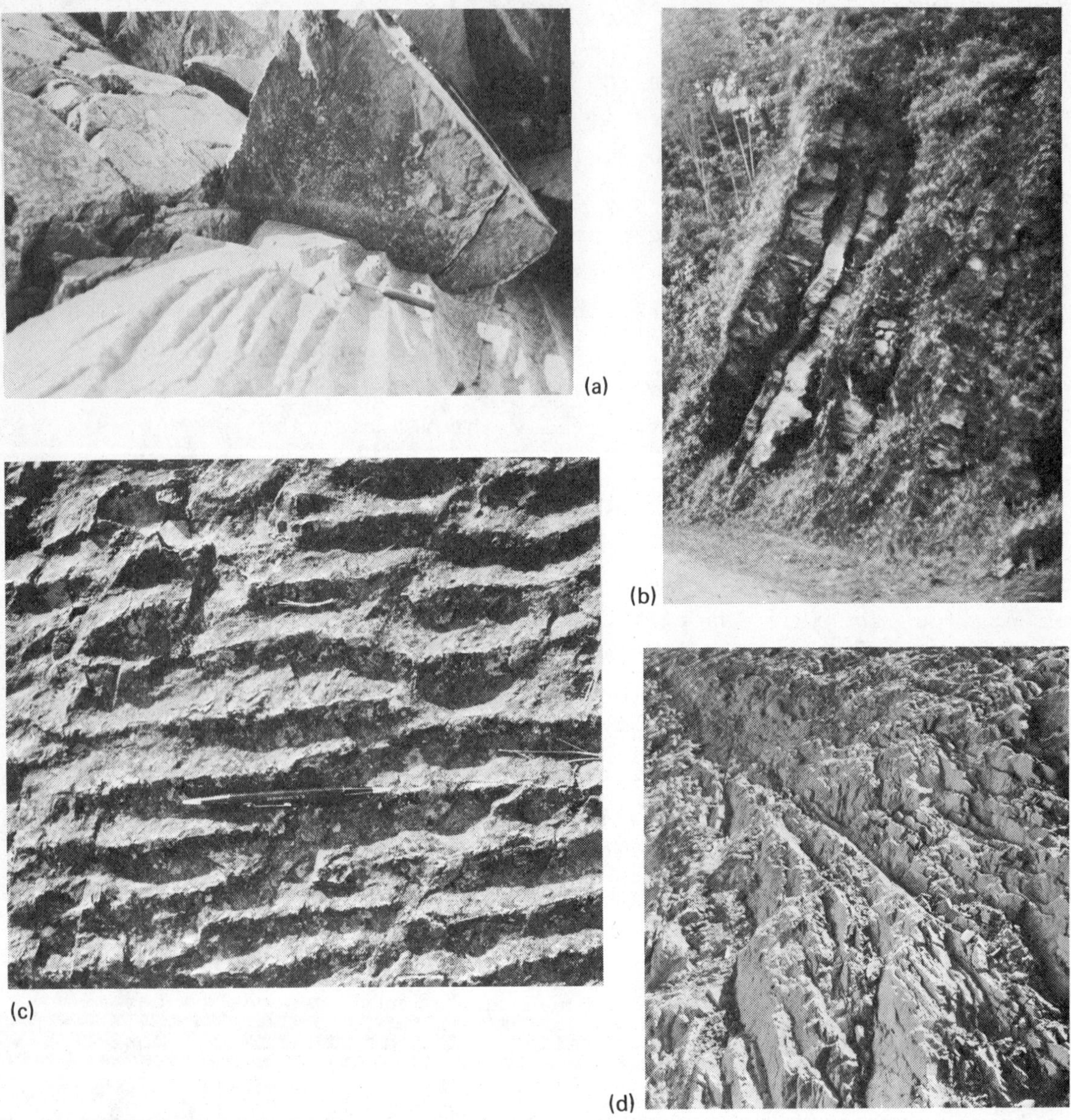

Figure 2-8. (a) A very rough fracture surface in granite; note the perfect mating of the joint blocks across the fracture. (b) A very rough bedding plane in limestone; voids between beds resulted from oversliding of asperities accompanying mass movement downslope. (c) A rough bedding plane surface; the roughness is formed by ripple marks preserved from the depositional surface. Photo by Dennis Lachel, (courtesy of the Corps of Engineers). (d) A rough joint surface; the roughness is created by the intersections of the joint with bedding; erosion has accentuated the relief. The rock is an argillite. Photo by Dennis Lachel, (courtesy of the Corps of Engineers). (e) A smooth surface whose mean plane parallels bedding. Slight roughness exists because the surface wanders from one argillite bed to another. Photo by Dennis Lachel, (courtesy of the Corps of Engineers). (f) A very smooth, wavy surface of discontinuity undercut in the spillway excavation of figure 1-6. This is probably an ancient shearing surface.

(e)

(f)

index of the strength of the wall rock is provided by Schmidt hammer readings on the joint surfaces. The scratch hardness value of the wall rock is also indicative of its strength. An appropriate scratch hardness scale devised by Ege (1968) uses grades of scratchability by a steel nail and the fingernail. Very hard rock surfaces cannot be scratched by a steel nail; hard corresponds to a faint scratch impression while intermediate hardness means a distinct scratch trace can be left. If the material can be scratched by the fingernail, it is soft.

The shapes and sizes of rock blocks formed by the system of discontinuities depend entirely on the orientations and spacings of the various joint sets, (see figure 5-1). Joint blocks formed mainly by one set of surfaces, as for example, layers in bedded rock, may be termed "Tabular". "Columnar" joint blocks are formed by two sets of joints (Burton, 1965) or, as in volcanic flows, by any number of sets parallel to a given axis. "Prisms" are formed by three very regular orthogonal sets. "Wedges" and "slivers" are formed by irregular jointing systems, as in metamorphic and granitic rocks. The sizes of the joint blocks obviously depend on the spacing of the various joint sets. It is usual to speak of joint spacing, rather than block size, although both numbers need to be appreciated. Table

TABLE 2-7

Terminology for Spacing of Discontinuities (after Watkins, 1970)

Descriptive Term		Spacing	
Bedding	Jointing	inches	mm
thinly laminated	fissured*	< 0.24	< 6
laminated	shattered	0.24 to 0.8	6 - 20
very thin	very close	0.8 to 2.4	20 - 60
thin	close	2.4 to 8	60 - 200
medium	moderate	8 to 24	200 - 600
thick	wide	24 to 80	600 - 2000
very thick	very wide	> 80	> 2000

* Called "comminuted" by Watkins.

2-7 after Watkins (1970) presents recommended terminology for various spacings of bedding and jointing. Figure 2-9 shows rock masses with various styles of jointing.

A description of the space between blocks includes the dimension of opening between the rock walls (aperture) and the completeness of filling with a description of the character and permanence of the filling materials. Joints and faults may be "unfilled", "healed"

Figure 2-9. (opposite) (a) Sheet joints in granodiorite, Yosemite National Park, California. (b) Bedding with tight joints in other directions. The bedding was undercut in excavation for a tunnel portal; rock reinforcement was installed from the catwalk above the top of the excavation. Ruedi Dam, U.S. Bureau of Reclamation. Photo by Dr. F. Heuze. (c) Closely jointed argillite with prismatic blocks; near Chivor Dam, Colombia. (d) Bedding and two joint sets defining tabular and prismatic blocks of relatively small size and very much larger wedge shaped blocks; near Libby Dam, Montana. Photo by Dennis Lachel, (courtesy of the Corps of Engineers). (e) A metamorphic rock mass with three regularly spaced, tightly closed joint sets and well developed schistocity. One well developed joint set (J1) forms the moderately rough surface inclined towards the observer. The traces of two additional joint sets and of the schistocity form three sets of parallel lines across the surface of J1. Near Don Pedro Dam, California. (f) Regular columnar jointing, and completely broken, irregularly jointed volcanic flow rock (latite). The columns formed from shrinkage during cooling, with their long axes perpendicular to the isotherms. The broken rock represents columns that rode along on still molten material underneath. Near Tullock Dam, California.

(a)
(b)
(c)
(d)
(e)
(f)

TABLE 2-8

Classes of Fault Gouge Materials (after Brekke and Howard, 1973)

Dominant Material in Gouge	Potential Behavior of Gouge Material in Tunnels: At Tunnel Face	Later
Swelling clay	Free swell, sloughing	Swelling pressure and squeeze against support or lining; free swell with down-fall or wash-in if lining inadequate.
Inactive clay	Slaking and sloughing caused by squeeze; heavy squeeze under extreme conditions.	Squeeze on supports or lining where unprotected. Slaking and sloughing due to environmental changes.
Chlorite, talc, graphite, serpentine	Ravelling	Heavy loads may develop due to low strength, particularly when wet.
Crushed rock fragments or sand-like gouge	Ravelling; standup time may be extremely short.	Loosening loads on lining; running and ravelling if unconfined.
Porous or flaky calcite, gypsum	Favorable condition	May dissolve, leading to instability of rock mass.

(cemented) or "filled". Brekke and Howard (1973) distinguish the five classes of fault fillings (gouge materials) listed in Table 2-8 in discussing the influence of fault gouge on tunnel stability. Swelling clay fillings are potentially the most troublesome class. Other clay fillings, and the sheet-silicate minerals chlorite, mica, talc, serpentine and graphite, can introduce extremely low shear strength, particularly if the thickness of filling is greater than the roughness amplitude (Goodman, 1969). The clay : quartz ratio of the filling has been demonstrated to be a sensitive parameter of joint shear strength in British coal measure rocks (Taylor, 1973).

Crushed rock fillings, and incomplete quartz or calcite fillings offer potentially high permeability; they can erode and undermine the adjacent rock in exceptional cases. This is particularly troublesome in unlined water tunnels.

CONTINUOUS AND DISCONTINUOUS ROCK MASSES

From a geological point of view, rock masses are divided into field rock units, e.g. formations, members, and zones. When such units are defined so as to lump similar lithologic units or groups, they are useful engineering divisions, mainly because they are coherent, mappable entities. From an engineering point of view, it makes sense to search for additional means of classification, motivated by pragmatic interests along functional lines. An enormous literature reflects the timeliness, if not the frustrations, of this pursuit.

A first order division can be made on the basis of the degree of importance of discontinuities. If a rock formation is described adequately by the rock material alone, it may be called a <u>continuous</u> rock mass. More often, in connection with civil engineering projects, the mass behavior is controlled by the discontinuities and the rock substance description is almost irrelevant; such rocks should be termed <u>discontinuous</u>. A formal classification can be made on the basis of the relative importance of the system of discontinuities on the significant properties; one such measure might be the ratio of formation deformability to rock deformability as given in Table 2-9; (or the ratio of shear strengths or permeabilities could be used if more appropriate and if they could be conveniently measured.) In jointed hard rocks, such as slate, gneiss, granite, quartzite, and marble, the rock is so strong, so indeformable, and so impermeable, that only the discontinuities contribute effectively to displacements, to stability problems, and to water flow. Such rocks should be treated as discontinuous. Unlike granular soil which is a discontinuum that can be approximated by a homogeneous continuum for many purposes; the discontinuous rock mass has very low porosity and any appreciable deformation requires a dramatic increase in open space through dilatancy. A "soft" rock on the other hand may belong to

TABLE 2-9

Classification of Rock Mass Continuity

Name	$E_{material}/E_{field}$*	Field fracture porosity	Typical materials
Continuous	1 to 1.3	< 1%	Some granites, massive sandstones, and massive limestones; many Tertiary and Quarternary sediments and argillaceous sediments of all ages.
Intermediate	1.3 to 2	< 1%	Many granitic and metamorphic rocks; iron oxide or calcite-cemented sandstones.
Discontinuous	> 2	1% to 5%	Highly jointed granitic and metamorphic rocks, quartzites and silica-cemented sandstones; volcanic flow rocks.
Loosened	> 5	Voids and cavities between joint blocks which are in edge to face contact.	Rock slides; fault zones.

* E_{field} means here "modulus of elasticity" as determined by using total deformation measurements from a field test in a formula based upon the theory of elasticity.

the class of continuous rocks, for its intrinsic rock deformability, strength, or permeability ("matrix permeability") may override the respective contributions from the system of discontinuities. Most sediments of Tertiary or Quaternary age, many older argillaceous sediments, soluble evaporites, chalk, and friable sandstones can be considered continuous. Weathered rocks should be classified separately through reference to the weathering profile as previously discussed. Some hard rocks are almost free of discontinuities and can be classed as continuous, e.g. some granites, massive sandstones, and massive limestones.

The behavior of a particular joint depends not only on its own properties but on the initial stress and water pressures along it. The same formations will change from discontinuous to continuous with increasing depth. Since a joint is very thin, it can be treated as essentially two dimensional; its state of stress therefore is in equilibrium with the stress state of the adjoining rock blocks. We cannot use the concept of joint stress in discussing a rock mass that has been displaced by sliding because block forces are then transferred partly through edge-to-face contacts across joints. A rock mass with open joints and with blocks in edge-to-face contacts is as different from a rock mass with mating joint surfaces as rock is different from gravel; such a rock mass, characterized by open joints, load transferrence along point and line contacts, and interblock cavities, shall be called loosened.

ENGINEERING CLASSIFICATIONS OF ROCK MASSES

Table 2-10 gives references to some engineering classification systems for rock material and rock masses. In continuous rocks, it is sufficient to classify the rock material alone and we can apply classifications of the type proposed by Coates (1964), motivated by interest in mining problems underground, or Deere and Miller (1966), motivated by an interest in standards for laboratory testing. All of the references to general purpose classification given in Table 2-10 are deficient in omitting explicit reference to micro-fissuring.

General purpose rock mass classifications for discontinuous rocks have been proposed by numerous authors, a sampling of whom are listed in Table 2-10.

Omitting reference to the weathering profiles, previously discussed, a functional classification must attempt to overlay a classification of discontinuities on one or more indices of rock material behavior. One typical scheme, for example, by Franklin, et al (1971) superimposes divisions of fracture spacing and rock strength (figure

TABLE 2-10

References to Some Engineering Classification Systems for Rock

Object	For general purpose	For a special purpose
Rock Material	Coates (1964) Coates and Parsons (1966) Deere and Miller (1966) and Deere et al (1967) Underwood (1967) - shales	Bergh-Christensen and Selmer-Olsen (1970) - resistance to blasting Selmer-Olsen and Blindheim (1970) - drillability
Rock Mass	John (1962) Onodera (1970) Iida et al (1970) Muller and Hoffman (1970) Franklin et al (1971)*	Terzaghi (1946) - tunnels Lauffer (1958) - tunnels Bieniawski (1974) - tunnels Barton et al (1975) - tunnels Kruse et al (1969) - tunnel liner design Ege (1968) - tunnels in granitic rocks Obert and Duvall (1967) - mining Goodman and Duncan (1971) - rock slopes Caterpiller Tractor Co. (1966) rippability

* Best applied to rippability classification.

2-10). The rock strength is to be determined either by unconfined compressive strength tests or by the point load index, previously discussed. This approach can be useful for organizing case experiences.

A number of workers have considered the specific problem of rock mass classification for tunnel excavation and supports. Barton,

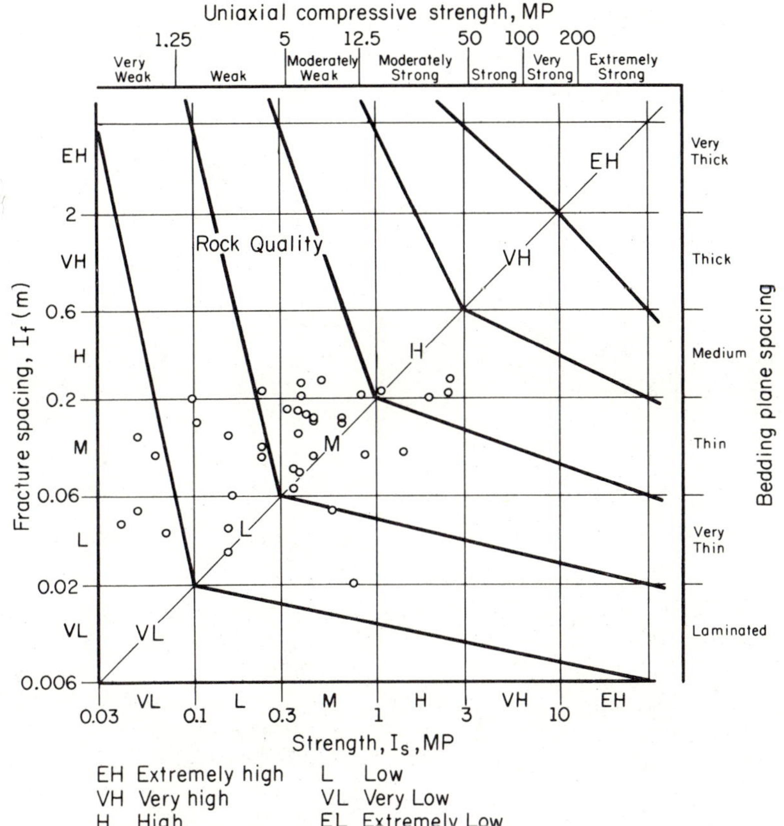

Figure 2-10. Franklin's rock quality classification; Franklin et al. (1971). The strength I_s is the point load index.

Lien and Lunde (1975), for example, adjusted six parameters by means of detailed study of 200 underground case histories, in proposing a single numerical tunneling index -- Q.

$$Q = \frac{RQD}{J_n} \cdot \frac{J_r}{J_a} \cdot \frac{J_w}{SRF} \tag{5}$$

RQD refers to the percent modified core recovery (Deere, et al., 1967), calculated from drilling logs by deleting from the "recovered" catagory all pieces of core less than four inches long*. A minimum

* Barton, Lien and Lunde state that Norwegian Geotechnical Institute geologists have found they can estimate RQD values in jointed, hard, clay-free rocks from field estimates of the number of joints per cubic meter (J_v). RQD = 115 - 3.3 J_v ; (RQD $\leq$ 100)

RQD of 10 is used in evaluating Q. The other terms in Equation 5 evaluate the number of joint sets and the roughness, alteration, water and stress conditions according to Table 2-11.

TABLE 2-11

Values of the Parameters in Barton, Lien, and Lunde's Classification

A. Number of sets of discontinuities

	Jn
massive	0.5
one set	2.0
two sets	4.0
three sets	9.0
four or more sets	15.0
crushed rock	20.0

B. Roughness of discontinuities

	Jr*
non-continuous joints	4.0
rough, wavy	3.0
smooth, wavy	2.0
rough, planar	1.5
smooth, planar	1.0
slick, planar	0.5
"filled" discontinuities	1.0

* add 1.0 if mean joint spacing exceeds 3 meters

C. Filling and wall rock alteration

	Ja
a) essentially unfilled	
healed	0.75
staining only; no alteration	1.0
silty or sandy coatings	3.0
clay coatings	4.0
b) filled	
sand or crushed rock filling	4.0
stiff clay filling <5 mm thick	6.0
soft clay filling <5 mm thick	8.0
swelling clay filling <5 mm thick	12.0
stiff clay filling >5 mm thick	10.0
soft clay filling >5mm thick	15.0
swelling clay filling >5 mm thick	20.0

Table 2-11 (continued)

D. Water conditions	Jw
dry	1.0
medium water inflow	0.66
large inflow with unfilled joints	0.5
large inflow with filled joints which wash out	0.33
high transient inflow	0.2 - 0.1
high continuous inflow	0.1 - 0.05

E. Stress reduction class	SRF*
loose rock with clay-filled discontinuities	10.0
loose rock with open discontinuities	5.0
rock at shallow depth (<50m) with clay-filled discontinuities	2.5
rock with tight, unfilled discontinuities under medium stress	1.0

* Barton et al also define SRF values corresponding to degrees of bursting, squeezing, and swelling rock conditions.

TABLE 2-12

After Barton, Lien, and Lunde (1975)

Q	Rock mass quality for tunneling
<0.01	exceptionally poor
0.01 - 0.1	extremely poor
0.1 - 1.0	very poor
1.0 - 4.0	poor
4.0 - 10.0	fair
10.10 - 40.0	good
40.0 - 100.0	very good
100.0 - 400.0	extremely good
>400.0	exceptionally good

Barton's analysis of case histories yielded a relationship for the maximum safe span (D) for an unsupported underground opening as a function of Q:

$$D = 2.1\ (Q)^{0.387} \qquad (6)$$

where D is in meters, and Q is in the range $0.001 \leq Q \leq 1{,}000$. Other functions of Q are given to select supports for different types of openings.

For example, consider the rock masses in Figures 2-9c and 2-9e with respect to tunneling at 40 meters depth. In the former case, assume the RQD is found to equal 30% and in the latter 75%. Assuming there will be no water inflow, we might estimate Q for each case as follows. For the rock of figure 2-9c:

$$Q = \frac{30}{9} \cdot \frac{1.0}{2.0} \cdot \frac{1.0}{2.0} = 0.83 \qquad (7)$$

According to Table 2-12, this classifies as <u>very poor rock</u>; the maximum unsupported span according to (6) is about two meters. For the rock of figure 2-9e:

$$Q = \frac{75}{15} \cdot \frac{1.5}{1.0} \cdot \frac{1.0}{1.5} = 5 \qquad (8)$$

This qualifies as <u>fair rock</u>; the maximum unsupported span is about four meters.

Barton's classification scheme has considerable potential for engineering for underground works as well as for generalization of experiences in other areas of engineering. A somewhat similar classification, developed by Bieniawski (1974) is presented in Table 2-13. Of course, no classification system can assign a name as generally informative as a careful description of the geological environment, the rock material, the weathering profile, and the system of discontinuities. Table 2-14 summarizes those factors appropriate in a geotechnical description of a rock mass. Table 2-15 is a standardized data sheet developed by the South African Central Scientific and Industrial Research Organization (CSIRO) and is useful for providing input for Bieniawski's classification.

TABLE 2-13

Geomechanics Classification of Jointed Rock Masses

A. CLASSIFICATION PARAMETERS AND THEIR RATINGS

	Parameter								
1	Strength of intact rock material	Point-load strength index	8 MPa	4 - 8 MPa	2 - 4 MPa	1 - 2 MPa	Use of uniaxial compressive test preferred		
		Uniaxial compressive strength	200 MPa	100 - 200 MPa	50 - 100 MPa	25 - 50 MPa	10-25 MPa	3-10 MPa	1-3 MPa
	Rating		15	12	7	4	2	1	0
2	Drill core quality RQD		90% - 100%	75% - 90%	50% - 75%	25% - 50%	25%		
	Rating		20	17	13	8	3		
3	Spacing of joints		> 3 m	1 - 3 m	0,3 - 1 m	50 - 300 mm	50 mm		
	Rating		30	25	20	10	5		
4	Condition of joints		Very rough surfaces. Not continuous No seperation Hard joint wall rock.	Slightly rough surfaces. Separation < 1 mm Hard joint wall rock	Slightly rough surfaces. Separation < 1 mm Soft joint wall rock	Slickensided surfaces OR Gouge < 5 mm thick OR Joints open 1-5 mm. Continuous joints	Soft gouge 5 mm thick OR Joints open 5 mm. Continuous joints		
	Rating		25	20	12	6	0		
5	Ground water	Inflow per 10m tunnel length	None		25 litres/min	25 - 125 litres/min	125 litres/min		
			OR		OR	OR	OR		
		Ratio $\frac{\text{joint water pressure}}{\text{major principal stress}}$	0		0,0 - 0.2	0.2 - 0.5	0.5		
			OR		OR	OR	OR		
		General conditions	Completely dry		Moist only (interstitial water)	Water under moderate pressure	Severe water problems		
	Rating		10		7	4	0		

B. ADJUSTMENT FOR JOINT ORIENTATIONS

Strike and dip orientations of joints		Very favourable	Favourable	Fair	Unfavourable	Very unfavourable
Ratings	Tunnels	0	-2	-5	-10	-12
	Foundations	0	-2	-7	-15	-25
	Slopes	0	-5	-25	-50	-60

C. ROCK MASS CLASSES AND THEIR RATINGS

Class No	I	II	III	IV	V
Description	Very good rock	Good rock	Fair rock	Poor rock	Very poor rock
Rating	100 ← 90	90 ← 70	70 ← 50	50 ← 25	25

D. MEANING OF ROCK MASS CLASSES

Class No	I	II	III	IV	V
Average stand-up time	10 years for 5 m span	6 months for 4 m span	1 week for 3 m span	5 hours for 1.5 m span	10 minutes for 0.5 m span
Cohesion of the rock mass	300 kPa	200 - 300 kPa	150 - 200 kPa	100 - 150 kPa	100 kPa
Friction angle of the rock mass	45°	40° - 45°	35° - 40°	30° - 35	30
Caveability of ore	Very poor	Will not cave readily Large fragments	Fair	Will cave readily Good fragmentation	Very good

TABLE 2-14

Some Factors to be Considered in a Geotechnical Description of a Rock Mass

A. Rock material

Petrologic description -- rock name, texture, fabric, principal and accessory minerals; nature of cement; alteration effects. Presence of alterable minerals such as gypsum, pyrrhotite, etc. should especially be noted.

Classification as "rock," "weathered rock" or "soil-like rock" according to results of simple tests (see Table 2-1).

Weatherability according to slake-durability or other test.

Mechanical properties according to an index test -- e.g. Schmidt hammer, point load test, or scratch hardness.

Degree of weathering according to laboratory index tests or mineralogic criteria.

State of fissuring, determined from polished sections or thin sections or by results of wave velocity measurements, tension tests, volumetric compression, or radial permeability tests.

Micro structures in the hand specimen -- bedding, foliation, etc.

B. Weathering Profiles

Description and classification of all the intermediate weathering products and their spatial arrangement together with results of laboratory tests indicative of their mechanical properties.

Description of joint properties in the different stages of weathering.

C. Discontinuities

Preferred orientations and spacings of each set, structural name, (e.g. bedding, joint) for each set; roughness angles versus wave length and description of wall rock as wavy, rough, smooth, or slickensided; note roughness anisotropy.

Wall rock scratch hardness expressed by a standard terminology, or strength as measured by Schmidt hammer.

Table 2-14 (continued)

Filling material: thickness; completeness of filling; compactness; composition; % clay and soil properties; classification as: swelling, erodible, soluble, or or stable.

lnterlocking and tightness of fit: healed, close, open, cavernous (or loosened).

Other features: estimate of relative extent; chemistry of water; wlll rock alteration.

TABLE 2-15

Input Data Form: Geomechanics Classification of Jointed Rock Masses

INPUT DATA FORM: GEOMECHANICS CLASSIFICATION OF JOINTED ROCK MASSES

Name of project:
Site of survey:
Conducted by:
Date:

STRUCTURAL REGION

ROCK TYPE AND ORIGIN

DRILL CORE QUALITY R.Q.D.

Very good quality:	90 - 100%
Good quality:	75 - 90%
Fair quality:	50 - 75%
Poor quality:	25 - 50%
Very poor quality:	< 25%

NOTE: R.Q.D. Rock Quality Designation in accordance with the method of Deere

WEATHERING

Unweathered
Slightly weathered
Moderately weathered
Highly weathered
Completely weathered

GROUND WATER

INFLOW per 10 m of tunnel length litres/minute
or
WATER PRESSURE, kPa
or
GENERAL CONDITIONS (completely dry, moist only, water under pressure, severe problems)

STRENGTH OF INTACT ROCK MATERIAL

Designation	Uniaxial compressive strength	Point-load strength index
Very high:	Over 200 MPa	> 8 MPa
High:	100 - 200 MPa	4 - 8 MPa
Medium	50 - 100 MPa	2 - 4 MPa
Low:	25 - 50 MPa	1 - 2 MPa
Very low:	10 - 25 MPa	< 1 MPa
	3 - 10 MPa	
	1 - 3 MPa	

SPACING OF JOINTS

		Set 1	Set 2	Set 3	Set 4
Very wide:	Over 3 m				
Wide:	1 - 3 m				
Moderately close:	0,3 - 1 m				
Close:	50 - 300 mm				
Very close:	< 50 mm				

NOTE: These values are obtained from a joint survey and not from borehole logs. Provide data for each joint set.

STRIKE AND DIP ORIENTATIONS

Set 1	Strike: (average)	(from	to	)	Dip: (angle) (direction)	
Set 2	Strike:	(from	to	)	Dip:	
Set 3	Strike:	(from	to	)	Dip:	
Set 4	Strike:	(from	to	)	Dip:	

NOTE: Provide data for each joint set. Refer all directions to magnetic north.

CONDITION OF JOINTS

		Set 1	Set 2	Set 3	Set 4
CONTINUITY					
Not continuous, no gouge					
with gouge					
Continuous, no gouge					
with gouge					
SEPARATION					
Very tight joints:	Less than 0,1 mm				
Tight joints:	0,1 - 1 mm				
Moderately open joints:	1 - 5 mm				
Open joints:	More than 5 mm				
ROUGHNESS					
Very rough surfaces:					
Rough surfaces:					
Slightly rough surfaces:					
Smooth surfaces					
Slickensided surfaces:					
JOINT WALL ROCK					
Hard rock:					
Medium hard rock					
Soft rock					

NOTE: Provide data for each joint set

MAJOR FAULTS OR FOLDS

Describe major faults and folds specifying their locality, nature and orientations.

GENERAL REMARKS AND ADDITIONAL DATA

If gouge is present specify its type, thickness, continuity and consistency.
Describe waviness of joints.
Assess regional stresses.

NOTE: The data on this form constitute the minimum required for engineering design. The geologist should, however, supply any further information which he considers relevant.

3
principles of stereographic projection and joint surveys

CONFORMAL STEREOGRAPHIC PROJECTION

Problems in geology and engineering which involve relationships of lines and planes in space can be solved or simplified through reference to stereographic projection. This is a method of mapping the surface of a sphere onto a plane, used in crystallography, cartography, navigation, structural geology, geophysics and other fields. By using the stereographic projection alone, one can solve problems involving orientations of lines and planes such as determining the plunge of the intersection of discontinuities. Where the position as well as the orientation of a line or a plane is involved, such as in analysis of overturning in a potential rock slide, the stereographic projection must be supplemented by other constructions. In this chapter we will consider the basic constructions necessary for applications in geological engineering and an introduction to statistics with reference to joint surveys. Applications in photogrammetry, the kinematics and statics of rock blocks, orientation of structures in drill cores, and assessment of the state of stress in rocks will be considered in later chapters. With these applications in mind, this chapter provides a thorough consideration of the properties and techniques of stereographic projection, including basic constructions.

Figure 3-1 illustrates the geometric principles of stereographic projection. Any point on the surface of a sphere is projected onto a diametral plane of the sphere (the projection plane) by means of construction lines radiating from a focus; the focus is fixed at one of the ends of the perpendicular to the projection plane. In figure 3-1, the projection plane is horizontal and the focus is at the bottom of the sphere. Points in the upper hemisphere will therefore appear inside the projection of the horizontal plane, which forms the "primitive circle", a name derived from crystallography. Line ON_R, for example, which pierces the sphere at N_R in the upper hemisphere projects to point $\hat{n}_R$ on the horizontal diametral plane. An inclined diametral plane, which intersects the sphere as a "great circle", projects as a circular arc as shown inside the

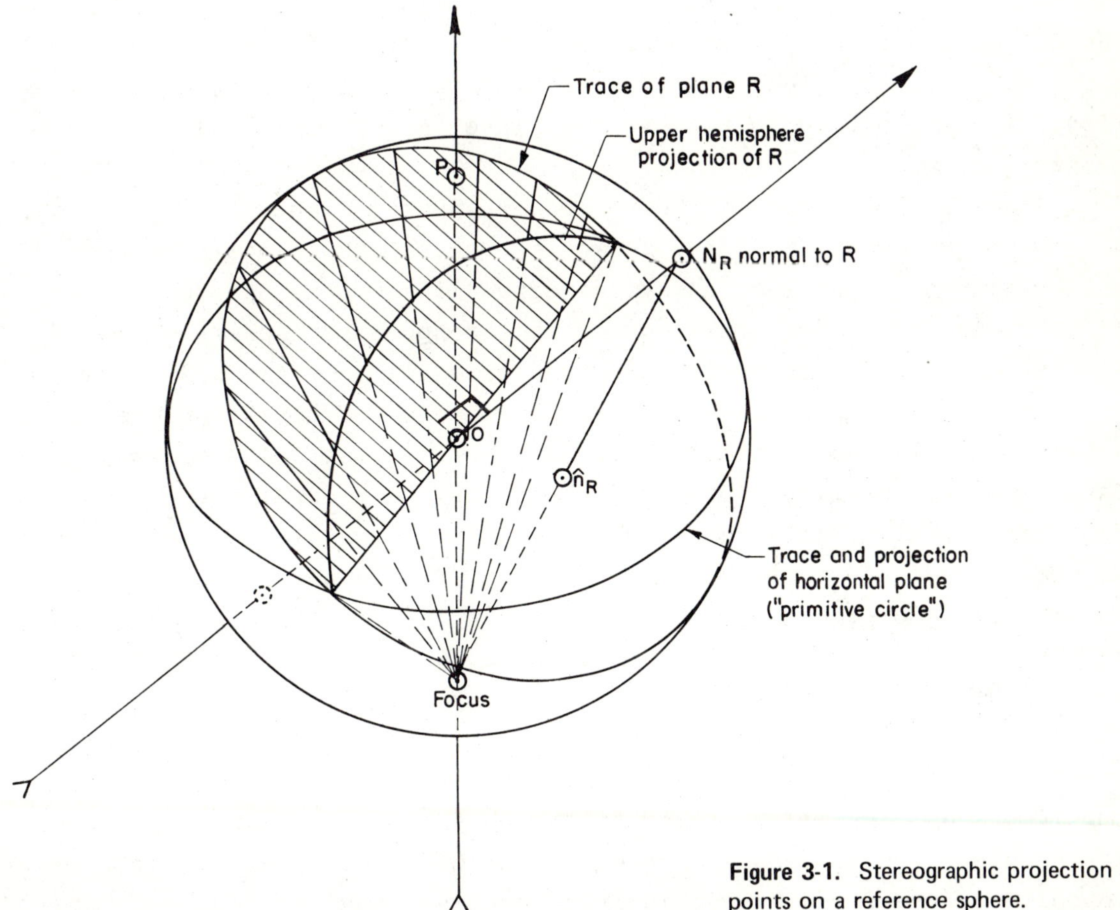

Figure 3-1. Stereographic projection of points on a reference sphere.

primitive circle. This arc represents the half plane in the upper hemisphere; the completion of the circle in the projection plane beyond the primitive would represent the projection of the half plane in the lower hemisphere. This fact derives from the theorem, proved in Phillips' text on Crystallography (1946) that the projection of a circle on the reference sphere is a circle on the projection plane. The projection is conformal; angles measured between lines on the sphere are preserved by the projection. As will be seen later, a variant of the projection exists that loses this quality (the "equal area" projection).

In dealing with problems of orientation only, i.e. where the spatial positions of lines and planes are not considered, one may move a line or plane parallel to itself to pass through the center of a single reference sphere. Any plane, then, can be described by the projection of a great circle, and any line by the projection of a piercing point on the reference sphere. A great circle, by convention, will be assumed to represent a plane, and a point will represent a line.

A vertical section of the reference sphere, figure 3-2a, provides a complete description of the geometric relationships basic to the method of projection. An upward directed line OP through the center of the reference sphere will appear in the upper hemisphere projection at a radial distance from the center equal to $r \tan \alpha/2$ in the direction of its bearing, where r is the radius of the reference sphere and α is the complement of the angle of rise. The opposite (tail end) of the line (-OP) projects to point -p outside the primitive circle, at a radial distance equal to $r \cot \alpha/2$ in the opposite direction. If -OP represents the dip vector of a plane, points p and -p are points on the great circle representing its stereographic projection. The bisector of line (-p) (p) is the center of this great circle which may be constructed with a compass, figure 3-2b. In this figure, the plan view of the projection plane and the vertical section of the reference sphere in the direction of dip have been superimposed. It will be found that the center, q, is the same point as the stereographic projection of a line plunging in the direction opposite to the dip at an angle from vertical equal to twice the dip

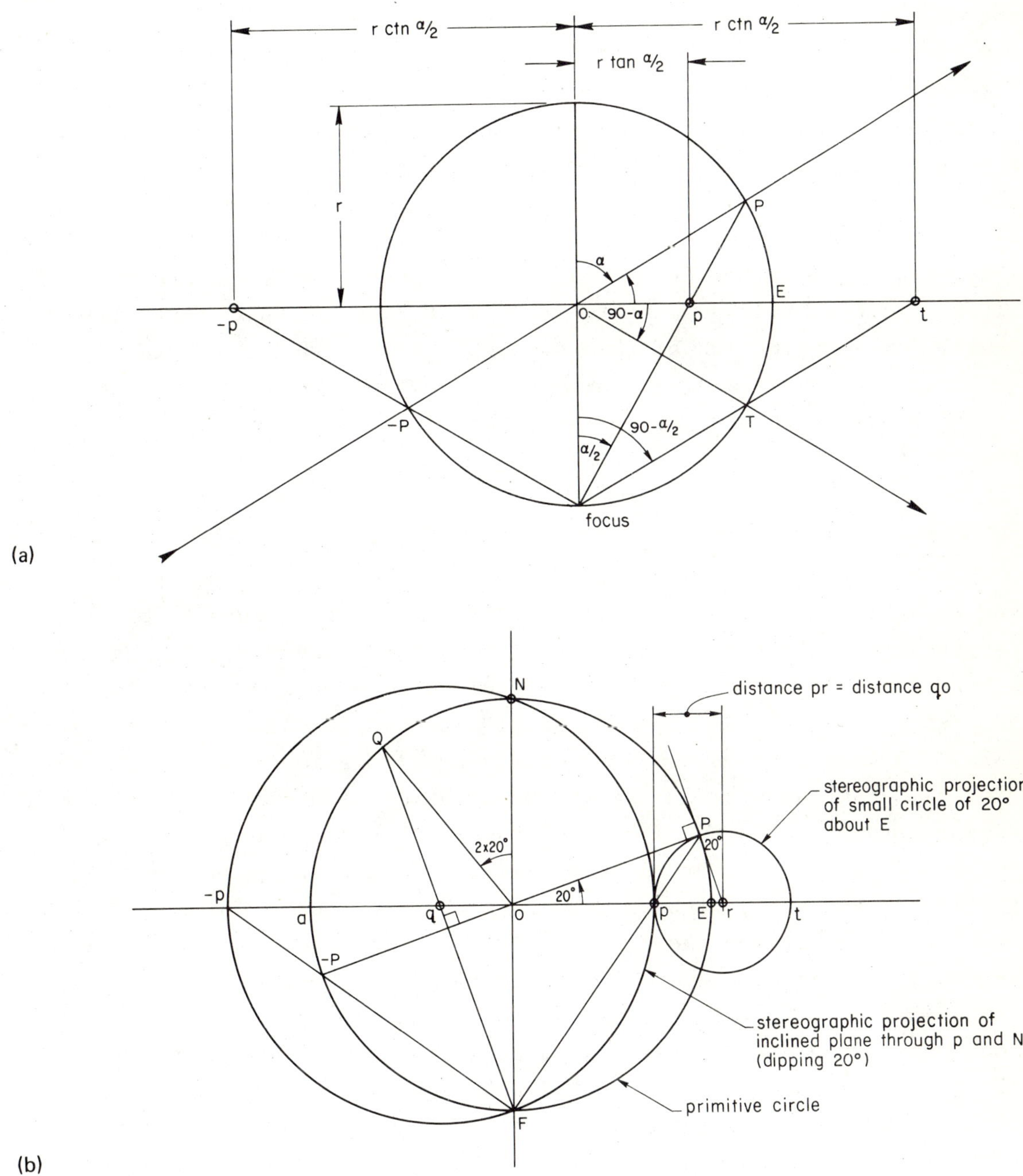

Figure 3-2. (a) Vertical section of the reference sphere through line PO. (b) Upper hemisphere stereographic projection (plan) superimposed on the vertical section of the reference sphere through line PO.

angle. The center can also be located graphically, as shown in figure 3-2b, by the intersection of the projection plane (OE) and a perpendicular to the dip vector through the focus (line FQ). It will also be useful to project small circles on the reference sphere formed by the double cone of the lines making a fixed angle with a given axis. For example, in figure 3-2a, we consider the cone formed by lines making an angle of 90-α with the horizontal (OE); two of the lines of this cone are OP, and OT. The bisector of the projections of these lines, points p and t, is point r (figure 3-2b) which is the center for construction of the required circle. When the axis of the small circle is horizontal, the center r can be found as the point of intersection of the projection plane OE and a tangent to the reference sphere at P.

Stereonets

Making use of the construction for the centers of great circles and small circles about a horizontal axis given in figure 3-2b, it is possible to construct a family of inclined planes having a common horizontal intersection and a family of small circles about this intersection, creating an "equatorial" stereonet. (Wulff net, figure 3-3a). If one regards the axis of the small circles (the intersection of the great circles) as the polar axis, (i.e. the focus of projection is on the equator) then the family of great circles can be viewed as the projection of the lines of longitude and the small circles as the projections of the lines of latitude. The equatorial net is convenient for the tracing of planes and the measurement of angles from tracings, as shown in figure 3-6. When dealing only with lines in space, it may be easier to use a net constructed about a focus on the pole. The resulting "polar" stereonet shows the lines of latitude as circles constructed about the center, while the lines of longitude are straight lines, i.e. vertical planes, through the center. By marking any point on the primitive circle as north, a line may be plotted by the intersection of the required bearing line and small circle of plunge. For example, in figure 3-3b, point A is 20 degrees from horizontal to the North 20 E. (The angle is above or below the horizontal according to the choice of hemisphere). It

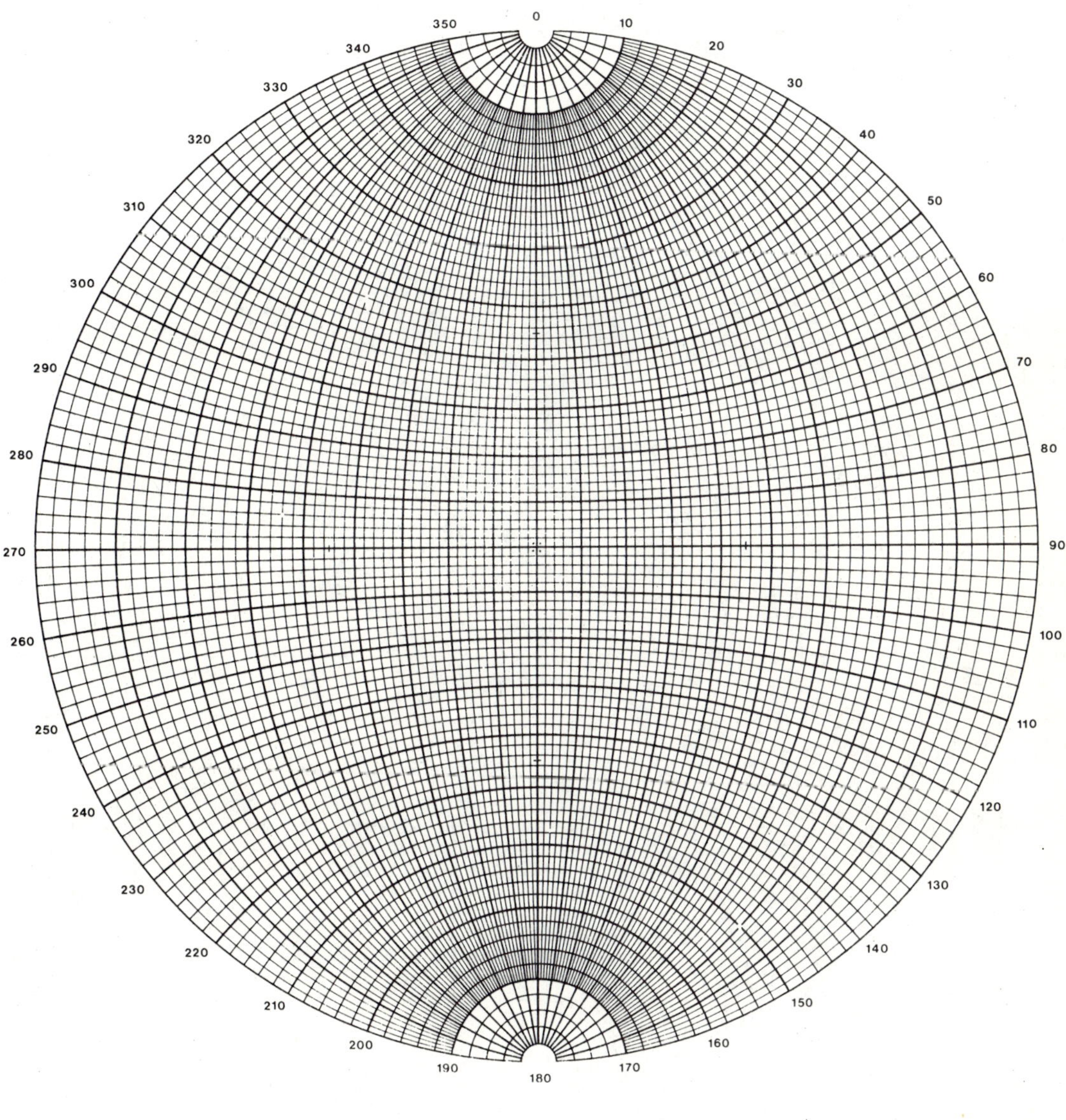

(a)

Figure 3-3. (a) Equatorial conformal stereonet; (computed by Dr. C. St. John; reproduced with permission). (b) Polar conformal stereonet; (computed by Dr. C. St. John; reproduced with permission).

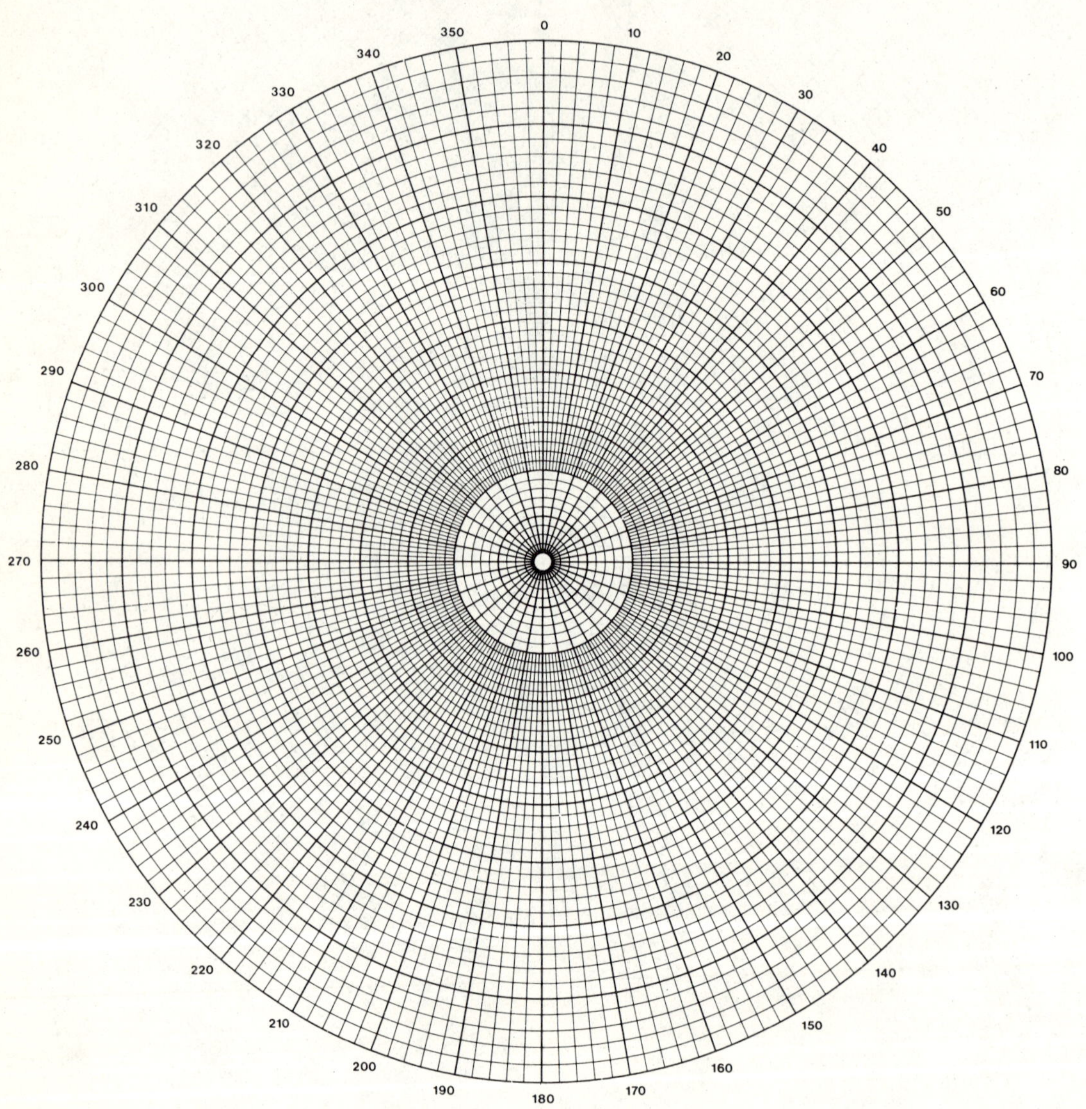
0
10
20
30
40
50
60
70
80
90
100
110
120
130
140
150
160
170
180
190
200
210
220
230
240
250
260
270
280
290
300
310
320
330
340
350

(b)

is useful to print a supply of tracing sheets having a polar conformal net lightly drawn upon them for use as overlays on the equatorial net. Lines can then be read directly from the tracing. (Until familiarity is gained with the basic constructions, however, it might be better to use a clear tracing to reduce the number of lines and arcs).

For statistical discussions of orientations, it is preferable to use a modification of the stereographic projection which produces equal areas for a given solid angle subtended anywhere on the sphere. It will be noted, in figure 3-3a and 3-3b, that a "square" bounded by two degree variation of latitude and longitude grows smaller towards the center of the projection. An equal area projection can be developed from the stereographic projection as shown in figure 3-4. Imagine a focus at the bottom of the reference sphere and the projection plane tangent to the reference sphere at the top. The stereographic projection referred to this plane becomes larger than the one referred to the diametral plane, but remains geometrically similar to it. The equal area projection of line OP, is obtained by swinging an arc from N (the top of the sphere) to point P, giving

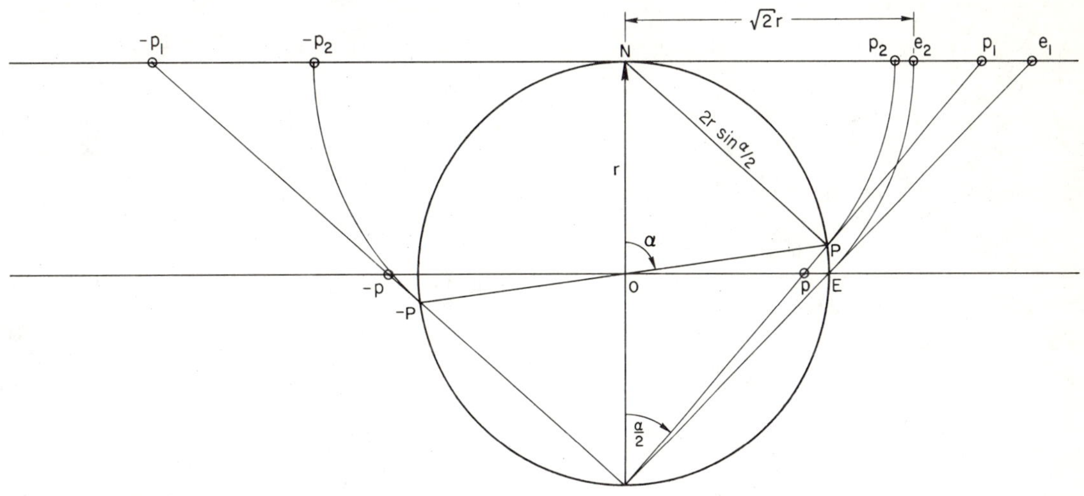

Figure 3-4. Equal area (Lambert) versus equal angle (Wulff) projections. The distance from the center of the reference sphere to the projection of OP (referred to plane OE) is r tan $(\alpha/2)$ in stereographic projection and $\sqrt{2}\, r \sin(\alpha/2)$ in equal area projection.

point p_2. So that the radius of the resulting primitive circle will be r, it is necessary to divide the projected lengths by $\sqrt{2}$ giving a distance from the center of $\sqrt{2}$ r sin ($\alpha/2$) for a line plunging at 90-α from horizontal. Equatorial and polar equal area nets (Lambert nets) are given in figure 3-5a and b. The equal area projection is not conformal and projections can not be made using circular constructions as for the stereographic projection. However, all the basic constructions required can be performed on a tracing by sketching lines from the nets given, exactly as presented for the stereographic projections in examples to be discussed.

BASIC CONSTRUCTIONS

In a number of examples basic manipulations with stereograms will be demonstrated, first by construction, and then by use of a tracing on the conformal equatorial stereonet, as depicted in figure 3-6.

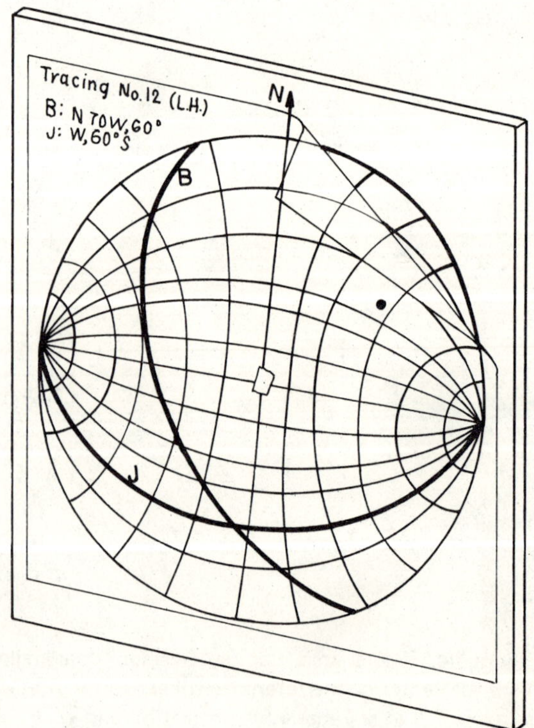

Figure 3-6. Use of the stereonet with a tracing.

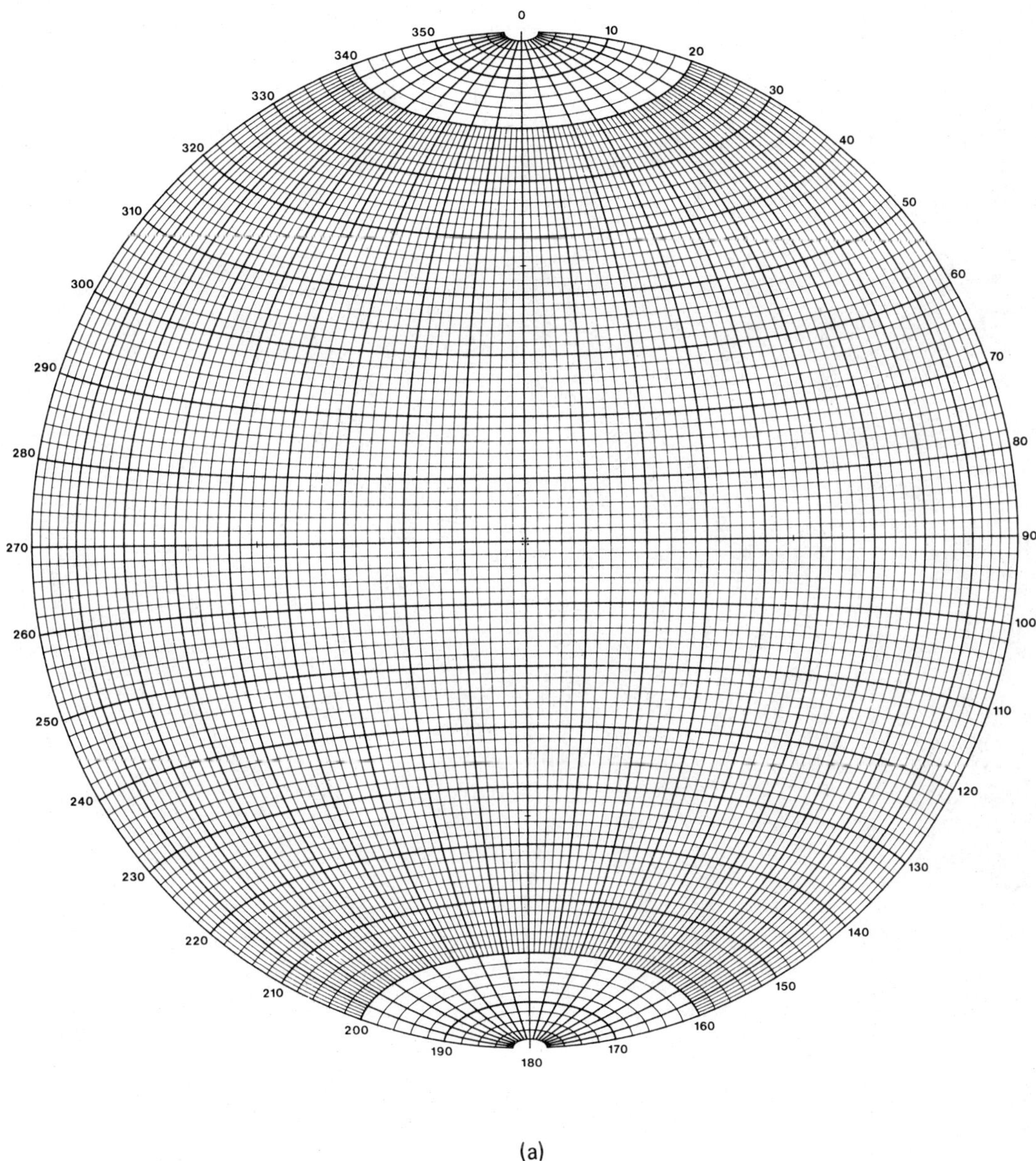

Figure 3-5. (a) Equatorial equal area stereonet; (computed by Dr. C. St. John; reproduced with permission). (b) Polar equal area stereonet; (computed by Dr. C. St. John; reproduced with permission).

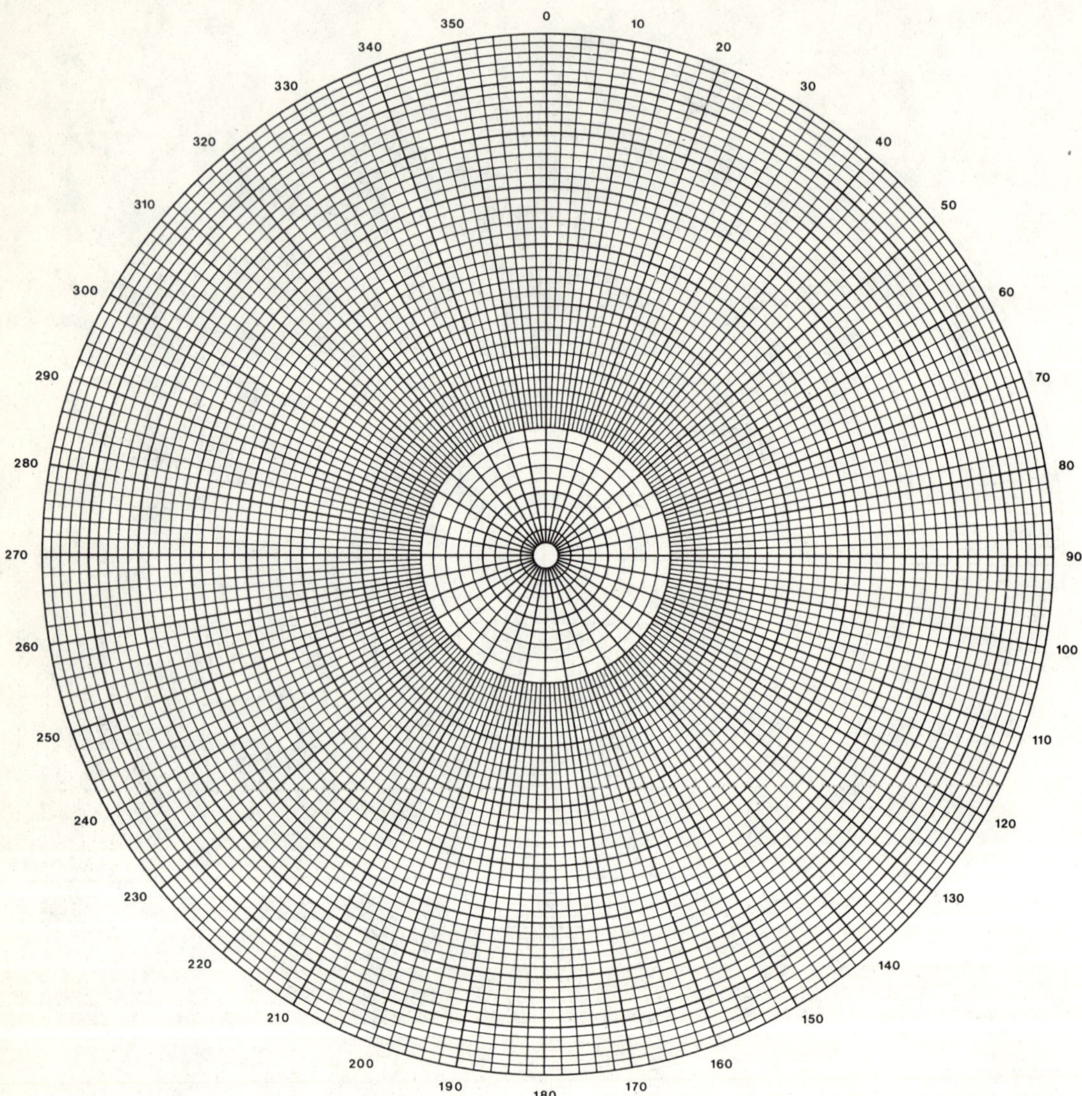
0
10
20
30
40
50
60
70
80
90
100
110
120
130
140
150
160
170
180
190
200
210
220
230
240
250
260
270
280
290
300
310
320
330
340
350

(b)

Construction of a Line in the Upper or Lower Hemisphere

Figure 3-7a reviews the steps for projection of a line. Given a line rising 20 degrees to the N30 E, draw a circle of convenient radius, mark north and locate the bearing on the primitive circle, point B, by measuring 30 degrees from north, subtended at the center O. For purposes of construction, draw a line through OB and a perpendicular to OB intersecting the primitive at F. We may view F as the focus for an upper hemisphere projection (an upper hemisphere projection was chosen because the required line is in the upper hemisphere). OB is the trace of the projection plane and the required line OP can be drawn 20 degrees above OB. Line FP intersects the projection plane at p, the required point. The opposite to OP can be plotted similarly by continuing OP in the opposite sense to point -P on the primitive circle, whence line (F)(-P) yields the required point -p as it intersects the projection plane. Points near the focus will project an inconvenient distance beyond the primitive and the focus itself can not be plotted. However, this is really no limita-

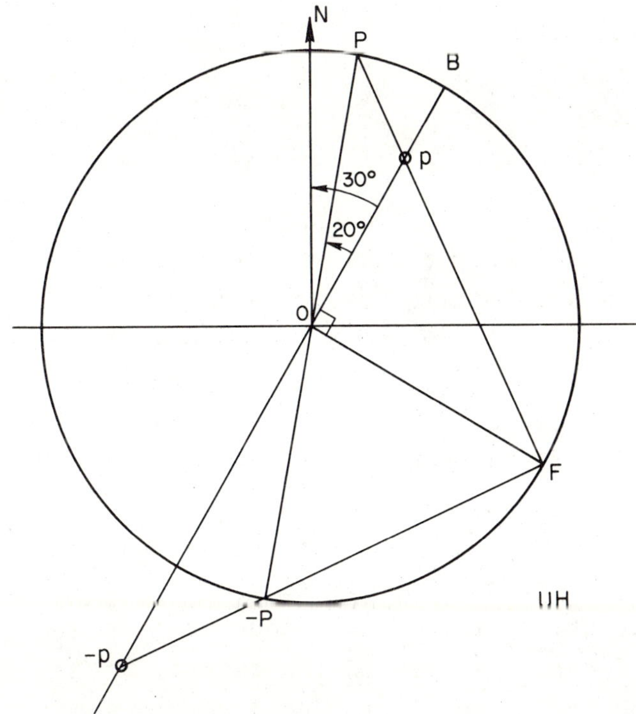

Plot line p rising 20° to N30E

Plot opposite -p

Figure 3-7. (a) Stereographic projection of a line and its opposite.

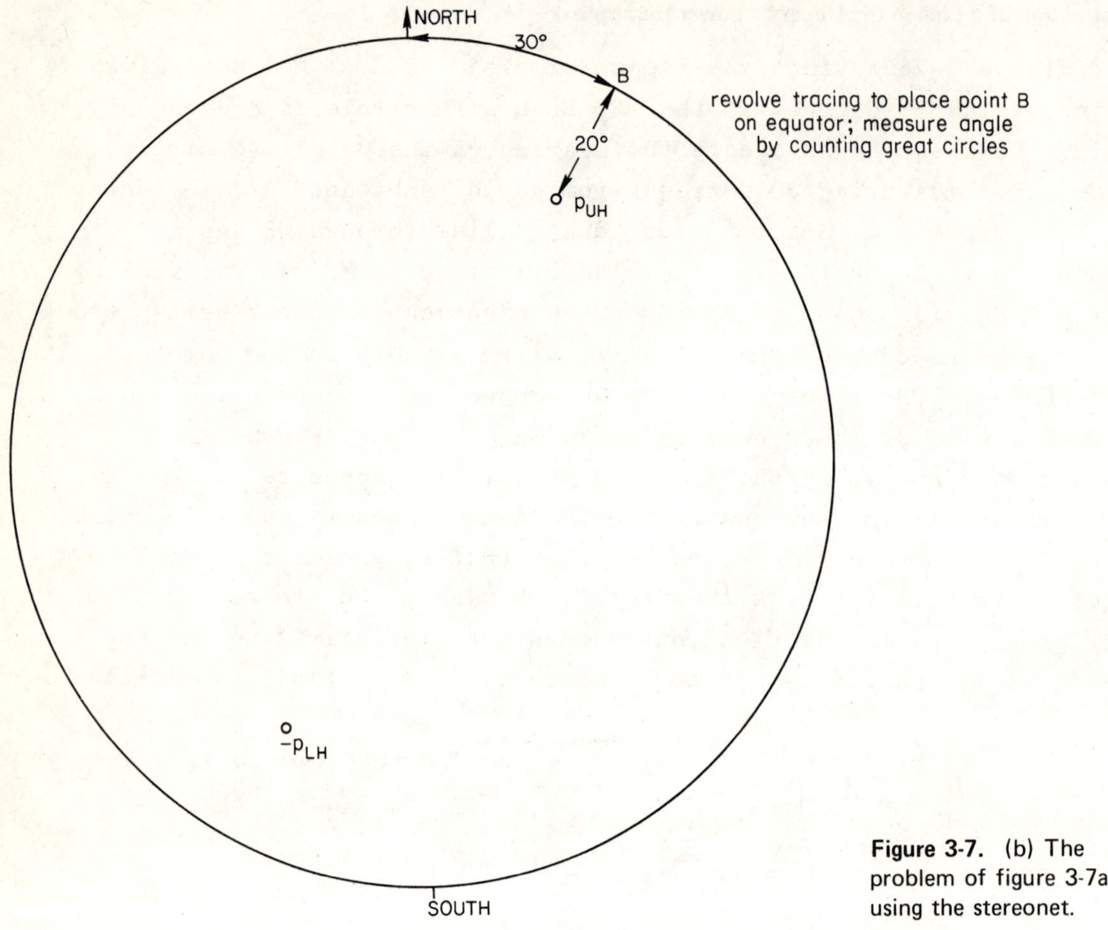

Figure 3-7. (b) The problem of figure 3-7a using the stereonet.

tion because one can switch the hemisphere, or work with the opposites of inconveniently located points without difficulty. Figure 3-8 shows the relationship between upper and lower hemisphere plots of a point and its opposite. In many problems of interest, as for example in joint surveys, structural geology calculations, and core orientation calculations, we do not distinguish a point from its opposite and all required constructions can be confined to one hemisphere. However, in statics, where we associate a direction with certain lines, the entire sphere must be mapped. In this case, as will be shown in chapter 6, we may use a single continuous projection with focus selected carefully to reduce the size of the figure, or we may use two side by side projections, one for the upper hemisphere,

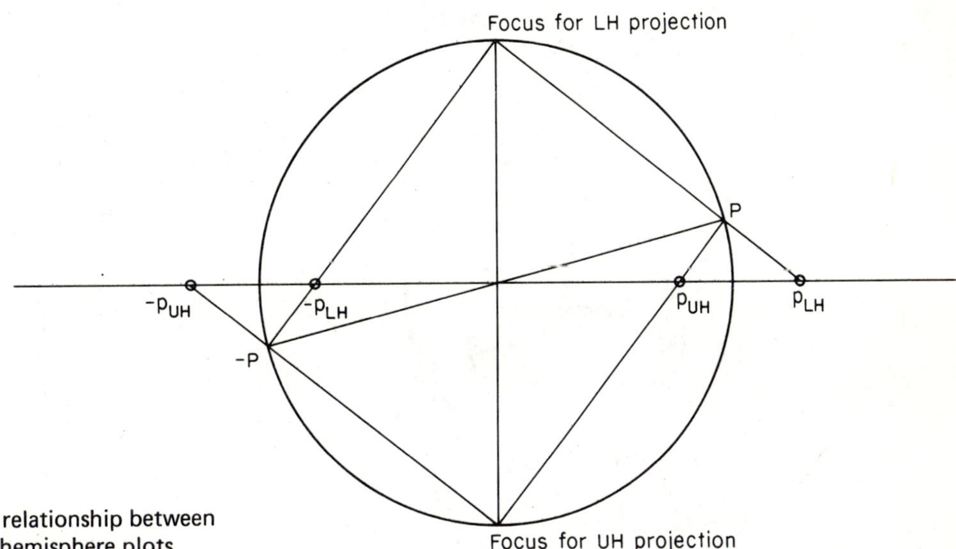

Figure 3-8. The relationship between upper and lower hemisphere plots.

marked U.H., and the other for the lower hemisphere, marked L.H. Figure 3-7b shows how the tracing can be used to facilitate the plotting of a line. Mount the stereonet on cardboard and push a thumb tack (drawing pin) through the center so that the tracing may be revolved about the center (protect the tracing with a piece of transparent tape to prevent enlargement of the hole with use). Mark point B 30 degrees from the arbitrarily selected north position and revolve the tracing to superimpose B on the equator. The angle of rise, 20 degrees, is measured along the vertical plane represented by the equator, counting ten great circles. The opposite to point p can not be plotted using the net (the opposite to the lower hemisphere projection of p is shown).

Plotting a Plane and its Normal

Figure 3-9a shows how to plot a plane given its strike and dip.* Plane R strikes N 20 W and dips 13 degrees to the NE. The strike vector, S_R, is horizontal so lies 20 degrees west of north along the primitive circle. The other end of the strike vector ($-S_R$) is on the opposite end of the diameter through S_R. The dip vector D_R and

---

* The terms strike and dip are discussed in chapter 4, page xxx

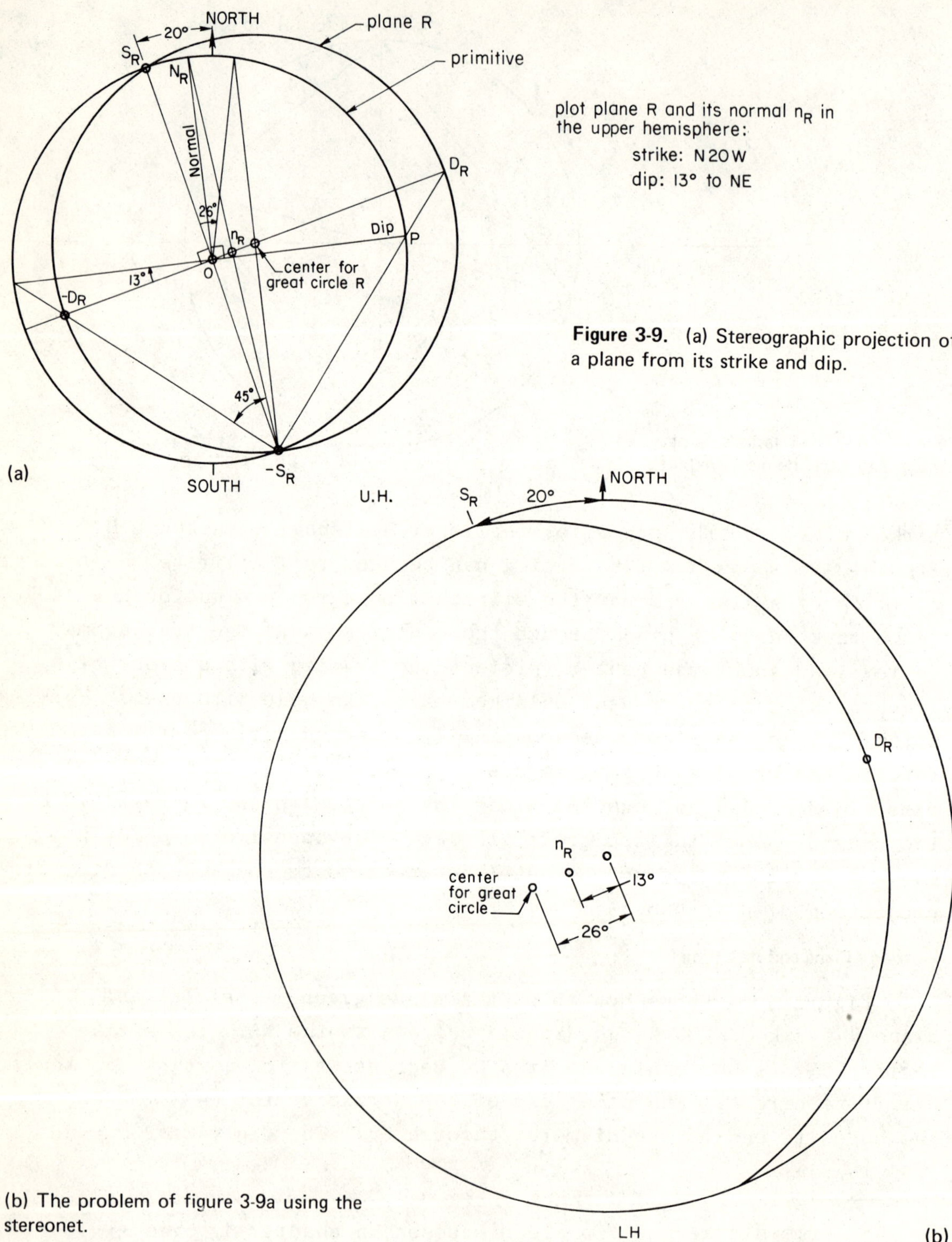

Figure 3-9. (a) Stereographic projection of a plane from its strike and dip.

(b) The problem of figure 3-9a using the stereonet.

its opposite $-D_R$ are constructed as in figure 3-7a. To construct the plane, simply draw a circle through S_R, $-S_R$, D_R, and $-D_R$. It will be helpful to remember that the center for this circle is at the intersection of: the trace of the projection plane (OD_R); and the perpendicular to the line of dip (OP) through $-S_R$ as shown.

The normal to a plane R is frequently required. This may be done by finding the dip vector D_R in the plane, recovering the line of dip in vertical section through OD_R, and projecting the piercing point N_R of the perpendicular to it erected at O. For example, in figure 3-7a, if we had been given point p in the stereographic projection, we could have constructed point F, and traced line Fp to determine point P, in this way recovering the actual line OP which point p represents. A simpler way to plot the normal to plane R with strike and dip vectors projecting at points S_R and D_R, figure 3-9a, is as the intersection of diameter OD_R and a line making an angle of 45 degrees with D_RS_R or $-D_R-S_R$ as shown.

To plot a plane and its normal using the stereonet and tracing, mark the strike and dip vectors as in figure 3-7b, and trace the great circle passing through both; since S_R is horizontal, this great circle will be found by revolving the tracing till S_R is over the polar position. To improve accuracy in problems involving construction of great circles, the circle should be constructed with a compass rather than traced free hand. The center is inside the net for planes dipping less than 45 degrees; as shown, the center is located as a point 26 degrees from vertical in the direction opposed to the dip. (The angle from vertical to the center is always twice the amount of dip.)

The Angle Between Two Lines

The angle between two lines (remembering the convention that all lines pass through the center of the reference sphere) is measured in their common plane. Given two lines p and q, (figure 3-10a), find the opposite to each and construct the great circle pq (passing through all four points). Construct the normal n_{pq} to plane pq as in the previous example, figure 3-9a. Now draw straight lines n_{pq} p and n_{pq} q, intersecting the primitive circle at points p_1 and q_1

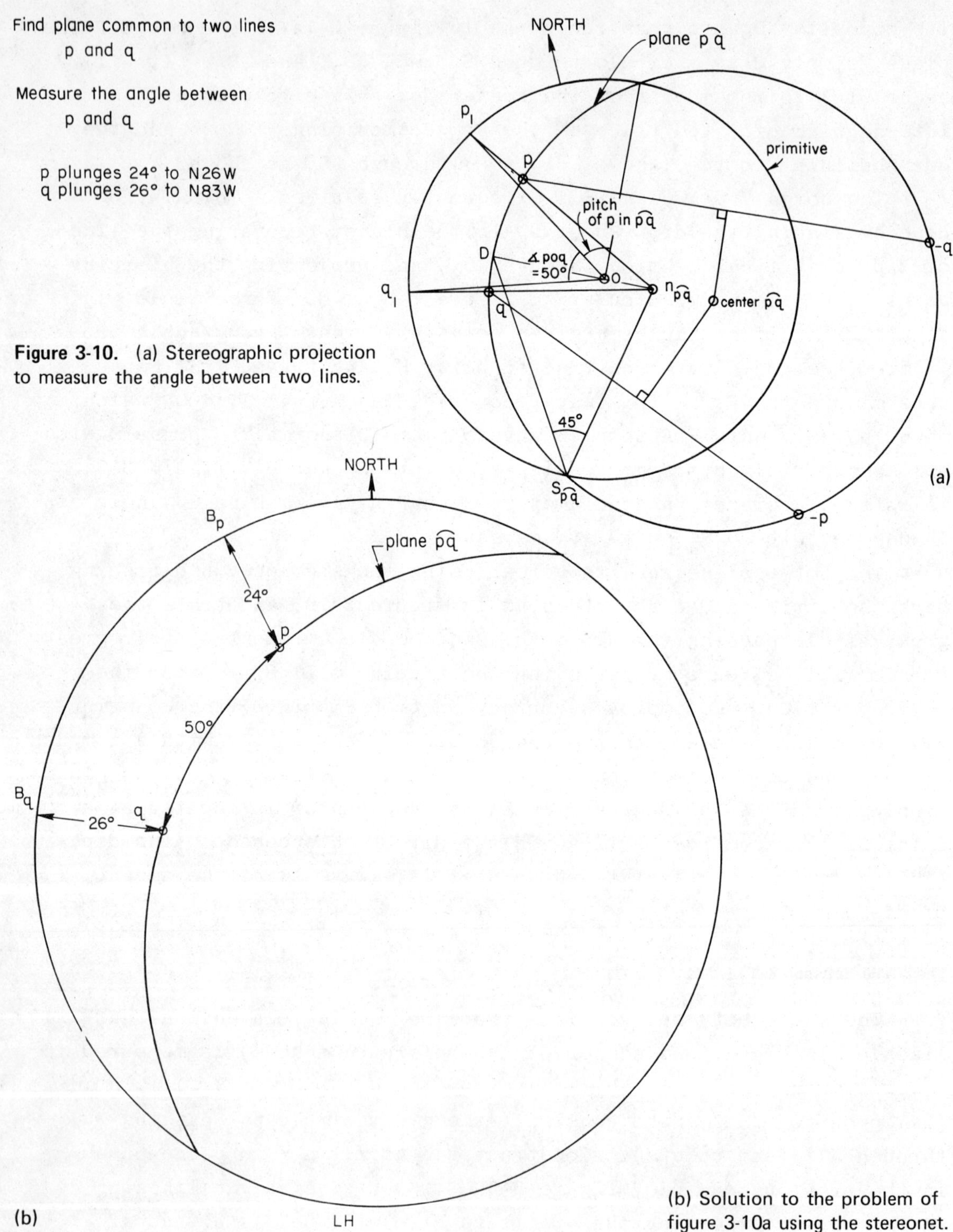

Figure 3-10. (a) Stereographic projection to measure the angle between two lines.

(b) Solution to the problem of figure 3-10a using the stereonet.

respectively. The required angle from p to q, is the angle p_1Oq_1, subtended at the center of the primitive circle.

This problem is easier on the stereonet, figure 3-10b. Plot points p and q and revolve the tracing until they lie along a common great circle. Angle pOq is measured along this great circle, which is calibrated by its intersections with small circles.

The Line of Intersection of Two Planes

Two planes will project as two circles, which intersect as two points, figure 3-11a. These are the intersection lines in each hemisphere. On the stereonet, figure 3-11b, only one of the intersections will be given, (the other being its opposite).

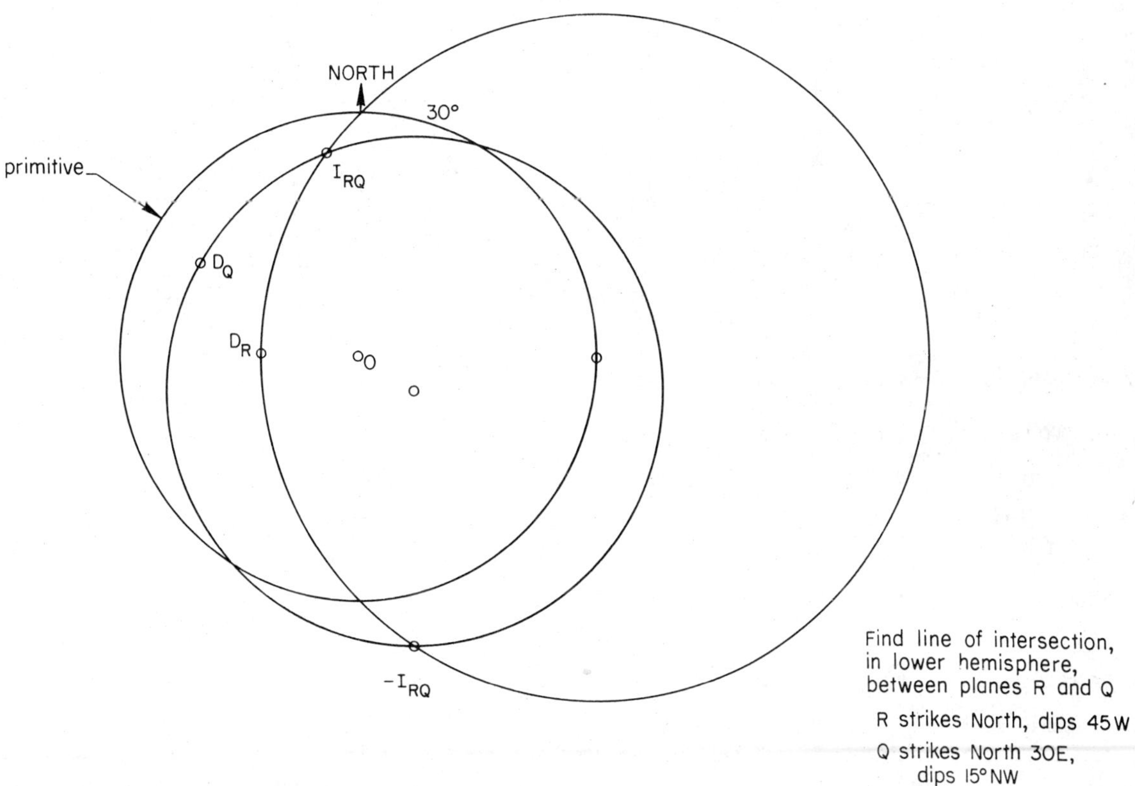

Figure 3-11. (a) Stereographic projection to find the line of intersection of two planes.

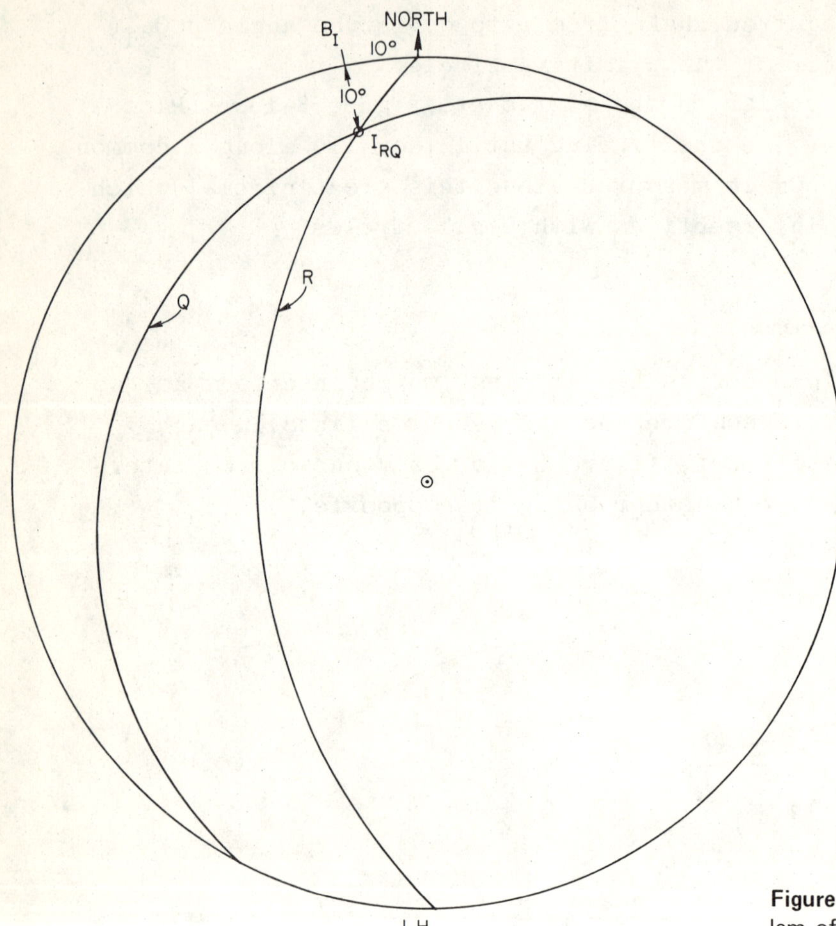

Figure 3-11. (b) Solution to problem of 3-11a using the stereonet.

Orthographic Projection of a Line on a Plane

This construction is used to determine angles in perspective views, and in other applications as illustrated in later chapters. Figure 3-12a shows the method of solution, given line p and plane R. Construct the normal to R (n_R). The orthographic projection of p in R (line J) is the intersection of the great circles n_Rp and R. The attitude of J is most often referred to plane R; in this case J pitches* 16 degrees below horizontal from the north in plane R. The solution on the stereonet is given in figure 3-12b.

---

* An angle of pitch, or rake, is an angle measured in a non-vertical plane.

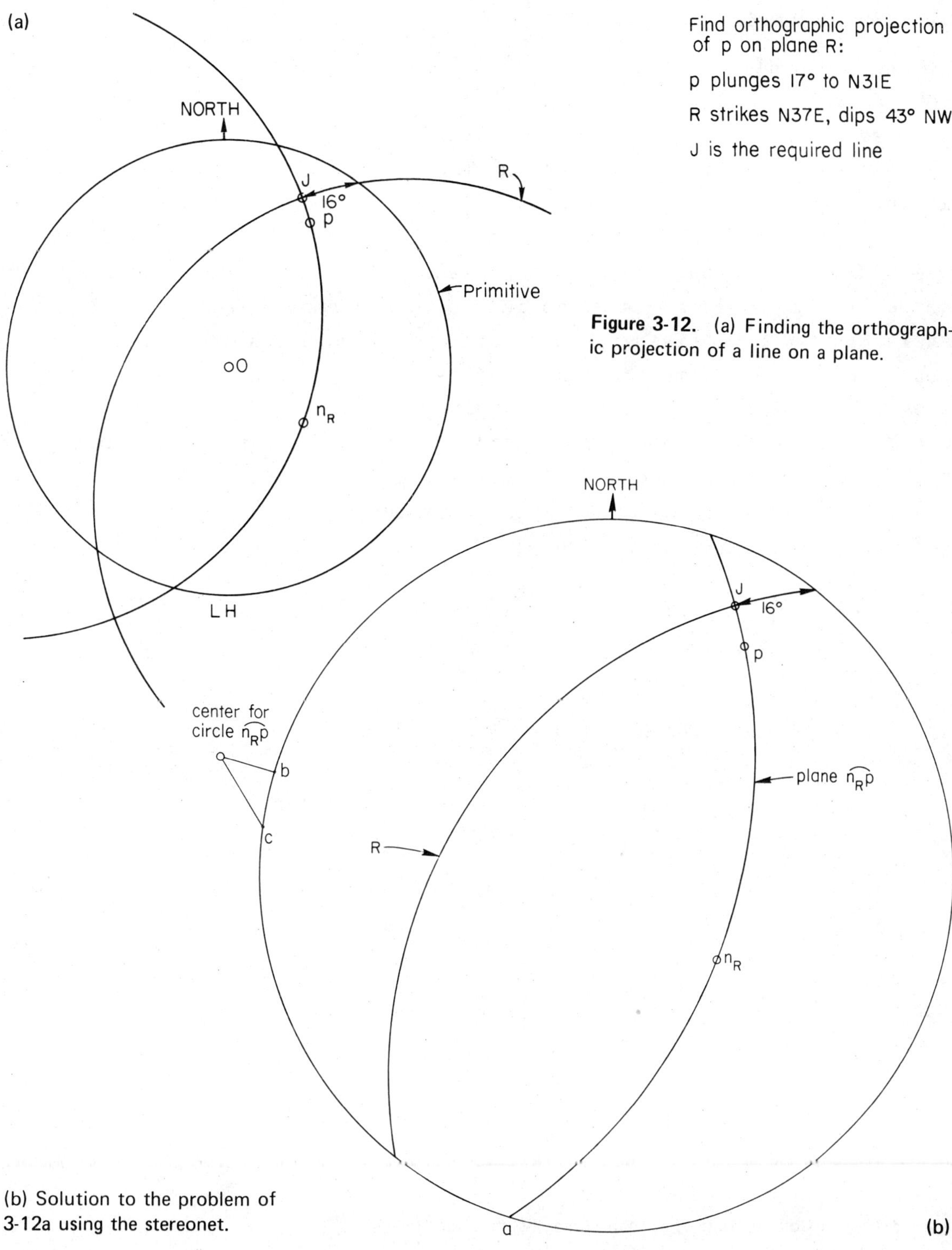

Figure 3-12. (a) Finding the orthographic projection of a line on a plane.

(b) Solution to the problem of 3-12a using the stereonet.

Construction of a Small Circle and Its Opposite

We use small circles in two ways. Problems involving rotation about an axis involve movement of points along small circles constructed about the axis. Problems with friction involve construction of the cone of static friction about the normal to a surface of sliding. The projection of a small circle is shown in figure 3-13a, where we desire a small circle of radius 15 degrees about the axis p. Draw a diameter through p and its perpendicular through O to locate the upper hemisphere focus F, as in previous examples. The line Fp intersects the primitive at P so that line OP is the axis of the double cone which forms the small circle on the reference sphere. Lines 15° above and below OP yield points P_1 and P_2, which project to p_1 and p_2 along Op. The required small circle is drawn with center at the bisector of p_1 and p_2; (the center coincides with the axis only if it is vertical). The opposite of the small circle about p is constructed similarly as a circle through the opposites, $-p_1$ and $-p_2$.

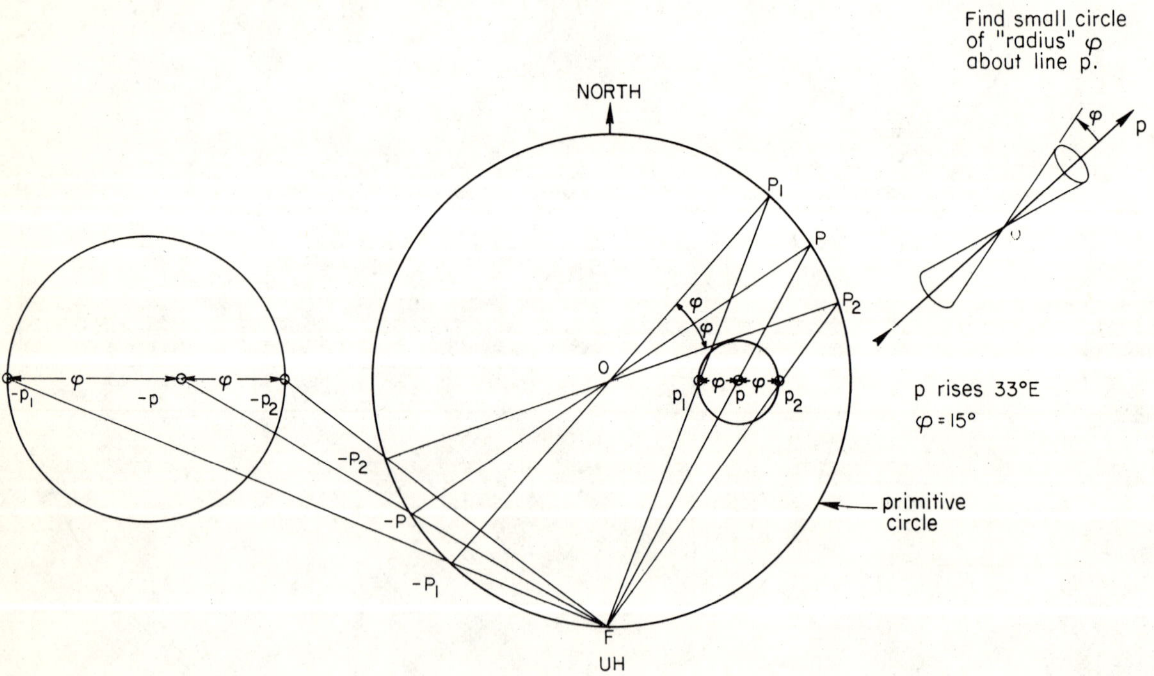

Figure 3-13. (a) Construction of a small circle and its opposite.

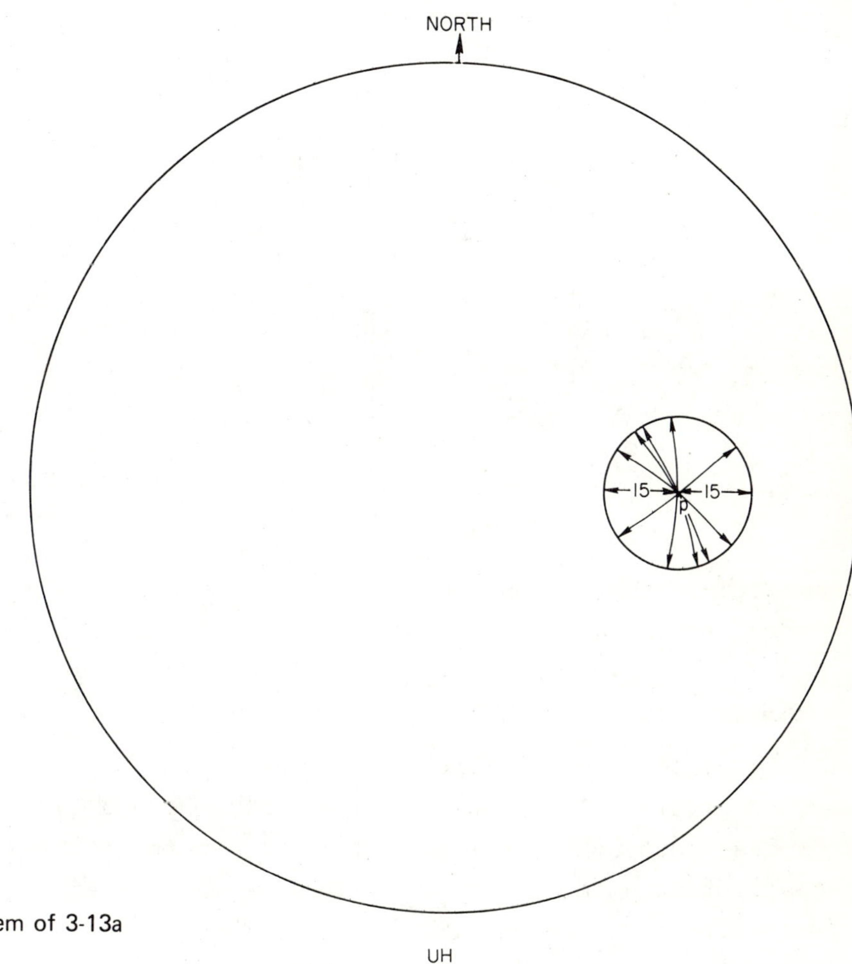

(b) Solution to the problem of 3-13a using the stereonet.

A small circle and its opposite may range over any portion of the plane.

The construction of a small circle on the stereonet tracing is illustrated in figure 3-13b. Points on the required small circle are obtained by revolving the net to pass a sequence of great circles through p, counting the required number of degrees in each direction from p in each great circle in turn. Of course, if the conformal stereonet is used, one can superimpose the construction of 3-13a on the tracing to improve the accuracy of the drawing. If the small circle lies partly in each hemisphere, as in figure 3-13c, it may be satisfactory for some applications to show two arcs, one being the

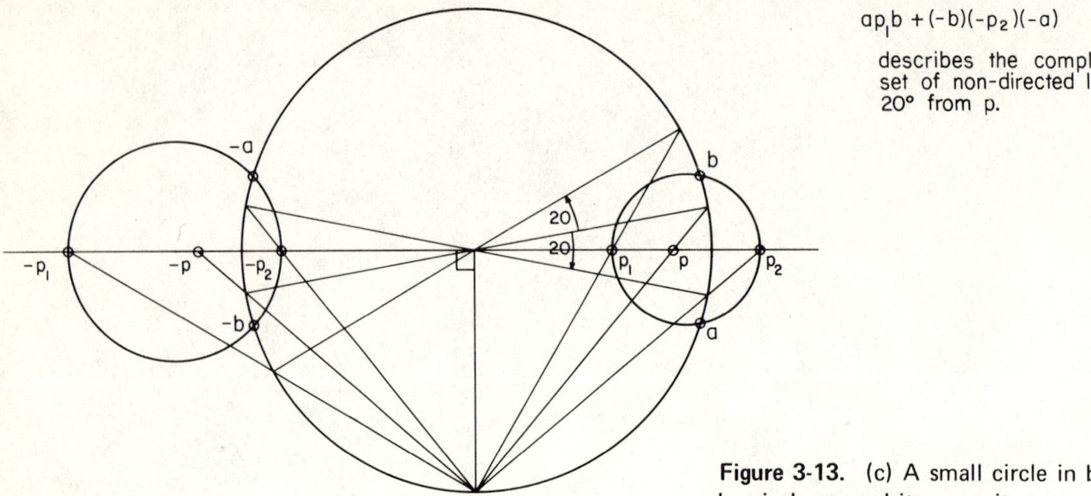

Figure 3-13. (c) A small circle in both hemispheres and its opposite.

opposite of the incomplete portion of the other (this is not satisfactory for problems in statics).

Rotation of a Line About an Axis

A small circle defines the locus of rotation of a line about an axis. Figure 3-14a shows an example in which a line p is rotated 30 degrees about a horizontal axis q. The path of rotation is found by constructing a small circle about q as in previous examples. This small circle can be calibrated as follows. Construct the plane R perpendicular to q. Locate point 1 in this plane by its intersection with the great circle common to p and q. Angles of rotation about q are measured in plane R. The required point p_r is found by locating m 30 degrees from 1 in plane R, and constructing the intersection of great circle mq and the small circle. The construction is identical when the axis of rotation is inclined, as in figure 3-15a.

On the steronet, one can immediately trace the path of rotation using the family of small circles, as shown in figure 3-14b, when the axis of rotation is horizontal. When the axis is inclined, the construction of figure 3-15a can be performed on the tracing, or the problem can temporarily be rotated so that the rotation axis becomes horizontal. The latter technique is illustrated in figure 3-15b.

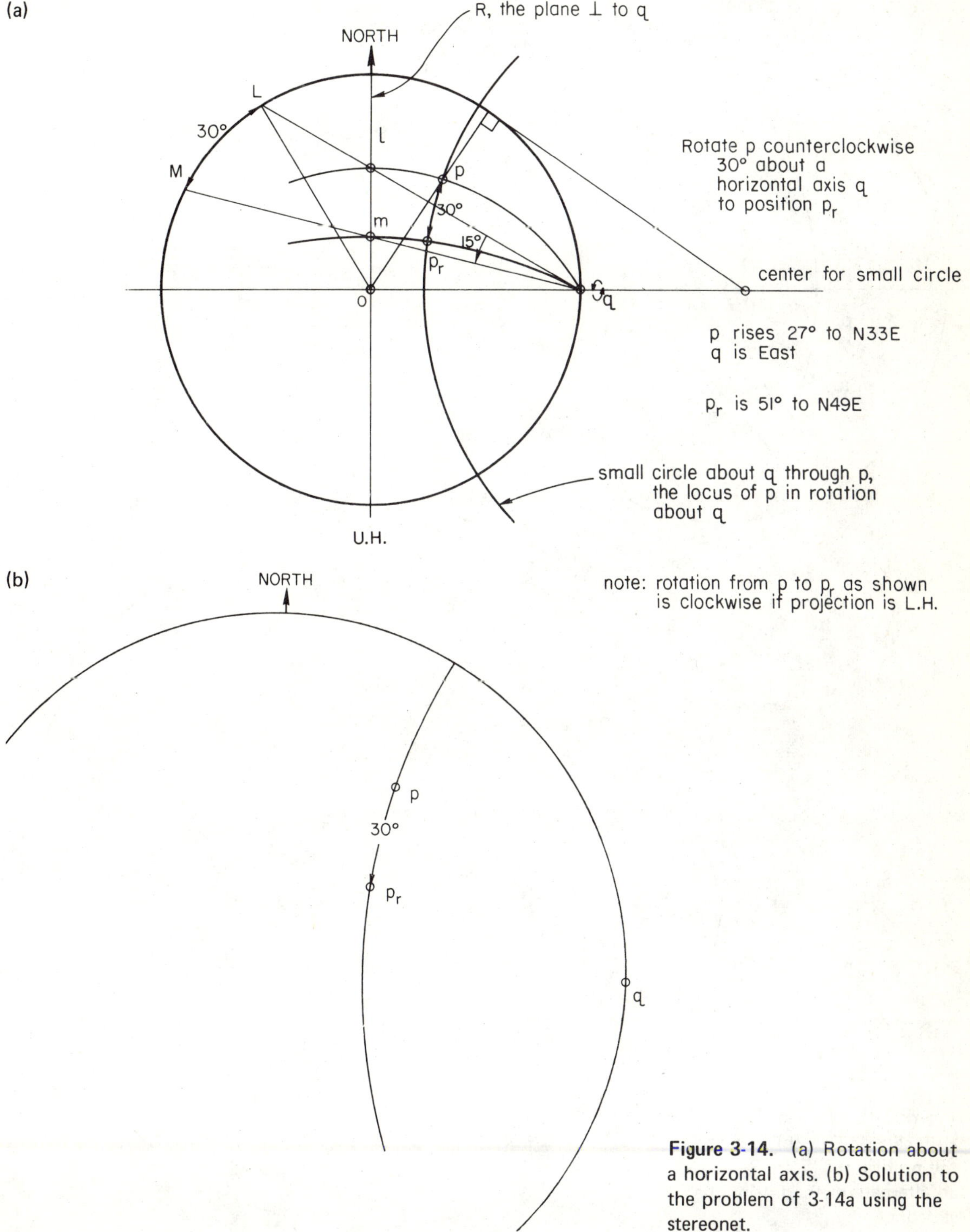

Figure 3-14. (a) Rotation about a horizontal axis. (b) Solution to the problem of 3-14a using the stereonet.

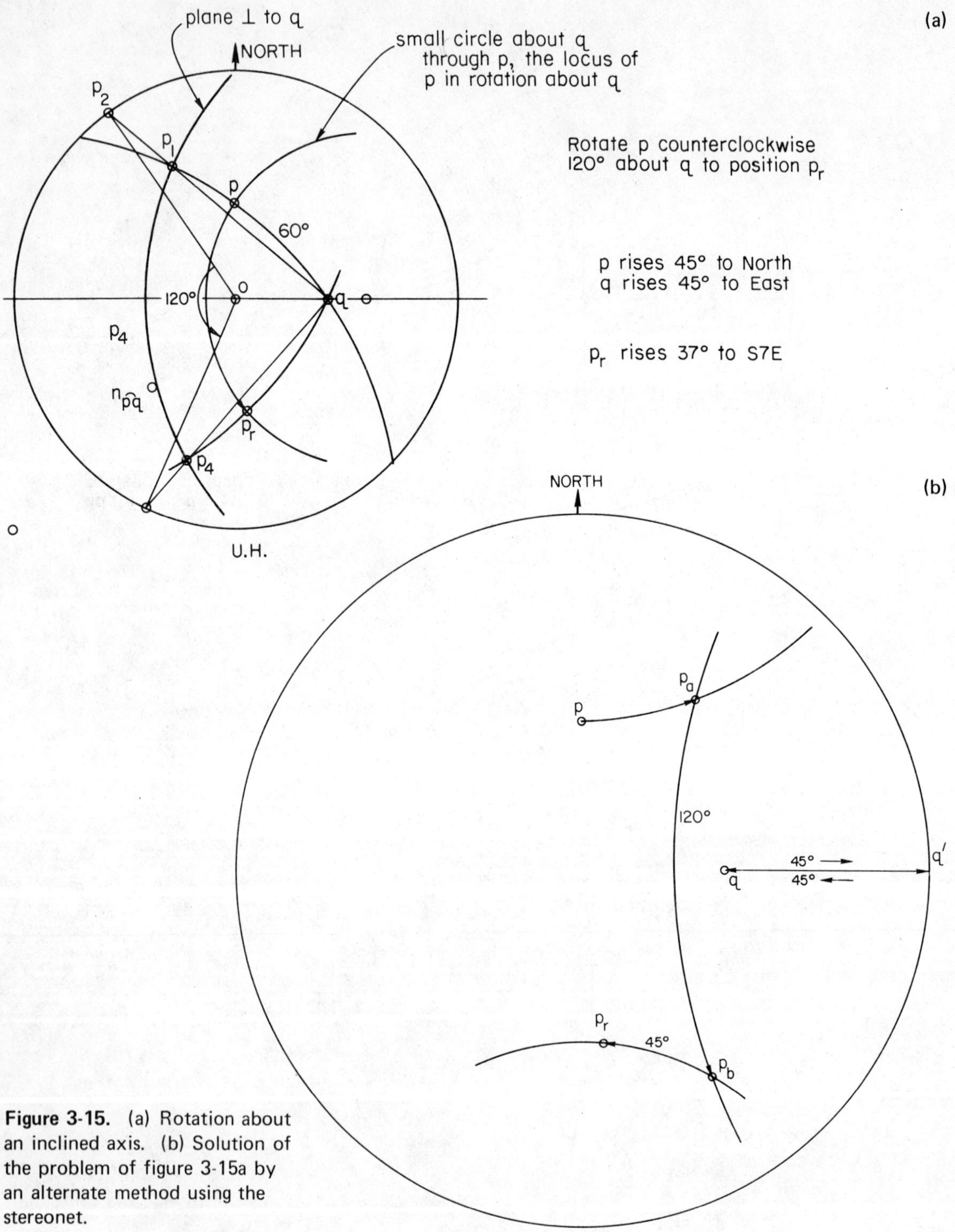

Figure 3-15. (a) Rotation about an inclined axis. (b) Solution of the problem of figure 3-15a by an alternate method using the stereonet.

First q is rotated to horizontal (q') along the equator carrying p to p_a, along the small circle through p. Then the tracing is revolved, to place q' over the polar position and the required angle of rotation about q is traced along the small circle through p_a, carrying it to point p_b. Finally, p_b is carried to its correct position p_r as q' is returned to q.

Problems involving rotation of data arise in solution of geological structures, and in analysis of symmetry of fabric data. An example will be discussed later (figure 3-16) in connection with joint surveys.

JOINT SURVEYS AND STATISTICS ON THE SPHERE

Field measurements of joint attitudes will define a scatter of orientations which can be interpreted statistically if plotted on an equal area steronet. Turner and Weiss (1963) discuss various means of contouring the resulting "scatter diagram" on the equal area diagram to highlight the symmetry of the discontinuity system. The "preferred orientations" of joint sets are defined by "highs" of the contour diagram. If many points are to be plotted and contoured, it might be appropriate to use a computer program with an automatic plotter routine, such as given by Jeran and Mashey (1970). Plotting and contouring can be done by hand using a polar net for plotting and a counting circle with radius equal to 1/10th of the net radius. The number of points inside the counting circle, when placed arbitrarily on a scatter diagram, is written in the center of the circle. By ranging over the hemisphere, pole density values are defined to a detail necessary to sketch contours. If two counting circles are provided exactly one net diameter apart, one circle will be the opposite of the other and points on the edge of the net can be counted by adding the number in the two circles.

Statistical distributions on the sphere have been discussed by Arnold (1941), and Fisher (1953) and applied to contoured joint diagrams by Pincus (1953), McMahon (1972), Mahtab et al (1972) and others. A problem basic to applications in geological engineering is finding the probability of occurrence of a joint normal in a given orientation, once the preferred orientations have been determined. For this type of work, the precision of the counting circle technique of contouring is questionable and one may consider using a vector approach as discussed by Mahtab et al (1972). The mean orientation of a group of unit vectors (joint normals) clustered within a cone is determined by the orientation of the resultant $\vec{R}$ of all the unit vectors in the cone. If all the joint normals are exactly parallel, the magnitude of $\vec{R}$ will equal the total number of normals in the cluster N; as the joint directions become dispersed, the length of $\vec{R}$ becomes less than N. Thus, as an index of dispersion of orientations, one can use a coefficient K defined as follows*

$$K = \frac{N}{N - |\vec{R}|} \tag{1}$$

K approaches infinity as the dispersion of joint normals approaches zero. This definition of K fits Arnold's and Fisher's hemispherical normal distributions which, for K greater than 6, yield the probability formula:

$$\cos \psi = 1 + (1/K) \ln (1 - P) \tag{2}$$

P is the probability that a normal to a joint occurs within ψ degrees of the mean orientation. It will be called the "probability of occurrence". The standard deviation of the distribution is $\delta = (K)^{-\frac{1}{2}}$

A rigorous statistical analysis must include an examination of the degree to which the hemispherical normal distribution fits the data; the application of the chi square test for goodness of fit is discussed by Irving (1964), and Mahtab et al (1972). A simple approach can be based upon the comparison of dispersion coefficients, K, fitting different probability contour levels. Suppose we are given a cluster of normals; first, estimate the orientation of the mean attitude as the high point of the set of contours drawn using

---

* The absolute value of $\vec{R}$ is used to permit replacing a pole by its opposite in the calculation.

a counting circle on an equal area plot.* On a separate tracing, figure 3-16 sketch a curve about the mean enclosing all but 5 percent of the number of points in the cluster; this is an estimate of the ninety-five percent probability curve (P = 0.95). Then sketch a

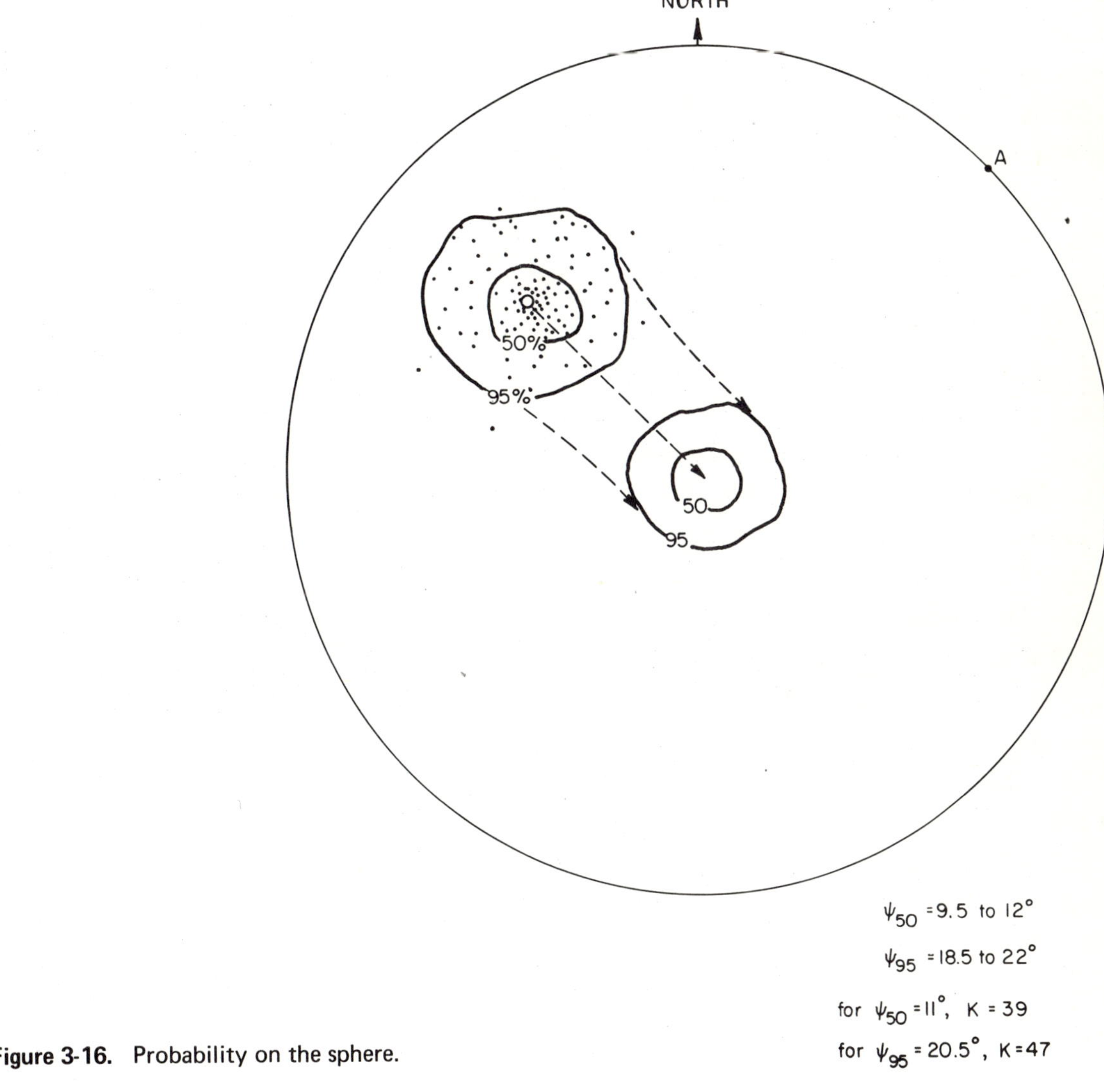

Figure 3-16. Probability on the sphere.

---

* The example of figure 3-16 is on a conformal net, but an equal area net might have been used.

closed curve about the mean orientation containing one half of the total number of points, estimating its position in such a way that the excluded points are equally distributed around the curve, if possible; this is one estimate of the fifty percent probability curve ($P = 0.50$). Rotate the diagram to orient the mean vertically (figure 3-16). If the shape of the curves after rotation is very different from a circle, an anisotropic Fisher distribution can be used, i.e. one in which K varies with direction (see Shanley and Mahtab, 1975). If the shape approximates a circle, as in figure 3-16, measure the angle ψ from the mean orientation to the 50 percent and 95 percent probability curves and compute K corresponding to each value from figure 3-17, derived from the expression

$$K = \frac{\ln (1 - P)}{\cos (\psi) - 1} \qquad (3)$$

If the values of $K_{.50}$ and $K_{.95}$ are not greatly different, the hemispherical normal distribution can be used to predict angular distances from the mean corresponding to other probability values -- (the positions of the 50 and 95 percent probability curves as drawn can be used directly whether or not the distribution fits the hemispherical normal distribution).

In the example of figure 3-16, ψ_{50} is in the range 9.5° to 12° while ψ_{95} is in the range 18.5° to 22°. For $\psi_{50} = 11^{\circ}$ and $\psi_{95} = 20.5^{\circ}$, figure 3-17 gives values of K respectively equal to 39 and 47. If we retain the value $K = 39$, we could predict for example that the 99% probability angle (ψ_{99}) is $\cos^{-1} (1 + \ln (0.01/39)) = 28^{\circ}$. Given the mean joint orientation in a design problem involving potential sliding, one might then consider the "design joint" to be inclined 28° steeper than the mean attitude. The suggested approach will be, however, to retain the mean joint orientation in calculations and reduce the angle of friction accordingly, as discussed in Chapter 6.

BIAS IN MEASUREMENT OF JOINT ORIENTATIONS FROM DRILL HOLES AND OUTCROPS

Ruth Terzaghi (1965) demonstrated that joint spacings inferred from measurements of joint attitudes in boreholes or from planar

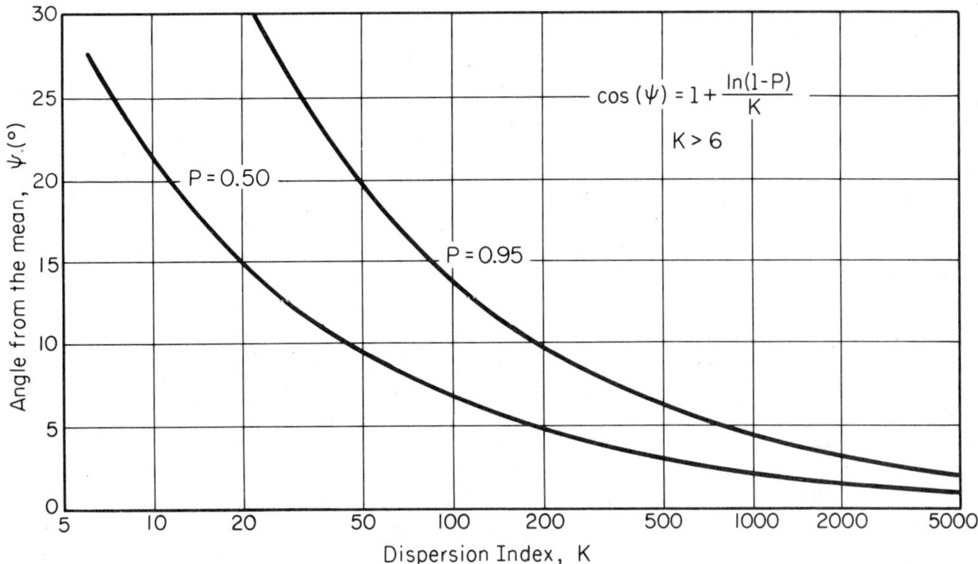

Figure 3-17. Angles of 50% and 95% probability poles from the mean of a hemispherical normal distribution After Arnold (1941), Fisher (1953), and Mahtab et al (1972).

outcrops are biased because joints parallel to the surface or the axis of the borehole are not seen. Let α be the angle between the normal to a joint and the axis of a borehole or the normal to a surface outcrop,(figure 3-18). The actual number of intersections with the outcrop or borehole by a series of parallel joints will depend on α. Only when α equals 90° will the true spacing equal the outcrop length divided by the number of joints; similarly, only when α equals 0° will the spacing equal the drill hole length divided by the number of joints. However, the data can be corrected as shown in figure 3-18. Let N_α be the number of joints observed at angle α. We can then calculate a "correct" number of joints N_c as follows.

For borehole measurements:

$$N_c = N_\alpha / \cos\alpha \qquad |\alpha| \leq 70^\circ \qquad (4a)$$

and for measurements from a planar outcrop:

$$N_c = N_\alpha / \sin\alpha \qquad |\alpha| \geq 20^\circ \qquad (4b)$$

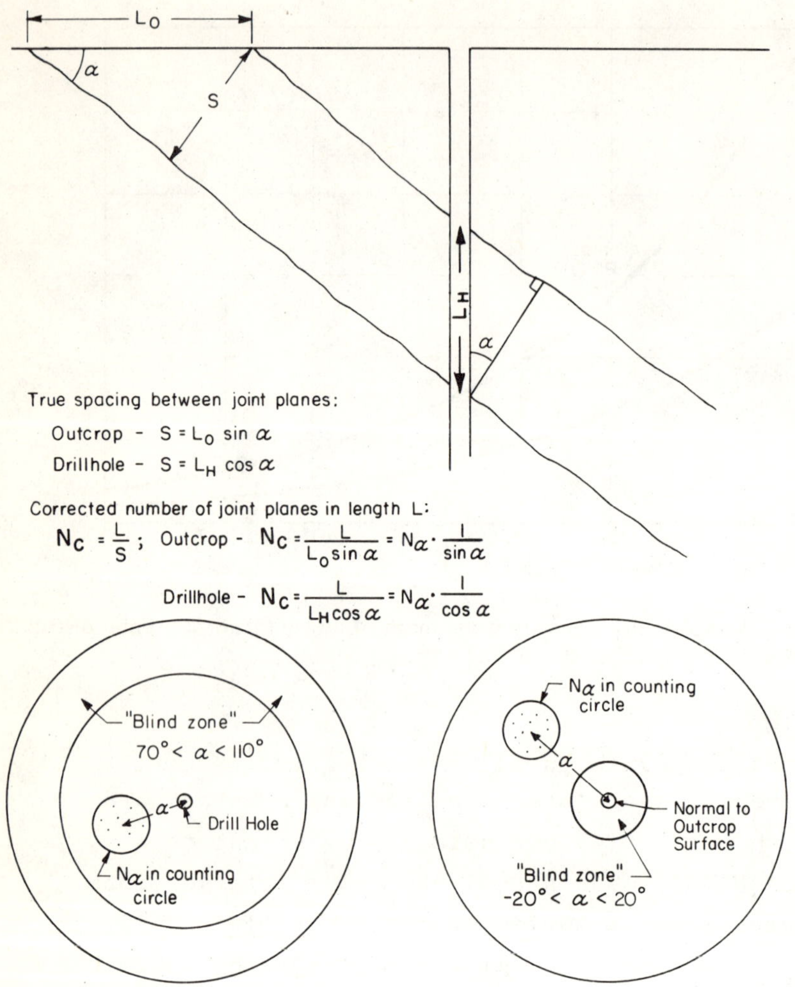

Figure 3-18. Correction for bias in joint surveys; after Ruth Terzaghi (1965).

The indicated limits on application of these corrections suggest that the measured data may be too sparse in the "blind zones" of the borehole or outcrop (figure 3-18). Normally, joint observations are made from compound surfaces and there is no absolutely blind zone. However, there still may be bias. The spacing of sheeting joints in granitic rocks and massive sandstones, for example, are difficult to appreciate from outcrops.

THE DIRECTIONALITY OF A JOINTED ROCK MASS

Talobre (1957) suggested a simple construction to determine

fundamental directions of a jointed rock mass. If the preferred orientations of joint sets and other discontinuity surfaces are plotted as poles on the stereonet, one can try to find a position which makes an angle with each pole of less than the friction angle ϕ (figure 3-19). Such a direction will be in the area common to small circles of radius ϕ about each pole (each joint set may present a different angle ϕ).

If an area common to all the small circles exists, it will contain the strongest directions of the rock mass because a rock loaded in such an orientation can mobilize rock strength without joint slip.

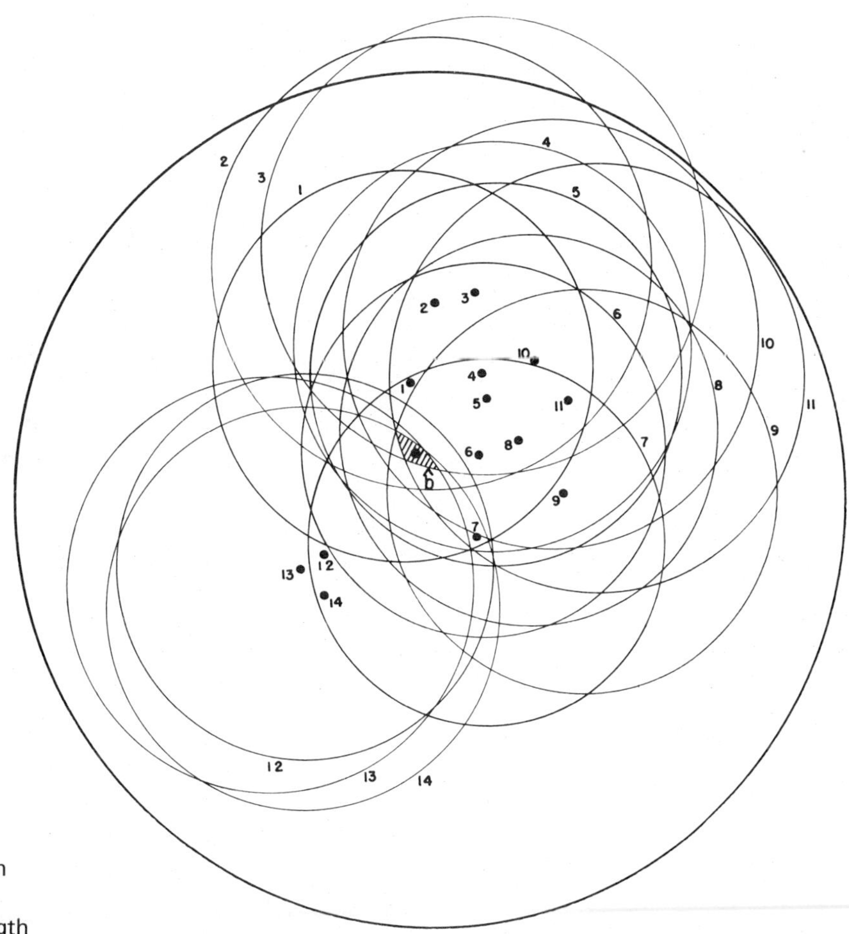

Figure 3-19. Construction for the direction ($\hat{b}$) of greatest compressive strength in a jointed rock mass; after Talobre (1967), chapter 10.

Let $\hat{b}$ signify such a direction. A drill hole along $\hat{b}$ will intersect a great number of joints, and $\hat{b}$ will accordingly be a good direction in which to drill drain or grouting holes. If a tunnel is oriented in direction $\hat{b}$, the walls will be in the least deformable directions of the rock mass, while the face of the tunnel will be in the most desirable orientation for mining. See also Lokin (1974).

4
exploration of rock conditions

Most construction in rock entails adapting to existing terrain and rock conditions. With adequate description of the morphology and properties of the different rock classes, the attitudes and characteristics of discontinuities, the hydrologic and sometimes tectonic regimes, it is possible to choose feasible sites and to find workable designs for engineering developments in discontinuous rocks. This chapter concerns the engineering geology methods used to obtain the required knowledge -- geological mapping, air photos and geophysics, and exploratory excavations and drill holes. Attention will be directed to methods pertinent to characterizing and mapping discontinuities. The review of principles presented here can be supplemented by reading in Simpson (1968), Donn and Shimer (1958), Badgeley (1959), Blyth (1965), Ragan (1974), or other works in structural geology and map interpretation.

GEOLOGICAL MAPS AND THEIR INTERPRETATION

Existing Maps

Exploration should always begin with a thorough review of existing geological maps and reports. It must be appreciated, of course, that all such maps are to a greater or lesser extent interpretive, the degree depending upon the scale and purpose of the map. Maps at scales larger than 1:500 can be almost entirely observational and

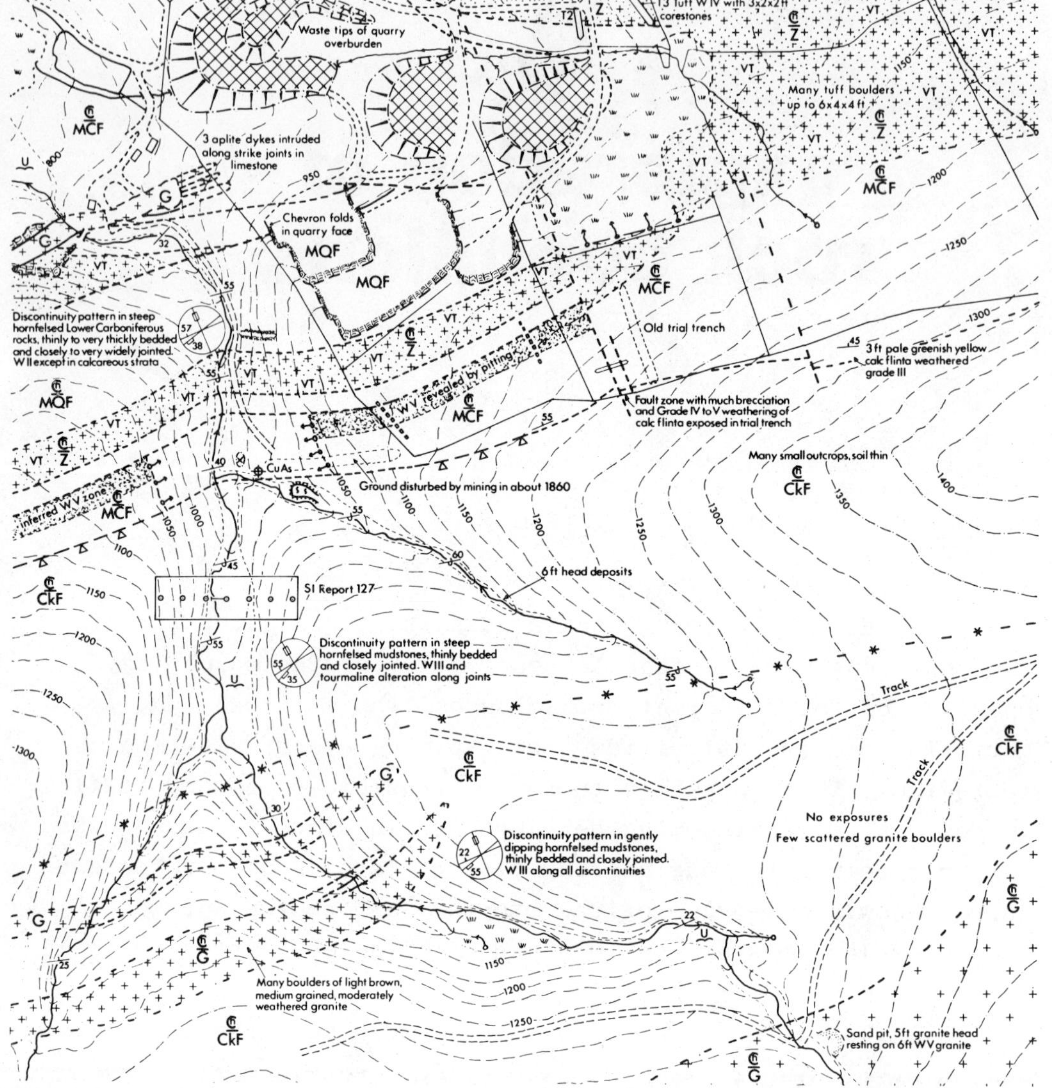

Figure 4-1. (a) An engineering geology map derived from a 1:10,500 general purpose geologic map. From the Geol. Soc. of London, Engineering Group Working Part Report, 1973; (courtesy Prof. W. Dearman). (b) A portion of the legend for figure 4-1a.

grade into logs, but these rarely will be found. Available general purpose geological maps show the mapper's opinion, based upon incomplete evidence, on the distribution of mappable bedrock formations and structures below the mantle of soil. The surface distributions of mappable formations are shown by means of the surface traces of their contacts with adjacent units. By means of structural symbols and supplementary cross sections, subsurface information is presented as well. Such maps are basic for all rock engineering.

(b)

SOLID FORMATIONS
CARBONIFEROUS
Upper Carboniferous (Namurian)

CkF	**CRACKINGTON FORMATION** Dark to very dark grey, very fine grained, thinly bedded to thinly laminated, very closely jointed, slightly to moderately weathered, poorly cleaved SHALE, weak, impermeable except along open joints. Interbedded with very subordinate grey to dark greenish grey, fine-grained, very thinly bedded, thinly laminated and cross-laminated, closely jointed, slightly to moderately weathered SILTSTONE, moderately strong, and dark greenish grey medium grained, very thinly to medium bedded, with closely to widely spaced joints slightly to moderately weathered, SANDSTONE, strong. The shale slakes on exposure and is suitable for brick making.
S	**SANDSTONE** It has been possible to map groups of beds in which SANDSTONE predominates. Beds are usually less than 12 in. thick and are separated by very thin beds of siltstone and shale. Sandstones are suitable for aggregate production. Within the contact metamorphic aureole of the granite, dark grey, very pale orange to dusky yellowish brown, fine to medium grained, thinly bedded, closely jointed, slightly to moderately weathered, hornfelsed SHALE and SANDSTONE, strong, impervious except along open joints. Locally with fine grained black tourmaline developed as selvedges up to 1 in. wide along discontinuities and with irregular quartz veins up to 2 in. wide.

Geological maps can be obtained from federal, state and local governments, geological societies, theses and journals. The U.S.G.S. offers indexes by states. State geological surveys also supply their own indexes to maps. Also, the Commission for the Geological Map of the World publishes periodic reports on worldwide mapping in the *Geological Newsletter*, (International Union of Geological Societies). Such maps can be helpful for engineering planning. Sometimes more detailed engineering geological maps exist and can be obtained from government or private engineering organizations.

Geological Mapping

Detailed geological mapping from surface exposures, reinforced by trenching, and other subsurface exploration, is the basic vehicle for developing the model of rock conditions. The mapping scale should be selected to delineate necessary details. Should the resulting map size be too large, supplementary notes can be incorporated, as for example in figure 4-1 from the Geological Society of London's Engineering Group Working Party Report (1972)*. The map was prepared

* *Quarterly Journal of Engineering Geology*, Vol 5, no. 4.

by annotation and expansion of detail from a previously existing published geological map at a scale of 1:10,560; potentially ambiguous names were replaced by objective descriptions and annotations using standard terminology and symbols defined in the Working Party report. Such a procedure minimizes the risk of misinterpretation, which can prove costly in engineering construction projects. Figure 4-2 is a portion of a bedrock geological map prepared by an engineering geologist (originally at 1:2400 scale with 10 foot contours) for design of a large earth dam. The accurate location of contacts, faults, landslides and structural attitudes in relation to project features required special mapping reinforced by coring and trenching through the soil.

Site maps and logs for engineering design should show the limits

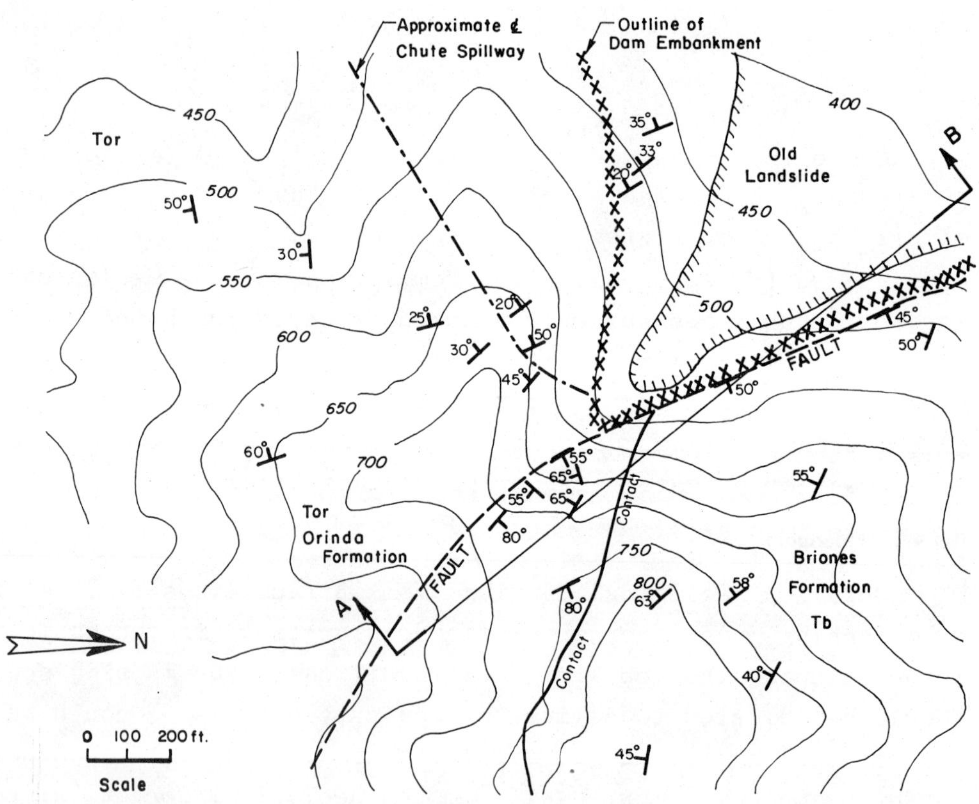

Figure 4-2. Geologic map of a portion of the left abutment of an earth dam; (courtesy Kaiser Engineers, Oakland).

of outcrops and the locations of borings and trenches to separate, wherever possible, the interpretative from the observational data. However, even directly observed features are interpretive owing to the nature of the geologic mapping process which entails fixing not only the location but the classification of the stratigraphy and structure. A geological map is prepared by charting the correct geographic location for sets of points on known geologic horizons and structures. For example, given a succession of layers consisting of (A) sandstone, (B) limestone, and (C) shale, the limestone is mapped by plotting the trace of the contacts AB and BC. If a fourth layer (D) consisting of limestone underlies the shale, a decision will have to be made at every outcrop of the shale/limestone contact to insure that BC and not CD has been located. Usually a contact such as BC can be located only approximately as a line (dashed) between proximate outcrops of B and of C. The quality and usefulness of a geological map depend markedly on the previous experience of the maker, who needs to be accurate simultaneously in his geographic and geologic "locations".

Special surveying is rarely necessary for general geological mapping but for site mapping required accuracy may demand survey of some suitably located control points. Rarely will accuracies in location of contacts be demanded better than one foot in plan or elevation, as geological uncertainties and variations will usually be more limiting.

The basic field tool is a combination compass and clinometer which is used to measure the attitudes of rock planes, to resect field positions by horizontal sightings, and to estimate elevations from vertical triangulation. The Clar compass*, allows simultaneous reading of strike and dip directions of planes in contact with the instrument. It is less versatile and more expensive than the better known Brunton compass** (or "pocket transit") which permits horizontal and vertical sightings as well as attitude measurements on planes in

---

* Breithaupt Co., Kessel, W. Germany

** "Brunton compass" of Ainsworth and the "pocket transit" of K&E are two examples.

contact with the case. By installing a protractor along the side to indicate the rotation of the cover it will be possible to read strikes and dips simultaneously. Golder & Brawner & Associates, Vancouver, B.C., and the U.S. Bureau of Mines (Bolstad and Mahtab, 1974), both have developed non-magnetic dip and strike instruments with azimuth measurements referenced by sightings on a control point. This is necessary in some iron ore mines and quite frequently in underground construction projects.

As additional data become available through exploratory excavations and drill holes, the geological map is improved. During construction its details can be checked as the subsurface is exposed to view. An opportunity to gain experience in understanding geological mapping is afforded by the collection of instructive British maps prepared by Blyth (1965).

Mapping Planar Features

Outcrop patterns of planar discontinuities and non-folded strata reflect the geometry of intersection of a plane and an uneven surface. The orientation of a plane is determined by the bearing and plunge of any two mutually perpendicular lines it contains; one of these lines --the strike vector -- is usually taken along the horizontal direction and the other -- the dip vector -- is then pointed down the plane (figure 4-3). The bearing of a line is the azimuth of its orthographic projection in a horizontal plane; the plunge of a line is the vertical angle between the line and its projection in the

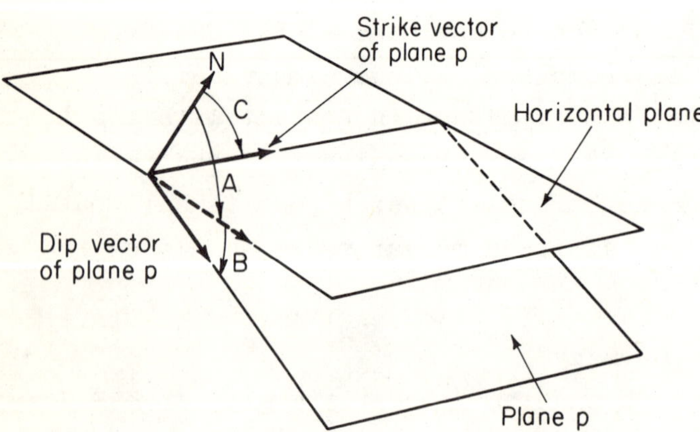

Figure 4-3. The convention for strike and dip.

horizontal plane. It is sufficient to state: the bearing of the strike vector (the "strike"), and the plunge of the dip vector (the "dip") together with an indication of the quadrant in which its bearing lies, for example: strike N30E, dip 20° SE. Table 4-1 presents strike and dip symbols for various planar features; when located on the map, these symbols fix a plane in orientation and in position. Some prefer to represent the plane by the bearing and plunge of the dip vector; others report the bearing and plunge of a plane's normal (the "pole"). In both cases, the user must understand that the given line represents a plane.

Representation of a discontinuity surface by a plane allows projection of data away from points of observation. Four types of projections arise: (1) the strike and dip of a planar feature are known at a point on the surface and the trace of the feature across the uneven ground surface is to be determined; (2) the surface trace of the plane is known and the strike and dip are to be determined; (3) the strike is known but not the dip, and the surface trace is incompletely known, for example only at two outcrops; and (4) the

TABLE 4-1

Symbols for Geological Maps (courtesy U.S. Geol. Survey)

- Contact, showing dip
- Contact, vertical (left) and overturned
- Contact, located approx. (give limits)
- Contact, located very approx.
- Fault, showing dips
- Fault, located approx. (give limits)
- Fault, existence uncertain
- Fault, projected beneath mapped units
- Possible fault (as located from aerial photographs)
- Fault, showing trend and plunge of linear features (D, down-thrown side; U, upthrown side)
- Fault, showing relative horizontal movement
- Thrust faults; T or sawteeth in upper plate
- Pault zones, showing ave. dips
- Normal fault; hachures on downthrown side
- Strike and dip of bedding
- Strike and dip of overturned bedding
- Strike of vertical bedding
- Horizontal bedding
- Undulatory or crumpled beds
- Strike and dip of bedding, uncertain
- Strike of bedding certain but dips uncertain
- Strike and dip of foliations
- Strike of vertical foliations
- Horizontal foliations
- Strike and dip where bedding parallels foliation
- Strike and dip of joints (left) and veins or dikes
- Strike of vertical joints (left) and veins or dikes
- Horizontal joints (left) and veins or dikes

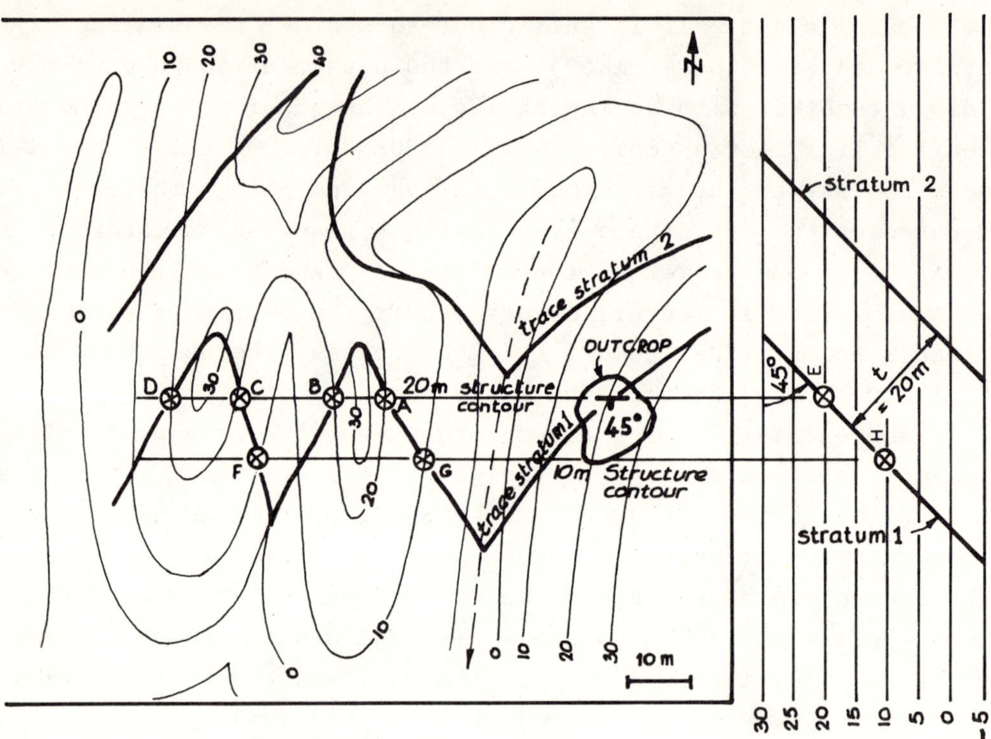

Figure 4-4. Projection of data from an outcrop.

amount of dip is known but not the strike, and the surface trace is known only at several points. Figure 4-4 shows a strike and dip symbol plotted at the outcrop of a clay seam. Based upon the measured plane, denoted stratum 1, one can estimate the trace of the clay seam across the ground surface, which can then be checked by excavations through the soil. Since a line parallel to the strike of a plane is a contour on the plane, the clay seam should also underlie the surface at A, B, C and D where the strike line intersects the 20 meter surface contour. A vertical section in the direction of the dip is shown on the right of the map. The line A, B, C, D intersects this section at elevation 20 (pt E) as shown. The edge of stratum 1 can be drawn in this section because the dip angle, 45°, has been given. Line A, B, C, D, E may be termed "the 20 meter structure contour". Structure contours at other elevations can be drawn from the cross-section, and their intersection with the appropriate surface contours define additional points along the trace of the clay seam across the surface.

Suppose geological mapping had revealed sufficient actual outcrops to define the trace just constructed yet without yielding anywhere a precise measurement of strike and dip. If the trace passes more than once through the same contour, the strike is determined by the trace. Line DA is parallel to the strike. Interpreting line DA as a structure contour at 20 meters and line FG as a structure contour at 10 meters elevation, the plane can be constructed in vertical section by extending line EH determining the dip.

Occasionally, some points along the surface trace are known but too few to allow the construction above. If the strike is known, the dip can be determined, allowing completion of the outcrop trace. For example, in figure 4-4, if only A and G were identified as points on the trace of stratum 1, knowledge of the strike would allow construction of points E and H and the balance of the solution follows. Following exploration by drilling, one usually known the amount of dip (the complement of the core/bedding angle in vertical drill holes) but not necessarily the strike. In this case one must select one of two possible solutions, as shown in figure 4-5. Here stratum 1 outcrops at G and B. The known dip of 45° gives a 10 meter horizontal offset distance (corresponding to the 10 meter difference in elevation) which is laid off as a circle about B. The strike line is a tangent to this circle drawn through G. It is uniquely defined as line GC if we know the direction of strike is approximately EW.

If 3 points are located on the stratum at different elevations, the strike and dip are obtainable. Points G, B, and K on the trace of stratum 1 (figure 4-5) present such a 3 point problem. Interpolate between the highest and lowest points (K and G) to find the position where the stratum will have an elevation corresponding to the middle point (B) This yields point X (the stratum is 10 m above the ground at X, meaning it has been eroded off). Line BX, the 20 meter structure contour, defines the direction of strike. The dip is determined by the projection of G or K into the cross-section as before or from the perpendicular distance d from G to BX, (since the elevation difference from line BX to line CG is known to be 10

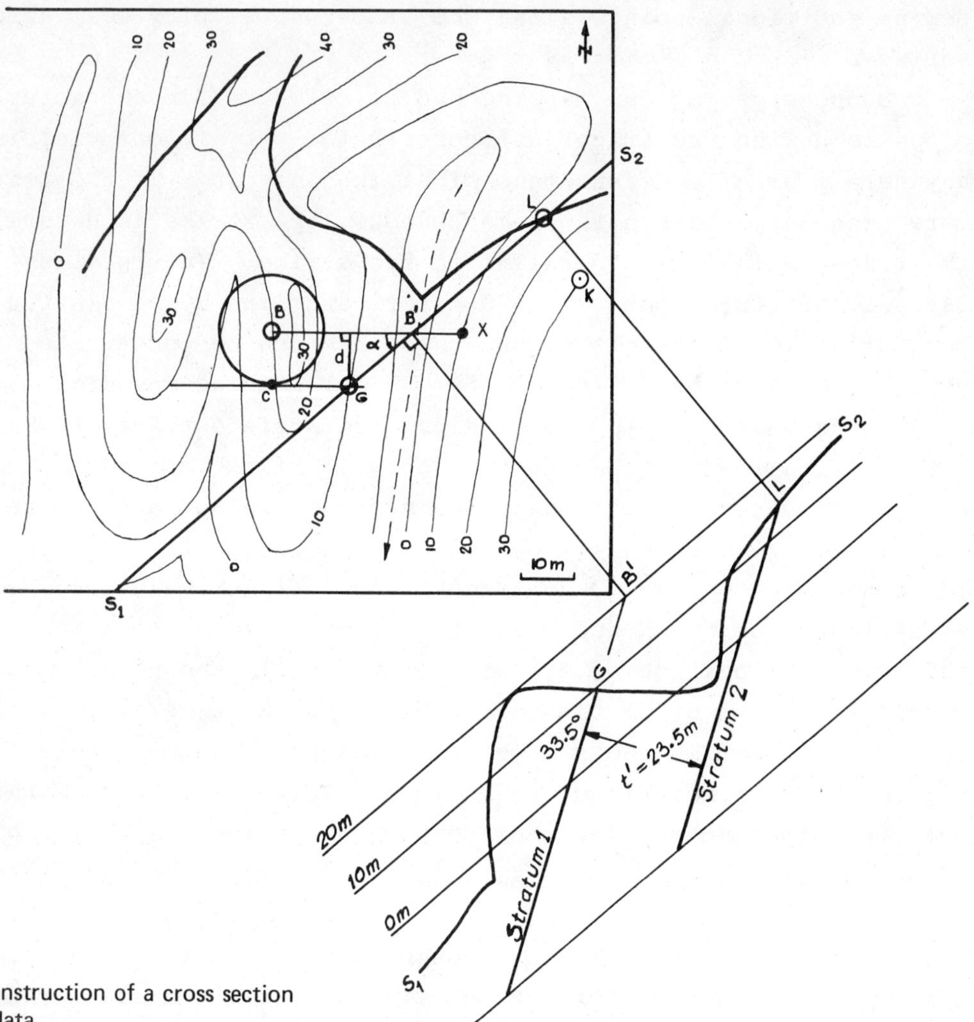

Figure 4-5. Construction of a cross section from outcrop data.

meters).

In these examples one observes how the trace of an inclined plane forms a "V" pointing in the direction of dip upon crossing a valley. The trace of a horizontal plane will follow the contours while a vertical plane will pursue a straight course in any terrain. A plane inclined more gently than the relief provides a more complex and often multiple trace across the surface.

Cross-sections to convey the geological structure can be drawn in any direction by projecting surface points parallel to the strike into the desired section. In figure 4-5, a vertical section has been drawn through S_1 and S_2. The apparent angle of dip in the section diminishes as the angle (α) between the section and the strike approaches 0. Section S_1 S_2, in figure 4-5 with α = 42 gives an apparent dip (AD) of 33.8°. It could also have been calculated from the relationship derived from figure 4-6.

$$\tan (AD) = \tan (D) \sin (\alpha) \qquad (1)$$

where α is the angle between the strike and the section.

The spacing (t) between parallel planes is defined in the direction of their normals. This direction is contained in only one vertical section, i.e. the one perpendicular to the strike. The direction perpendicular to the edge of planar features in any other vertical section will give an apparent spacing t', derived from figure 4-6 as follows:

$$t' = t/(\cos^2 D + \sin^2 D \sin^2 \alpha)^{\frac{1}{2}} \qquad (2)$$

In figure 4-4, a second planar surface 20 meters beneath stratum 1 has been drawn on the true dip section and the surface trace of stratum 2 has then been constructed as shown. The area between the two traces maps the area underlain by the stratigraphic interval between 1 and 2. In figure 4-5 the trace of the lower surface is seen to intersect the line of the oblique section at L. The apparent spacing, t', found by constructing an apparent dip through L, is 23.5 m, which agrees with equation 2, (α = 42° and D = 45°).

Application of these simple principles will enable you to predict where excavations and foundations are likely to encounter important planes of weakness and to lay out site plans so as to minimize the difficulties such discontinuities can invite.

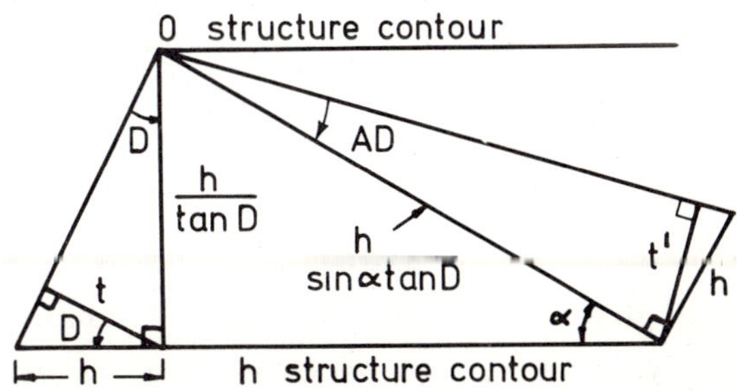

Figure 4-6. Apparent spacing (t') and apparent dip (AD) in vertical sections not parallel to the direction of dip.

AERIAL PHOTO INTERPRETATION

The opportunity to observe, and even measure, discontinuities in the three dimensional model generated by viewing overlapping vertical aerial photographs with a stereoscope can provide information on the relative orientations, spacings, and extents of discontinuity surfaces. Figure 4-7 presents several examples to illustrate how such material can be obtained. (These figures should be viewed with a lens stereoscope.) Figure 4-7a shows tilted sedimentary strata -- Mesozoic sandstones and siliceous shales -- in the Colombian Andes. The zig-zag pattern reflects resistant beds and the rule of "V's" discussed in the previous section can be used to determine the direction of dip. A set of transverse joints (T) crosses the beds without displacing them. Figure 4-7b shows sedimentary strata in a desert environment in North Africa. Again a set of recurrent fractures crosses the bedding, but in contrast to figure 4-7a, these do offset the strata, showing that they are minor faults. In both examples, the spacing of the fractures can be established if the scale is known. The scale ratio, e.g. 1:20,000, for any point equals the focal length of the lens (f) divided by the height of the camera above the point (H). In figure 4-7b, the average distance between shears is of the order of 0.01 feet and the photograph was taken with a 6 inch lens from a height of 30,000 feet, giving a spacing of 600 feet. A detailed study will indicate a distribution of spacings.

Figure 4-7c contains a dip slope defended by sandstone from which a bed of shale has been almost completely stripped away. The beds are widely spaced but two sets of major, apparently open fractures spaced at 100 to 200 feet (the scale is 1:20,000) disrupt the continuity of the formation. The fracture orientations are regular but there is a gradual swing in the preferred orientation of one of the joint sets from one side of the photo to the other. At (1) in figure 4-7c a joint surface is exposed on which a strike and dip measurement may be obtained. Presumably the joints visible in the landscape are only a small fraction of the total population, which must include members spaced several feet apart. However, the

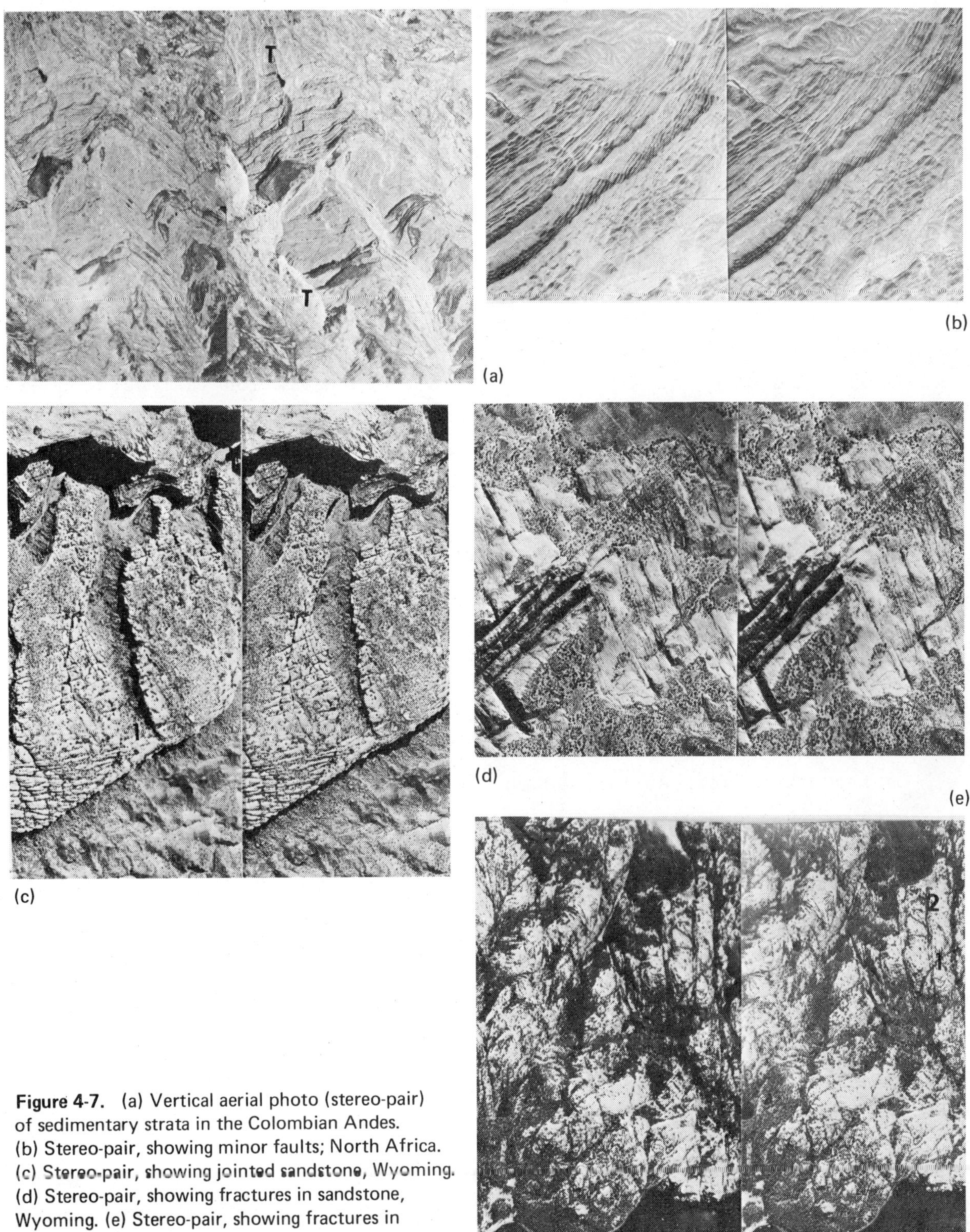

Figure 4-7. (a) Vertical aerial photo (stereo-pair) of sedimentary strata in the Colombian Andes. (b) Stereo-pair, showing minor faults; North Africa. (c) Stereo-pair, showing jointed sandstone, Wyoming. (d) Stereo-pair, showing fractures in sandstone, Wyoming. (e) Stereo-pair, showing fractures in granite, South East Alaska.

100 foot spacing figure may represent a reasonable estimate for the spacing of key discontinuities. Field study would tell. Observe that the lengths of particular individual fractures are more than 1400 feet. Figure 4-7d presents a similar example but one in which spacings are more variable. Again, two approximately orthogonal sets are expressed in the terrain. The Northeast trending joints are deeply etched in a closely spaced zone. Figure 4-7e presents a fracture system representative of granitic terrains. There are at least three different sets with considerable scatter about the mean orientation of each, as well as considerable variation in length. Study of aerial photographs may be the only reasonable way in many instances, to gain data about joint lengths.

Measurements of Attitudes of Planar Features in Aerial Photos

Strikes and dips of discontinuities like those shown in figure 4-7 can be estimated from their outcrop trace and surface expression, but one must be careful about the exaggeration of relief, usually between two and three times, when making such estimates. From the discussion of geological mapping it will be evident that any discontinuity lineament which presents a straight trace across uneven topography such as (2) in figure 4-7e must be dipping very steeply. On the other hand, surfaces such as (1) in figure 7e may be inclined because their outcrop trace across the terrain is not straight.

The geometry of the vertical aerial photograph is such that points on a horizontal plane are in correct map position with respect to one another. The scale of the map changes with elevation but horizontal lines are not rotated. This makes it easy to measure strikes of line elements in the photograph with respect to a reference direction. The usual reference -- the x axis -- is the line connecting the principal point (the lens center) and the image of the principal point of the adjacent photograph (the conjugate principal point). (The best way to view two overlapping photographs stereoscopically is with their x axes aligned). Figure 4-8 shows the image of two points R and Q in adjacent, overlapping vertical photos. To determine the bearing of the line RQ relative to the x axis, simply read the angle with a protractor. Determination of the plunge of the line between

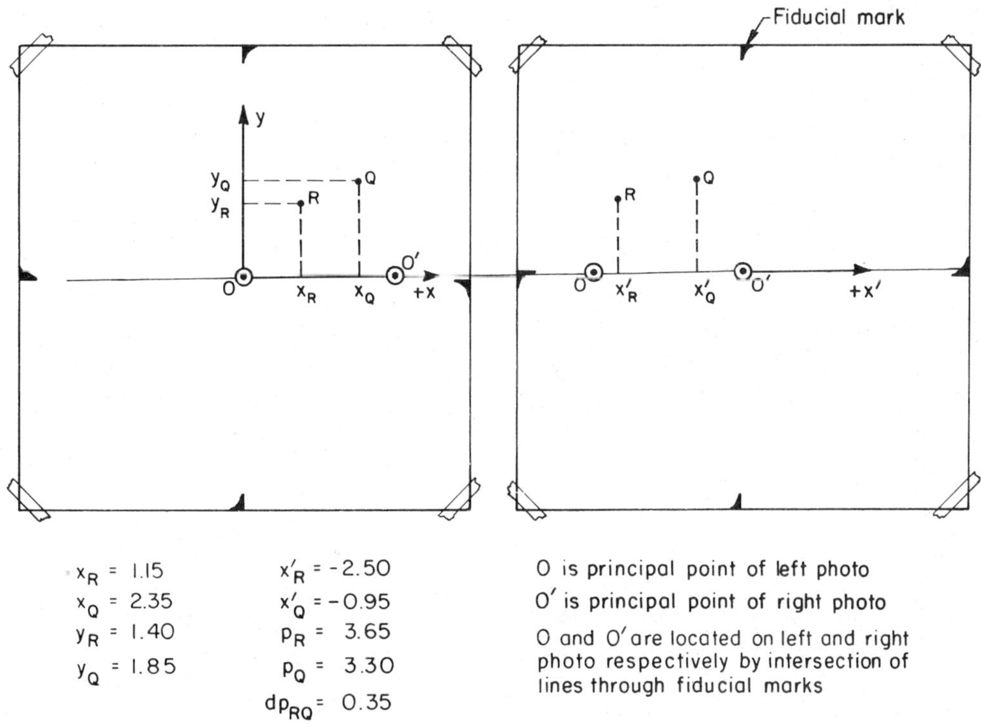

Figure 4-8. Parallax and coordinates of points in overlapping vertical aerial photographs.

points R and Q requires that we first calculate their parallax difference, dp_{RQ}

$$dp_{RQ} = p_R - p_Q = (x_R - x_R') - (x_Q - x_Q') \tag{3}$$

where, as shown in figure 4-8, the primes denote points in the right photograph and the unprimed coordinates refer to points in the left photograph. The elevation difference between R and Q is (Moffitt, 1959)

$$dh_{RQ} = \frac{H_Q dp_{RQ}}{p_Q} \tag{4}$$

where H_Q is the flying height above point Q. An approximation, usually sufficiently accurate for strike and dip measurement is:

$$dh_{RQ} = \frac{H \cdot dp_{RQ}}{b + dp_{RQ}} \tag{5}$$

where H is the camera height above the principal point, and b, the <u>photo base</u>, is the distance between the principal point and the conjugate principal point (OO'). Both H and b are averaged from the two photographs. Let S_{RQ} be the distance from R to Q measured on the photograph. The ground distance between R and Q is approximately equal to $S_{RQ}(H_Q/f)$. Thus, the plunge (S) of RQ is given by

$$\text{Tan } S = \frac{(f)\ (dp_{RQ})}{(S_{RQ})\ (p_Q)} \tag{6}$$

Using the average photo base b and flying height H, if the point Q is at about the same elevation as the principal point,

$$\text{Tan } S = \frac{(f)\ (dp_{RQ})}{(S_{RQ})\ (b+dp_{RQ})} \tag{7}$$

Since the determination of parallax difference, dp, involves the difference in measurements of larger quantities, the measurements can not usually be performed with sufficient precision using an engineer's scale. The precision of standard stereometers, such as the parallax bar, Coutour Finder, and Stereo Comparograph, is of the order of 0.01 mm. These instruments utilize the <u>floating dot principle</u>, in which the fused image of dots over the right and left photograph appears to rise or fall as the separation is changed while viewed under a stereoscope. By fixing the dot "on the ground" at each point along the opposite ends of a dip vector, the difference in parallax can be read from the difference in absolute readings of the instrument. It is easier for some people to use a <u>floating line</u>, illustrated in figure 4-9, with a "parallax wedge height finder"*, consisting of two non-parallel lines with a separation in the range that includes

---

* Felsenthal Instruments, Chicago

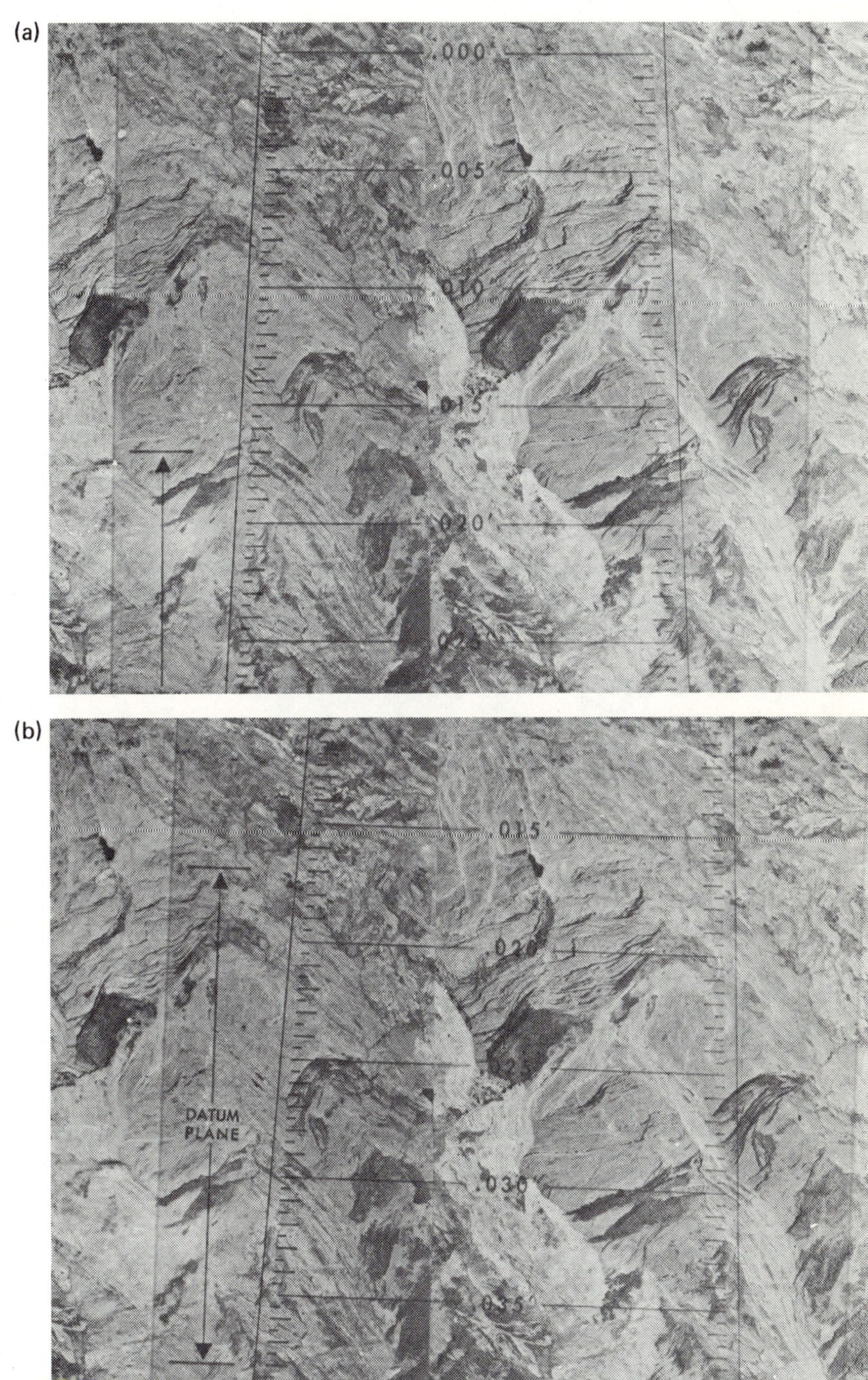

Figure 4-9. Measurement of dip using a parallax wedge: (a) the floating line pierces the ground at the bottom of the slope; (b) the floating line pierces the ground at the top of the dip slope.

the average person's eye-to-eye spacing. When placed over the stereo pair and viewed with a lens stereoscope, the lines fuse to give the appearance of one sloping line in space. In figure 4-9a, the sloping line appears to pierce the ground at point Q at the top of a dip slope. The separation of the lines on the wedge is 0.0130 feet. In figure 4-9a, the wedge has been moved to make the floating line pierce the base of the slope at R; the separation is now 0.0210 feet (estimating the last digit) giving a parallax difference dp_{RQ} of 0.0088 feet. The photo distance S_{RQ} is 0.043 feet. The photos were taken through a lens having a six inch focal length and the photo base, b, averaged from the two photos is 0.317 feet (this can not be verified on the stereopair of figure 4-9 as not enough of the image area has been copied). Solving equation 7, the bed is dipping at an angle of 17.5 degrees.

If the angle between the lines on the parallax wedge can be varied while the model is viewed, the inclination of the line in the model will appear to change. This is the principle of the "parallax ladder"* for measurement of slopes. "Hackett's method" employs an analogue device to measure directly the exaggerated slope angle in the stereo-model; the exaggeration factor is then obtained by a graphical procedure as discussed by Ray (1960).

The most accurate measurements of attitudes will be obtained using a plotting instrument such as a Kelsh plotter or Multiplex, in which a tilt free stereo-model is created by projecting through film or glass plate diapositives of the aerial photographs. The attitude of a planar surface can be obtained by mathematical plane-fitting to coordinates of three or more points on the surface. Ross-Brown (1972) in fact even determined joint roughness by computing the scatter of planes fitting four or more points in a stereo-model of terrestrial photographs. Attitudes of discontinuity surfaces in the model can also be measured by tilting the plotting platen into conformity with the surface to be measured and reading the attitude of the table with a clinometer just as it is done on the outcrop (Ray, 1960).

---

* Photogrammetry, Inc., Silver Spring, Maryland

While it is difficult to estimate dips reliably owing to vertical exaggeration, the eye can be trained by first measuring several reference slopes in the model. For fracture analyses, it may be satisfactory merely to classify each of the joint sets by strike and approximate dip, e.g., nearly vertical, steeply, moderately, or gently dipping, or nearly horizontal. The principles of geological mapping reviewed in the previous section will enable such estimates to be made from the outcrop trace. In fact, non-parallel lines in a planar surface can be combined with an apparent dip solution, using the stereographic projection (Wallace, 1950).

Types and Sources of Aerial Photos

Black and white vertical aerial photos like those in figure 4-7 have been taken of nearly every part of the United States by agencies of the U.S. government and are attainable at modest cost. Table 4-2 lists the principal sources of photographic coverage in the United States. An order may be expedited by visiting the offices of the agency and identifying the photo exposure numbers on an index mosaic. Alternatively, the area desired in photos can be marked on a topographic map and forwarded with the order. Faster response can be obtained through private aerial photo companies. Local government agencies, such as the County Tax Assessor's Office, may possess recent aerial photos, flown for them on contract, which might be borrowed. For analysis of fracture patterns, high altitude photographs, e.g. 30,000 or 20,000 feet, are often more desirable than the lower altitude photos, e.g. 5000 feet, used to make contour maps for engineering design.

In military and space research, there has been considerable interest in expanding the scope of photo interpretation outside of the visible spectrum through the use of special films and artificial imagery techniques (Cassines, 1972). Infra-red photographs bring out contrast of wet and dry terrain. Color photography, though rather expensive, can point out zones of alteration. Thermal imagery produces a visual record from natural heat radiating sources, such that warmer areas appear lighter in tone; thermal imagery can sometimes show up seepage along underground paths, subaqueous discharge,

TABLE 4-2

Sources of Existing Government Aerial Photos

United States Department of Agriculture, Agricultural Stabilization and Conservation Service, Aerial Photography Division, 45 French Broad Avenue, Asheville, N.C., 28802 (for eastern United States) or 2505 Parley's Way, Salt Lake City, Utah, 84109 (for western United States).

United States Department of Agriculture, Forest Service, Division of Engineering, Washington, D.C., 20250. (This is the address for general information. Orders are processed through regional offices.)

United States Department of Agriculture Soil Conservation Service, Cartographic Div., Federal Center Building, Hyattsville, Md., 20783.

United States Department of the Interior, Geological Survey, Map Information Office, Washington, D.C., 20242. (This is the address for general information. Orders are processed through regional offices.)

United States Department of Commerce, National Ocean Survey, Washington Science Center, Rockville, Md., 20852. (Formerly the Coast and Geodetic Survey)

National Archives and Record Services, Cartographic Branch, Washington, D.C., 20250. (Old government photographs are obtainable from this source.)

United States Department of the Interior, Geological Survey, EROS Data Center, 10th and Dakota Avenues, Sioux Falls, S.D., 57104. (This office maintains information on ERTS satellite photographs and imagery.)

National Air Photo Library, Surveys and Mapping Buildings, Room 130, 615 Booth Street, Ottawa 4, Ontario, CANADA. (This is the centralized source for Canadian air photos)

Note: The military authorities in most countries have extensive photo coverage which engineering organizations with a "need to know" can sometimes borrow. (Figure 4-7a is a U.S. Air Force photograph.)

and geothermal anomalies. Radar imagery (SLAR) is perhaps the most valuable remote sensing technique for study of fracture patterns (Barr, 1969). In this technique, reflections from a radar beam directed obliquely from an airplane are displayed on a cathode ray tube, and photographed. Radar (and thermal) images will be obtained even at night and through clouds. The low angle of incidence, compared to conventional high sun angle photography accentuates linear features such as faults. Cluff and Slemmons (1971) demonstrated that conventional photographs taken in the early or late hours, when the sun angle is low, will highlight linear features (figure 4-10). In any of these displays, the resolution of fracture lineaments can be enhanced by viewing the photo through a film with fine parallel rulings, e.g. 200 lines per inch, held at a distance from the photo; for when the ruled lines are perpendicular to a linear feature, it can be seen to the exclusion of lines of all other orientations. Information on sources of radar imagery pictures and other remote sensing techniques can be obtained by writing to the EROS data center, U.S. Geological Survey, (see Table 4-2). High altitude and satellite photographs taken as part of the American space program can also be consulted through the Earth Resources Program of NASA, Lyndon Johnson Space Center, Houston, Texas.

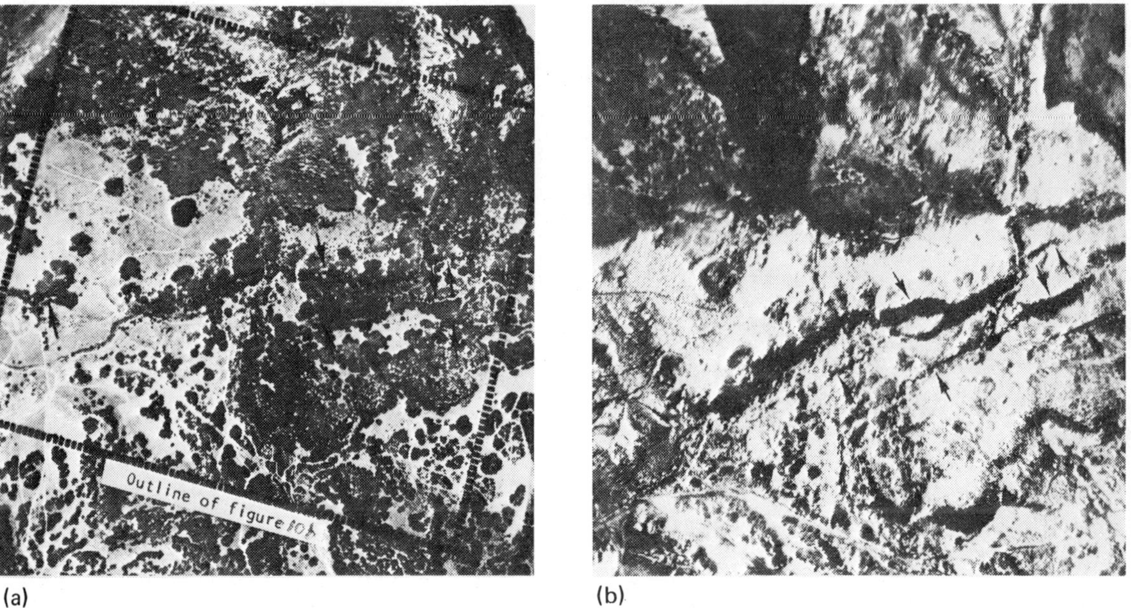

(a) (b)

Figure 4-10. Low sun angle aerial photography will display shadows along prominent discontinuities, such as the Wasatch Fault scarp photographed here. (a) A conventional vertical aerial photograph of a portion of the Wasatch front, Utah, photographed at mid-day. (b) A portion of the same area, photographed early in the morning. Reproduced from Cluff and Slemmons (1971).

TERRESTRIAL PHOTOGRAPHS

Photographs can be used to measure the orientations and spacings of discontinuities in inaccessible outcrops and in transient exposures such as cleaned rock foundations, mines, and exploratory trenches.

Measurements on Photographs

Using a 150 mm phototheodelite and good survey control, coordinates of points 100 meters from the camera are determinable with a precision of the order of 2 centimeters (Ross Brown and Atkinson, 1972). This is adequate for mapping open pit mines and will usually permit sufficiently accurate determination of joint orientations at inaccessible points by "3 point" determination. (In practice, Ross-Brown, et al. (1973) use 4 points.) Many engineers do not realize that accurate measurements can also be made from photographs taken with a hand-held camera. In connection with geological mapping onto a suitable available topographic base and logging of exploratory excavations, ordinary snapshots can supplement quantitative observations made in the field with a precision consistent with the geological mapping process. An inspiring book on techniques appropriate for analysis of photographs taken with a hand held camera was written by J.C.C. Williams (1969).

A photograph records an infinite family of convergent sightings from the camera point to all points within the field of view. A line from the lens center (O) perpendicular to the film plane establishes the focal distance (f) (figure 4-11), and intersects the photo at the principal point (p) found on an untrimmed photo by the intersection of diagonals connecting corners. The nadir point (n) lies beneath the lens center on the extension of the positive plane in the case of a photo below the horizontal. (The zenith point (Z) lies above the lens center on the extension of the positive plane on an elevated photograph.) The obliquity of the photo is the angle Onp (or Ozp) while its complement is termed the angle of rise or depression (θ). The horizon is a horizontal line in the positive plane at the elevation of the camera; a perpendicular from the horizon through p will pass through n (or z) and defines the trace of the principal vertical.

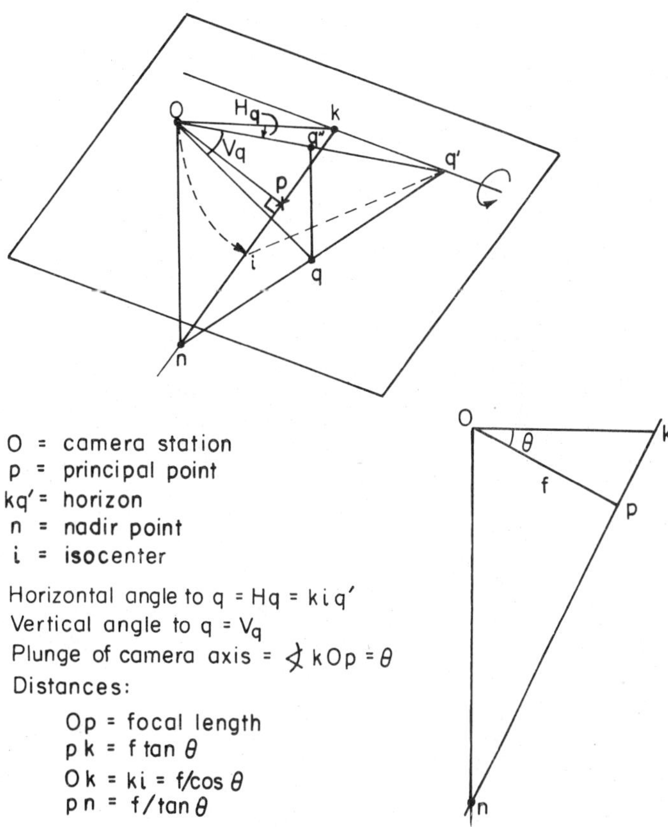

Figure 4-11. Geometry of a photograph.

All points on the ground along the same ray to the lens are superimposed on the positive. Distances can be found by fixing points through resection from two photos. Figure 4-12a, adapted from Moffit (1959), defines the co-ordinates of a point on a photo. The x-axis is a line parallel to the horizon through the principal point, p; the principal vertical is the y axis. (A camera <u>depression angle</u> is defined to be negative, giving a negative value y = f tanθ for the principal point, p.) Let q be a point on the positive. Point q' is the projection of q on the horizontal plane through the camera. The horizontal angle, H_q, is kOq' given by

$$\tan H_q = \frac{x_q}{f \sec\theta + y \sin\theta} \tag{8}$$

the vertical angle, V_q, is qOq' given by

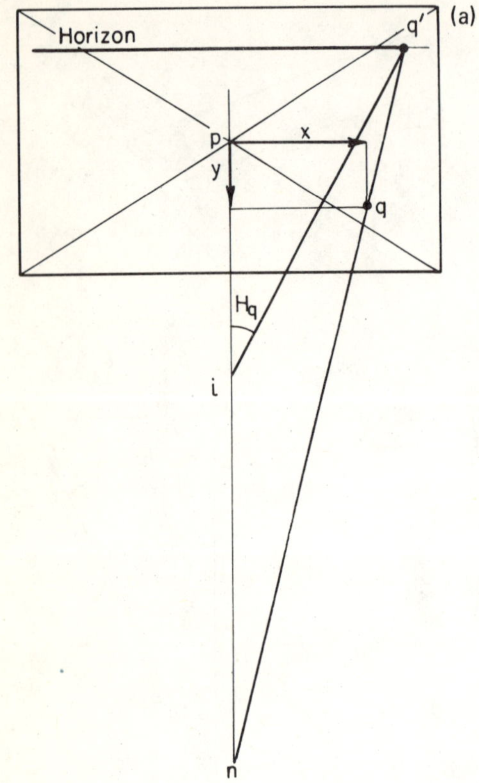

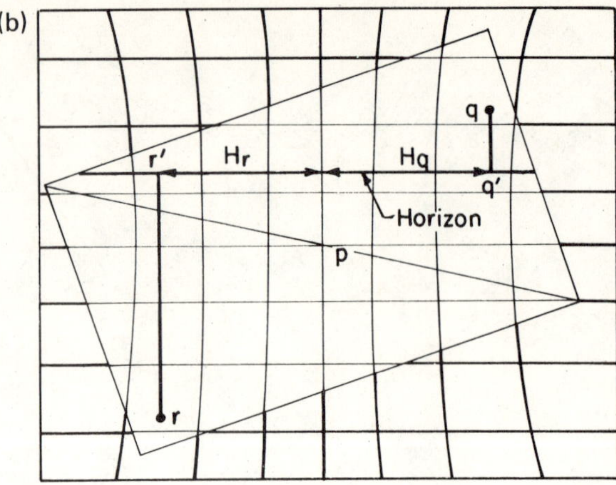

Figure 4-12. (a) Measurement of horizontal angles on a photograph by construction; points i and n are described in figure 4-11. (b) Measurement of horizontal angles on a tracing overlying a gnomonic net.

$$\tan V_q = \frac{y_q \cos\theta \sin H_q}{x_q} \qquad (9)$$

The angle H_q may also be found graphically as shown in figure 4-12a. Moffitt gives a simple construction for vertical angles as well.

A quick method of measuring angles on photos is through the use of a gnomonic net. A photograph is a gnomonic projection wherein the principal distance of the lens (f) equals the radius of the reference sphere at p. Recognizing this, Wallace (1950) proposed superimposing a suitably enlarged gnomonic net overlay centered on the principal point of a photo to measure angles graphically.* Figure 4-13b

* Wallace also showed how to solve for the attitude of a plane, using the stereographic projection, from the angular relations of lines in the plane as seen from two different perspectives. This method is often inaccurate in practice, however.

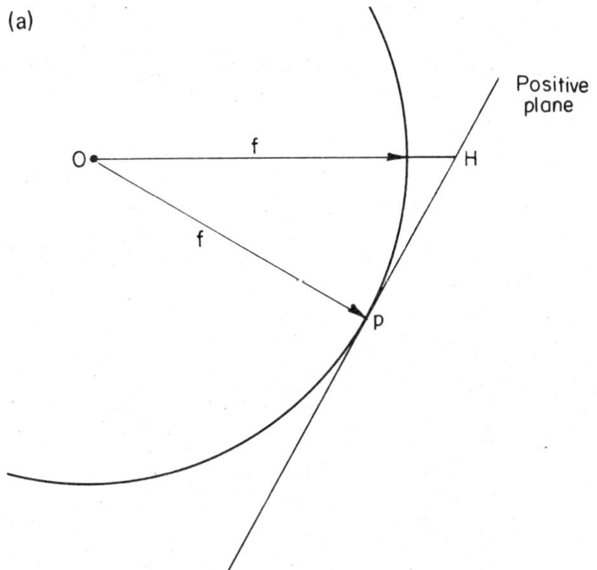

Figure 4-13. (a) A photograph as a gnomonic projection. (b) A one degree gnomonic net corresponding to a focal length of 84.5 mm.

(b)

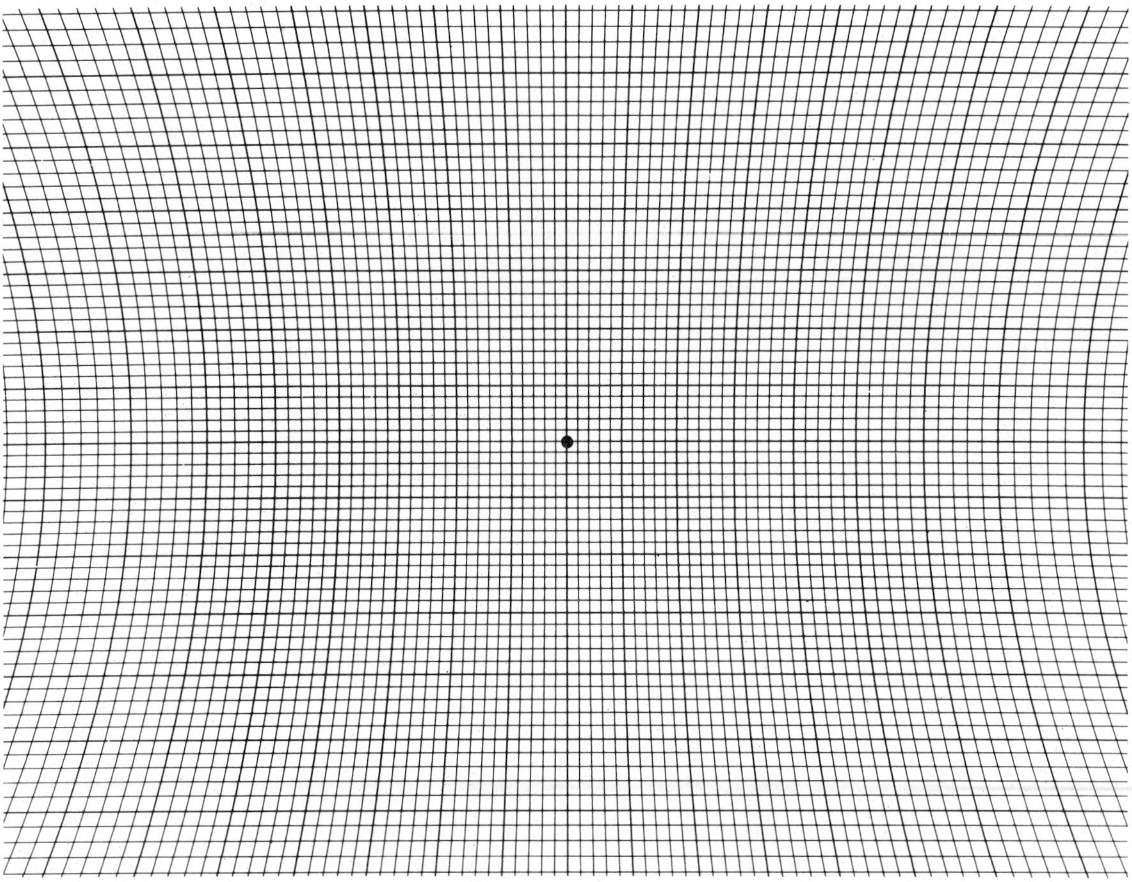

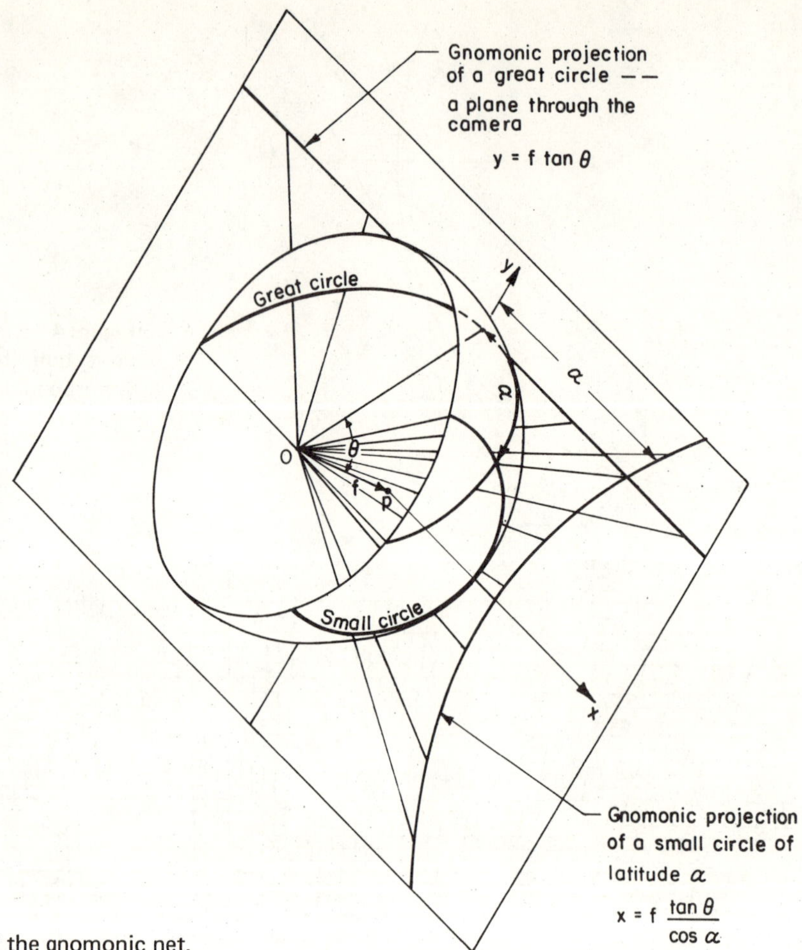

Figure 4-13. (c) Explanation of the gnomonic net.

is a 1° gnomonic net which can be superimposed on a photo if enlarged or reduced accordingly. This net, plotted by computer, corresponds to a focal length of 84.5 mm. (The focal length after any enlargement will be found by the relationship $f = y/\tan\theta$ where y is the distance along the equator corresponding to an angle θ.) As shown in figure 4-13c, the straight lines are the gnomonic projections of lines of longitude -- great circles of the reference sphere having a common intersection; they can represent any plane passing through the point O. The curved lines are the projections of lines of latitude, i.e., small circles of the reference sphere; they graduate the straight lines. To find the angle subtended at the camera between any two image points, revolve the gnomonic overlay about p until one of its straight lines (planes) passes through both points and count the number of latitude curves, each of which marks an angle of 1°. Angles are always measured along the straight lines by counting the curved lines. To measure the horizontal angle H_q to an image point, q, once the true horizon has been drawn on the photograph, revolve the overlay about the principal point p to orient the straight lines parallel to the horizon and count the angular distances H_q as shown in figure 4-12b. Then rotate the overlay 90° and measure V_q. The horizon may be established using the gnomonic net if the horizontal and vertical angles of two points in the photo field are known, e.g., points r and q in figure 4-14.

The above constructions require you to know the focal length (principal distance) of the camera. The focal length may be determined precisely by photographing a line of known length and position with respect to the camera. An Asahi Pentax Spotmatic with a nominal 50mm lens focused at infinity gave (f) = 50.5mm. A wide angle lens, nominally 35mm, gave (f) = 34.4mm.

Applications of Terrestrial Photogrammetry

In reconnaissance geological mapping, where there is rarely enough time to measure all the observations from each vista, an

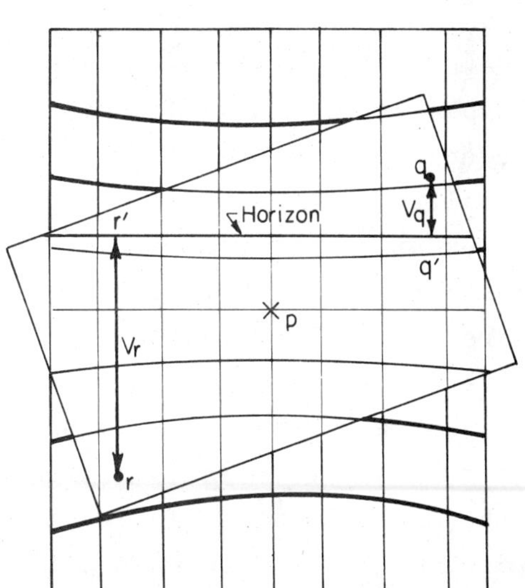

Figure 4-14. Finding the horizon from known vertical angles to two points r and q. Rotate the tracing on the gnomonic net, about point p, to find an orientation giving correct values of v_r and v_q. Then check the horizontal angles H_r and H_q. (see figure 4-12b).

oriented snapshot is valuable. Measurements may be desired later when an idea of the project layout is proceeding and these can be obtained without the necessity of revisiting every site. If a topographic base map is available, outcrop limits, geological contacts, springs, faults, prominent discontinuities, and other features seen in the hillsides from a vantage point can be mapped onto the base using the horizontal and vertical angles from one camera station, or the horizontal angles from two camera stations. Photos from helicopters and light planes can also be used for this purpose if a base map is available and if prominent points can be recognized in the photos. In this case, the camera point can be found by a trial and error resection using a tracing. If one is fortuitous in being able to recognize a sufficient number of images corresponding to points at known map positions, a method of alignments, discussed by Williams, can be used to resect the plan position of the camera; any two points on the same line perpendicular to the horizon establish a ray to the camera and two such rays intersect at the camera point. With the plan position of the camera point known, the vertical angles to images recognized on the topographic map will establish the elevation of the camera and the orientation of the principal point can be established, etc. Thus, one can transfer geological details from random photos to a topographic map.

In mapping for design stage studies, terrestrial photogrammetry can be used in two ways: logging of exposures; and mapping geological details from inaccessible places such as cliffs, rocks and walls. Exploratory excavations should be logged in complete detail; a procedure making use of systematically taken photos will improve the quality of the logging. So many styles of exploratory excavation present themselves in practice that the geologist will have to establish the photo logging system for his own purpose. Deep trenches are difficult to photograph effectively as a great many exposures are required to cover the entire surface. Nevertheless, photographs are relatively inexpensive, and the opportunity they afford for later study and measurement supports their routine application. It should be possible to arrange for multiple parallel photography of the walls (each wall alternatively) of a trench or adit, by mounting the camera on a trolley running parallel to the axis of the excavation on a wooden track; one can trip the shutter using a long cable release.

During construction and operation, terrestrial photogrammetry can be helpful in recording geological details during the short time when new exposures become accessible. Measurements can be made later as needed if the location and orientation of the camera, and the focal length of the lens, have been recorded.

Standard aerial photogrammetric techniques have been adapted to terrestrial phototheodelite photos for measurements as a basis of payment to contractors* and for mapping joint orientations in mines and quarries (Ross-Brown, 1972 and 1973; Rengers, 1967; and Linkwitz, 1963).

Savage (1965) used aerial photo plotters with enlarged overlapping stereo photographs taken with an ordinary snapshot camera from a levelled plane table. Stereo photography is a helpful practice even in those cases where the parallax measurements will not form the basis for the photogrammetry because inspection of the stereo pairs will help in identification of image points and in interpretation of geological data.** The simple graphical methods discussed here complement the standard photogrammetric techniques for geological measurements at all stages of engineering projects.

Example

Figures 4-15a and 4-15b are photographs of a road cutting taken with a Pentax Spotmatic camera from two stations, O_1 and O_2. The original focal length, f = 34.4mm, has been enlarged to 84.5 mm. The horizontal and vertical angles to several recognizable points in the

---

* T.A. Lang described the use of such techniques in the Snowy Mountains project in unpublished notes on Rock Mechanics written in 1965. He referred to work by W.A.G. Mueller (1959) in the Australian Surveyor, Vol. 17, No. 7.

** Williams points out that one must be careful in the field to identify which photo is the one to which the control observations apply.

(a)

(b)

Figure 4-15. (a), (b) Photographs of a road cut taken from the ends of a surveyed base line. Horizontal and vertical angles to a number of points in each photo were measured with a Brunton compass when the photos were taken. The enlarged focal length (ef) is 84.5 mm. The map of figure 4-15c was made from enlargements to 248.5 mm. (c) Plan and cross section of the cut slope shown in figure 4-15a and b by triangulation from the two photographs.

photo field were read with a Brunton compass mounted on a tripod,* and the distance and direction between O_1 and O_2 was measured with

---

* A transit would be preferable, if available, as the quality of control data limits the map.

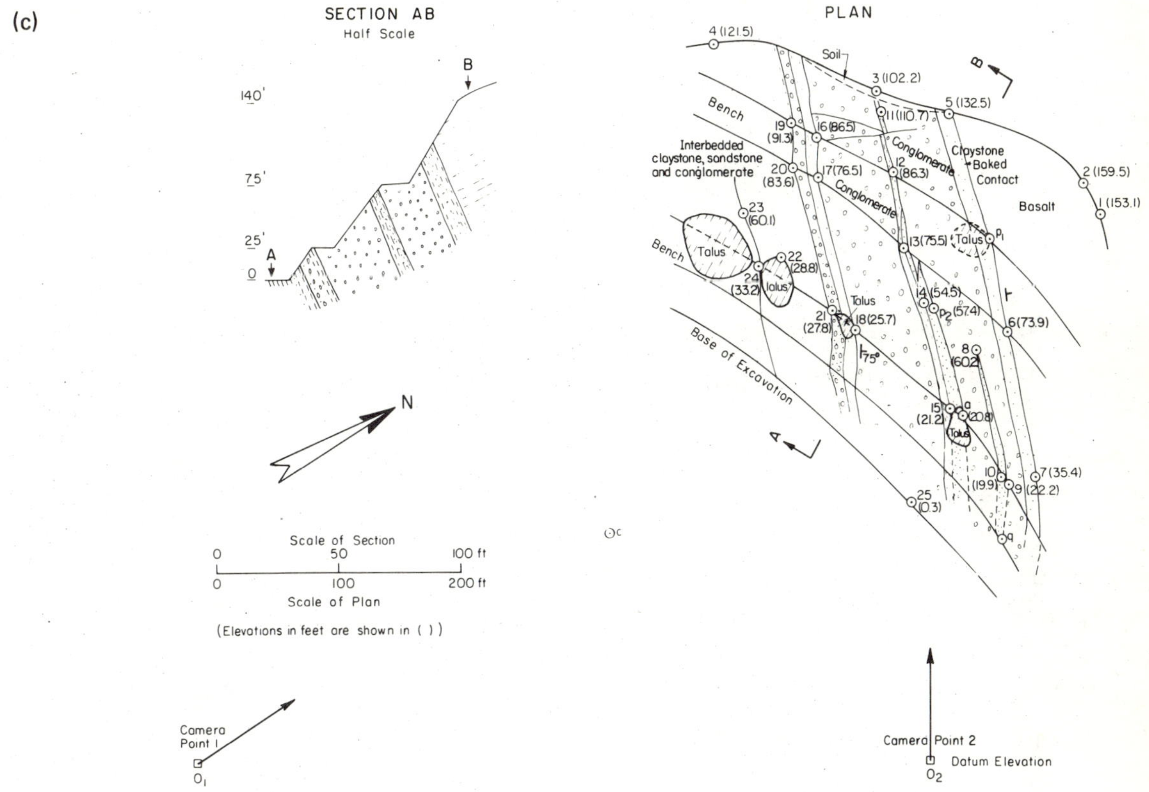

a steel tape. A number of key points, recognizable in each photo, were labelled on individual tracings; Using the gnomonic net centered on the principal point, which was found from intersecting diagonals, the tracings were oriented as previously shown (figure 4-14); then horizontal angles to all the marked points were read, and the plan (figure 4-15c) was developed by triangulation from O_1 and O_2. Next, vertical angles on photo 2 were read for all the labelled points, and the distances from O_2 to each point were scaled from the plan, yielding relative elevations. Finally, geology was interpreted and sketched between the plotted points.

GEOPHYSICAL METHODS

Geophysical surveys conducted from the ground surface can supplement the data of geological mapping. A number of methods of geophysical exploration are discussed in books by Grant and West (1965),

Dobrin (1960), Heiland (1946), Jakosky (1950), Van Nostrand and Cook (1966) and others. Many of these methods have application in geological engineering as summarized by Griffiths and King (1965). Here it will be sufficient to list those techniques which help to describe geometrical and mechanical properties of discontinuities in the rock.

Geophysical surveys usually yield two types of results. First, they evaluate a physical property at different points within the rock mass -- properties like the velocity of sound waves, magnetic susceptibility, and density. This result may be a direct product of the instrument readings, but more usually it is calculated from the data. Secondly, the geophysical exploration will usually reveal the distribution of the measured quantities over the map; interpretation of these data, based upon a suitable model of the geologic structure, will often permit the geometric constants of the subsurface structural model to be determined. For example, data might be interpreted to yield the depth of each layer in a multi-layered configuration, or the width of a fault zone in a model with a tabular fault between continuous walls.

The physical quantities measured by geophysical methods may be of direct value in a design problem. The longitudinal wave velocity, for example, is used in assessing potential damage from blasting operations. But more frequently, the real value in knowing the measured quantity lies in its associations, such as with the rippability of rocks; rocks with sonic velocity less than 7000 feet per second are usually excavatable by ripping, whereas rocks with velocities greater than 10,000 feet per second will usually require blasting (Caterpillar Tractor Co., 1966). Figure 4-16 shows how the sonic velocity correlates with the characteristics of weathered granite, including the percent of "core stones" and the porosity (Polak, 1963). The next section considers briefly some methods which are relevant to the problem of mapping and assessing discontinuities in a rock mass.

Seismic Methods

Several geophysical methods utilize the signature and relative

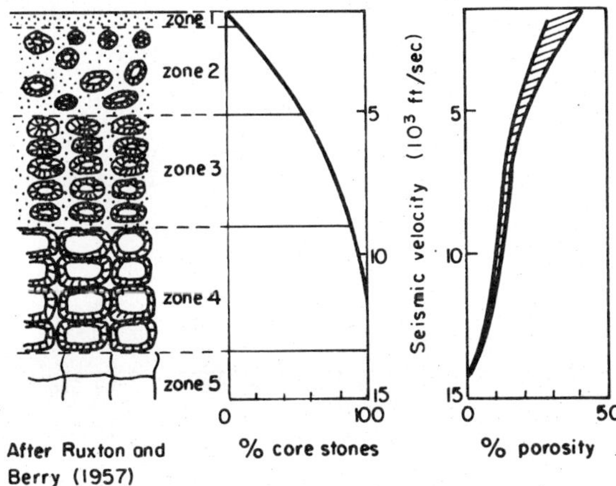

Figure 4-16. The variation of sonic velocity in the weathering profile of granite. From Polak (1963).

arrival times of elastic waves travelling between a shallow "shot point" and one or more transducers. If an underlying rock or soil layer has a higher sonic velocity than the overlying layer, it will serve as a refracting surface. In a specific range of distances from the shot point, the time versus distance data of first arrivals from the shot will represent waves travelling along this surface. Within this range, the rate of change of distance with first arrival time, $\partial x/\partial t$, is equal to the longitudinal wave velocity in the refracting layer, if it is horizontal and regular. If it is irregular or inclined, a simple graphical interpretation can reveal the velocity, as well as the topography and depths, as shown in figure 4-17. When the refracting surface is a stratigraphic horizon, a fault is indicated by a sudden topographic change along it, as in figure 4-17. Joints and shear zones will rarely be seen in refraction profiles, but they may be recognizable in continuous overwater reflection profiles, using a boat to advance a transducer and a repeating shock source, such as a "sparker" (Fanshawe and Watkins, 1971). Other methods of seismic measurement are sometimes used in engineering investigations. Cross hole shooting, with seismic source in a borehole and receivers in other boreholes, is used to map the velocity variations in the subsurface. Faults and close jointing will reduce the net travel time and yield apparently low velocities. Wave front diagrams, constructed from data with the shot point in a drill hole and the transducers in a line on the surface, will supplement data

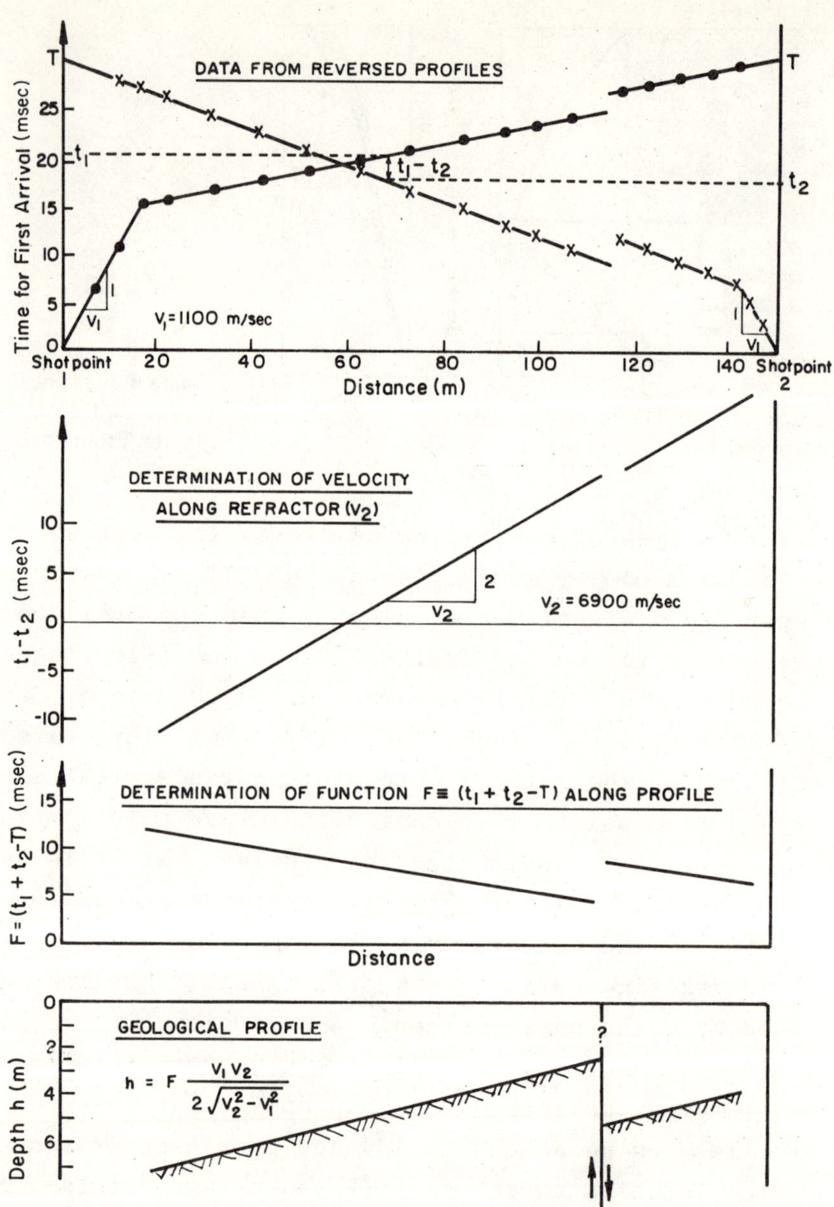

Figure 4-17. Approximate graphical method for locating faults from seismic refraction "reversed" profiles. Based upon Griffiths and King (1965)

from surface profiles, especially where a lower velocity layer lies below a higher velocity layer (Meissner, 1961). If the whole signature of the wave arrivals are recorded rather than just the first arrivals, additional information about discontinuities in the rock can be determined from the wave attenuation, as discussed by Polak (1971).

Resistivity Surveys

Since most rocks are themselves nonconductive, the electrical resistivity of a rock derives mainly from salinity in the ground water occupying pores and fractures. Accordingly, rock formations will differ in resistivity because of porosity and jointing differences. Faults that act as contacts may, therefore, be mappable by resistivity. In faults and shears, the water content may be higher than in the country rock, and anomalously low resistivity will be measured. Occasionally, in porous country rock, a fracture will act as a drain and appear as an anomaly of high resistivity (Stahl, 1973). Field measurement may be accomplished by traversing along a line with a set of four electrodes at fixed spacing, ("Fixed Wenner" array). Batteries or an AC current source are connected to set up a current flow (I) between the outer electrodes, while the potential drop (ΔV) between the inner electrodes is measured with a millivoltmeter. The resistivity (r) is calculated from the relationship

$$r = 2\pi a \Delta V/I \qquad (10)$$

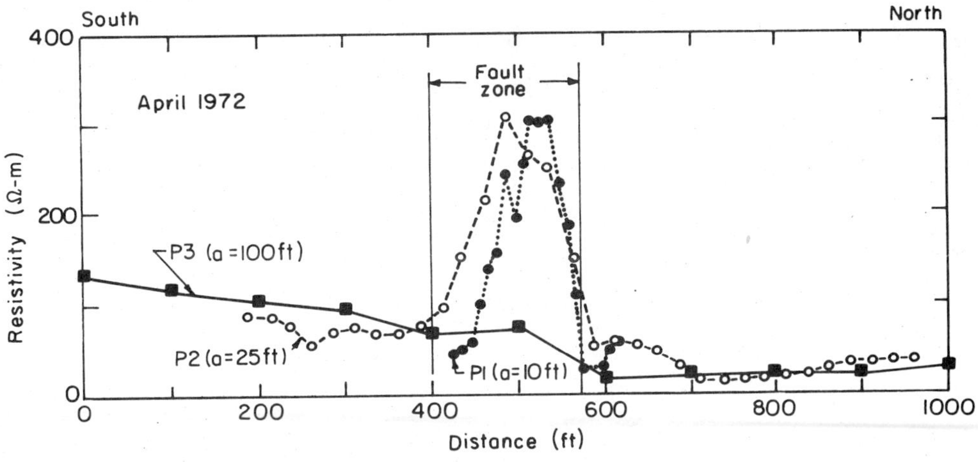

Figure 4-18. Results of a fixed Wenner resistivity profile across a fault zone. From Stahl (1973)

Figure 4-18 shows the resistivity variation in moving fixed Wenner arrays, with electrode spacings of 10, 25, and 100 feet, across a resistive fault zone. Interpretation of resistivity surveys is more complicated than interpretation of seismic profiles; consequently one usually checks the interpretations by selected trenching or boring. (This is a good practise in engineering geophysics work in general.)

Gravity Surveys

Measurement of changes in the gravity field from point to point may be used to map changes in geology because density changes between different formations produce anomalies. The subject is complicated and has not been applied extensively in engineering practice. Very precise gravimeters now available will respond to 1 mm changes in elevation and are being used to detect vertical rock movement. An approach for characterizing the rock mass porosity in-situ is to measure the apparent density of a rock mass in which the rock specimen density is known (Eaton et al, 1964). If the mass density of a sample of the rock is ρ_s while the bulk mass density deduced from the gravity survey is ρ_b, the additional porosity of the formation beyond that represented in the specimen is $\Delta n_j = 1 - \rho_b/\rho_s$. In saturated rock, $\Delta n_j = (\rho_s - \rho_b)/(\rho_s - \rho_w)$ where ρ_w is the mass density of water. If this porosity is created by a given pattern and spacing of joints, it should be possible to deduce the average apertures of each set; or if the apertures are assumed, the average spacing of each joint set can be calculated. Interpretation of field gravity measurements, with excellent levelling control, will yield density values for different rock bodies if the true geometric properties of each body are determined by drilling or otherwise.

Magnetic Methods

The intensity of the earth's magnetic field at a point on the surface is the sum of two vectors -- the induced magnetic intensity derived from the earth's magnetic field acting on rocks of given magnetic susceptibility, and remanent magnetism in the rocks. Since rocks vary in their magnetic susceptibility and remanent magnetism, magnetometer surveys can map the boundaries of formations. Rocks derive magnetic susceptibility from the minerals magnetite, hematite and pyrrhotite, and as these minerals are more prevalent in the basic igneous rocks, the magnetic highs correspond to occurrences of such rocks. The California Department of Water Resources used aeromagnetic surveys successfully to locate bodies of serpentine in connection with the planning of water tunnels and the Corps of Engineers used surface magnetometer surveys to map buried serpentine bodies at a dam site. Fracturing and weathering of rocks demagnetizes them, so one can expect faults and highly jointed igneous rock masses to show up as anomalies of low magnetic intensity. Wiebenga and Polak (1969) used a "micro-magnetic" method to reveal joint trends in an area of rock thinly masked by weathered debris. An area of about 100 feet by 100 feet was covered with a grid of measuring points on ten foot centers.

Unfortunately, it is often easier to make good geophysical measurements than to provide good interpretations. However, geophysical techniques are generally relatively inexpensive and can yield useful information under the proper circumstances.

DRILL HOLES

Geological field mapping rarely yields sufficiently detailed data for design because the surface exposures give incomplete details of the stratigraphy and structure. Even in the planning stage, it is usual to invest a significant sum in bull-dozer cuts or trenches and some drill holes to determine definitive answers to crucial questions centering on the geology. Geophysical measurements, aerial photographic interpretation, and shallow excavations and test pits must eventually be supported by drill holes, and sometimes, exploratory tunnels, in which tests and observations can be made and from which samples of the rock can be studied. We will consider methods of drilling and sampling rock and the important question of orienting the core so that the attitudes of discontinuities can be measured.

Drill holes are expensive, c.a. $25.00 per foot for diamond drilling, with pressure testing, boxing of samples, and logging of the hole and core. The amount of drilling can be minimized if planning of each hole considers previous geological knowledge, including previous drill holes. Evaluation of the adequacy of a drilling program requires appreciation of the probabilities of intersecting the possible targets of search with the given pattern of holes. If the sampling of a site consists of a series of parallel planes such as deep trenches or geophysical profiles, then the probability of encountering a specified target is approximately the ratio of the length of the target, normal to the line of the profile, to the spacing between profiles. When linear samples or observations are made, as with parallel drill holes on a regular grid, the probability of encountering a target is related to the ratio of its area normal to the sample lines to the area of the unit cell of the exploration grid. When little is known about the geological structure, the greatest probability of intersecting and finding targets of search will be obtained from a square grid of holes. Figure 4-19, from Slichter (1955), gives the probability of encountering a target of circular or rectangular area normal to holes on a square grid with spacing S. Given a particular spacing of drill holes, this figure

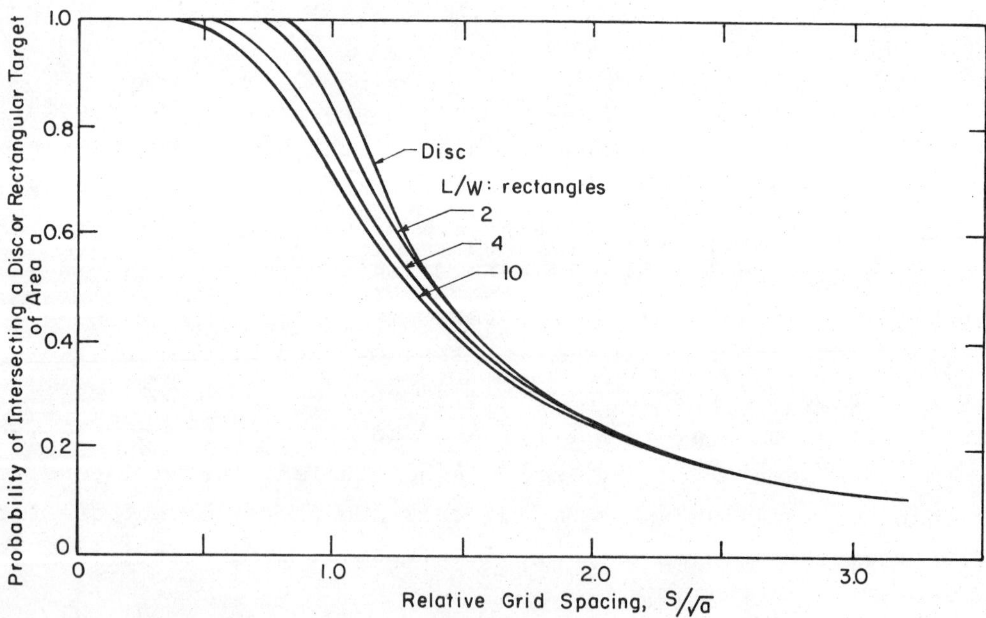

Figure 4-19. The probability of a single intersection of disc or rectangular target using a square grid of drill holes with spacing s; after Slichter (1955).

will permit you to determine the degree of certainty with which you can be assured of having found a geological feature, such as a fault, a hydrothermally altered zone, a cavity, a buried valley, etc. For example, with rectangular targets of length to width ratio l/w equal to 2, and a grid spacing S equal to $2(lw)^{\frac{1}{2}}$, the probability of encountering the target is only 25%. A drilling program can not be expected to find targets appreciably smaller than the spacing between drill holes.

Methods of Drilling

There are numerous ways to drill rock, each developed for a special combination of factors. For the small diameter exploration holes of the engineering geologist, only coring methods, mainly diamond drilling, provide a satisfactory rock sample; this then is the principal approach used to explore details of the subsurface structure. Less expensive non-coring methods can satisfy other exploration needs, for example: to provide relatively cheap access to a point where coring is to begin; to emplace seismometers, piezometers, tilt-meters and other instruments; to measure water levels or rock temperatures; to map the top of rock surface; or to provide blast holes for seismic exploration.

Non-coring drilling is performed by percussive or shearing methods individually or in combination (Table 4-3). Soil exploration methods by alternate drive sampling and wash boring, combining a chopping action with erosion by water, are not often appropriate for rocks because the "split spoon" can not be driven while the wash boring produces a valueless sample and a damaged hole. Drilling by raising and lowering a chisel (churn drilling) is no longer as economical as pneumatically driven percussion drills, which are driven from the top or by down-the-hole hammers. Churn drilling is still used in the absence of compressed air. In soft rocks, augering and combined percussion-auger drills can be used, very cheaply. However, rotary drilling with carbide blades (drag bits) is faster, even in rocks of medium hardness (e.g. coal, some shales and sandstones, evaporites, marls and some serpentines and limestones). Rotary drilling using toothed rolling cones ("tricone" bits) can

TABLE 4-3

Drilling Methods

Data from McGregor (1967) and other sources

Augering - Disturbed samples of soils and soil-like rocks; 2 to 24" diameter holes, maximum depth 60 to 100 feet; holes usually vertical.

Wash Boring - Used for making a hole, without return of a sample, in soils and in soft rocks; only in vertical holes.

Center Sample Rotary Drilling (Becker CSR) - Chips or cores of all rocks; holes 3½" and larger, up to 750 feet deep; holes may be inclined up to 45 degrees

Churn Drilling - Used for making a hole, without return of a sample, in soft and medium hard rocks. Holes commonly 2" to 15" in diameter, up to 4000 feet deep; usually vertical.

Track or Wagon-Mounted Percussive Drilling - Used for making a hole, without a sample, in hard rocks. Holes commonly 1-3/4" to 4-1/2" diameter, up to 100 feet deep, in any orientation.

Down the Hole Percussive Drilling - For making holes 4" to 6" in diameter without a sample, in hard rocks. Usual depth range 120-200 feet. Can be drilled in any orientation.

Rotary Drag Bit Drilling - Very fast drilling in soft to medium hard rocks, but without a sample. 2-3/8" to 7-1/8" diameter holes in any orientation. Maximum depth in the range 100 to 250 feet.

Rotary Tricone Drilling - Small diameter core sample obtained in the larger hole sizes in all but the very hardest rocks. Holes up to 3" to 12" in diameter to any depth; rigs for drilling up to 200 feet in any orientation are common.

Shot Drilling - A continuous core sample in all but the hardest rocks with holes up to 6 feet in diameter; rather slow. Small diameter shot drilled holes have been drilled up to 1000 feet deep vertically. Calyx holes rarely are drilled deeper than 100 feet.

Diamond Drilling - Continuous core sample of all rocks except highly fractured rocks of great hardness. Common hole diameters 15/16" up to 7-3/4". Maximum depths of rigs are 200 - 1500 feet. Wireline equipment can be used in steep holes.

penetrate all but the hardest rocks. This method was developed by the petroleum industry where high bit pressures can be provided by the weight of the drill string. For shallow drilling, very heavy rigs are needed. A continuous rock core is obtained using diamond coring bits (figure 4-20) or shot drilling methods. The latter work by turning a slotted steel cylinder on continuously supplied abrasive steel shot; this slow expensive method is used only on very large holes or for smaller holes in situations where diamond bits are unobtainable. Czech engineers, for example, have used shot drilling to explore the foundations of dam sites because they could not purchase diamond bits. Large diameter shot drilled holes (36" to 60" "calyx" holes) were used by the T.V.A., the Corps of Engineers, and other engineering agencies to permit downhole inspection of the rock by a geologist.* Oriented core techniques, and to a lesser extent bore-hole cameras and television devices, have superseded calyx drilling for exploration in all but very special circumstances.

Good drilling practice reduces costs and is vital in coring for civil engineering work since the quality and completeness of the sample are sensitive to drilling efficiency. Thrust, rotation and flushing action must be properly adjusted according to equipment and rock conditions: Thrust applied to the bit forces the mechanical work to be done on the rock and not on the drill. In diamond drilling it is the force per carat which controls; too high a value breaks diamonds and too low a value polishes them. Flushing the bottom with fluid, water or compressed air, transports the cuttings from the rock breakage region and cools the cutting tools; an insufficient flushing rate risks plugging of the annulus, indirectly causing blocking of the drill core. The speed of rotation affects the cutting action of the bit; it varies according to the bit design and rock type. The optimum speed of rotation (in the range 250 to 1500 rpm) is determined by the minimum thrust. Since the controlling variable is the tangential velocity of the diamonds, larger diameter bits must be turned more slowly than smaller ones.

For maximum core recovery, diamond drills for civil engineering geology should be "hydraulic feed" types, in which rotation speed

---

* Calyx hole inspection is unacceptably hazardous unless an air line and a protective man car are provided.

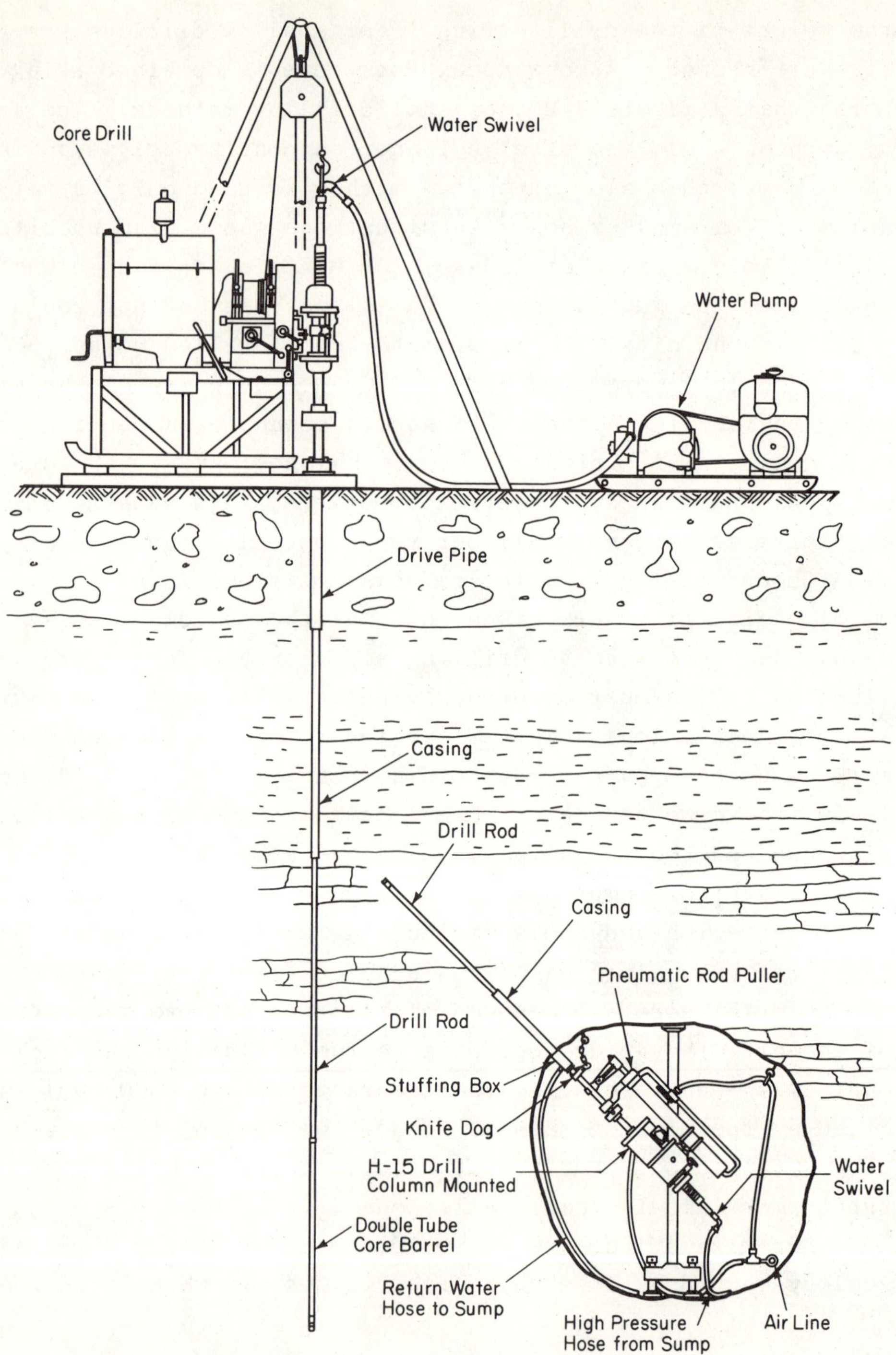

Figure 4-20. Typical diamond drilling set-ups; (courtesy Joy Manufacturing Co.).

and bit pressure can be varied independently. Large diameter hard rock drilling requires relatively slow rotation with high pressure whereas friable soft rocks want a moderate speed of rotation, with low pressure. The hydraulic feed is usually helpful in layered formations of variable hardness, where high core loss will result from improperly controlled drilling owing to the breakability of the individual layers.

Core Barrels

Sizes of diamond drill holes were standardized in 1930 by the Diamond Core Drill Manufacturers Association (X series)*. Later, the need for larger diameter holes in geological exploration for engineering works led to a "large diameter series" (Table 4-4). The core bit and casing sizes in standardized series designations are "nested" (Table 4-4); i.e. if drilling begins with an NX casing bit, (3.6 inch diameter hole), and at some point steel casing is installed to prevent caving of weathered or seamy material, drilling can continue with an NX bit (3 inch hole) which admits BX casing and so on. Metric equipment does not correspond in size to the X series.

Large diameter core barrels with sophisticated design features are usually required in civil engineering exploration to prevent core loss or damage. Core loss arises from grinding, breakage and erosion. With a single tube core barrel, essentially a section of casing following the bit, the wash water travels alongside the core and discharges at the bit, returning in the annulus outside the barrel. Rock chips eroded by the wash water can jam the bit, blocking free entrance of new core into the barrel and leading to core breakage. Erosion of friable materials seriously damages the sample and washing out of seams destroys the continuity of the sample, exposing it to breakage inside the barrel. The double tube core barrel was designed to isolate the core from the water stream (figure 4-20); water flows down between the inner and outer barrels, discharges

---

* The Canadian Diamond Drilling Association standardized similarly in 1950 (XT series); in 1956 the coupling and thread details were modified in a new standard (W series) but diameters of holes of the X series given in Table 4-4 were retained.

TABLE 4-4

Standard Sizes of Drill Holes and Typical Sizes of Drill Cores

TYPICAL EUROPEAN SIZES

Hole Diameter (mm)	Core Diameter (mm) thick walled bits	Core Diameter (mm) thin walled bits	
146	120		Swedish Diamond Rock Drilling Co.
131	105		
116	90		
101	75		
86	58	72	Craelius Co.
76	48	62	
66	38	52	
56	34	42	

U.S. AND CANADIAN SIZES

Name	Hole diameter (in)	Usual core diameter (in)	
6 x 7-3/4	7-3/4	6	"Large Series" double tube swivel barrels.
5-1/2 x 4	5-1/2	4	
3-7/8 x 2-3/4	3-7/8	2-3/4	
WX (NX casing bit)*	3-5/8	2-13/16	
NX	3	2-1/8	
BX casing bit*	3	2-3/16	
BX	2-3/8	1-5/8	
AX casing bit*	2-3/8	1-3/4	
AXT	1-7/8	1-9/32	
AX	1-7/8	1-3/16	
EX casing bit*	1-7/8	1-3/8	
EXT	1-1/2	1-5/16	
EX	1-1/2	7/8	
1-1/4" BH	1-1/4	3/4	
XRT	1-3/16	3/4	

WIRELINE CORE BARREL SIZES (Longyear Co.)

Name	Hole Diameter in.	Hole Diameter mm.	Core Diameter in.	Core Diameter mm.
PQ	4-53/64	122.6	3-11/32	85.0
HQ	3-25/32	96	2-1/2	63.5
NQ	2-63/64	75.8	1-7/8	47.6
BQ	2-23/64	60.0	1-7/16	36.5
AQ	1-57/64	48.0	1-1/16	27.0

TABLE 4-4 (continued)

MINDRILL SERIES S CORE BARREL SIZES

Name	Hole Diameter* in.	Core Diameter in.
7S	6.290	4.375
6S	5.318	3.750
5S	4.347	3.063
4S	3.630	2.625
HXS	3.906	2.875
NXS	2.980	2.155
BXS	2.360	1.655

* Dimension given is outside diameter of reaming shell.

Figure 4-21. Several types of diamond drill bits; (courtesy of Mindrill, Ltd.).

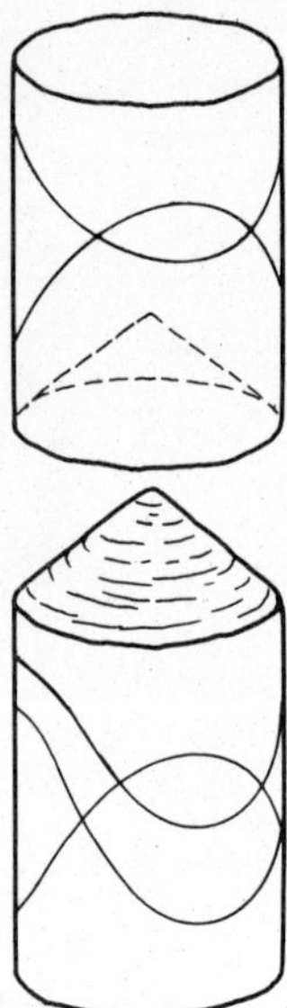

Figure 4-22. Ground ends of core.

against the core inside the bit and returns outside of the outer barrel. In erodible or seamy formations, the water stream can be further separated from the core by using bottom (face) discharge bits (figure 4-21). Longyear Co's "L" series core barrels (e.g. NXL) and Joy, and Sprague and Henwood's "M" series (e.g. NXM) reflect improved designs which among other things discharge the wash water closer to the bit.

Blocking and wedging of core inside the barrel can cause core to break, especially if it is of small diameter. Ideally, core barrels should not be smaller than NX size to explore discontinuous rocks for civil works.

Grinding in the barrel occurs when previously drilled core turns against fixed core still attached to the bottom (figure 4-22). Many feet of core can be lost in this way. Grinding can be minimized by withdrawing ("pulling") core frequently so that the barrel is never allowed to fill up. However, frequent interruption of drilling and withdrawal of all the rods to empty the barrel proves costly. In vertical and in steeply inclined holes, "wire line" equipment can be used (e.g. Longyear's Q series) in which the inner tube can be uncoupled from the core barrel and pulled to the surface by a wire cable inside the string of special hollow drill rods. While the tube is being emptied, a second inner tube is lowered and drilling continues. If core separates along natural or new fractures it can wedge against the sides of the barrel, blocking entrance of new core. With Mindrill's S barrels and the L series barrels, circulation of water cuts off as soon as a block occurs to warn the driller. To prevent grinding, the core should be prevented from rotating by means of a free swivel joint between the top of the inner tube and the head of the barrel. Ball bearing swivel type heads are now generally available for all sizes of double tube core barrels, e.g. the L, M and "Large" series mentioned above, and should be selected for geological engineering.

Ideally, core should be removed from the sample tube (inner tube) without disturbance to preserve mating across natural fractures and to prevent loss of filling material or small rock pieces. Longyear, Mindrill and other manufacturers provide a core barrel either with a split inner tube or as an additional tube sample holder within the inner tube (figure 4-23a), wherein the split tube is extruded hydraulically from the inner tube and separated to expose the core without disturbance (figure 4-23b). A triple tube barrel can be made by inserting a split plastic tube inside the inner core barrel of a double tube type.

Good core recovery can be difficult to achieve in friable materials. Christensen Diamond Products Co. developed a rubber sleeve core barrel which encases the core in a tight fitting neoprene sleeve as it comes into the barrel. This is particularly useful when drilling friable rocks with drilling mud which can invade and damage the core. Another special barrel is Christensen's "pressure core barrel", which tries to preserve the original pore fluid in the core sample by sealing the inner tube as soon as it is occupied.

Good core recovery is also difficult to achieve in materials of variable quality, e.g. where hard rock and soil occur in the weathering profile. For uniformly soft intervals, large diameter design core barrels have been equipped with soil sampling extensions on the inner tube which work like drive samplers when pushed ahead of the coring bit. Berents (1961) reported a design by the Snowy

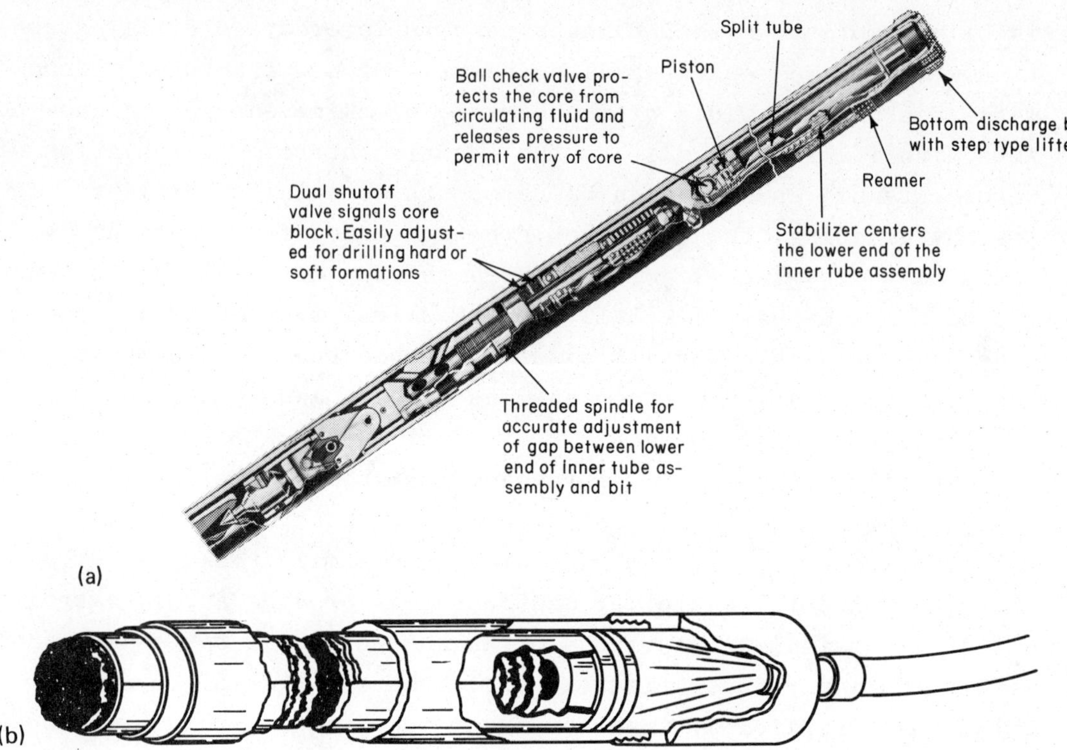

Figure 4-23. (a) Longyear triple tube wireline core barrel (Q3). (b) Extruding the split inner tube of the Q3 barrel; (courtesy Longyear Co.).

Mountain Hydroelectric Authority* for use in decomposed granite containing layers too hard for the soil sampling extension on the NM barrel. A spring was placed between the back of the inner tube and the drill string; when fully extended, the soil sampling extension protruded beyond the bit; but a thrust of four hundred pounds, the minimum for drilling soft weathered granite in this case, would compress the spring sufficiently to expose the diamond bit.

It can be appreciated from the above that the sampling of rock for exploration of structural features demands careful use of precision equipment. It is unlikely that the available drilling contractors at a remote site will have access to the most advanced

---

* In conjunction with Triefus Industries, Australia

tools, such as the Mindrill S series, Longyear Triple Tube (V3 and Q3) and comparable products, nor be experienced in their use. It is a good idea for an engineering organization to purchase its own core barrel and to become proficient in operating it. Some suppliers of good core barrels are: Acker; Atlas-Copco; Boyles Bros.; Christensen; Joy; Longyear; Mindrill; and Sprague and Henwood.

Logging of Drill Holes and Core Samples

The drilling log, made by the drilling inspector or the driller himself, indicates the kind of equipment used at each step, the rate of drilling, the location and orientation of the hole, the depth of water loss, caving, excessive vibrations, blockages and other incidents, as well as any other technical details related to the work. The geologist's log, on the other hand, should give complete, descriptive details about the core and observations and measurements in the borehole, e.g. water pressure tests. Ideally, the logs are made routinely by a geologist as he opens the core barrel; this is especially important in shales and seamy rocks which change appearance and character markedly when the core dries out. In such rocks, an effort should be made to preserve the field moisture content by sealing the lengths of core, or at least a representative selection, in plastic bags. The core should be placed in sturdy wooden boxes with hinged covers and divided in compartments of the same length as the missing sections and the entire box photographed routinely before being placed in storage. A tendency exists to economize on storage requirements by closing up gaps in the core record; photographs of such core are misleading to users, as is a log with broken lines to shorten monotonous section. The core boxes should be stored on frames allowing any box to be removed without first displacing those on top, and a current index map to the core shed should be displayed.

The excellent logging procedure used by the Snowy Mountain Hydro-Electric Authority was described by Moye (1967). A columnar section was made down the center of each sheet and corresponding graphic response logs were drawn to show the number of fractures per foot (fracture log), the percent core recovery, and the water loss

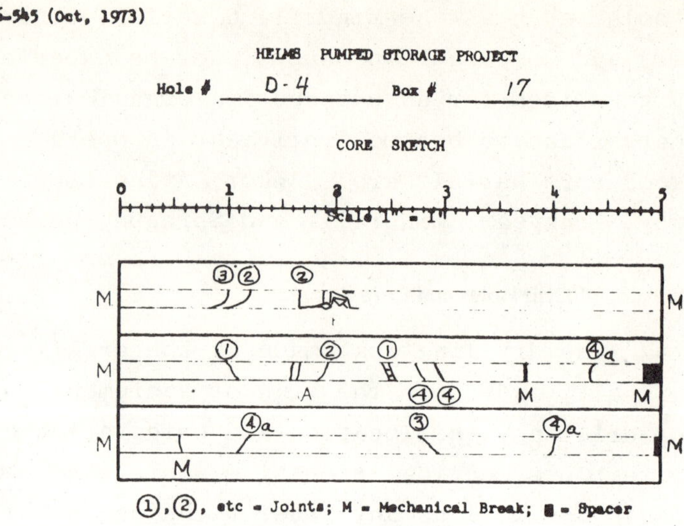

PHOTO

Figure 4-24. Format for geological log of diamond drilling introduced by Prof. Tor Brekke; the log was prepared by Mr. Larry Myer; (courtesy of the Pacific Gas and Electric Company, San Francisco).

in pressure tests. Considerable space was provided to describe petrology, weathering, and discontinuities.

A somewhat different format for presenting the results of exploratory core drilling was developed by Professor Tor Brekke of the University of California for the Pacific Gas and Electric Company and is reproduced in figure 4-24. Each core box is represented by two facing pages in the log book. On the bottom left is a photo-

76-545 (Oct, 1973)
A

HELMS PUMPED STORAGE PROJECT
FIELD CORE LOG

Hole # D-4 Box # 17

DEPTH INTERVAL OF BOX: 235.6 to 250.4
CORE RUNS INCLUDED: 234.5 to 244.5
244.5 to 254.5
to

INFORMATION FROM DRILLERS' LOG
Drilling Rate: .3-.4 ft/min
Water Loss: 100%
Remarks: Water return at end of core run

CORE RECOVERY: 100%

ROCK
Rock type, color, degree of weathering: Fresh grey granodiorite

Texture, structure/foliation: Medium grained, foliation indistinct at 44-50°

CHARACTER OF JOINTS	#1	#2	#3	#4
Inclination:	59°	30°	71°	35°
Shape & Roughness:	Rough Subplanar	Smooth Planar	Smooth Planar	Rough Subplanar
Surface Weathering:	Fresh	Trace (rust)	Fresh	Trace to Light (rust)
Coatings:	Clean	Grey Talc Coat	Chloritic or Green Talc Coat	Light Coat Yellow Clay
Filling Material:				

Remarks:
#1 Subparallel foliation
#2 Perpendicular foliation
#3 (3) Exhibit close joint spacing even though almost axial in orientation
#4 Orientation varies 20°-35°
(4a) - has light to moderate weathering (rust)

OTHER DISCONTINUITIES
(A) - Broken zone due to close jointing- surfaces of pieces chloritic
(B) - Broken zone due to close jointing and mechanical breakage.

REMARKS

Date Logged 8/7/74 By Larry R. Myer

graph of the core box, preferably in color; above it is an annotated sketch of the core as seen in the photograph. The symbols used in the core sketch are fully described on the right page together with identifying information and highlights of the drillers' log. The main joint sets are classified by orientation and character, and additional individuals are noted under the section "other discontinuities". This method of logging and displaying the information from the core box not only presents the data vividly, but provides an interpretation and classification of the discontinuities that will be appreciated when entering into analysis and decision making.

The potential penalties for careless or faulty logging procedures are high, as the logs are often included as legal entities available to bidders and become a part of the contract. One legal case concerning rock excavation centered around a marginal notation "probably needs blasting" scribbled on a log sheet in the field. In Zambia, a construction company blamed a faulty log of a boring at the site of an underground power house for a chain of events which led ultimately to the forfeiture of the contract and the bankruptcy of the company*; in this case it was alleged that biotite schist bands within a biotite gneiss, logged uniformly as "sound gneiss", created a falling rock condition.

ABSOLUTE ORIENTATION OF STRUCTURAL FEATURES IN DRILL CORE

A drill core is considerably more valuable if the true attitudes of planar discontinuities can be logged. The orientation of a discontinuity is often its most immediately significant attribute; furthermore, the use of the sub-surface data for correlating structures from hole to hole is greatly enhanced when the core is absolutely oriented. When a distinctive discontinuity is recognized in the core of non-parallel drill holes, one can calculate its orientation. In the majority of cases where this can not be accomplished, absolute orientation can be determined by reference to other structures in the core whose orientation is known, either through knowledge of the regional geology, or from pointwise down hole orientation measurements. Such measurements can be made using: the Craelius core orienter; bore hole periscopes, cameras, or television devices; paint marks on the hole bottom; or paleomagnetic determinations. It is also possible to perform continuous orientation of the entire drill core by scribing it with knives in the core barrel by using

---

* "The Story of Borehole 4, Kariba North", New Civil Engineer, 8th February, 1973, pp 10,11.

the "integral sampling technique" and by down hole "dip meter" surveys.

Core Orientation by Reference to Known Structural Attitudes

Rosengren (1970), showed how lengths of core can be placed in a consistent relative orientation. Though tedious, this can be rewarding if core recovery has been excellent. The core pieces are laid on a "V" trough in proper sequence and rotated as necessary to fit all the mating pieces together. The entire length of core is thus divided into a series of continuous sub-lengths; a continuous reference line is painted down each such sub-length, and the points where it must be interrupted are clearly marked. Planar features will intersect the core to form ellipses (figure 4-25a). The angle, α, between the core axis and the major axis of an ellipse of intersection can be measured with a contact goniometer (figure 4-25c).

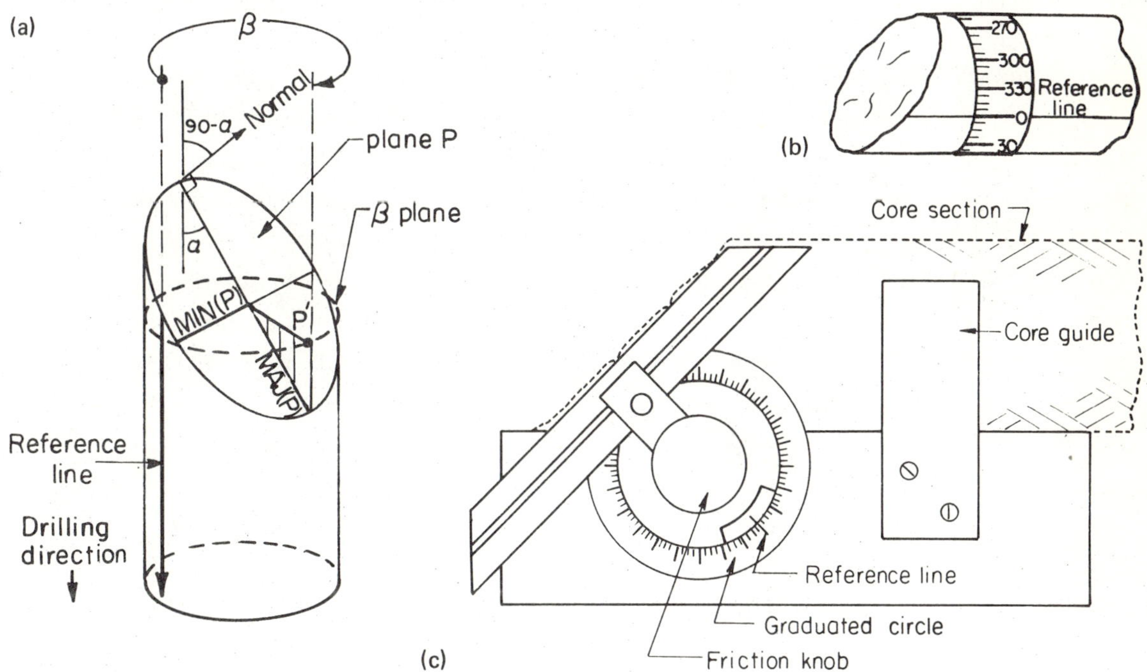

Figure 4-25. (a) Rosengren's reference line and the angles α and β. (b) A simple method of measuring β; β = 250° (courtesy of Golder and Brawner Associates, Vancouver). (c) A goniometer for measurement of α; (courtesy of Golder and Brawner Associates, Vancouver).

Figure 4-26. Core being referenced and described; note the continuous reference line on the mated lengths of core laid in "v" troughs (courtesy of Dr. K. Matthews, Mt. Isa Mining Co., Australia).

The downward end of the major axis of the fracture ellipse meets the core circumference at an angle β from the reference line (measured clockwise looking in the direction of drilling). This can be measured routinely using a circumferential band calibrated in degrees (figure 4-25b). Figure 4-26 shows core being referenced as above. The angles α and β determine the orientations of any fracture in coordinates local to the particular reference line. If the absolute orientation is known or can be determined for any planar feature (other than one perpendicular to the drill hole), all other planes can be oriented absolutely.

The stereographic projection facilitates the solution. Consider the ellipse formed by the intersection of an inclined plane of discontinuity and a vertical cylinder (figure 4-25a). The minor axis of the ellipse -- the strike of the discontinuity when the core is vertical -- is the line of intersection of the plane of discontinuity and the plane perpendicular to the axis of the core. Since the latter is the plane in which β is measured, we shall call it "the β plane". The major axis of the ellipse of intersection projects into the β plane at 90° from the minor axis (the β value for a plane is measured from the reference line to the major axis). Respecting the sign convention given, the major axis will be in the lower hemisphere if the hole is downward. Figure 4-27a shows a stereographic projection relating the angles β_R and α_R for a reference plane (R) of known attitude in a drill hole (H) of known orientation. The β plane is constructed as the plane normal to H, and its intersection with R, together with the measured value of β_R determines the position of $\beta = 0$. In this illustration, the drill hole H is inclined 70° to the north, while the reference plane strikes N40E and dips 40° to the southeast. $\beta_R = 7^o$ and $\alpha_R = 60^o$.

Now that the orientation of the reference line in the β plane has been determined, any other plane defined by values of α and β may be oriented absolutely. For example,(figure 4-27b) a plane P, in the same section of core as the reference line of figure 4-27a, has values $\alpha_P = 45^o$ and $\beta_P = 65^o$. First mark the point P' at 65° from $\beta = 0$ along the β plane. Then, in the plane HP', determine the position of the major axis of the ellipse MAJ(P) located 45° from H. Next plot one of the minor axes of the ellipse MIN(P) at an angle 90° from P' along the β plane (the sign does not matter). Plane P is defined by the great circle common to MIN(P) and MAJ(P) which in the example strikes N75E, and dips 64° to the South.

This solution, as well as others to be discussed, requires a knowledge of the orientation of the drill hole. Boreholes, especially inclined ones, tend to wander from their initial orientation after a hundred feet or more, depending upon the nature of the rock; the problem can be acute when the angle (α) of foliation or bedding becomes small. Wherever core orientation is being considered, it is good practice to order a survey for the inclination of the drill hole. This can be done by a variety of methods as discussed by Cumming (1956).

Approach Using Non-parallel Holes

If a planar feature maintains constant orientation throughout a volume of rock, and has a distinctive character making it recognizable in different boreholes, it can be oriented absolutely from knowledge of its α values in non-parallel holes. Bedding, cleavage, schistocity, foliation, any joint set with a strongly expressed

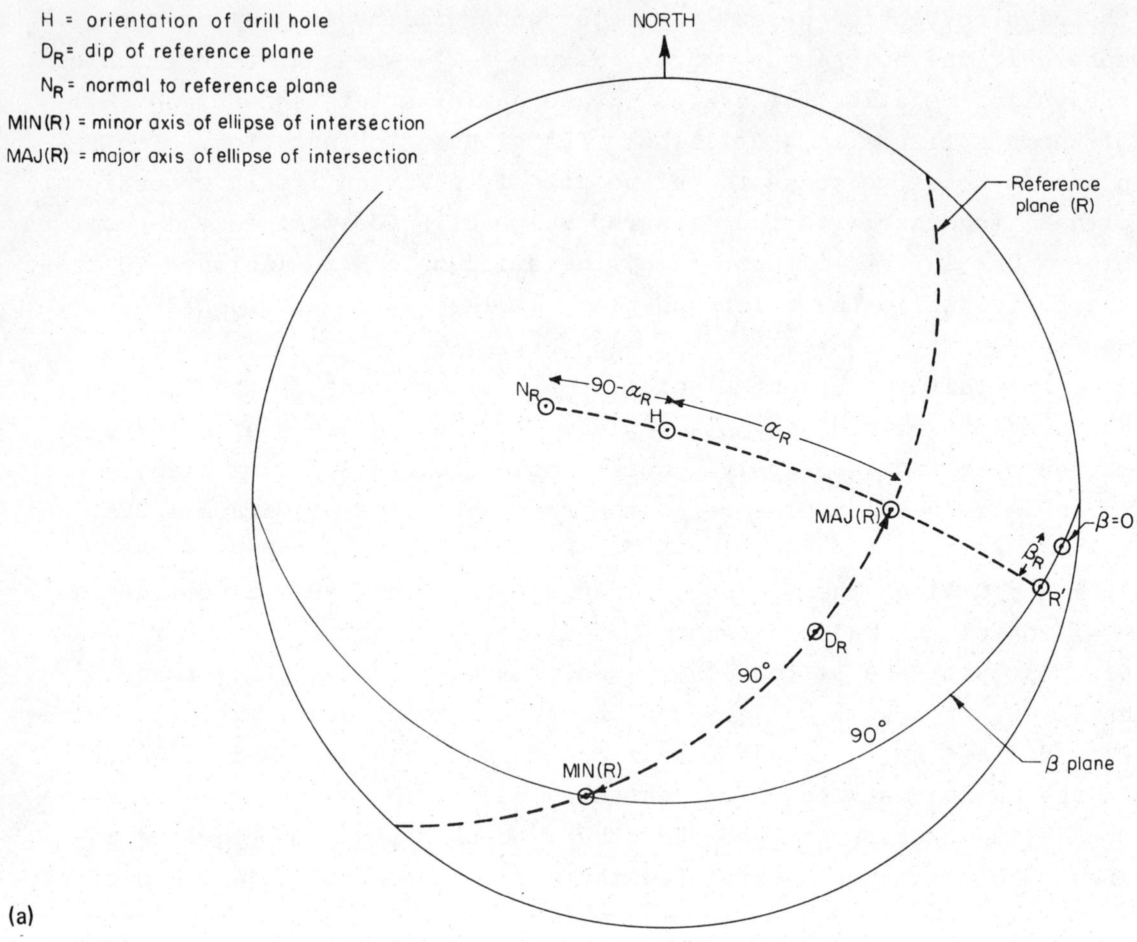

Figure 4-27. (a) Absolute orientation of the reference line in a lower hemisphere stereographic projection. (b) Absolute orientation of plane P given its α and β values. $\alpha_P = 45°$ and $\beta_P = 65°$

statistically preferred orientation, faults, and other structures can be used for this purpose as long as they can be demonstrated to maintain planarity. Two non-parallel holes yield as many as four possible solutions, which can often be reduced to a single acceptable solution from minimal knowledge of the region. If one of the holes is vertical, its α value gives the true dip and only the strike is in question. For a unique solution, three holes are required. The basis for the solution, as discussed by Phillips (1971), is the

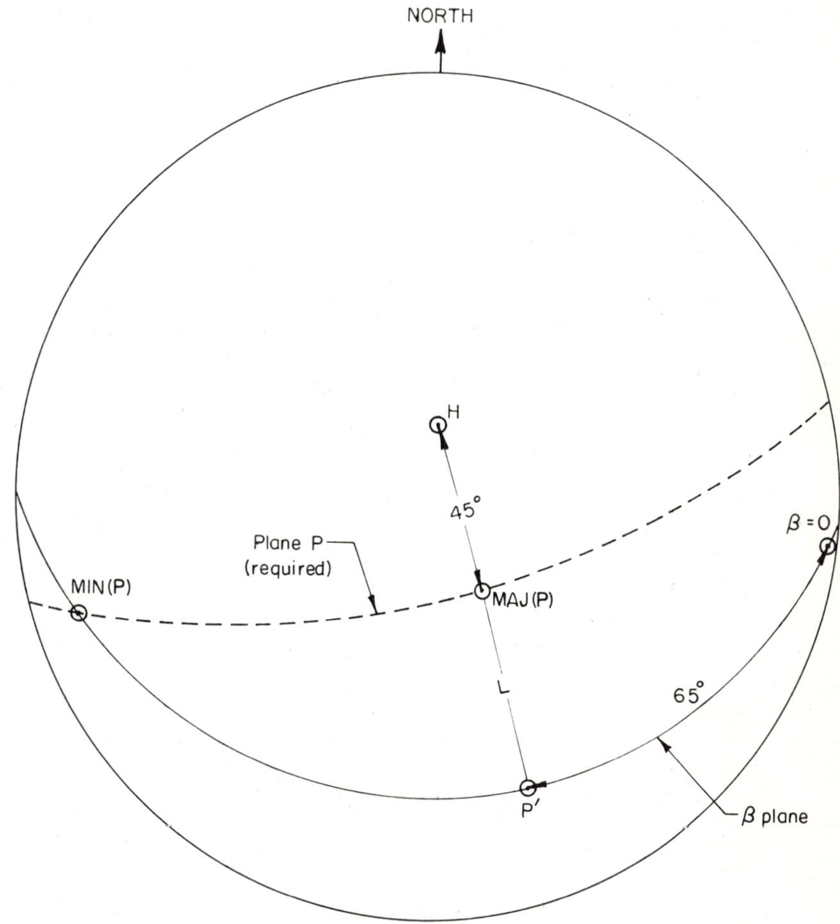

(b)

observation that the true orientation of the normal lies in a double cone of possible orientations of radius (90 - α) centered about the drill hole axis and its opposite. The solution is given by the unique intersection of cones. Since we attach no directionality to the normal to a plane, for the purposes of this exercise, a complete mapping on one hemisphere is sufficient. Figure 4-28 presents an example. The core-bedding angle was measured in three non-parallel drill holes as follows:

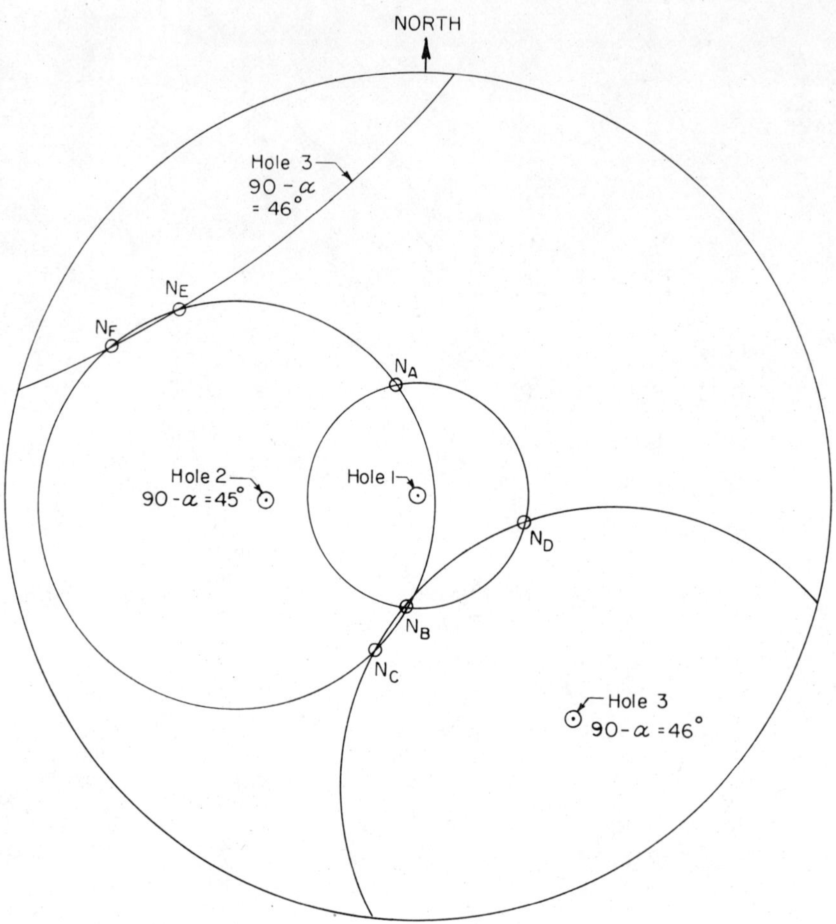

Figure 4-28. Absolute orientation of discontinuities from α angles (figure 4-25a) measured in three non-parallel drill holes.

hole number	orientation	core - bedding angle (α)	core - normal angle (90-α)
1	vertical	60°	30°
2	50° to the west	45°	45°
3	24° to the S33E	44°	46°

Small circles of required radius are constructed about the hole axes yielding 6 intersections. (In the case of drill hole 3, a part of the cone about the upward direction of the hole is required as the downward axis defines a cone extending into the upper hemisphere.) NB, the only simultaneous intersection of all three cones, is the solution, giving the strike of the plane as N82W, and dip 30° to the North. Had only holes 1 and 2, or 2 and 3 been used, there would have been two possible solutions, in this instance with vastly different strikes.

Frequently a few parallel planar features will recur in a number of holes. If the core recovery has been excellent, so that few reference line interruptions occur, these features can be used to orient the reference line, as shown in figure 4-27a, thereby allowing all the core to be oriented without any special surveys. If this is not feasible, then core orienting devices can be used during the drilling operation.

Core Orientation Devices

In shallow inclined boreholes at Mt. Isa, Australia, Rosengren recorded the absolute orientation of core simply by breaking a bottle of paint against the end of the hole. The paint ran down the core stub, marking the dip of the β plane on the end of the core. In holes inclined more than about ten degrees from the vertical, a simple and effective mechanical orientation device -- the Craelius core orienter (Atlas Copco) -- can be used in hard rock. This device (figure 4-29a) is a cylinder, of about the same diameter as the core, with six locking extension feet. At the beginning of each core run, it is inserted on the front of the core barrel with the feet fully extended. It is lowered with the core barrel and when it hits the bottom, the feet are depressed differentially until they lock into position, while a free aluminium ball is impressed against an aluminium marking disc, causing an indentation defining the true dip of the β plane. The orienting device rides into the barrel as coring progresses. When the core barrel is emptied, the top of the core is laid in an alignment cradle (figure 4-29b) against the Craelius device and rotated to find the proper fit of the feet against the rough top-of-core. The angle β_o between the dip of the β plane and the reference line establishes the orientation of the reference line for the contents of the core barrel. When using the Craelius core

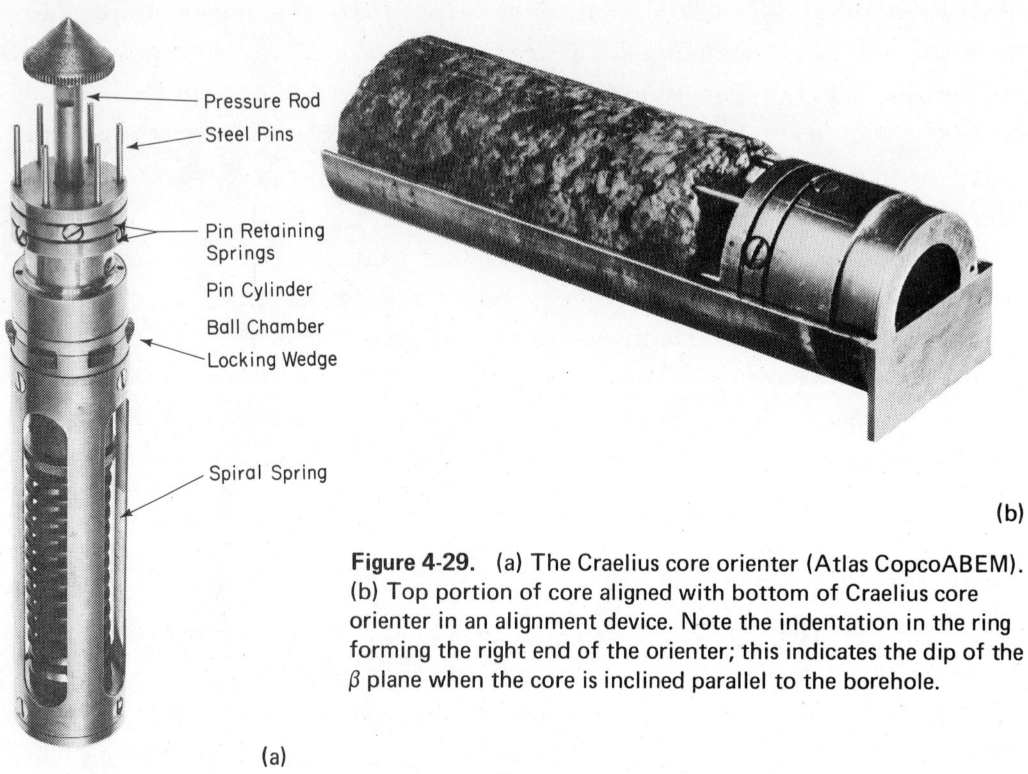

Figure 4-29. (a) The Craelius core orienter (Atlas CopcoABEM). (b) Top portion of core aligned with bottom of Craelius core orienter in an alignment device. Note the indentation in the ring forming the right end of the orienter; this indicates the dip of the β plane when the core is inclined parallel to the borehole.

orienter, it would be most convenient to draw the reference line at the dip of the β plane.

The Christensen-Hügel barrel contains three knives on a shoe mounted on the end of the inner barrel, so that as the core enters the core holder, three grooves are cut longitudinally (figure 4-30). The barrel also has an Eastman, Multishot directional survey instrument, which photographs a compass giving the bearing and plunge of the hole, and the orientation (the angle β) of a marker oriented relative to one of the scribing lines. The determination of attitudes of planar features in the core can be made using the method previously given, or by placing the core in a goniometer (figure 4-30). Voloshin, et al (1968) described application of this equipment by the California Department of Water Resources for investigation of stability for surface excavations in shale. This approach is more expensive than the Craelius device but better suited to the softer rocks.

The "Integral Sampling Method"

This method, developed by the National Civil Engineering Laboratory of Portugal (LNEC), (Rocha, 1971), returns a complete sample in perfect relative orientation by overdrilling pre-reinforced core. This is accomplished by cementing a reinforcing bar in a co-axial hole of smaller diameter pre-drilled in the bottom of the borehole (figure 4-31). The core orientation is known if the orientation of the positioning rods is measured at the time the reinforcing bar is installed. The method is expensive because the drilling must be interrupted until the cement hardens and the orientation through the intermediary of positioning rods is probably not reliable beyond a maximum depth of about 300 feet. However, as a special technique addressed to specific questions, e.g. concerning the reason for core loss at a critical location or of the true nature of a known fault zone, this is a particularly important tool. For

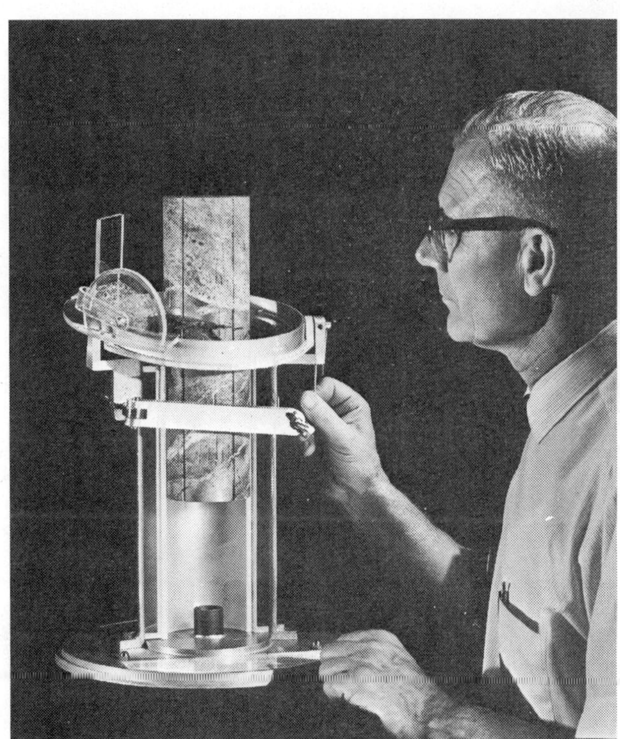

Figure 4-30. Orientation of core, grooved with the Christensen-Hugel barrel, being reconstructed by means of a core goniometer (courtesy of Christensen Diamond Products Co., Salt Lake City).

Figure 4-31. (a) The Integral Sampling Method; from Rocha (1971). (b) An integral sample of decomposed granite; from Rocha (1971). (c) An integral sample taken across a gouge-filled fault.

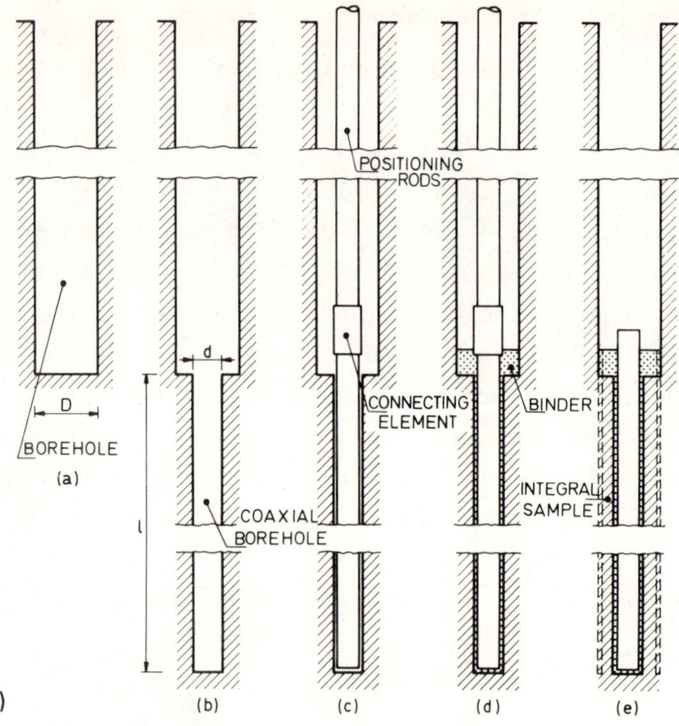

(a)

(b)

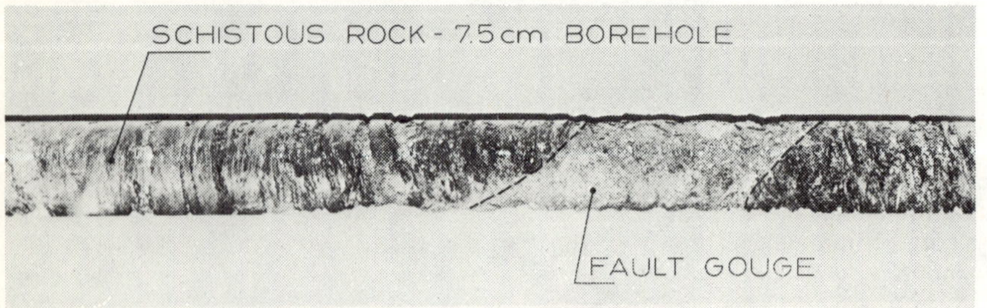

(c)

example, Rocha and Barroso (1971) upgraded the evaluation of a dam foundation in fractured hornfels after examining the continuity, smoothness, and filling of fractures in an integral sample. The integral samples allow measurement of joint apertures; in figure 4-32, from Rocha (1974), such data are usefully presented on a polar equal area projection by using different symbols for joint normals in each of four aperture ranges. The application of integral sampling to site investigations will greatly increase our knowledge and appreciation of discontinuous rock masses.

Geophysical and Optical Orienting Devices

Core orientation is possible with a photographic record of the borehole wall obtained with a borehole camera. For example, the NX borehole camera manufactured by Republic Engineering and Manufacturing Co., St. Paul, Minnesota, photographs the image displayed on a conical mirror, which looks through a cylindrical window. The film drive is timed with a strobe light as the camera is lowered, exposing frames every 3/4 inch as the camera is hoisted up the hole. Interpretation of the resulting photos for planar orientation was described by

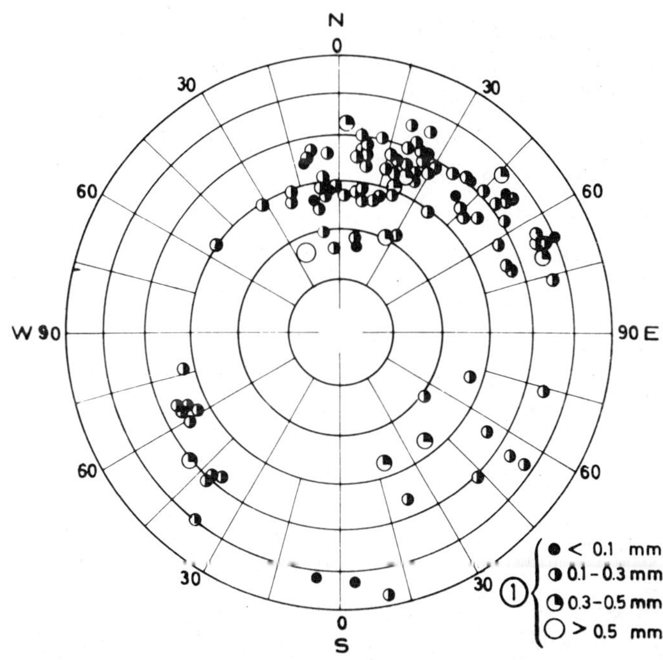

Figure 4-32. Joint attitudes and apertures measured in schists at a Portugese dam site using the integral sampling method; reproduced from Rocha (1974).

Trantina and Cluff (1963). Borehole television devices, e.g., the Eastman F.B. 400*, allow detailed mapping of the borehole walls in the field. Portability may be more restricted than with a borehole camera, and surveys consequently may be more expensive. Borehole periscopes, e.g., Eastman BP34, allow measurement of planar orientations down to maximum depths of the order of 100 feet. Direct viewing optical instruments such as this are particularly valuable for study of discontinuities behind the walls or roof of an underground opening or at shallow depth beneath a foundation. The disadvantage of all optical instruments is that they do not work in holes filled with muddy water. Geophysical loggers with orienting capability can overcome this difficulty.

A borehole logging probe which permits orientation of planar features is the Birdwell Seisviewer** described by Myung and Baltosser (1972). It contains a rotating acoustical transducer, pulsed 2000 times per second. Reflections are received by a detector and transmitted, together with a north marker provided by a built-in magnetometer. On each revolution, the north marker triggers a sweep on the viewing oscillograph, and the sweep is intensity modulated by the reflected acoustic signal. Advance of the logging tool down the borehole produces additional horizontal traces separated from each other in depth. A typical picture, photographed from the visual display, is presented in figure 4-33; it can be thought of as an unrolled borehole with the longitudinal cut at magnetic north. Because damaged rock gives poorer reflections, major fractures appear as dark curves. Geophysics can also aid in core orientation through the measurement of magnetic remanence stratigraphy (Zimmer, 1963) and by means of Dipmeter logs. The Dipmeter is a borehole probe providing continuous resistivity versus depth logs simultaneously along three radii at 120° angles. When the instrument passes an inclined bed or significant discontinuity of anomalous resistivity, distinctive patterns repeat on each trace at different depths. The

---

* Eastman International Co., GMBH, Hannover-Westerfeld, Germany.

** Birdwell Div., Seismograph Service Corp., Tulsa, Oklahoma; the instrument was previously called the "televiewer".

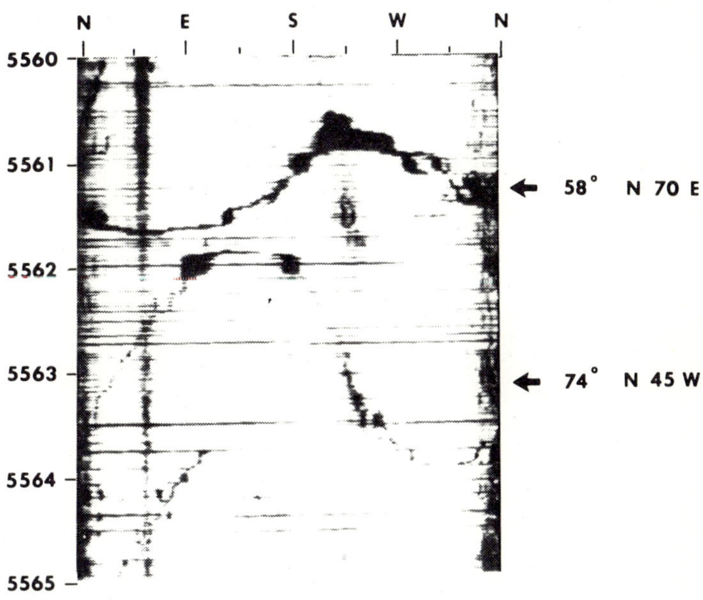

Figure 4-33. The image of high angle fractures intersecting a borehole as logged by the Birdwell Seisviewer; reproduced from Myung and Baltosser (1972).

depth separations of an anomaly allow calculation of two apparent dips which determine the attitude of the plane, as discussed by Phillips (1971) and DeChambrier (1953).

Other Downhole Measurements

The investigation of dam sites frequently includes pump-in water pressure tests, from which the rate of steady flow (q) from a packed off section of the borehole can be plotted against the water pressure (Δp), as discussed by Lugeon (1933), Sabarly (1965), and Maini (1971). Sabarly suggested programming the test for a series of steady flow measurements at successively higher pressures and then at successively lower pressures, as shown in figure 4-34. In this figure, q is the flow rate (e.g. cubic meters per second) and Δp is the differential pressure -- the difference between the downhole pressure inside the borehole (corrected for head losses) and the original static water pressure in the rock at the test section. The shapes of the graphs of q(Δp) help to identify the phenomena occurring. Laminar flow gives a response like that of figure 4-34a. Turbulent flow, possibly explained by an open fracture, or by a leaky packer, may give a curve resembling figure 4-34b. Washing out or opening of a fracture, or rupture of the packer, may give results

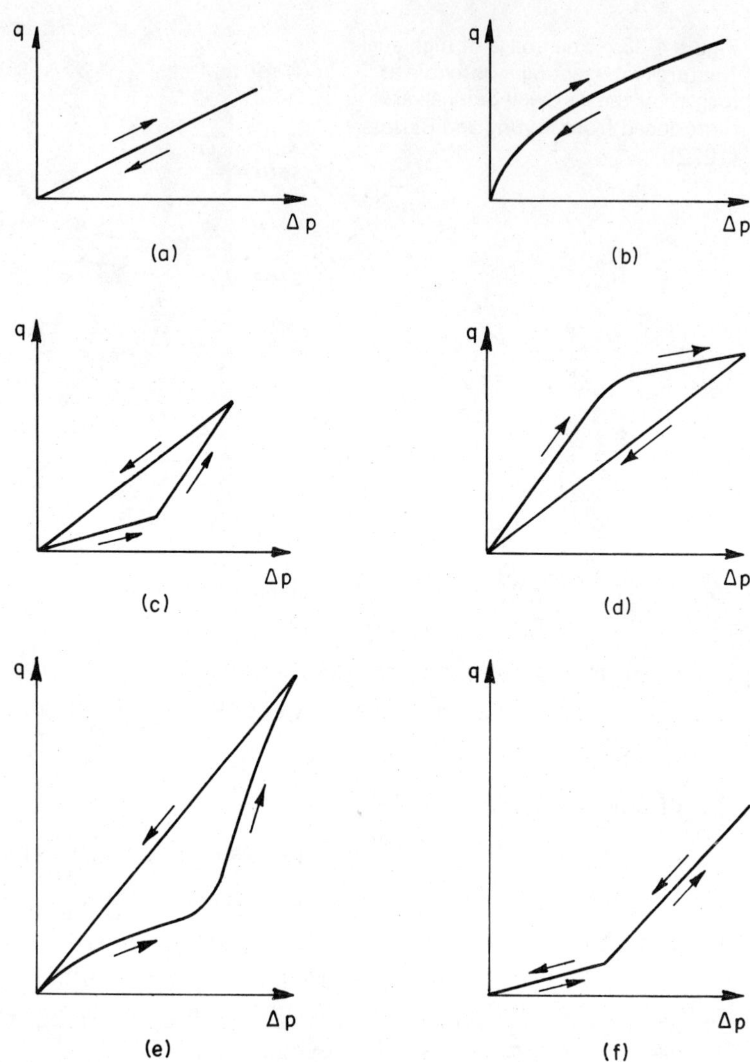

Figure 4-34. Different types of response to pump-in water pressure tests and their interpretation; after Sabarly (1965).

like those of figure 4-34c. Figure 4-34d can be explained by plugging of fractures or pores in the rock by fines in the water. It is important to use clean water for these tests. Figure 4-34e may be explained by plugging of the flow paths in the rock at low pressures, followed by opening of fractures or washing out of fillings at higher pressures. Reversible opening of a fracture as the normal stress is reached and exceeded by the down-hole pressure may explain figure 4-34f. A proportion of such tests will intersect no conducting fracture, and consequently will yield no flow at any pressure. Snow

(1968) used the percentage of no-flow tests according to the Poisson distribution to calculate the mean spacing between permeable fractures.

Borehole logging by geophysical techniques can reveal important discontinuities and permit their correlation from hole to hole. Equipment is now available to use standard techniques of the petroleum industry in NX boreholes. For example, the Widco model 1200 resistivity and spontaneous potential logger is a portable, battery operated unit having a sonde 1-5/8 inches in diameter. By a hand operated winch, it can operate to 1200 foot depths. The Birdwell "3-D Velocity Logger" is available with a 2-1/4 inch diameter sonde which encompasses a seismic source and receiver as close as one foot apart, (Geyer and Myung, 1971). It can also be used in a hole to hole configuration. The device produces a record of amplitude versus time of the transmitted energy at successive depths and can be interpreted to yield the longitudinal and shear wave velocities of the rock around the borehole. The relative amplitudes of shear and compressive signals can be used to evaluate the mechanical properties of fractures, especially if individual fractures are separately logged, e.g., by a Seisviewer or other fracture orientation instrument.

It is also possible to apply static loads to the walls of the borehole and measure directly the deformability and elasticity of the wall rock. Some instruments, e.g., the LNEC dilatometer, (Rocha, et al., 1969), apply a radial pressure inside a rubber tube and measure the radial expansion of the borehole. Borehole jacks, such as the "Goodman Jack," force steel plates to expand along a diameter and monitor the resulting borehole stretch (Goodman, Van, and Heuze, 1972). If the hole is logged to reveal the number of fractures in each test section, with either type of device it is possible to study the variation of rock mass deformability with severity of fracturing in-situ.

The developing technology of exploration makes it possible to extract a great deal of information about the locations, orientations, spacings, and properties of discontinuities in rocks. The remainder of this book is concerned with how description and properties of discontinuities can be introduced into engineering analysis.

5
mechanical properties of discontinuities

DETERMINATION OF PROPERTIES

Discontinuous rock masses can be analyzed in either of two ways. The weakening and softening influence of the network of discontinuities can be accounted for implicitly in calculations by modifying the strength and deformability properties assumed for a large body of rock. Or, the actual properties of individual discontinuities can be introduced explicitly in the analysis as discrete elements of the rock mass. In order to characterize each joint set, physical properties may be estimated on the basis of detailed descriptions. Or, after a program of field sampling, they may be measured in laboratory experiments. It is sometimes feasible to characterize joint properties through tests conducted in the field. In special cases, properties may be "back calculated" from deformations observed in jointed rocks in the field.

Samples of Joints

There are several ways of obtaining samples of joints for laboratory testing. Suitable specimens of natural joints might be found in the core box, and prepared for the shear test by trimming. Usually, however, special drilling will be needed expressly for sampling particular discontinuities. In a rock face, it is sometimes possible to orient a thin walled coring bit parallel to a discontinuity to provide a longitudinal joint sample (figure 5-1a).

It is more difficult to sample by drilling perpendicularly to a joint as the core tends to turn on any initially open crossing joint grinding off the asperities. Pre-bolting can be used for joints at shallow depth (figure 5-1b). For triaxial testing, as discussed later, it is desirable to orient joints at about 30 degrees to the core axis and this can sometimes be done satisfactorily in the field by orienting the core barrel appropriately; a triple tube barrel is especially useful for such an effort (see Chapter 4).

Another approach to sampling of joints is to extract a block containing an undisturbed joint (figure 5-1c). For soft seams, pre-bolting is necessary to retain the seating of the seam during extraction of the block. Extremely sensitive seams can be extracted by wire sawing ("S.E.I.L." method, Hoek and Bray, 1974).

An entirely different approach is to construct an artificial joint in the laboratory. One can introduce a rough or smooth joint in a sample of the actual rock by splitting the rock in a Brazilian test (figure 5-1e), diamond sawing (figure 5-1f), or producing a planar shear failure in a triaxial test. Or one can mold part of a rock outcrop with dental molding plastic (e.g. "Jeltrate") or

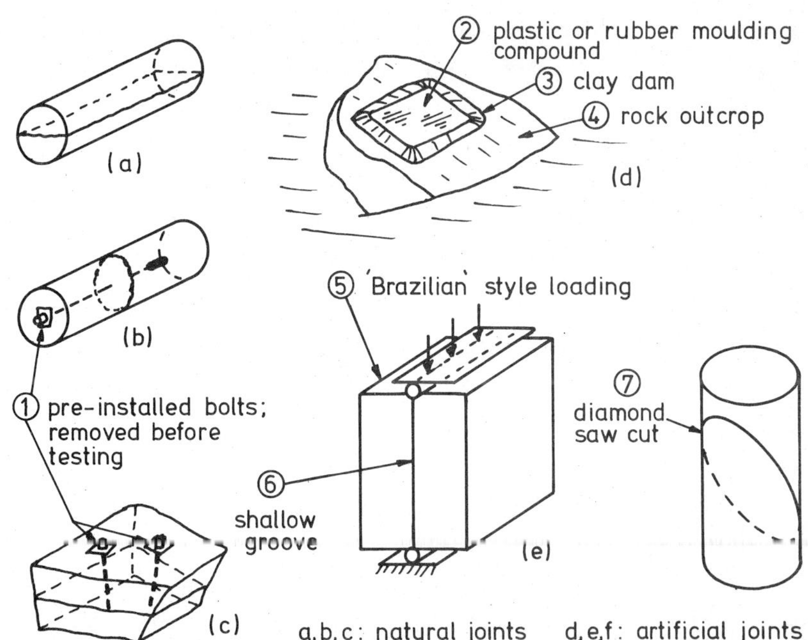

Figure 5-1. Types of joint specimens

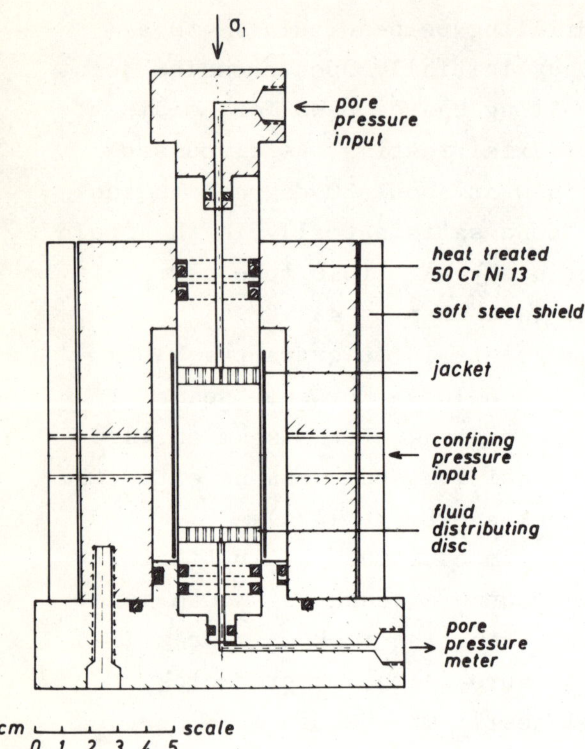

Figure 5-2. Triaxial test chamber; (courtesy of Dr. F. Rummel, Ruhr University, Bochum, Germany).

"potting compound" (e.g. "Dow Corning Silica Set No. 105"), after which the mating rough surfaces are cast with hydrostone, sulfur capping compound ("cylcap") or epoxy (figure 5-1d); this allows repeated virgin tests with different environmental conditions, for example different filling material characteristics (Schneider,1974).

Laboratory Triaxial Testing of Joints

The most generally available equipment for joint testing is a triaxial testing chamber. Figure 5-2 shows a triaxial cell with sufficient internal space for shearing of joints. The joint triaxial test was introduced by Jaeger (1959), and results of triaxial tests on joints have since been reported by Lane and Heck (1964), Handin and Stearns (1964), Raleigh and Paterson (1965), Byerlee (1967), Rosengren (1968), Heuze and Goodman (1967), Goodman and Ohnishi (1973) and others.

Consider the triaxial specimen with a joint inclined at ψ with the long axis (figure 5-3a). First, an all around pressure equal to

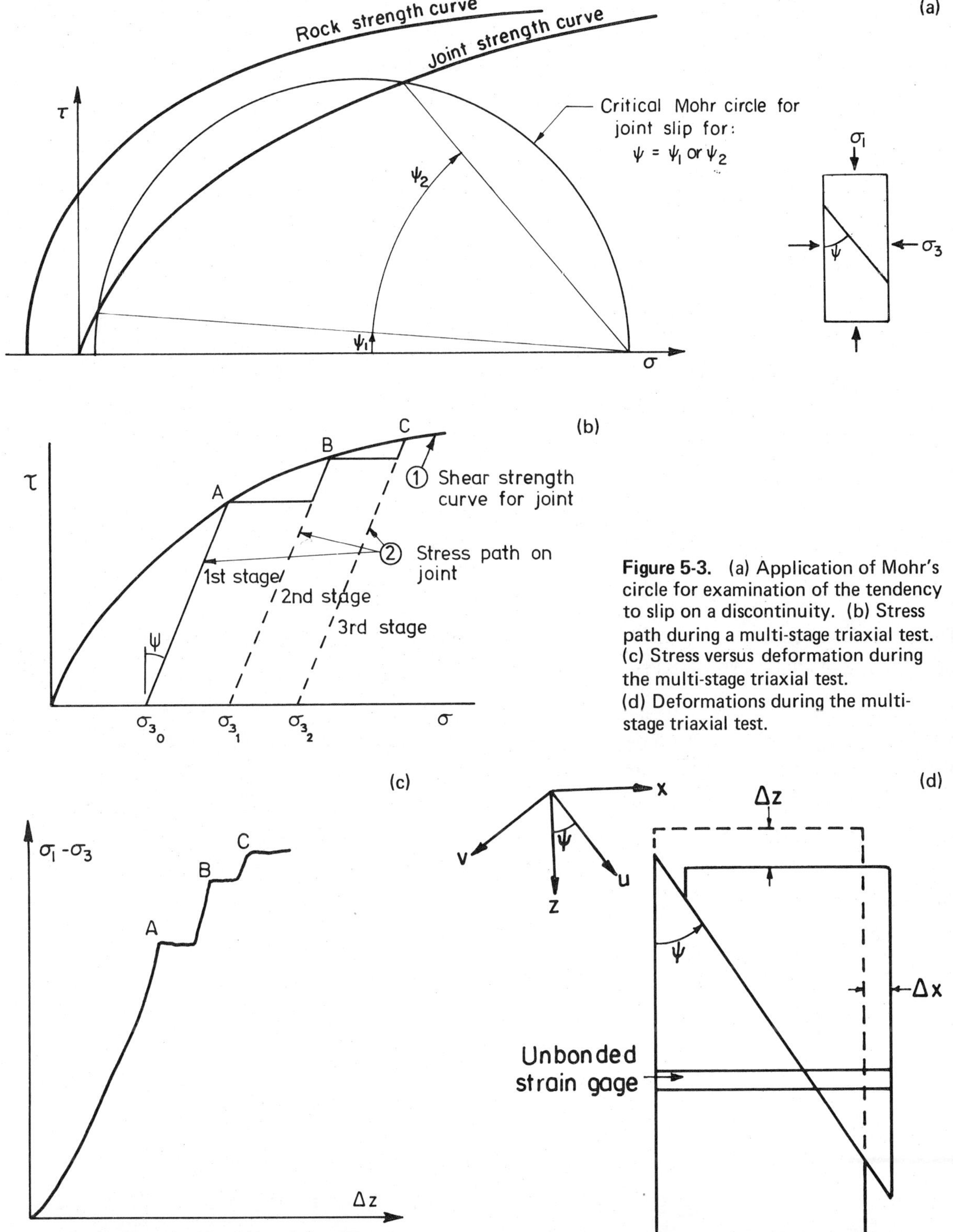

Figure 5-3. (a) Application of Mohr's circle for examination of the tendency to slip on a discontinuity. (b) Stress path during a multi-stage triaxial test. (c) Stress versus deformation during the multi-stage triaxial test. (d) Deformations during the multi-stage triaxial test.

σ_3 is applied, and then maintained as the deviator stress $(\sigma_1 - \sigma_3)$ is raised. Slip occurs when point A, representing the shear and normal stress on the joint plane first touches the shear strength curve for the joint. If ψ is near 0 degrees, or larger than about 70 degrees, the Mohr circle may become tangent to the Mohr envelope of the rock before the joint can slip, as discussed by Jaeger (1959), so that ψ must be controlled within definite limits. The normal and shear stresses on the joint are, theoretically,

$$\sigma = \sigma_3 + (\sigma_1 - \sigma_3) \sin^2 \psi$$

$$\tau = (\sigma_1 - \sigma_3) \sin\psi \cos\psi \qquad (1)$$

These equations, (1), can be combined to yield:

$$\tau = (\sigma - \sigma_3) \operatorname{cotan}\psi \qquad (2)$$

Thus the stress path during a triaxial test with a joint is as shown in figure 5-3b. When slip is initiated at A, the confining pressure is quickly raised from $\sigma_{3,0}$ to $\sigma_{3,1}$ and then the deviator stress is raised provoking slip at B, etc. This test may therefore be termed a "multistage" triaxial test.

The results of such a test can be useful only if the specimen is free to slide along the joint without frictional restraint from the ends. Rosengren (1968) showed that a friction coefficient K at the ends adds new stress contributions on the joint σ_F and τ_F, in the first loading stage, equal to:

$$\sigma_F = K \sigma_1 \sin\psi \cos\psi$$

and (3)

$$\tau_F = -K \sigma_1 \sin^2 \psi$$

When added to equation 1, the effect of friction proves unacceptable unless $K < 0.01$. This can be achieved using a pair of smooth, flat plates on each end of the specimen lubricated with molybdenite grease

(Rosengren, 1968) or polished to a fine finish (Wawersik, 1973), (see figure 5-2).

Vertical and horizontal displacements measured in the triaxial test are inclined with respect to local coordinates normal and parallel to the joint (figure 5-3d). The joint shear displacement Δu and normal displacement Δv are obtained from the vertical and horizontal movements Δz and Δx across the joint:

$$\Delta u = \Delta z \cos\psi + \Delta x \sin\psi$$

and (4)

$$\Delta v = \Delta z \sin\psi - \Delta x \cos\psi$$

where Δz equals the total specimen shortening minus that due to compression of the rock and Δx equals the total specimen lateral deformation minus that caused by lateral strain of the rock. (Only the volume change of the rock influences the volume of the confining fluid so the rock deformation can easily be separated from the total diameter expansion indicated by a circumferential unbonded extensometer.)

A correction is necessary to account for continuous change in the area of contact as a result of shear displacement. The true contact area across the joint, after Rosengren (1968), is:

$$A_C = \frac{D^2 (2\theta - \sin 2\theta)}{4 \sin\psi} \tag{5}$$

where $\theta = \cos^{-1} \frac{\Delta u \sin\psi}{D}$, $0 \leq \theta \leq \pi/2$

and D is the diameter of the cylindrical specimen. The arg of $\cos^{-1}$ $\simeq (\Delta x)/D \simeq \Delta z (\tan\psi)/D$

When $\Delta u/D$ is small, Rosengren suggested:

$$A_C \simeq \frac{\pi D^2}{4 \sin\psi} - D(\Delta u) \tag{5a}$$

As shear progresses, the area of the joint which is no longer in

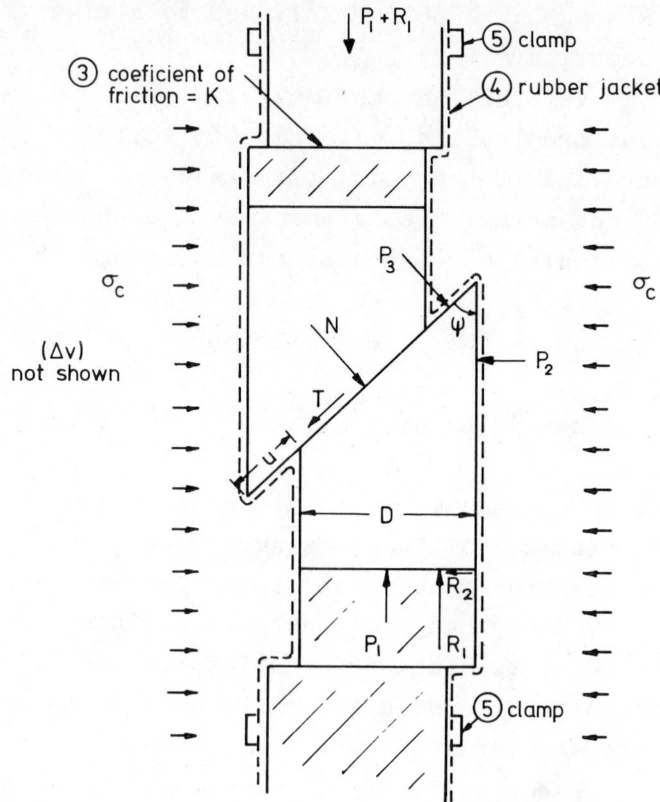

Figure 5-4. Joint triaxial test with large displacement; after Rosengren (1968).

contact becomes exposed to a normal pressure equal to the confining pressure (figure 5-4). The resulting curves of $\sigma_1 - \sigma_3$ versus specimen shortening, z, are consequently difficult to interpret. The test data should therefore be transformed to display τ and σ versus Δu and Δv. Consider the equilibrium of the bottom half of a jointed triaxial specimen under confining pressure σ_c after shear displacement Δu, as shown in figure 5-4. The normal stress σ and shear stress τ in the joint are given by

$$\sigma A_c = N = (R_1 + P_1)\cos\psi + (R_2 + P_2)\sin\psi - P_3$$

and (6)

$$\tau A_c = T = (R_1 + P_1)\sin\psi + (R_2 + P_2)\cos\psi$$

where

$$P_1 = \sigma_c \frac{\pi D^2}{4}$$

$$P_2 = \sigma_c D^2/\tan\psi$$

$$P_3 = \sigma_c \left(\frac{\pi D^2}{4 \sin\psi} - A_c\right) \simeq \sigma_c D (\Delta u)$$

R_1 = axial deviatoric force (load read on testing machine dial minus P_1), and

R_2 = friction force on platens = $K(R_1 + \sigma_c A_c \sin\psi)$

(A_c is given by equation 5).

Laboratory Direct Shear Testing of Discontinuities

The direct shear test (figure 5-5) is a natural way to test properties of discontinuities especially at low normal pressures.

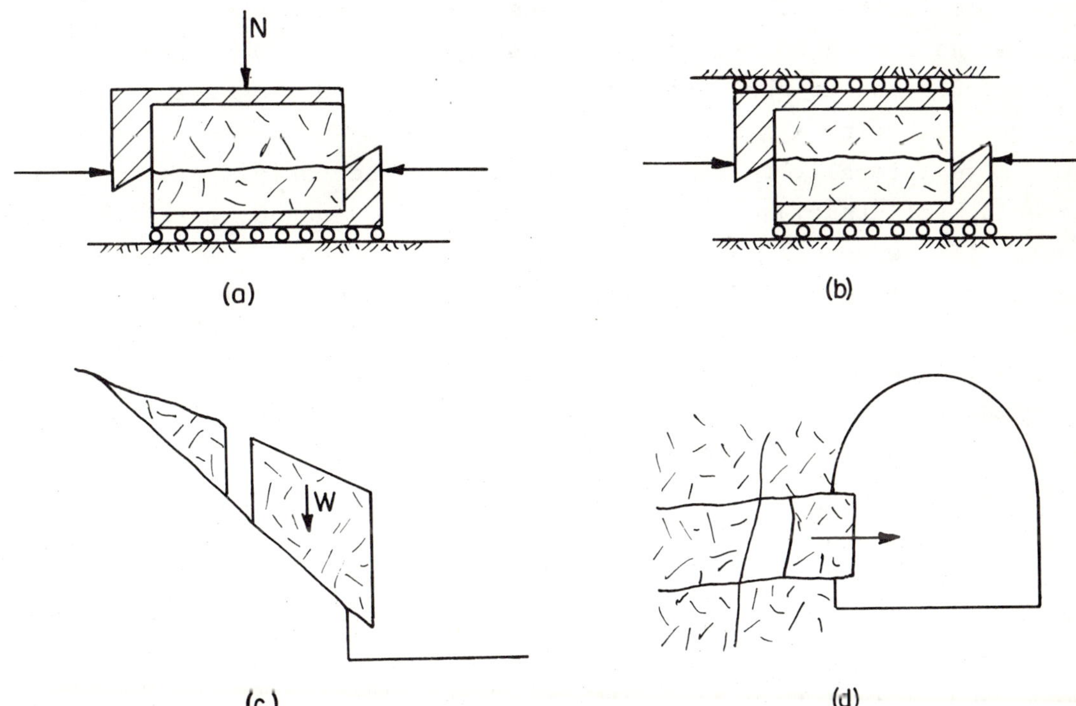

Figure 5-5. Controlled normal stress (a and c) and controlled normal displacement (b and d) shearing modes.

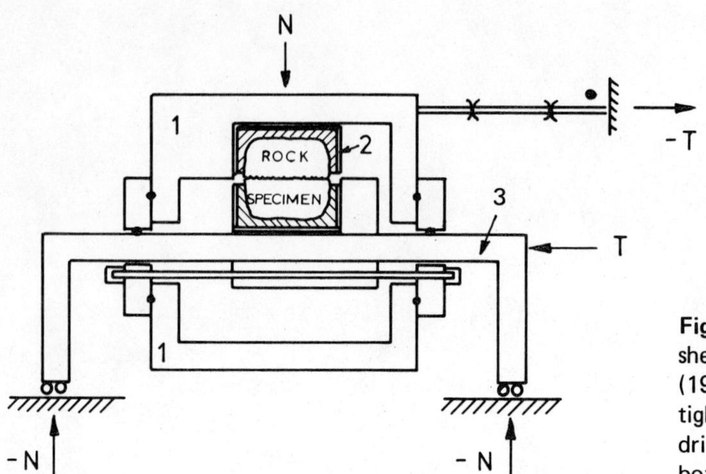

Figure 5-6. Scheme of the Berkeley direct shear machine; from Goodman and Ohnishi (1973). The test is conducted inside a water-tight chamber (1) by advancing a piston (3) driving the bottom half of an inner shear box (2).

A sample is cemented in a shear box, using hydrostone, sulfur capping compound (cylcap), or epoxy. The surface of discontinuity is coincident with the plane of shearing and a gap of about one centimeter (more if the joint roughness is great) is left unbonded between the upper and lower boxes. Normal load is applied by a hydraulic ram, air cushion, or screw and maintained while shearing load is built up, either by a screw or a hydraulic ram. The normal load should move laterally to remain centered over the contact area. Joint thickening (dilatancy) or thinning (contractancy) can be measured directly if the normal load is very soft (figure 5-5a). If the normal load is stiff (figure 5-5b), a tendency for dilatancy will be determined by monitoring the change in the normal load during shear. Most machines in use prohibit any specimen rotation during shear. Specific direct shear machines are described by Krsmanovic and Langof (1964), Lombardi and Del Vesco (1966), Evdokimov and Sapegin (1967), Locher (1968), Bernaix (1969), Hoek (1970), Rengers (1970), Burman (1971), and Goodman and Ohnishi (1973), (figure 5-6). An economical portable direct shear device developed at Imperial College, (Hoek and Bray, 1974) is available commercially.* Objections to the direct shear test on intact rock specimens arise from the unknown local stress conditions and failure mode by en-echelon tension cracks. However,

* Robertson Research, Llandudno, North Wales; and Golder & Brawner Assoc., Vancouver, Canada.

a finite element study of stress conditions in a shear box by Kutter (1971) showed that a uniform stress distribution exists over most of the joint surface in the case of tests on seams or preexisting joints.

The direct shear test yields directly the relationships between stresses τ and σ and displacements Δu and Δv necessary to characterize the deformability and strength properties. A typical test record for a sample with a joint is presented in figure 5-7. The parameters to be extracted from such tests will be considered later.

It is usual to program a direct shear test so that the normal pressure remains constant during shear as in figure 5-5a; this corresponds to sliding of a free block on a slope (figure 5-5c). However, such a test yields too low a shear strength for sliding of blocks constrained between parallel dilatant joints, as for example in an underground opening (figure 5-5d) or in the middle of a rock wall. A no-displacement or controlled stiffness condition in the normal load direction, which can be obtained by servo or manual feedback control, is the correct type of shear test for such design situations. The results of a normal displacement controlled shear test will depend not only on the stiffness of the normal load member, but upon the stiffness of the wall rock so the data require pro-

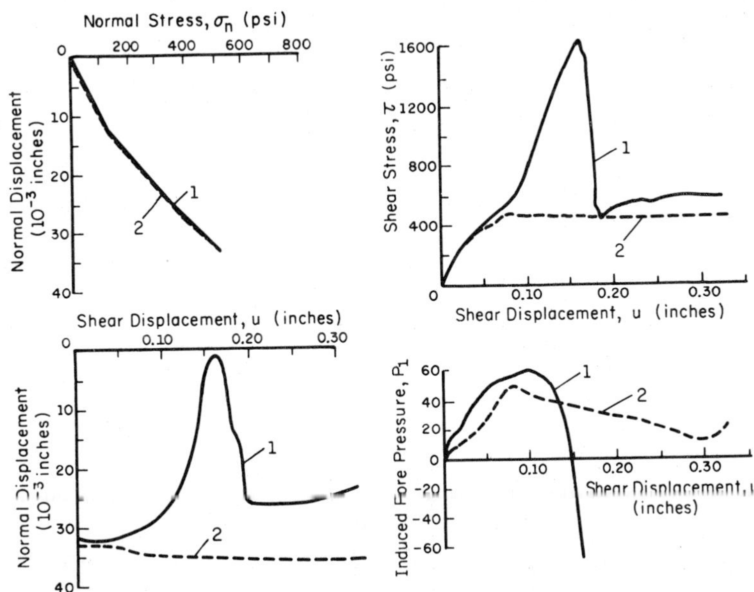

Figure 5-7. Direct shear results for an intact specimen (1) and a specimen with an initially open discontinuity oriented in the plane of shear (2). First the normal force was applied (upper left); thereafter the specimen was sheared at constant normal pressure. From Goodman and Ohnishi (1973).

cessing to scale to any given field case. One of the advantages of the direct shear test over the triaxial test is the greater ease in specifying and controlling the normal load boundary condition.

Field Shear Tests

The direct shear test can be conducted in situ (figure 5-8). The procedure involves selective excavation to isolate a test block on a plane of weakness, either in a gallery or on the surface. In the former case the walls and roof of the chamber provide reactions for the normal and shear forces, (see e.g. Serafim and Lopes, 1961). On the surface, a cable anchored beneath the center of the block, (Zienkiewicz, 1966; Haverland and Slebir, 1972) provides reaction

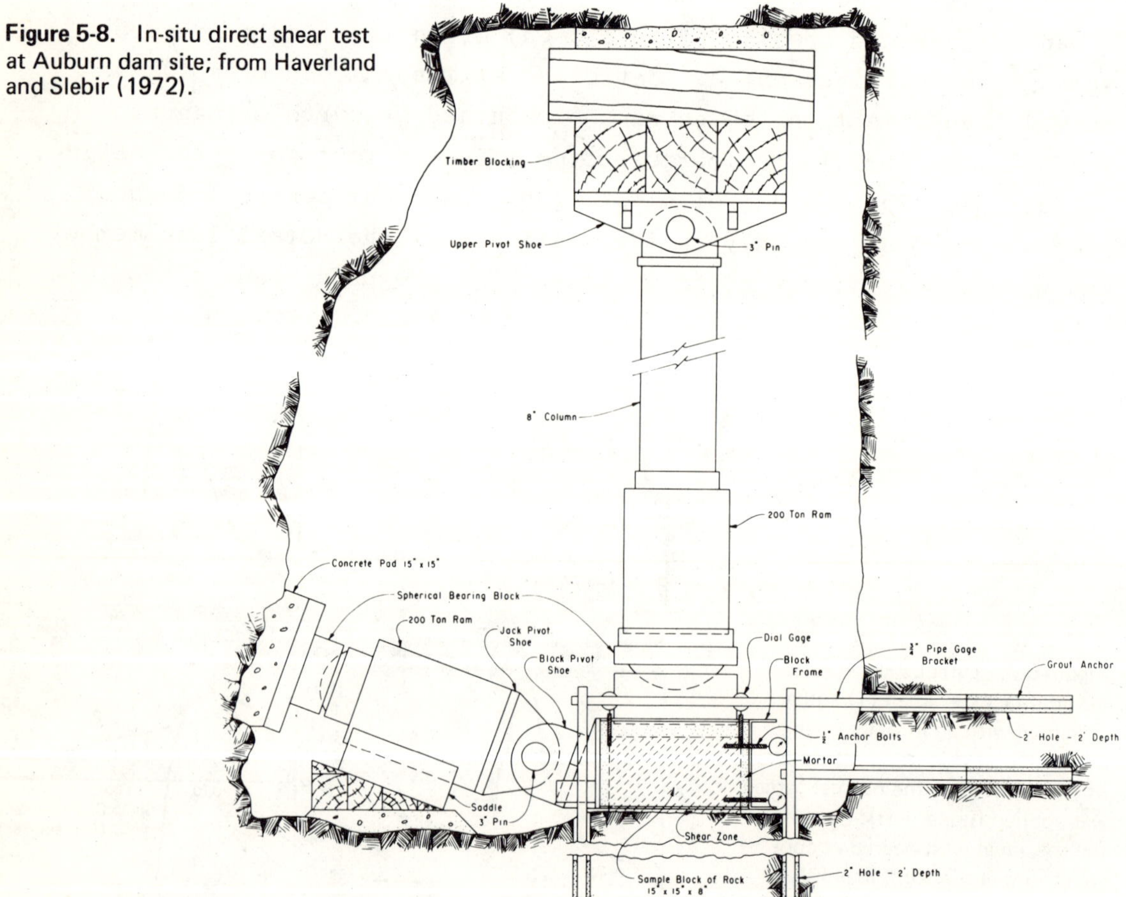

Figure 5-8. In-situ direct shear test at Auburn dam site; from Haverland and Slebir (1972).

for the normal force or the test is conducted with the block's self weight alone (Ruiz and Camargo, 1966). Shear forces can be arranged by jacks across a trench behind the rear face of the test block. In soft rocks such as coal, marl and soft shales, it may be possible to jack into place a shear box equipped with cutters (Brawner et al, 1972); this provides support for the shear block during its preparation overcoming the damaging effects of total decompression. Haverland and Slebir applied the normal force before freeing the sides of the test block to avoid total decompression. In situ shear tests are quite expensive; because of scatter in results a few in situ test results may have questionable worth unless supported by a laboratory test program.

Study of Case Histories

No laboratory or field test, however carefully controlled, can duplicate the scale and character of the loading, boundary and environmental conditions inherent to engineering service. One hopes through shear tests to evaluate components of the joint properties, which can later be integrated in an analysis or model study. But a relevant case history can help to insure that the extrapolation from test to field is basically correct. The extraction of properties of joints from "back calculations" presents a problem inverse to that of design and in general can not yield unique answers. Therefore the most useful case histories are those that resemble the work in question. For limiting equilibrium problems, natural and artificial rock slopes can provide considerable help in evaluating properties of joints. In this connection Hoek and Bray (1974) developed a series of charts and functions greatly facilitating back calculations of slope failures. As discussed later, it is possible to approximate the relation between residual shear strength and normal stress for joints sufficiently well for some applications by the Coulomb equation:

$$\tau = C_J + \sigma \tan(\phi_r) \tag{7}$$

The residual friction angle ϕ_r can be established by shear tests and field observations, but the cohesion, C_J can not. Therefore, determination of reasonable ranges for probable values of C_J may be the main object of case studies. On the basis of more than 40 case histories, Hoek and Bray were able to suggest a range of values for the cohesion of rock masses (Table 5-1); these values help to place in context cohesion values obtained for individual weakness planes.

Finite element analysis, model studies, and other analytical methods as well as graphical solutions are used in calculating field case histories. These methods are to be discussed in later chapters.

DEFORMATIONS IN JOINTS

Normal Deformations

When a block is placed lightly on a rough surface, the proportion of the surface area in actual contact is almost zero. The entire contact force is sustained at three or more point contacts. Under increasing normal load, the point contacts enlarge by elastic deformation, crushing, and tension cracking, while the deformation

TABLE 5-1

Order of Magnitude of Joint Cohesion for Rock Masses (after Hoek & Bray (1974) figure 70)

	C_J (psi)	C_J (Kg/cm^2)
Soil	< 56	< 4
Weathered soft rock; Discontinuities in hard rock	56 - 140	4 - 10
Soft rock masses or jointed hard rock disturbed by blasting or excess loading	140 - 280	10 - 20
Undisturbed jointed soft rock masses	280 - 420	20 - 30
Undisturbed hard rock masses	420	30

brings new regions into contact. It is possible to pursue these mechanisms mathematically and develop a theory of normal deformation under increasing normal load, as was done for metals by Bowden and Tabor (1964); however the system is so poorly defined that an empirical approach is more useful.

There are two physical constraints on normal deformations in discontinuities. First, an open joint has no tensile strength. Secondly, there is a limit to the amount of compression possible, -- a maximum possible closure, V_{mc}, which must be less than the "thickness" of the joint, e (figure 5-9a). But presently we can only guess the relationship between V_{mc} and e. Combining these two conditions demands that we fit the normal pressure-deformation relationship into a quarter space as shown in figure 5-9b. A simple relationship satisfying these conditions is the hyperbola:

$$\frac{\sigma - \xi}{\xi} = A \left(\frac{\Delta v}{V_{mc} - \Delta v}\right)^t \qquad (\Delta v < V_{mc}) \qquad (8)$$

where ξ is the seating pressure, defining the initial condition for measuring the normal deformation Δv. The continuous curvature of $\sigma = f(\Delta v)$ described by (8) can be observed experimentally as shown in figure 5-9c. Curve A, in the left half of the figure, shows the shortening of an intact cylindrical specimen (3.6 inches long by 1.75 inches in diameter) on its third cycle of loading. The first load cycle showed large hysteresis and inelasticity, but the second and third cycles of loading of the intact specimen produced almost identical, elastic compression curves. Then the specimen was turned on its side and compressed between knife edges creating a single, rough and wavy extension fracture parallel to its ends. The specimen was reassembled and recompressed yielding curve B (figure 5-9c). The difference between the compression curves for the jointed and intact specimens describes the compression of the mated joint; it is plotted in the right half of the figure.

At the end of these loading cycles, the wall rock and joint showed no visible damage. Then, the upper block was rotated to create a mismatched joint, with point contacts and mean aperture (e)

Figure 5-9. (a) Idealization of a joint. (b) Behavior of a joint in compression. ξ is the seating load. (c) Normal compression of an extension fracture in a granodiorite specimen.

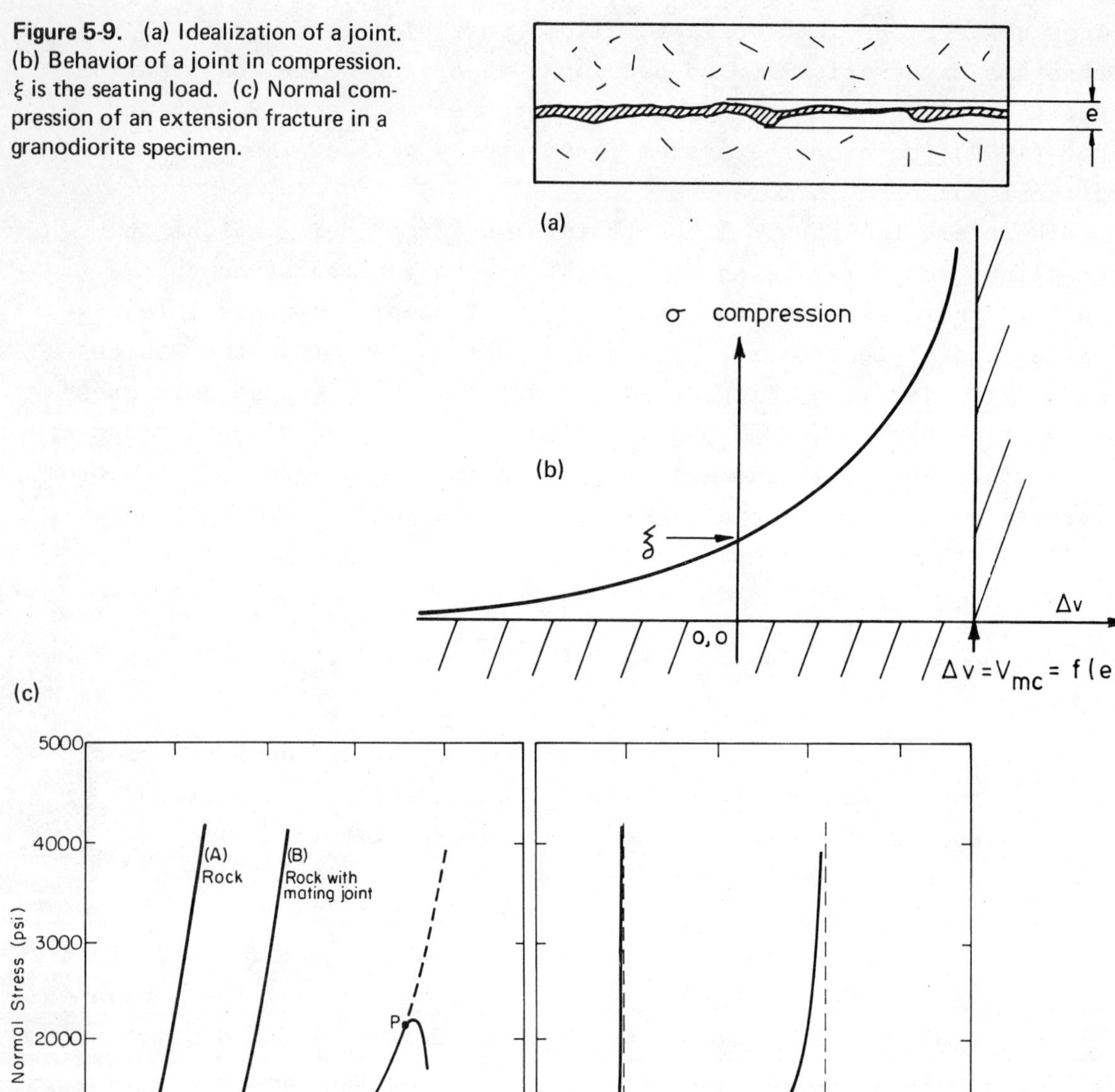

approximately 0.05 inches. Compression of this sample produced curve C; at point P on this curve, the rock began to split lengthwise. The test was discontinued; however the curve was extrapolated as shown. The difference curve, (C)-(A) describes the compression behavior for the non-mating joint. At the end of the test, about ten percent of the area of the joint showed the results of rock crushing.

The following equations (with Δv in inches and σ in psi) fit the joint compression curves (B)-(A) and (C)-(A) of figure 5-9c. For curve (B)-(A) representing deformation of the mating joint,

$$V_{mc} = 0.0047 \text{ inches, and}$$

$$\Delta v = -.0004 + .0007 \ln \sigma$$

For curve (C)-(A) representing deformation of the non-mating joint,

$$V_{mc} = 0.0152 \text{ inches, and}$$

$$\Delta v = -.0094 + .0031 \ln \sigma$$

The dimensionless form of the joint compression curve, equation (8), (with σ=67 psi arbitrarily established as the seating load ξ) can be fitted approximately to the data of figure 5-9c. For the mating joint: A = 3.00 and t = 0.605; for the non-mating joint A = 5.95 and t = 0.609.

The unloading cycles for the jointed specimens followed essentially the same path as for the intact rock signifying that the elastic portion of the normal deformation in a rock with a joint is entirely derived from the rock. Joint compression is essentially unrecoverable. The highly non-linear, and inelastic deformation of a discontinuity under compression accounts for similar non-linear and inelastic stress dependency for all properties which are linked with joint aperture, e.g. electrical resistivity and fluid permeability. The behavior of discontinuities in compression is discussed further in Chapter 8 in association with finite element analysis.

Shear Deformations

Krsmanovic and Langof (1964), Hoek and Pentz (1968), Rosengren (1968), Goodman (1970) and Coulson (1972) discussed the shear stress - shear deformation curves for discontinuities undergoing shear at constant normal stress. Rough clean (i.e. "unfilled") joints (curve

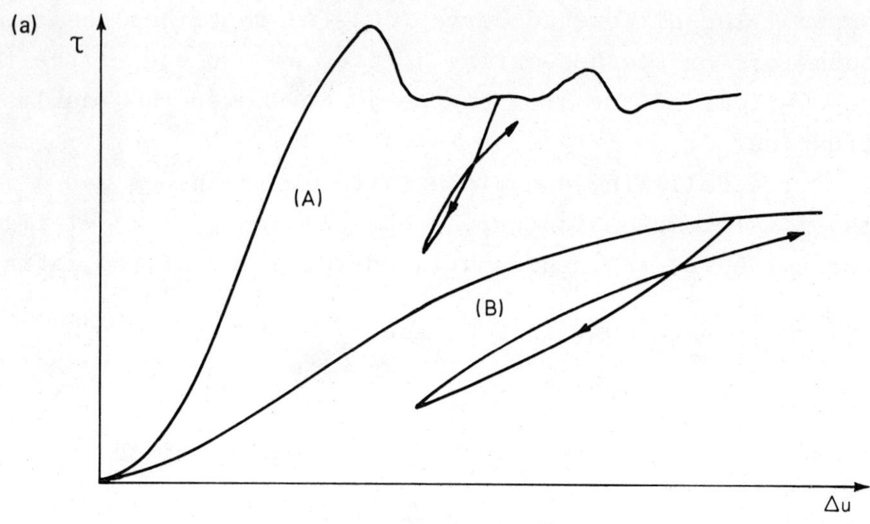

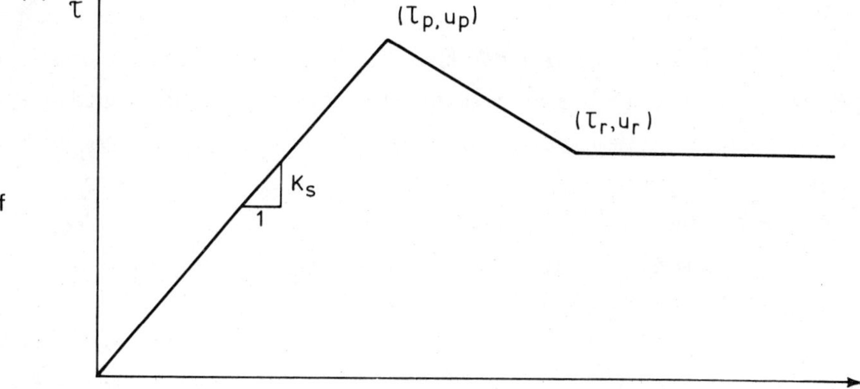

Figure 5-10. (a) Types of shear stress versus deformation curves at constant confining pressure. (b) Parameters of shear deformation at constant confining pressure.

A of figure 5-10a) show relatively rapid rise in shear stress to a peak followed by an irregular post peak history with considerable loss in load carrying capacity. The ratio, B, of residual to peak shear stress increases with normal pressure; it may be as low as 0.3 in previously healed or incipient joints at low normal pressure and 0.6 in open rough joints. Seams or "filled" joints (curve B of figure 5-10a) show a convex downward stress-deformation curve like that of clay, with the peak stress poorly defined and the slope continuously changing. In a sense, the type B curve is a product of a history of deformations, which may have started with type A. When the filling material becomes dessicated, the type B curve is replaced by a type A curve, but wetting reverses this response. Most in situ shear tests reviewed by Goodman (1970) proved to have type B curves, perhaps because one tends to authorize the great expense of in situ shear testing only for major seams.

The shear deformation versus shear stress curve for a test conducted under constant normal stress can be characterized by elastic, peak, and plastic resions as depicted in figure 5-10b. The peak shear stress (τ_p) is termed the shear strength while the minimum post-peak shear stress (τ_r) is the residual strength. The peak strength demonstrates a scale effect, while the residual strength may not (Bernaix, 1974). The slope characterizing the elastic region is termed the unit shear stiffness k_s (Goodman, Taylor, and Brekke, 1968). Joints with type B shear curves generally have lower stiffness than those having type A curves. Representative values of shear stiffness for different classes of discontinuities are not generally known and values are often assumed in analysis. The reason is that shear stiffness measurements are sensitive to the testing technique and apparatus, particularly the technique of gripping the specimen and the location of the displacement measuring instruments. Moreover, shear stiffness displays a strong scale effect as revealed by Barton (1972) (figure 5-11).

All of the parameters of joint shear behavior are greatly influenced by changes in normal stress. The variation of peak shear displacement and shear stiffness with changing σ can be simplified using a model having constant stiffness as in figure 5-12a, or a model having constant peak displacement as in figure 5-12b (John, 1970). Jaeger (1971) gave examples of direct shear test results with polished saw cuts in trachyte in which the first loadings fit the constant stiffness model; however reloading the worn surfaces more nearly matched the constant peak displacement model. The variation of peak shear strength with normal stress is described by the shear strength curve, as discussed later. As shown in figure 5-22, the τ_p-σ relationship for a rough discontinuity tends to be more highly curved than this relationship for the intact rock. For a restricted range of normal stresses, linear approximations to the

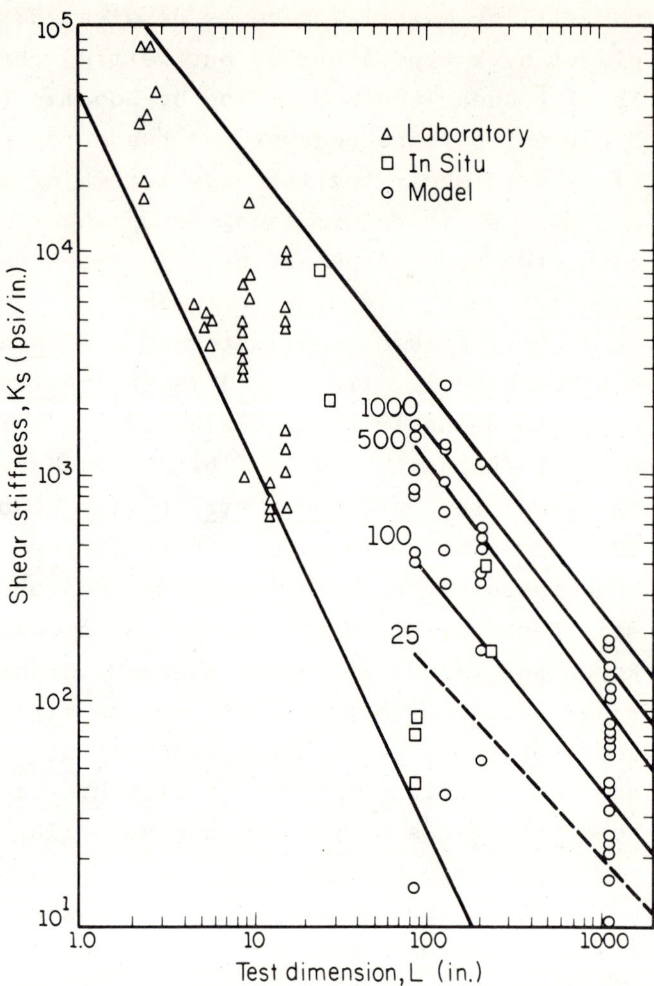

Figure 5-11. Joint shear stiffness as a function of the square root of the loaded area (L) and the normal stress (25, 100, 500, and 1000 psi lines are given); from Barton (1972).

peak and residual shear strength curves can be used with acceptable precision. In some cases, the ratio of residual to peak strength B approaches unity as σ increases.

Dilatancy of Continuous Rough Surfaces

A shear test conducted under restricted normal displacement conditions, (figure 5-13) curves B, will generally yield a considerably higher shear strength than one conducted under constant normal stress (figure 5-13) curves A. The reason for these strength differences is connected with dilatancy. Perfectly mating rough blocks can be forced to slide past one another only if they are free to move apart,

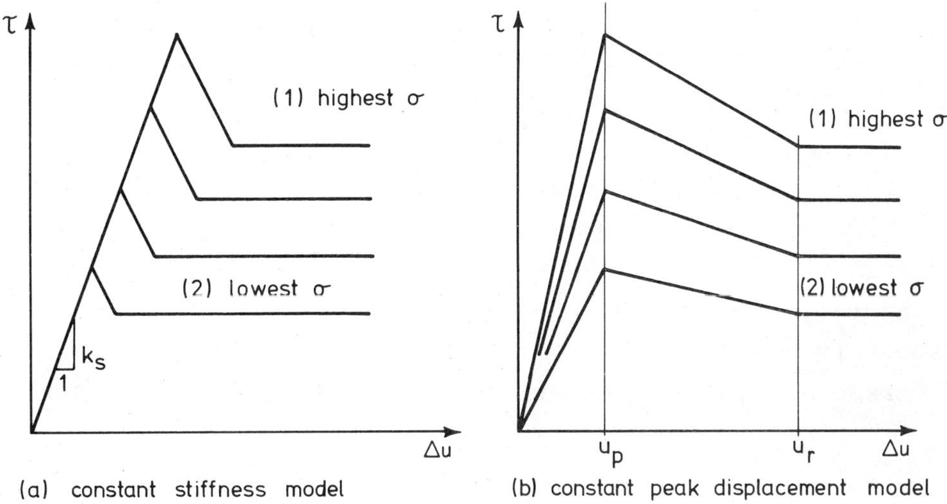

Figure 5-12. Shear deformation models.

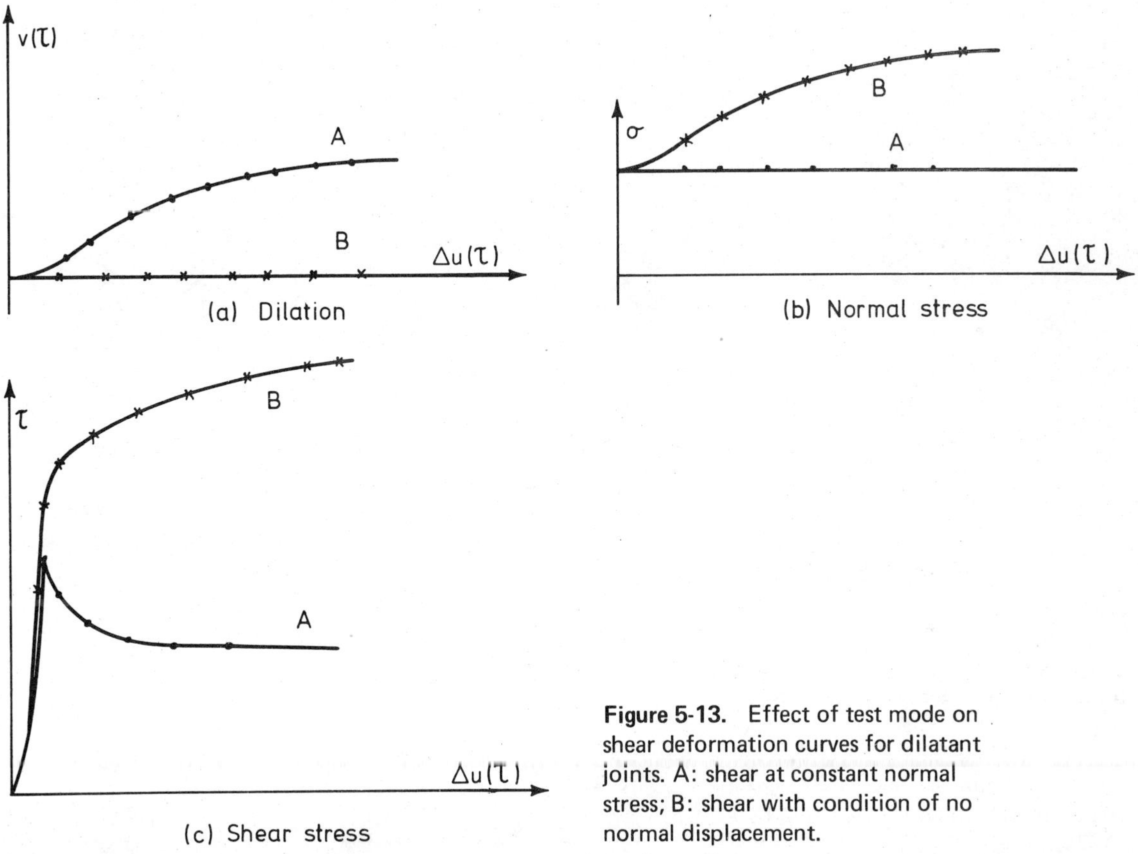

Figure 5-13. Effect of test mode on shear deformation curves for dilatant joints. A: shear at constant normal stress; B: shear with condition of no normal displacement.

(to "dilate") to work around asperities; if the blocks are confined, shearing is possible only if the asperities themselves break. Since dilatancy can very considerably strengthen a joint, it is an important property.

This mechanism of dilatancy, involving the over-riding of asperities, originates mainly from surface roughness. (Another mechanism -- rotation -- will be considered later.) Rengers (1970) measured roughness angles of natural joint surfaces over a band of steps from 0.01 to 1000 cm, using successively a variable focus microscope, a profilometer, and terrestrial photogrammetry. Corre-

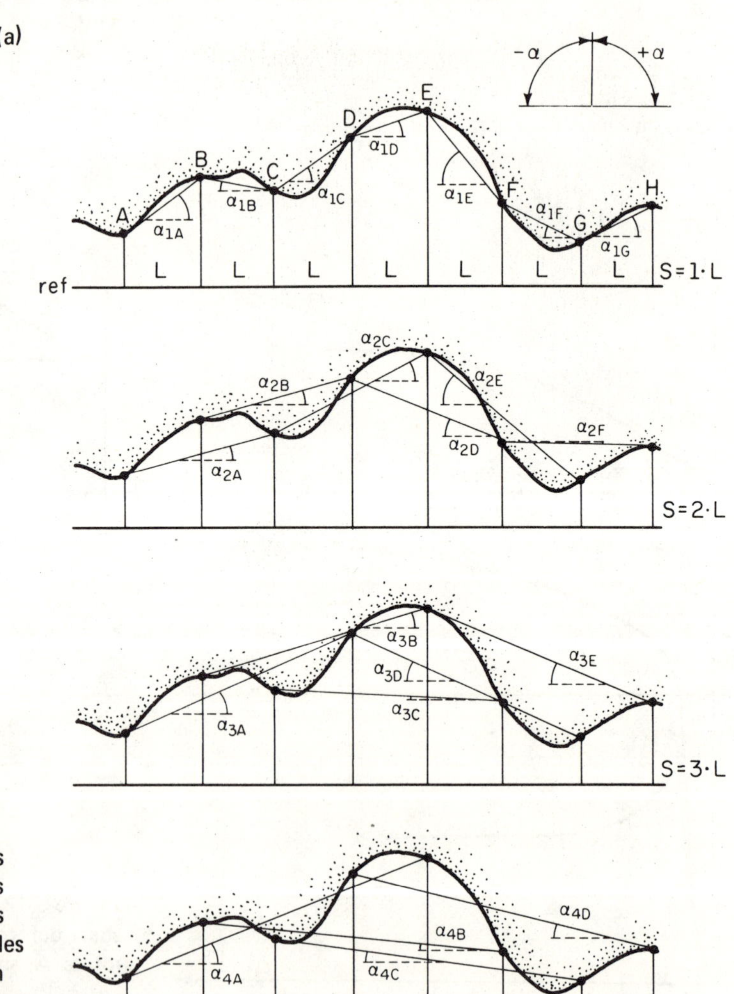

Figure 5-14. (a) Example of roughness angle calculation from digital roughness amplitude measurements; from Rengers (1971). (b) Envelope of roughness angles for sliding to the left and to the right in the example of figure 5-14a; from Rengers (1971).

sponding to each selected step size there is a distribution of roughness angles; for example, in figure 5-14a, corresponding to step size S = 1·L the surface presents angles $\alpha_{1,A}$ through $\alpha_{1,G}$, varying from +35° to −45°. Similarly, step S = 2L produces angles $\alpha_{2,A}$ through $\alpha_{2,F}$ varying over a smaller range and so on for S = 3L and S = 4L. Rengers plotted these angles corresponding to the value of S (figure 5-14b) and constructed envelope curves (solid lines), assuming that the steepest surface angle of contacting mating blocks always regulates dilatancy during shear with over-riding of asperities. The envelope of positive angles governs right lateral shearing (clockwise sense) while the envelope of negative angles governs left lateral shearing (counter clockwise sense). The actual dilatancy during shearing reflects the cumulative effect of movements along the rough surface. Since the effective roughness angle varies inversely with the distance of relative shear motion, the rate of dilatancy decreases progressively as shearing continues. In fact, the dilatancy curve $\Delta v(\tau) = f(\Delta u(\tau))$ is essentially the mirror image

(b)

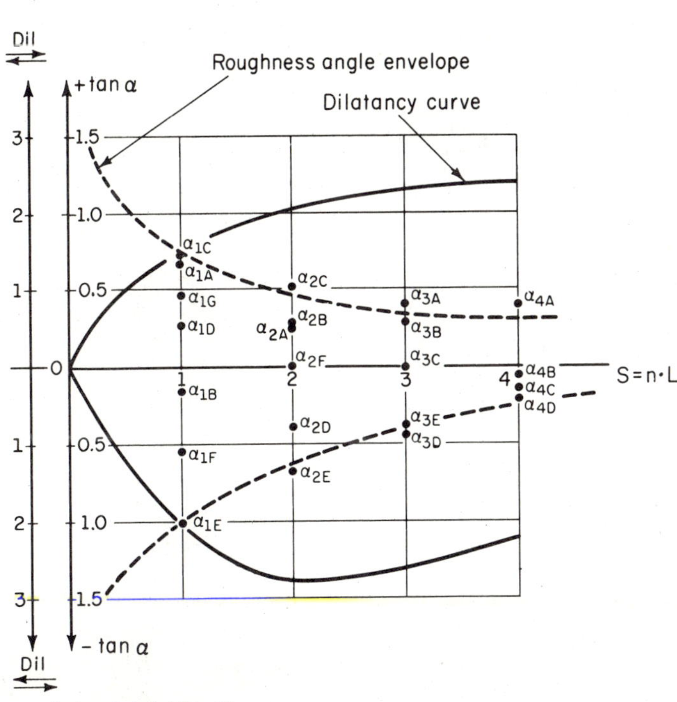

of the roughness angle envelope (figure 5-14b). Barton (1971) made a similar analysis for rough joints in a model material.

In the field the reference state from which dilatancy begins to be measured may reflect past shear displacement. In the case considered (figure 5-14b) a previous shear displacement equal in magnitude to L implies a remaining dilatancy of about one half that inferrable from the analysis of roughness.

Oversliding of asperities without rock breakage is unlikely, except at zero normal stress; when there is no normal stress or restraint, asperities presenting angles less than 90 - ϕ_μ can be over-ridden, where ϕ_μ is the friction angle for sliding of flat surfaces of the rock in question. But when the normal stress is high, the work required to dilate against normal force will exceed the work sufficient to shear through some asperities. Thus dilatancy decreases with normal stress and tends to become completely suppressed when the average normal stress reaches the order of magnitude

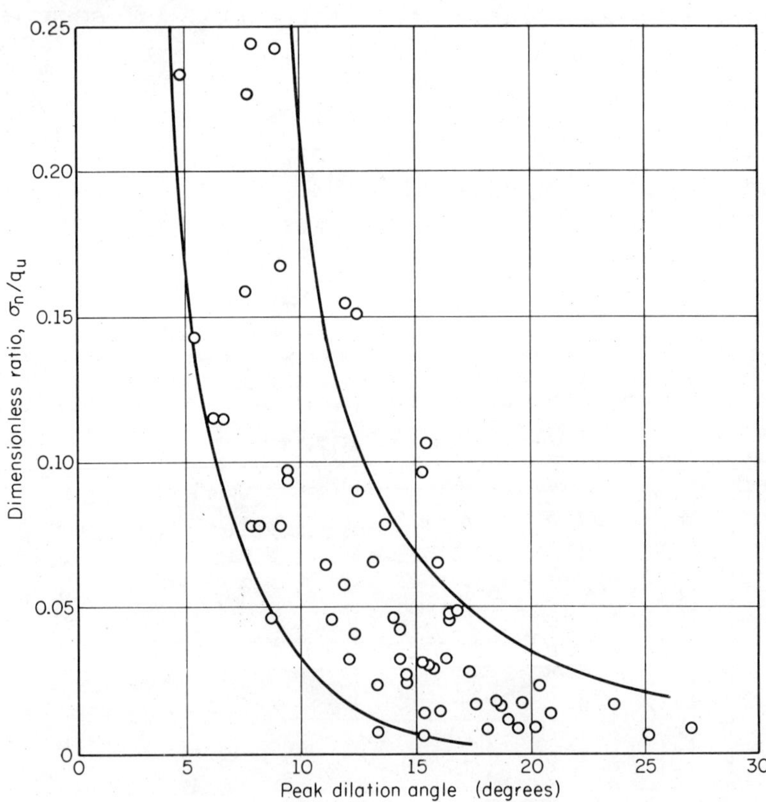

Figure 5-15. Peak dilatancy angle (i) as a function of the ratio of normal stress to compressive strength for model extension joints; from Barton (1971).

of the unconfined compressive strength of the asperities (Barton, 1971). The shear strength of an individual asperity varies with its width. Corresponding to each value of normal load, all asperities up to a given base width suffer the possibility of rupturing. Accordingly, the variation of dilatancy with increasing normal stress is governed by a function similar to the reduction of dilatancy with increasing step size, (figure 5-15).

Field measurement of roughness, as a function of base distance, can be made photogrammetrically (Ross-Brown et al, 1973; Patton, 1966), and by profilometer (Fecker, 1970). If repeated measurements of the attitude of an exposed surface are plotted on a stereographic projection, Fecker and Rengers (1971) showed that the roughness angles can be estimated from the extent of the scatter of poles. When a rigid plate is laid on the surface of discontinuity, its attitude will depend upon the relative locations of its contacts with the surface; thus repeated measurements taken with a field compass attached to a rigid plate will give a range of readings. The roughness angle, corresponding to a step size equal to the diameter of the plate, is estimated by measuring the angle between the mean orientation of the poles and the extreme orientation of the envelope to the set of poles. Figure 5-16a gives two examples as well as comparisons with profilograph data for two specific directions of sliding.

Use of a joint pole scatter diagram for analysis of roughness is accurate only for isotropic roughness. When roughness in a discontinuity is created by planar asperities, rather than by detached hills, the roughness angles are the apparent dips of the planar asperities in the direction of sliding.* To illustrate, (figure 5-16b) consider shear on a surface having roughness formed by planes 1 and 2 striking parallel to each other and dipping 30° and 40° in opposite directions. The "envelope" of normals is the line N_1 N_2 whereas the true roughness angle envelope is the boundary of the ruled area, constructed by finding poles to apparent dips in

---

* John Boyd, Imperial College, London; personal communication.

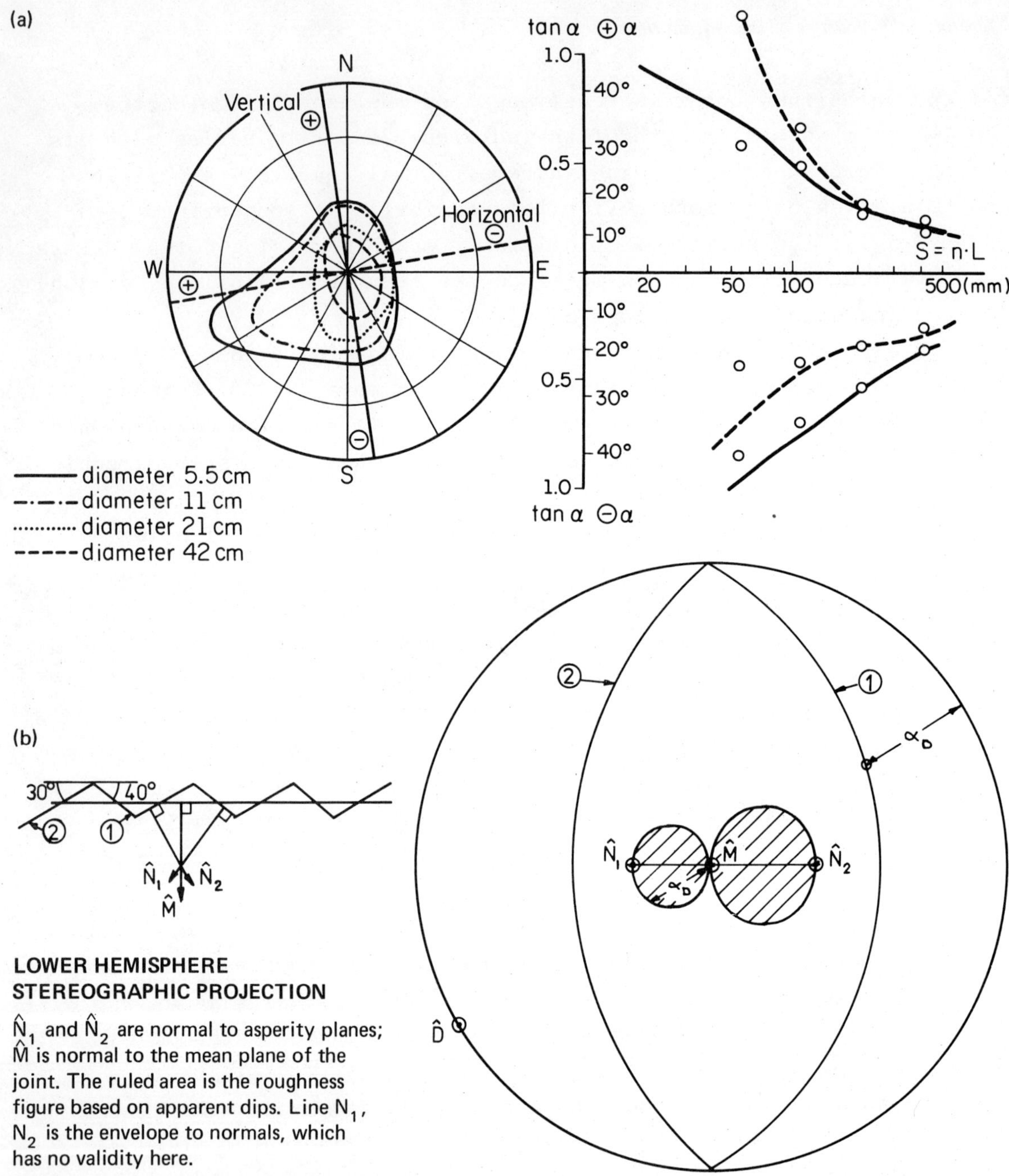

Figure 5-16. (a) Comparison of roughness estimate from the scatter of normals to a single rough discontinuity (left figure) and actual roughness as measured by surface profiles (right figure). The four scatter diagrams on the stereonet are envelopes to repeated measurements of attitude of a rough joint with the compass placed on a plate of 5.5, 11, 21, or 42 cm. diameter. After Fecker and Rengers (1971). (b) Roughness of a discontinuity formed of two sets of component planes when profiled in different directions.

all vertical planes. For example, for sliding parallel to D, the roughness angle is α_D as shown.

PEAK SHEAR STRENGTH

The "shear strength" of a discontinuity refers to the peak load in a test with constant normal stress (which as previously noted may be quitc conservative for dilatant joints under restricted normal deformation). Shear strength is sometimes called "friction" although as discussed by Nascimento and Teixeira (1971), not only surface friction, but wedging, rotation, and even rolling effects contribute normal stress dependent shear resistance, while cementation and interlocking can develop additional shear strength.

Surface Friction of Minerals and Rocks

The surface friction of smooth rock and mineral surfaces derives from micro-interlocking and adhesion, which may require rock breakage for sliding, as well as "ploughing" of harder minerals into a softer matrix. Jaeger (1971) reviewed motivations, methods, and data of friction measurements between rock surfaces. Friction measurements on smooth rock surfaces were reported by Jaeger (1959), Byerlee (1967), Jaeger and Rosengren (1969), Coulson (1972), and others; friction of individual minerals was measured by Horn and Deere (1962). Friction experiments of even relatively smooth rock surfaces, e.g. lapped with #400 grit (roughness ≈0.001 in) usually show considerable scatter and are generally more sensitive to changes in moisture conditions and roughness than to changes in mineralogy. The notable exception is the family of sheet silicate minerals, mica, chlorite, clays, talc and serpentine, which exhibit low friction, particularly when wet. Generally the coefficient of friction ($\tan \phi_\mu = \tau_p/\sigma$) varies in the range 0.4 to 0.8, but in the sheet minerals it can be as low as 0.2 ($\phi_\mu = 12^\circ$) and rocks composed largely of such minerals can have quite low friction angles. For example Richards (1973) reported $\phi_\mu = 20^\circ$ ($\tan\phi_\mu = 0.36$) for moist, smooth surfaces of slate. Drying increases the friction of sheet silicate minerals but, oddly, oven drying significantly lowered the friction of quartz, calcite and feldspar in Horn and Deere's tests at low confining pressure.

Most rock surfaces however are stronger when dry than when wet.

Most specimens tested by Coulson, which included granite, basalt, gneiss, sandstone, siltstone, limestone, and dolomite, showed higher friction after displacement of one to three centimeters, accompanied by secondary fracture of wall rock ("Riedel shears") and formation of gouge, figure 5-17, (see Lajtai, 1969b), particularly at normal pressures above 500 psi. Wear of rock surfaces accompanying continued shear displacement eventually causes the surface to be coated with crushed material, and it becomes a new kind of specimen. In the case of dry, unweathered rock surfaces, the new material can bring higher friction than polished surfaces, but in moist weathered rock surfaces, wear produces a clay film with a considerable drop in friction; for example Richards (1973) found residual friction of 15° in slightly weathered diorite having peak friction near 30°.

In summary, the friction properties of smooth rock surfaces vary with micro-roughness, normal load, weathering, environmental conditions, test apparatus, and testing procedure. A value for the coefficient of friction of 0.5 to 0.6 is a reasonable assumption in general, but values as low as 0.2 can be expected for rocks rich in mica or other platy minerals or whose discontinuities are weathered.

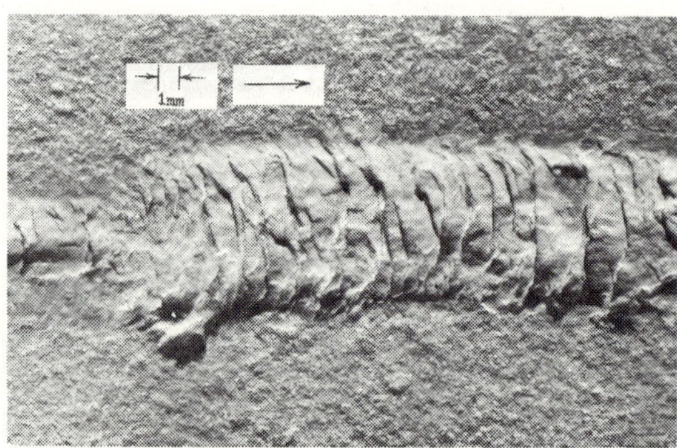

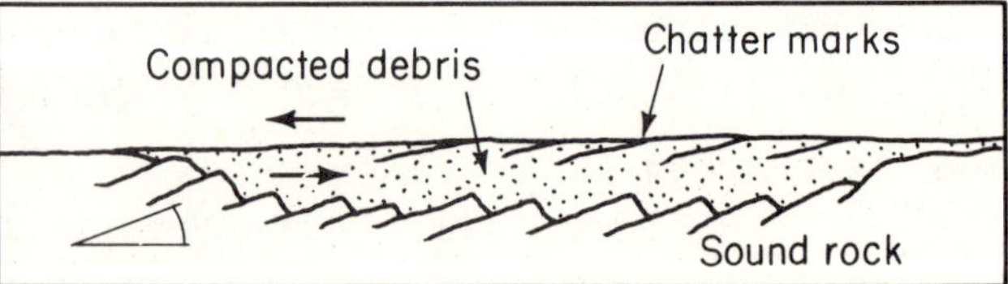

Figure 5-17. Characteristics of a gouge zone, artifically produced by shearing at 1177 psi normal pressure along sawed joints in Solenhofen limestone; from Coulson (1972).

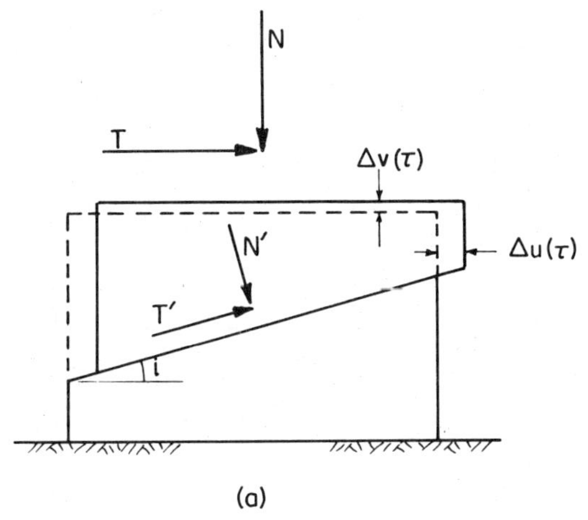

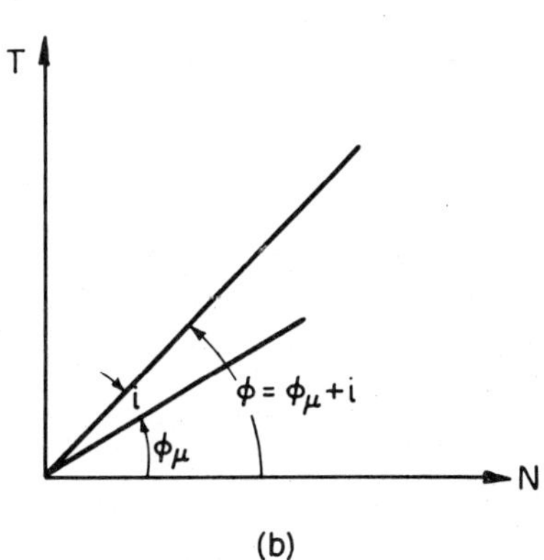

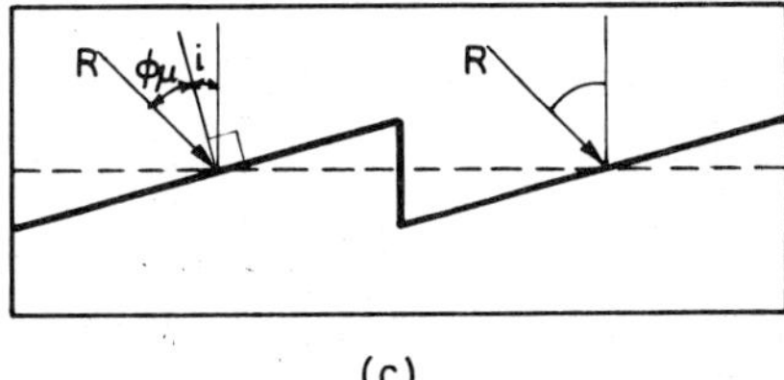

Figure 5-18. Sliding on inclined asperities.

Additional Resistance From Sliding on Inclined Wedges

We have seen previously that dilatancy can cause an increase in σ and thereby strengthen a joint. Sliding asperities can increase the frictional resistance of rock surfaces even when the dilatancy does not increase the normal stress, as discussed by Patton (1966). Consider an ideal wedge-shaped asperity (figure 5-18a) inclined i degrees above the direction of sliding. The friction angle on the sliding surface itself is ϕ_μ, i.e.

$$T' = N' \tan \phi_\mu \tag{9}$$

At the limit of sliding, T and N are connected by:

$$\frac{T}{N} = \frac{T' \cos i + N' \sin i}{N' \cos i - T' \sin i} \tag{10}$$

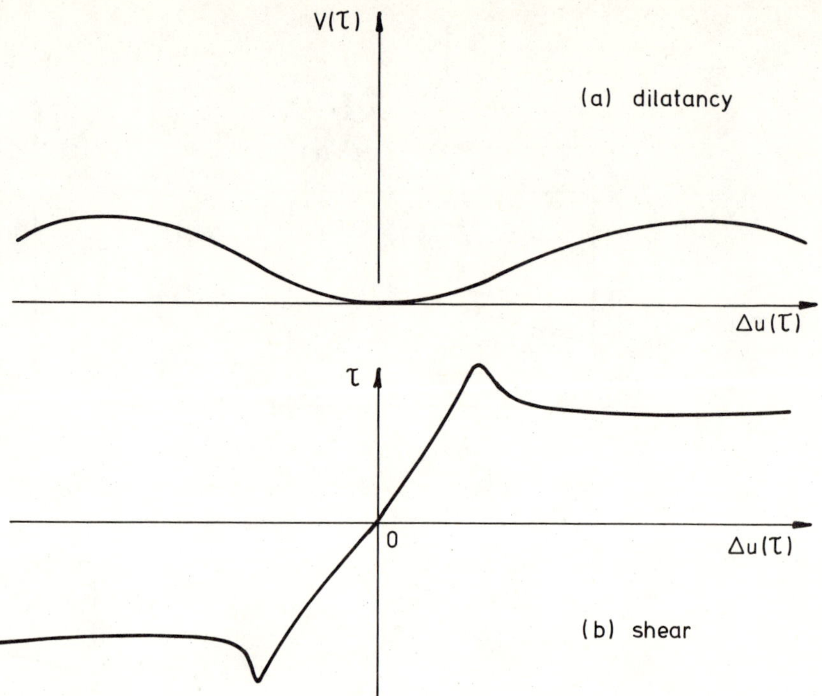

Figure 5-19. Shear behavior of bidilatant joints without rotation under constant normal stress.

introducing 9 and letting tan ϕ = T/N gives:

$$\tan \phi = \tan (\phi_\mu + i) \qquad (11)$$

The effect of regular asperities at a uniform angle i is therefore to increase the friction angle by i (figure 5-18b). This result is also apparent upon examining the inclination (ϕ + i) of the resultant force on the plane of sliding (figure 5-18c).

The wedge effect is associated with dilatancy,

$$\Delta v\ (\tau) = \Delta u\ (\tau) \tan i \qquad (12)$$

The inclination i is signed; that is, for left lateral shearing in figure 5-18, i is negative and (ϕ) = (ϕ - $|i|$) as the joint shears with contractancy. Contraction rarely occurs in practice since actual rough surfaces possess a distribution of both positive and negative angles causing dilatant behavior when sheared in either direction (figure 5-19). Only previously loosened surfaces or soft seams will exhibit contractancy; in the former case, once the surfaces displace they will lock and the friction angle will be restored. If limiting deformations govern design, contractant behavior may need to be considered. Iida and Kobayashi (1974), for example, discuss both dilatancy and contractancy effects for computations of stresses in dams on rock foundations.

Peak Strength as a Function of Normal Pressure

Equation (11) can not hold at high normal stress because the work required to shear through asperities is less than the work to override them. Patton (1966) found that a bilinear relation (figure 5-20) described his experimental data for shear of model joints with regular teeth. At normal stresses less than σ_T

$$\tau_p = \sigma \tan (\phi_\mu + i) \tag{12a}$$

while at normal pressures greater than σ_T

$$\tau_p = C_J + \sigma \tan \phi_r \tag{12b}$$

where $\sigma_T = C_J/(\tan (\phi_\mu + i) - \tan \phi_r)$,

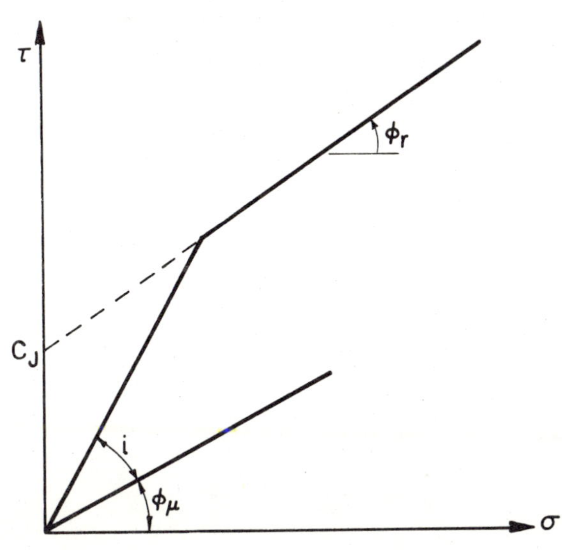

Figure 5-20. Patton's bilinear criterion for shear strength of joints.

C_J is the shear strength intercept ("cohesion") derived from the asperities and ϕ_r is the residual angle of internal friction of the rock comprising the asperities.

Actual rock surfaces obviously cannot be fit by such a simple model. Jaeger (1971) considered a continuously variable empirical shear strength equation which can be written:

$$\tau_p = C_J (1 - e^{-b\sigma}) + \sigma \tan \phi_r \quad (13)$$

When $\sigma = 0$ the strength curve rises at

$$\left(\frac{\partial \tau}{\partial \sigma}\right) = \tan \phi_r + C_J b$$

while when σ is large,

$$\tau_p = C_J + \sigma \tan\phi_r$$

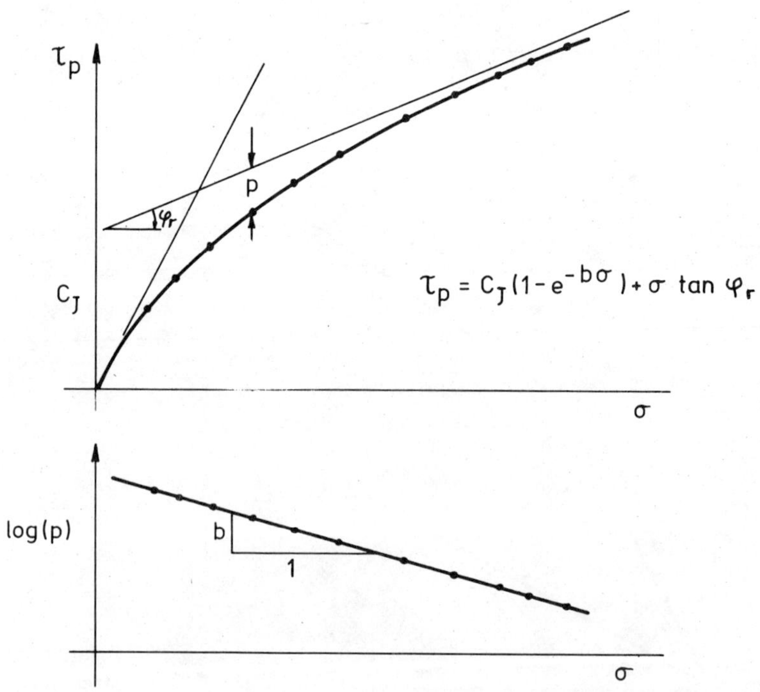

Figure 5-21. Jaeger's empirical shear strength relationship.

Figure 5-21 shows how such a formula smooths Patton's shear strength criterion (12) and suggests a simple graphical way of evaluating the parameters. Sketch a line asymptotic to the peak strength data plotted in τ-σ coordinates; this defines C_J and ϕ_r, and a value of p for each test: $p = C_J + \sigma \tan \phi_r - \tau_p$; then -b is the slope of the line of log (p) plotted against σ. Jaeger found, for example, for residual sliding of natural joints in andesite (in psi units):

$$\tau_p = 270 (1 - \exp (-0.0015 \sigma)) + 0.41 \sigma$$

Ladanyi and Archambault's Equation. While Jaeger's empirical equation should be satisfactory for a wide range of conditions, it is helpful to have an equation derived from identifiable properties of the joint and wall rock. Ladanyi and Archambault (1970) combined the friction, dilatancy and interlock contributions to peak shear strength to derive a general strength equation for discontinuities which has proved accurate in model studies. The peak strength is given by:

$$\tau_p = \frac{\sigma(1 - a_s) (\dot{v} + \tan \phi_\mu) + a_s s_R}{1 - (1 - a_s) \dot{v} \tan \phi_\mu} \qquad (14)^*$$

where a_s, $\dot{v}$, and s_R are the following functions of σ:

a_s is the proportion of joint area sheared through the asperities

$\dot{v}$ is the dilation rate at the peak shear stress (secant dilatancy rate) $\Delta v (\tau_p)/\Delta u (\tau_p)$, and

s_R is the shear strength of the rock composing the asperities.

Equation (14) reduces to (12a) at low σ when $a_s = 0$ and $\dot{v} = \tan i$. Putting $s_R = C_J + \sigma \tan \phi_r$, equation (14) reduces to (12b) at very high σ where all the teeth are sheared off, $a_s = 1$, and $\dot{v} = 0$. However Ladanyi suggested substituting Fairhurst's parabolic criterion for s_R:

---

* The derivation follows the work of Rowe (1962) and Rowe et al (1964). The equation given in (14) differs from that presented by Ladanyi in replacing Rowe's ϕ_f by ϕ_μ.

$$s_R = q_u \frac{\sqrt{1 + n} - 1}{n} (1 + n\,\sigma/q_u)^{\frac{1}{2}} \qquad (15)$$

where q_u = the unconfined compressive strength and n = the ratio of compressive to tensile strength, of the rock comprising the asperities.

Ladanyi and Archambault suggested power laws for $\dot{v}$ and a_s; for $\sigma < \sigma_T$,

$$a_s = 1 - \left(1 - \frac{\sigma}{\sigma_T}\right)^{K_1} \qquad (16a)$$

and

$$\dot{v} = \left(1 - \frac{\sigma}{\sigma_T}\right)^{K_2} \tan i_o \qquad (16b)$$

The suggested values of the exponents are $K_1 = 1.5$ and $K_2 = 4$ (Ladanyi and Archambault, 1972); a_s increases from 0 at $\sigma = 0$, to 1 at $\sigma = \sigma_T$ while $\dot{v}$ decreases from $\tan i_o$ when $\sigma = 0$, to 0 at $\sigma = \sigma_T$ (see figure 8-17). The transition pressure σ_T is the normal stress at which the joints cease to be weaker than the rock itself and in the absence of sufficient data it can be approximated by $\sigma_T = q_u$. With these conditions, equation 14 will define a curved peak stress criterion as shown in figure 5-22.

Barton's Empirical Shear Strength Equation. Barton (1974a) offered an empirical shear strength criterion for unfilled discontinuities accounting for the variation of dilatancy with normal stress and the shear strength of the asperities.

$$\tau_p = \sigma_n \tan\left(R \log_{10}\left(\frac{q_u}{\sigma_n}\right) + \phi\right) \quad (\sigma_n < q_u) \qquad (17)$$

The factor R expresses the influence of roughness, varying linearly from 0 to 20 over the range from perfectly smooth to very rough. In both expressions (14) and (17), the normal stress is the effective stress if the discontinuity contains a fluid under pressure p; i.e. $\sigma_n = \sigma_{total} - p$. The unconfined compressive strength q_u refers to the rock forming the asperities. Since weathering is often considerably more advanced along joints than through the body of rock, q_u may be considerably lower than values for unweathered rock and it should be obtained from results of tests on the wall rock, e.g. Schmidt hammer or scratch hardness tests. A comparison of equations (14) and (17) for rough joints ($i_o = 5^o$, R = 20) showed Ladanyi's peak strength to be higher than Barton's in the region $0.5 < \sigma_n/q_u < 0.7$, and lower elsewhere.

Filled Discontinuities

When soil material completely covers the walls of a discontinuity, burying all the asperities, shear entirely within the filling material is possible. If the thickness of filling, e, is less than the maximum asperity height, rock contact will occur after a displacement Δu_c approximated by (Barton, 1974),

$$\Delta u_c \quad = \quad e/\tan i \tag{18}$$

and thereafter the discontinuity will stiffen and strengthen. In filled, smooth discontinuities, this will not happen and, in fact, the peak strength may be less than that of the filling material when sheared alone at similar rate and confinement (Eurenius and Fagerstrom,

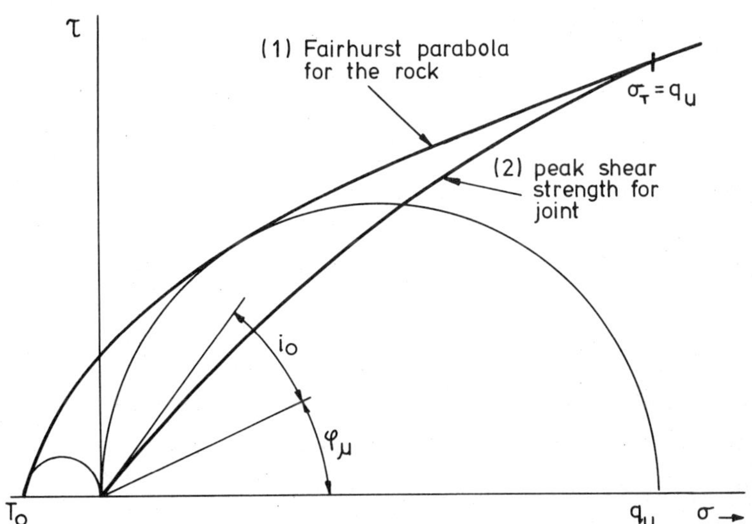

Figure 5-22. Ladanyi and Archambault's shear strength relation for rough joints.

Figure 5-23. Shearing though a discontinuous rock mass on compound surfaces.

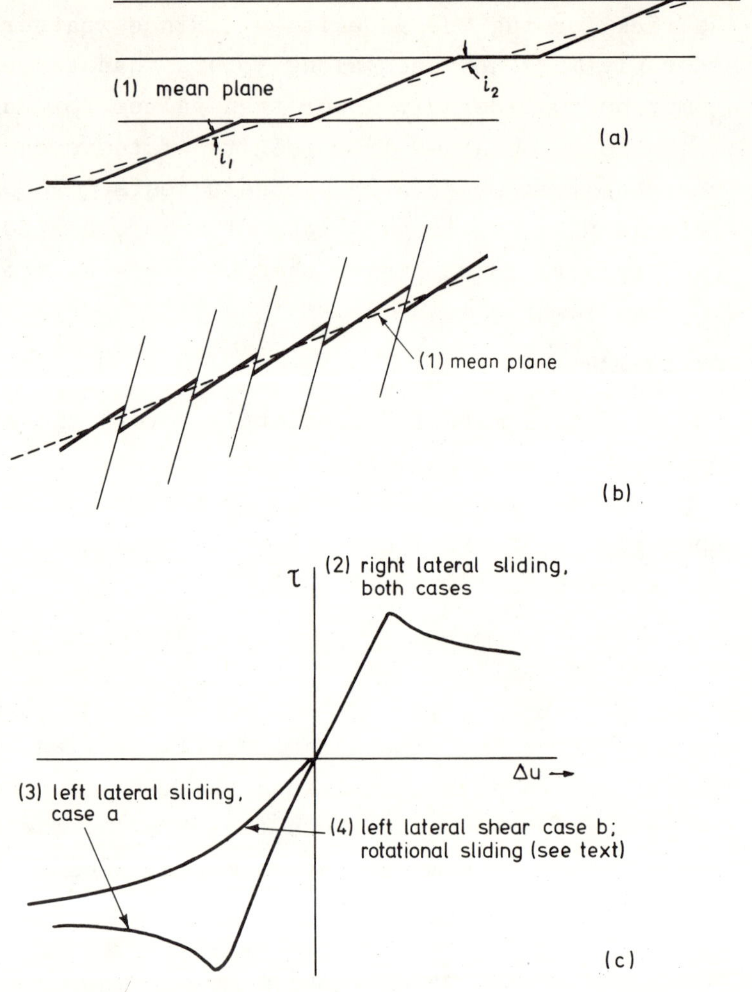

1969). This is because the shearing occurs at the rock wall where pore pressures can not be dissipated and where a small displacement creates sufficient strain to reach the residual shear strength. Since the shear strength of filled discontinuities varies over a range from as little as 7°, for montmorillonite fillings, to as much as 50°, for crushed rock or preconsolidated, dry gouge, shear tests are warranted when the strength needs to be known accurately.

Sliding on Compound Surfaces—Rotational Friction

The peak strength models discussed can describe sliding on a compound surface made up of sections of two different planes as discussed by Bray (1967), Jaeger (1971), Deere et al (1967), and Ladanyi and Archambault (1972). If the mean plane through such a surface is constructed, the actual surfaces along which sliding occurs can be considered to define wedge shaped asperities (figure 5-23). Then if there is shear without rotation of joint-blocks, the angles i_o are defined for sliding in either direction along the mean plane and either equation (12) or equations (14) through (17) can be applied. Figure 5-23b shows a case in which over-sliding of asperities is possible only in one direction along the mean plane due to the steep angle of imbrication; however, rotational sliding can occur in this case. Compound slide surfaces as in figure 5-23 occur in regularly jointed rock masses sheared at intermediate orientations. However, whereas equation (12) and (14) require continuous wall rock, a case where one set of layers or blocks slides on another implies a sliding system with additional degrees of freedom, the most important of which relate to individual block rotations. The jamming of a drawer when pulled eccentrically illustrates the large modification rotational tendencies can introduce. Nascimento and Texeira (1971) discussed rotational friction where shear occurs on two parallel cross-jointed sliding surfaces as in figure 5-24. In this figure a continuous row of blocks, three of which are shown, is subjected to left lateral shear by a force T applied at the upper platen. The overturning moment of T_1 on a block initiates rotation about the left bottom corner (O) with the upper right corner (P) tending to move along a circular arc P P', with dilatancy v. A small rotation γ may cause changes in N because of the dilatancy and shear forces develop along the sides of the blocks, ultimately inclining the side forces F_2 and F_3 at an angle ϕ to the normal with a back turning moment about O. The displacement and the rotation necessary to neutralize the moments depend upon the magnitude of T and the stiffness of the normal load mechanism. Thus corresponding to every T, up to a limit, there is an equilibrium γ. The limit is reached when overtoppling occurs at $\gamma = \delta$ where $\delta = \tan^{-1} (a/b)$; instability will be preceded by a flattening of the rate of dilatancy, which becomes stationary at the peak load, figure 5-24b.

If the blocks become separated so that side forces are nil, the

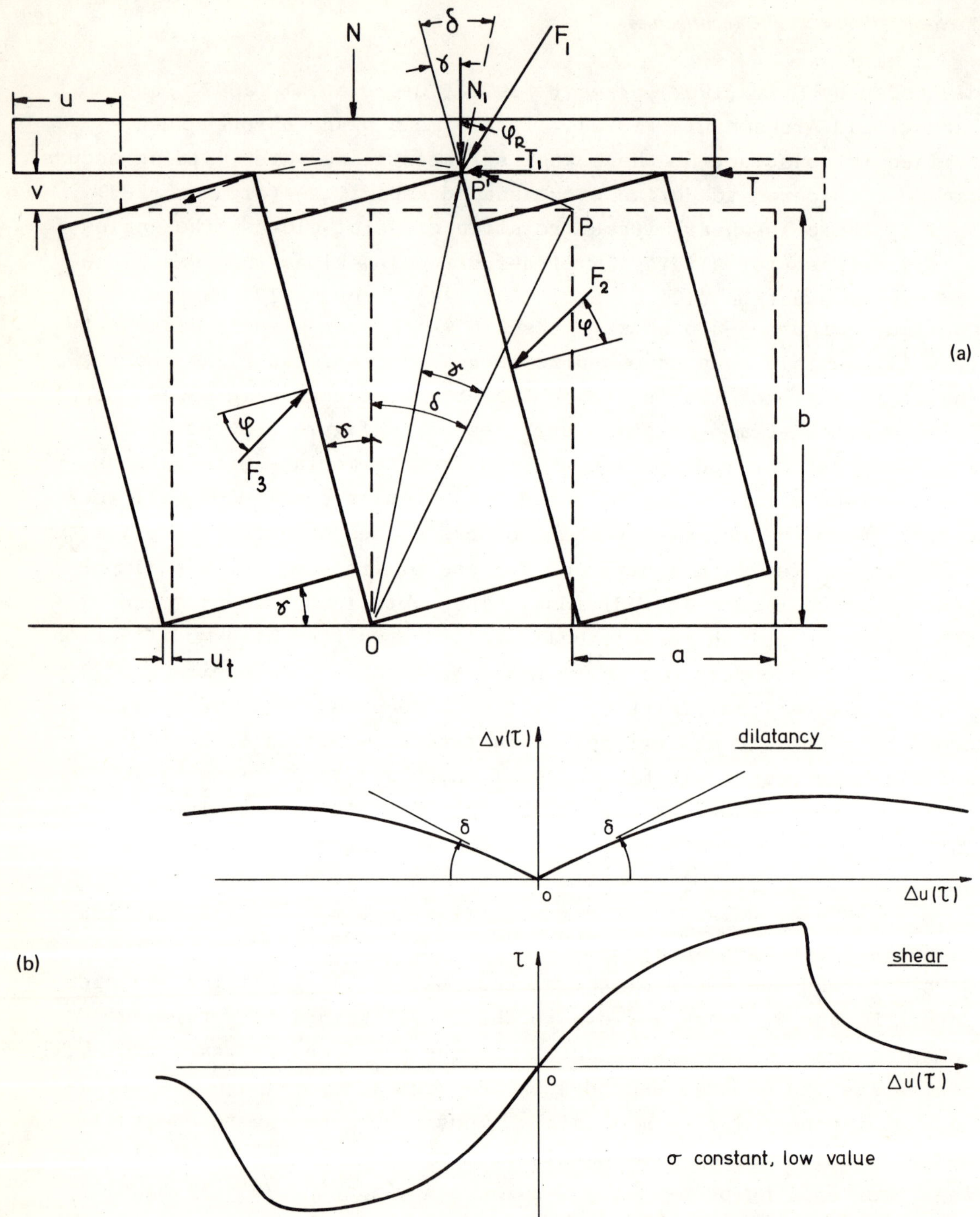

Figure 5-24. (a) Rotational friction; after Nascimento and Teixeira (1971). (b) Rotational shearing behavior of bidilatant joints at low normal pressure; compare with figure 5-19.

line of action of F_1 and its reaction at O must lie along the diagonal OP; in this case, the inclination of F_1 with the normal load is always $\phi_R = \delta - \gamma$. In the presence of side forces, the rotational friction angle ϕ_R will be greater than $\delta - \gamma$. If γ increases uniformly with increasing T, this implies that the rotational peak load variation with σ is convex downward resembling equation 14. Accordingly it is not unreasonable to include the effects of block rotation in that equation. Ladanyi and Archambault (1972) did this through an adjustment of exponents K_1 and K_2 in equations (16a) and (16b) as follows. Let n_r be the number of rows of blocks required to define the shearing zone; for example, in figure 5-24, $n_r = 1$. Results of shear tests with n_r in the range 2 to 5 could be simulated by (14) through (16) by putting in:

$$K_1 = \left(\frac{2}{n_r}\right)^3 \tan i_o \qquad (19)$$

and

$$K_2 = 5$$

$$(2 \leq n_r \leq 5)$$

Reducing K_1 makes the area of contact, a_s, smaller at any normal pressure, σ. The increased value of K_2 implies a faster loss of

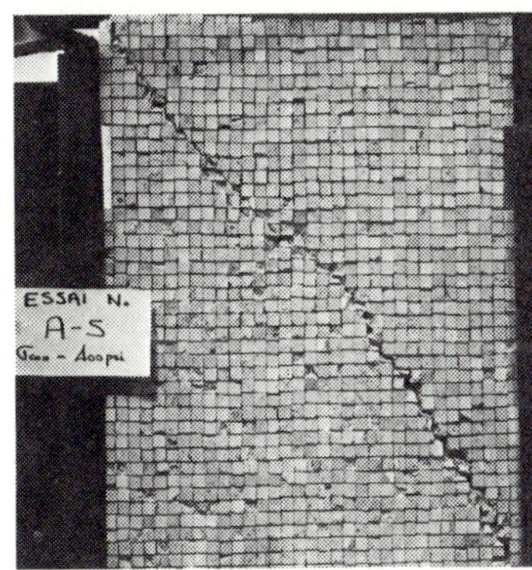

Figure 5-25. Rupture modes observed in biaxial model tests with imbricated joint blocks; from Ladanyi and Archambault (1972). (a) formation of a shear plane; $\theta = 0$; $\sigma_3 = 28$ Kg/cm^2 (b) formation of a shear zone; $\theta = 30°$; $\sigma_3 = 7$ Kg/cm^2 (c) formation of a kink band; $\theta = 60°$; $\sigma_3 = 35$ Kg/cm^2

dilatancy with σ, associated with rock crushing along the edge to face contacts. Reasonable agreement was found using this theory to calculate results of biaxial model tests on imbricated block systems. The blocks were of square cross-section 1.27 cm x 1.27 cm and 6.3 cm long; 1800 were used per model. Three modes of failure were observed depending on the orientation of the joint system:

(1) Shear along a well defined plane in general oblique to both joint sets as in figure 5-25a; $K_2 = 4$, $K_1 = 1.5$;

(2) Formation of a narrow zone of rotated blocks with $n_r = 2$, as in figure 5-25b; $K_2 = 5$, $K_1 = \tan i_o$; and

(3) Failure along a wide band of rotated and separated columns of blocks with $n_r = 3$, as in figure 5-25c; $K_2 = 5$, $K_1 = (2/3)^3 \tan i_o$. The authors used the term "kink band" to describe this mode of failure, which resembles kinking in metamorphic rocks as described by Ramsay (1967), and Paterson and Weiss (1966). Kink bands occur when sliding takes place on the discontinuous joints; movement of any block will then exert an eccentric load on its neighbor which overturns. This mechanism was called "block jacking" by Goodman (1972). The failure process is similar to toppling of slopes described by Hoek (1973), Goodman (1972), Cundall (1971), Barton (1971), Ashby (1971) and others.

As an illustration of calculation by Ladanyi's theory, consider the block model result of figure 5-26. The continuous joint set was inclined 15° with respect to the direction of σ_1. Failure occurred by formation of a kink band, two to four blocks wide, at a mean orientation of $\psi = 35^o$ with the direction of σ_1, as shown in figure 5-26. Instability occurred when the diagonal of the block columns rotated into the direction of σ (i.e. $\gamma = \delta = \tan^{-1}(1/3) = 18^o$). The confining pressure $\sigma_3 = 13.5$ kg/cm^2 was applied all around and then σ_1 was increased. At failure, σ_1 was 115 kg/cm^2. The normal stress, σ, and shear stress τ on the failure zone at failure were, from (1)

$$\tau_p = (\sigma_1 - \sigma_3) \sin 35 \cos 35 = 47.7 \text{ kg/cm}^2$$

$$\sigma = \sigma_3 + (\sigma_1 - \sigma_3) \sin^2 (35) = 46.9 \text{ kg/cm}^2$$

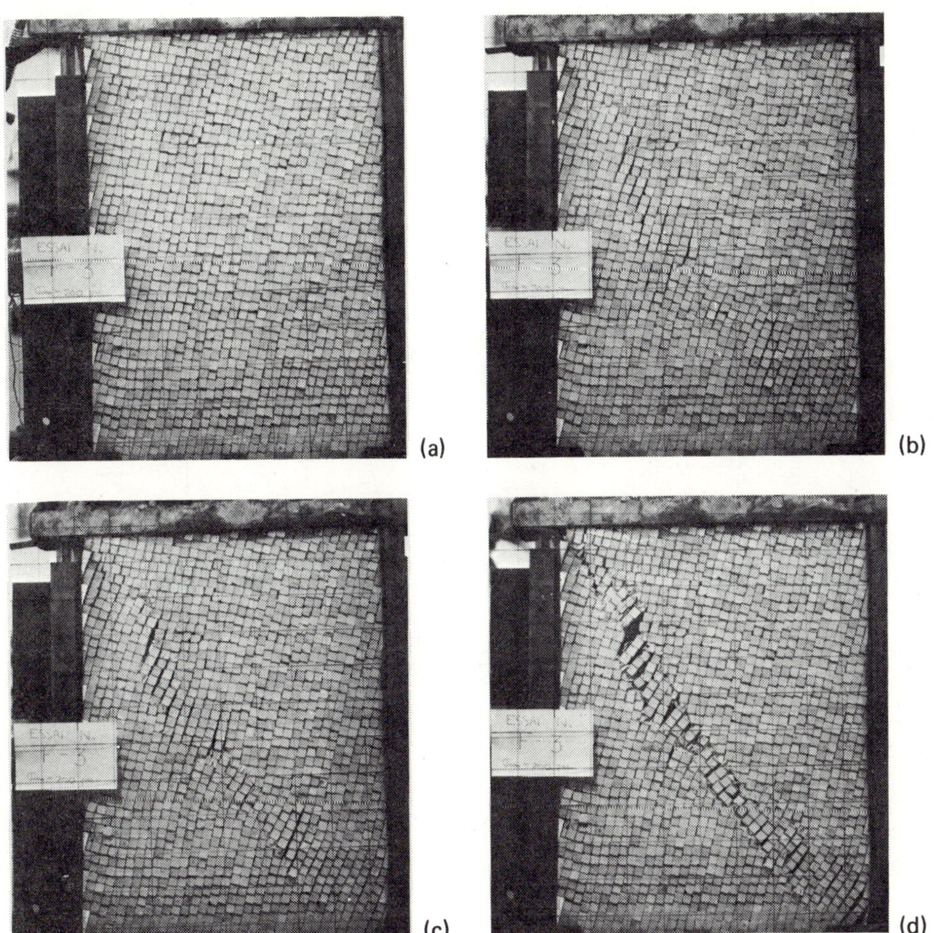

Figure 5-26. Development of a kink band; σ_3 = 13.5 kg/cm^2. From Ladanyi and Archambault (1972).

Corresponding to σ = 46.9, the peak shear stress will be calculated by equations (14) to (19) and compared with 47.7 kg/cm^2. Figure 5-27a shows the relative orientations of the two joint sets and the mean orientation of the kink band. The orientation of the blocks relative to the direction of shearing is more readily appreciated from figure 5-27b in which the region of the eventual kink band is shown as if in a direct shear test; the continuous joints dip in the direction of the shear load or in the negative direction according to the terminology of Hayashi and Kitihara (1970). It is easy to appreciate why block rotations occur in such a test. Hayashi also

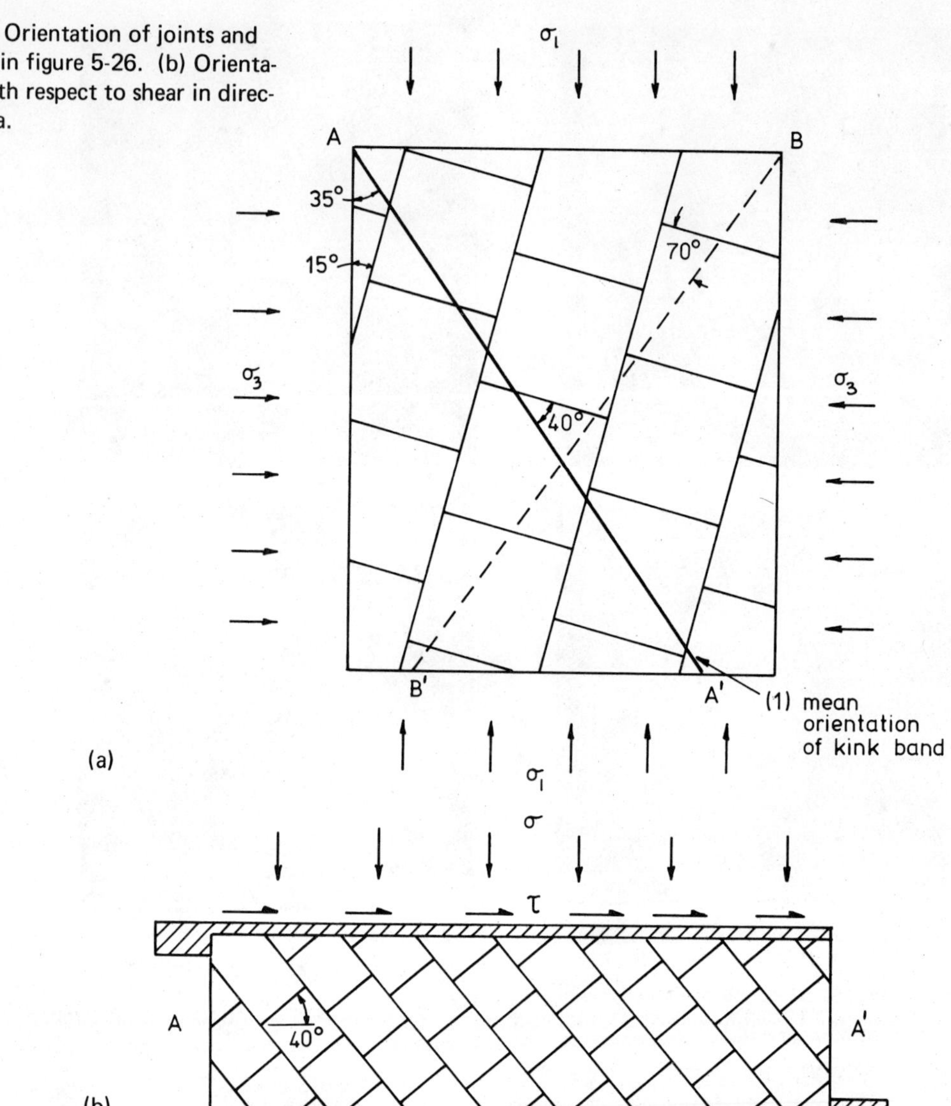

Figure 5-27. (a) Orientation of joints and kink band (AA′) in figure 5-26. (b) Orientation of blocks with respect to shear in direction AA′ of 5-27a.

found rotational modes in negative joint orientations giving reduced shear strength and increased dilatancy, (see also figure 54 of Bernaix, 1974).

Table 5-2 lists the parameters and results of calculation by Ladanyi's theory. Using the actual value of 39° for ϕ_μ of the model material, the calculated shear strength τ_p is too large. The actual and calculated values of τ_p agree for $\phi_\mu = 27^o$. This low friction angle -- less than the residual friction angle -- is not unreasonable

TABLE 5-2

Calculation of Peak Strength For Kink Band of Figure 5-27

Input

ϕ_μ = 39°, 30°, 27°
q_u = 254 Kg/cm^2
T_o = 28.2 (n = 9 in equation 15)
i_o = 40° (see figure 5-27)
n_r = 3
K_2 = 5
σ = 46.9

Intermediate Results

a_s = 0.057
$\dot{v}$ = 0.261

Peak shear Strength - compare with τ_p = 47.7

$\phi_\mu = 39^o$	$\tau_p = 71.$
$\phi_\mu = 30^o$	$\tau_p = 52.9$
$\phi_\mu = 27^o$	$\tau_p = 48.2$

in view of Nascimento's discussion of rotational friction (figure 5-24) which showed a continuous decrease of ϕ_R with shear strain.

The subject of rotational friction is very important in view of the low shear strengths associated with instability by buckling and kinking. Unfortunately our appreciation and understanding of these phenomena are just beginning.

The Role of Water

As previously discussed, water changes the chemistry of joint surfaces and therefore modifies sliding friction values; the presence of water also changes the consequences of wear during sliding. When shear occurs quickly or drainage is slow, excess pore pressures will develop in the wall rock by virtue of the stress changes during increasing shear. An additional increment of water pressure can build up in the joint itself before peak stress if the volume of the joint tends to contract. Rotational shear, in which large dilatancy precedes peak load, should drain the excess water pressure and raise the strength; but this has not yet been confirmed experimentally. In non-rotational shear of smooth joints, Goodman and Ohnishi (1973) observed continuous pore pressure rise until slip, with a net loss in strength because of induced pore pressure; in other words, dilatancy occurred too late to drain the pore pressure.

Handin and Stearns (1964) and others have demonstrated that the effective stress principle is valid for rock joints in shear. Thus we can write $\tau_p = f(\sigma - p)$ where p is the water pressure in the joint; that is, the peak strength equations discussed earlier should be considered to be effective stress criteria.

INFLUENCE OF JOINT ORIENTATION

A rock mass with parallel joints can support tensile stress in some directions but not in others. Some principal stress orientations are safe with respect to shear and others induce slip. Thus joints introduce anisotropy in rock masses.

The joint opening and shear failure criteria previously considered refer to local coordinates normal and parallel to the plane of the joint. Therefore they formulate global anisotropy implicitly, since the peak shear strength can be expressed as a function of the principal stresses and their direction through a transformation of coordinates. It is instructive to consider such a transformation for the very simple peak shear strength criterion:

$$|\tau_p| = \sigma \tan \phi_J \tag{20}$$

Such an assumption simplifies the discussion but is not as limiting as may at first appear since water pressure and cohesion C_J can be introduced through the substitution

$$\sigma = \sigma_a + \frac{C_J}{\tan \phi_J} - p \tag{21}$$

where σ_a is the total normal stress on the shear plane and p is the water pressure. Or, in terms of principal stresses, substitute

$$\sigma_1 = \sigma_{1,a} + \frac{C_J}{\tan \phi_J} - p$$

and

$$\sigma_3 = \sigma_{3,a} + \frac{C_J}{\tan \phi_J} - p \qquad (22)$$

where $\sigma_{1,a}$ and $\sigma_{3,a}$ are the total principal stresses.

Following Bray (1967), consider a discontinuity inclined ψ degrees with the direction of σ_1, as in figure 5-3a. Then

$$\sigma = \frac{\sigma_1 + \sigma_3}{2} - \frac{\sigma_1 - \sigma_3}{2} \cos 2\psi$$

and

$$\tau = \frac{\sigma_1 - \sigma_3}{2} \sin 2\psi \qquad (23)$$

Inserting equations (23) in (20), and introducing $K_f = \sigma_{3,f} / \sigma_{1,f}$ where $\sigma_{3,f}$ and $\sigma_{1,f}$ are the principal stresses to produce slip:

$$|(1 - K_f) \sin 2\psi| = (1 + K_f) \tan \phi_J - (1 - K_f) \cos 2\psi \tan \phi_J \qquad (24)$$

Consider ψ positive, solve for K_f and introduce trigonometric identities for sin (A + B), and (sin A ± sin B) to obtain the condition for limiting equilibrium:

$$K_f = \frac{\tan \psi}{\tan (\psi + \phi_J)} \qquad (25)$$

(σ_1 compressive)

This expression is unchanged if in place of ψ we substitute $\psi_1 = 90^\circ - (\psi + \phi_J)$. Finally, if $\psi_2 = -\psi$ is introduced, then in (24) only sin 2ψ becomes negative and the absolute value brackets restore the positive sign, so that we again obtain the result (25).

Thus we can conclude that four values of ψ (in each 180° range) produce identical values of K_f, i.e.:

$$K_f(\psi) = K_f(-\psi) = K_f(\pm(90 - (\psi + \phi_J)) \qquad (26)$$

Equation (26) shows that the principal stress ratio to produce slip on a joint with friction angle ϕ_J varies markedly with the orientation of the stresses with respect to the joint; the maximum value of K_f (minimum strength) occurs when $\psi = 45^\circ - \phi_J/2$. This is shown in Table 5-3, and in figures 5-28a and b, where K_f is plotted against ψ for $\phi_J = 20^\circ$ and 40° respectively. Any set of values of σ_3, σ_1, and ψ define a point by the polar coordinates $(K = \sigma_3/\sigma_1, \psi)$. Such a point is consistent with the criterion of slip only if it is outside of the shaded region of the Bray diagram. During a joint triaxial test, ψ is fixed and the test commences with $K = 1$. Then as $\sigma_1 - \sigma_3$ is increased, K decreases and slip occurs when $K = K_f$.

Within the complete band $-90 < \psi < 90^\circ$, negative values of K_f will be calculated from (21) when $\psi > (90 - \phi)$ or $\psi < -(90 - \phi)$. Since σ_1 is the algebraically greatest stress, with compression positive, a negative value of K occurs when σ_3 is tensile. A certain amount of tension is allowable as long as it is in a direction almost parallel to the joint plane, i.e. σ_1 almost perpendicular to the joint plane. For a given orientation ψ, slip will occur if $\sigma_3 \leq \sigma_1 K_f$ (see Table 5-3). A joint will open when $\cos 2\psi = (1 + K) / (1 - K)$. When both principal stresses are tensile, the joint will open for all values of K.

The shear strength criterion for the rock can also be plotted in the Bray diagram. For isotropic rock, $K_f = f(\sigma_3)$, which can be plotted as a series of concentric circles of radii $f(\sigma_3)$. For example, the Mohr-Coulomb criterion for the rock expressed in terms of principal stresses takes the form:

$$\sigma_{1,f} = q_u + \tan^2(45 + \phi/2)\,\sigma_3$$

or

$$K_f = \frac{\sigma_3/q_u}{1 + \sigma_3/q_u \tan^2(45 + \phi/2)} \qquad (27)$$

TABLE 5-3

Limiting Principal Stress Ratio $K_f = \sigma_3/\sigma_1$
For Slip on Joints in Different Orientations (ψ)*

ψ	ϕ: 20°	25	30	35	40	45	50	55	60
0	0	0	0	0	0	0	0	0	0
5	.1876	.1515	.1249	.1043	.0875	.0734	.0613	.0505	.0408
10	.3054	.2518	.2101	.1763	.1480	.1235	.1018	.0822	.0642
15	.3827	.3193	.2679	.2248	.1876	.1547	.1249	.0975	.0718
17.5								.0994	
20	.4338	.3640	.3054	.2549	.2101	.1697	.1325	.0975	.0642
22.5						.1716			
25	.4663	.3913	.3265	.2692	.2174	.1697	.1249	.0822	.0408
27.5				.2710					
30	.4845	.4042	.3333	.2692	.2101	.1547	.1018	.0505	0
32.5		.4059							
35	.4903	.4042	.3265	.2549	.1876	.1235	.0613	0	-.0613
40	.4845	.3913	.3054	.2248	.1480	.0734	0	-.0734	-.1480
45	.4663	.3640	.2679	.1763	.0875	0	-.0875	-.1763	-.2679
50	.4338	.3193	.2101	.1043	0	-.1043	-.2101	-.3193	-.4338
55	.3827	.2518	.1249	0	-.1249	-.2518	-.3827	-.5198	-.6660
60	.3054	.1515	0	-.1515	-.3054	-.4641	-.6304	-.8077	-1.0000
65	.1876	0	-.1875	-.3781	-.5746	-.7805	-1.0000	-1.2381	-1.5016
70	0	-.2404	-.4845	-.7362	-1.0000	-2.7475	-1.5863	-1.9238	-2.3054
75	-.3265	-.6581	-1.0000	-1.3584	-1.743	-2.1547	-2.6132	-3.1315	-3.7521
80	-1.0000	-1.5196	-2.0642	-2.6446	-3.2743	-3.9711	-4.7588	-5.6713	-6.7588
85	-3.7321	-4.1602	-5.3299	-6.5991	-8.0034	-9.5910	-11.4301	-13.6218	-16.3238
90	$-\infty$	$-\infty$	$-\infty$	$-\infty$	$-\infty$	$-\infty$	$-\infty$	$-\infty$	$-\infty$

* ψ is the angle between the plane of the joint and the direction of σ_1 (see figure 5-3a)

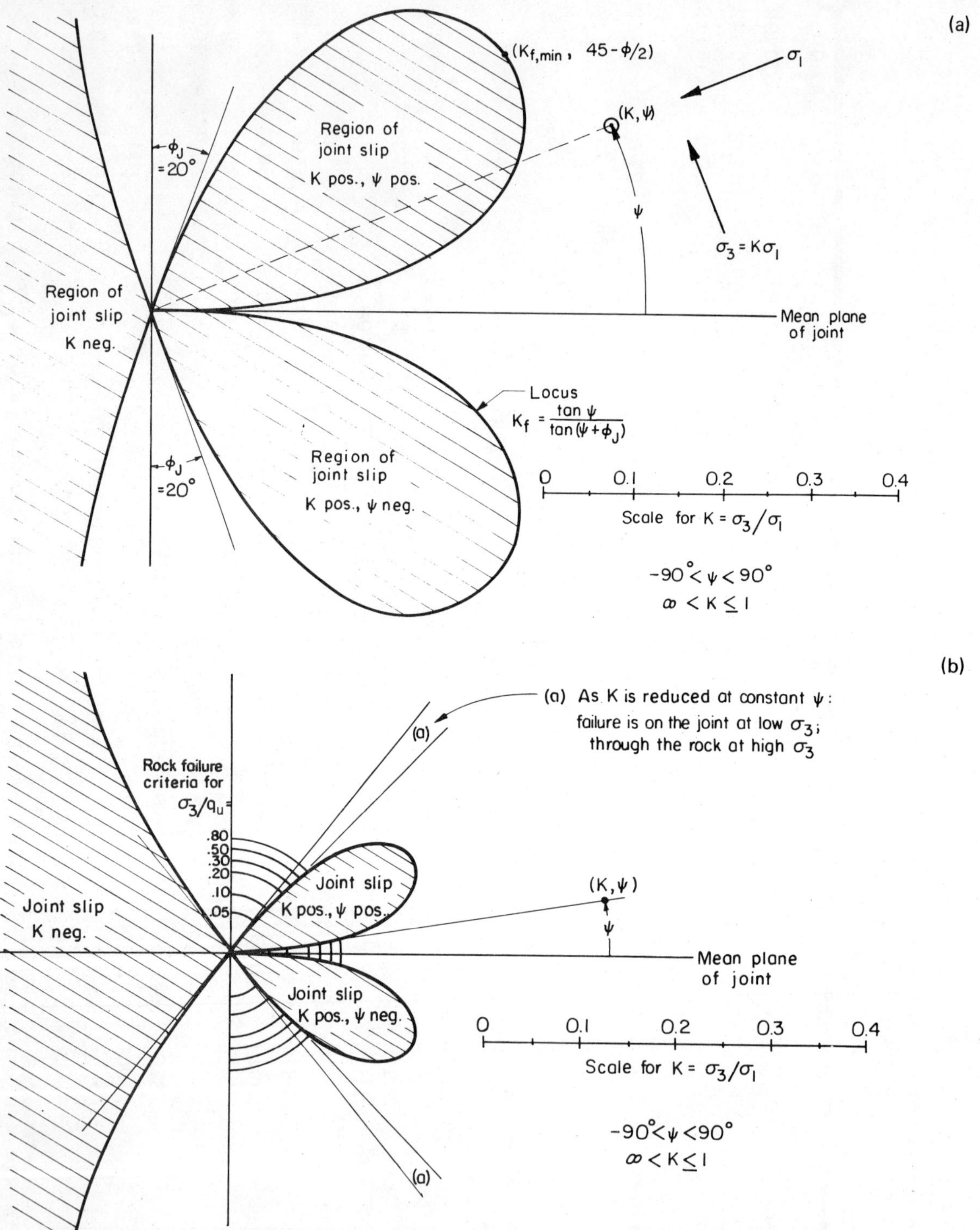

Figure 5-28. (a) Bray's diagram for the Coulomb criterion for joint slip for $\phi_J = 20°$; σ_1 compressive. (b) Bray's diagram for a joint with $\phi_J = 40°$ in rock with $\phi = 50°$.

Thus as σ_3 increases, K_f for the rock grows larger, as shown in Table 5-4 and in figure 5-28b where a circle for $\phi = 50^o$ has been superimposed on the Bray diagram for $\phi_J = 40^o$. We can see from this figure why, in certain orientations, the joint ceases to be weaker than the rock when σ_3 is above a given value. Figure 5-29 is an alternate mode of presentation in cartesian coordinates $x = \psi$, $y = (\sigma_1 - \sigma_3)_f$ for varying values of σ_3. Donath (1964), Jaeger (1960), Chenevert and Gatlin (1965) and others have plotted anisotropy this way, while Müller (1966) and John (1970) use a somewhat different form with polar coordinates ($(\sigma_1 - \sigma_3)_f$, ψ).

Bray (1967) used his graphical presentation to discuss superposition of slip criteria for multi-jointed rocks. He ingeniously traced the changing orientation and size of the "petals" through development of imbrication by slip on crossing joints. Of course

TABLE 5-4

K_f: Radius of Rock Failure in Bray Diagram

$$K_f = \frac{\sigma_3/q_u}{1 + (\sigma_3/q_u)\tan^2(45 + \phi/2)}$$

σ_3/q_u	$\phi = 40^o$	$\phi = 50^o$	$\phi = 60^o$
0	0	0	0
.05	.0406	.0363	.0295
.10	.0685	.0570	.0418
.20	.1042	.0797	.0528
.30	.1261	.0919	.0579
.40	.1409	.0995	.0609
.50	.1515	.1047	.0628
.60	.1596	.1085	.0641
.70	.1660	.1114	.0651
.80	.1710	.1137	.0659

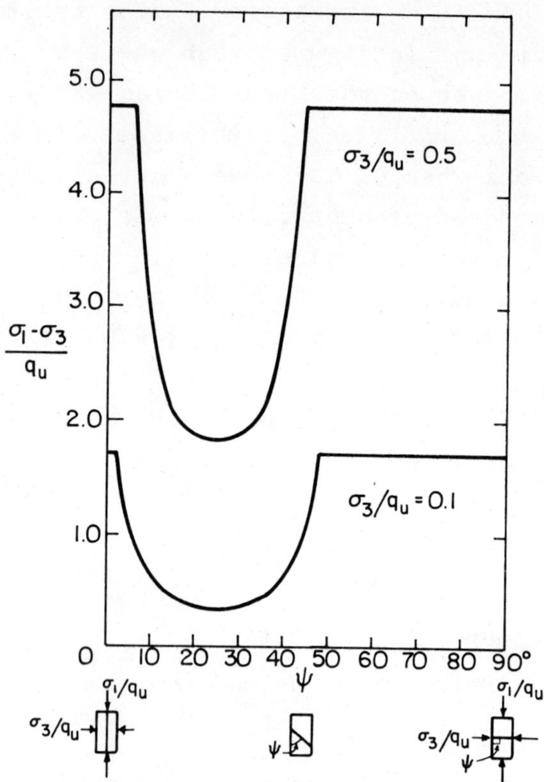

Figure 5-29. Anisotropy of a rock with one joint: $\phi_J = 40°$; $\phi_{rock} = 50°$

superposition can afford only a first estimate for the shear strength anisotropy of a rock mass cut by several sets of joints, since stress redistributions during slip introduce local block breakage as well as rotation and buckling. However, a closer approximation to the real behavior of multi-jointed rock systems is possible with physical model and finite element techniques, to be discussed later.

Although the discussion of strength anisotropy caused by a set of joints has been two dimensional, it must be appreciated that a peak stress criterion like (20) is truly three dimensional; for in place of (23) we can introduce the three principal stresses in (20) through:

$$\sigma = \sigma_1 l^2 + \sigma_2 m^2 + \sigma_3 n^2$$

and (28)

$$\tau = \sqrt{(\sigma_1^2 l^2 + \sigma_2^2 m^2 + \sigma_3^2 n^2) - \sigma^2}$$

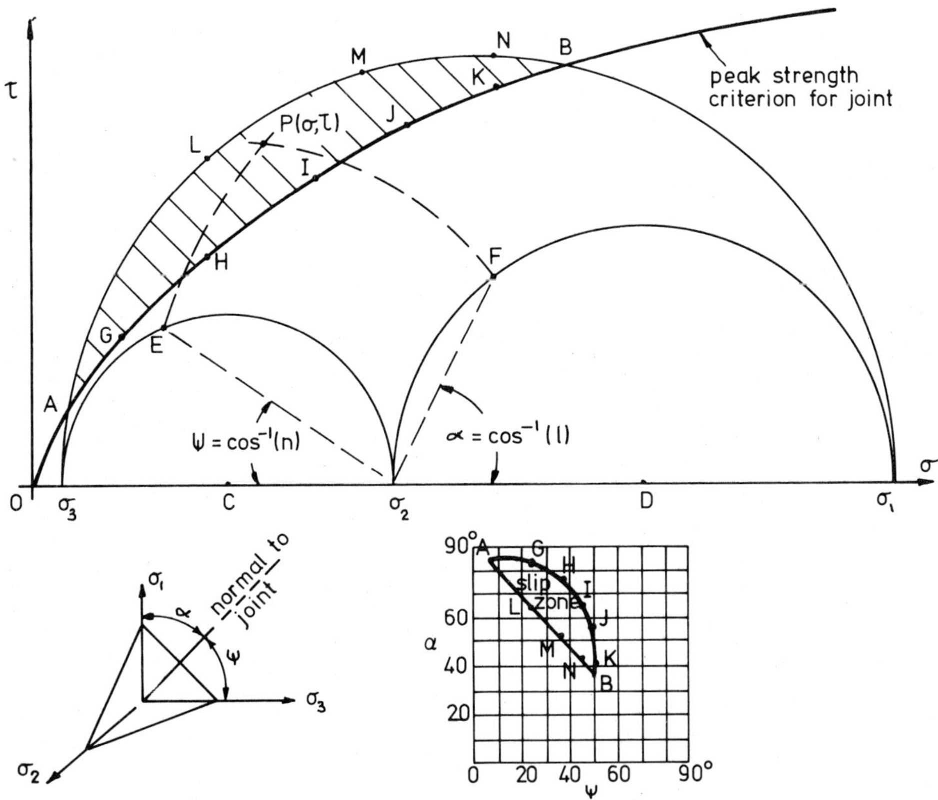

Figure 5-30. Application of Mohr's circles to test for slip on joints of varying orientation according to a given peak strength criterion.

where l, m, and n are the cosines of the angles the normal to the joint plane makes with the directions of σ_1, σ_2, and σ_3 respectively* (figure 6-37b). A quick appreciation can be gained using the three dimensional Mohr's circle diagram, figure 5-30. Given the directions and magnitudes of σ_1, σ_2, and σ_3, a joint whose normal has direction cosines l, m, and n has normal and maximum shear stress defined by the coordinates of point P determined as follows. From σ_2 lay off angles $\psi = \cos^{-1}(n)$ and $\alpha = \cos^{-1}(l)$ as shown in figure 5-30 defining points E and F. P is at the intersection of an arc through E with center D and an arc through F with center C, where D = $1/2\ (\sigma_1 + \sigma_2)$ and C = $1/2\ (\sigma_2 + \sigma_3)$. If P defines a point in the

---

* In the two dimensional case considered, ψ corresponds to $\cos^{-1}(n)$.

ruled area, the stress and strength data are incompatible, i.e. slip occurs. Using this construction, the locus AB of the peak strength criterion across the Mohr's circle can be transformed to cartesian axes (ψ, α) and the zone of safe joint orientations the Coulomb criterion, were presented by Jaeger and Rosengren (1969). These show that the relative magnitude of σ_2 exerts a significant influence on the relative orientations of safe joints.

6

applications of stereographic projection in mechanics of discontinuous rocks

INTRODUCTION

In the design of rock foundations and excavations, we become concerned with potential instability of rock blocks bounded by intersecting planes of weakness. The stability analysis of such blocks, whether they are subjected mainly to self weight, or to additional loads such as static or dynamic water forces, will usually begin with a qualitative investigation to define the morphology of the body and the constraints at the limit of equilibrium, together with description of the mode of failure. After this "kinematic" study, equations for stability are postulated and the degree of safety is calculated. For example, to determine the possible loads on the lining of a large tunnel it will be necessary first to determine possible shapes of wall or roof overbreak; then, knowing the frictional properties of the various rock surfaces and the unit weight of the rock, it will be possible to specify a lining to provide the passive support necessary to retain the rock in place. As another illustration, consider the stability analysis for the rock abutments of an arch dam. The thrust of the dam may tend to displace wedges and blocks, but countless deformational modes will exist, each with different possible effects on the dam. One needs a kinematical study of the system of discontinuities in relation to the dam in order to know where and how to draw "free body diagrams".

This chapter introduces stereographic projection procedures for both kinematical and equilibrium calculations in discontinuous rock masses. The method of stereographic projection is selected for simplicity, since the problems posed by any real case are invariably three dimensional.

As discussed in chapter 3, stereographic projection relates to operations on a unit reference sphere. In the applications to be discussed, the reference sphere will be considered to be the envelope of unit vectors radiating from a point. Operations with orientations of vectors can be done directly on the stereonet, but most operations relevant to statics demand an accounting of not only directions, but magnitudes and positions of vectors as well. However, by combining the stereographic projection with graphical or trigonometric operations, the necessary calculations can be completed. Thus the stereographic projection is useful not only in discussions of the shapes of rock blocks formed by intersecting weakness surfaces, but in analysis of stability of such blocks. Most of the examples discussed in this section relate to the stability of blocks in excavations at the ground surface. However, the problem of limit equilibrium of one or more rock blocks is a basic stability problem for jointed rock masses both at the surface and underground.

KINEMATICAL CONSIDERATIONS

"Kinematics" deals with the motion of bodies without reference to the forces which move them. In the context of rock masses, this implies the movement and modes of failure of the system of rock blocks and joint surfaces. Jointing in hard rocks produces blocks of varied shapes, some of which are illustrated in figure 6-1. Regular patterns of equally spaced joints will produce cubes or regular prisms. When the jointing directions are not perfectly specified but rather are dispersed about preferred orientations, tapered prisms arise. Slivers occur close to excavations where discontinuities lie at some small distance behind the free surface. They will also be found in dispersed joint systems. In sedimentary rocks, individual beds form flat slabs. The granitic rocks possessing exfoliation jointing (sheeting), present families of slightly

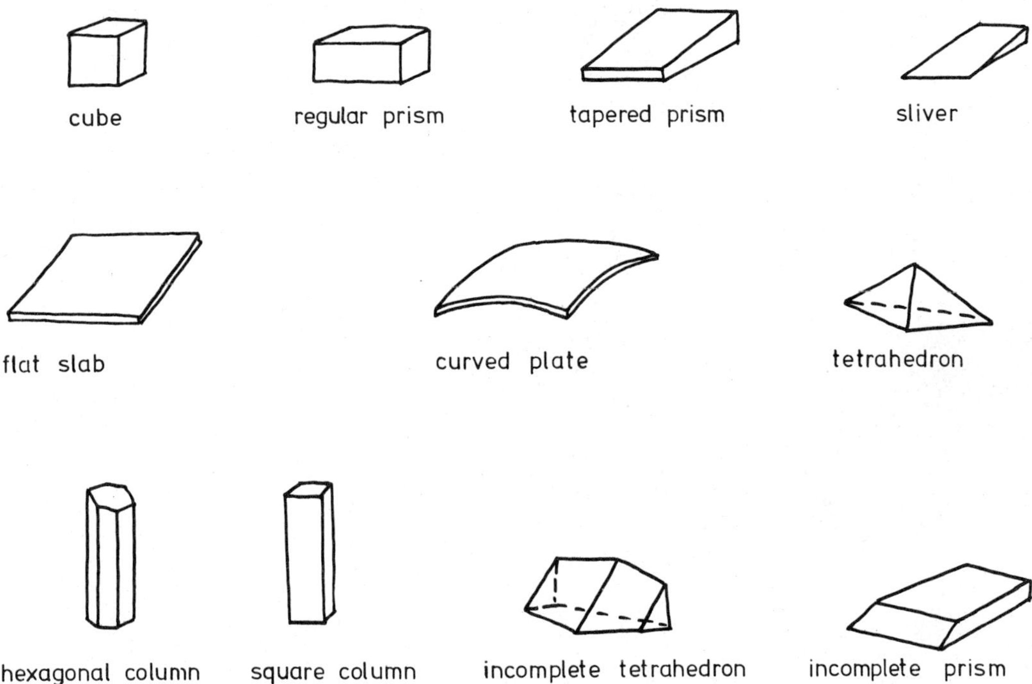

Figure 6-1. Some common block shapes.

curved sheets. Intersections of three discontinuity sets will produce tetrahedra; four sets will produce truncated tetrahedra incomplete tetrahedra or incomplete prisms. Finally, volcanic flow rocks frequently have columnar structure formed by joints parallel to a given axis. The undisturbed rock mass, as noted in Chapter 2, consists of a tight packing of such blocks essentially without void space (figure 6-2a). An excavation considerably larger than an individual block liberates a large number of differently shaped blocks which could potentially move into the excavation. It is impractical to analyze each individual block independently, but extreme combinations of members of each joint set can be considered. Here, discussions of kinematics are particularly relevant. On the other hand, one block alone is sometimes sufficiently large that it merits individual study as a single feature. In such a case, the stereographic projection permits a rigorous discussion of force and moment equilibrium.

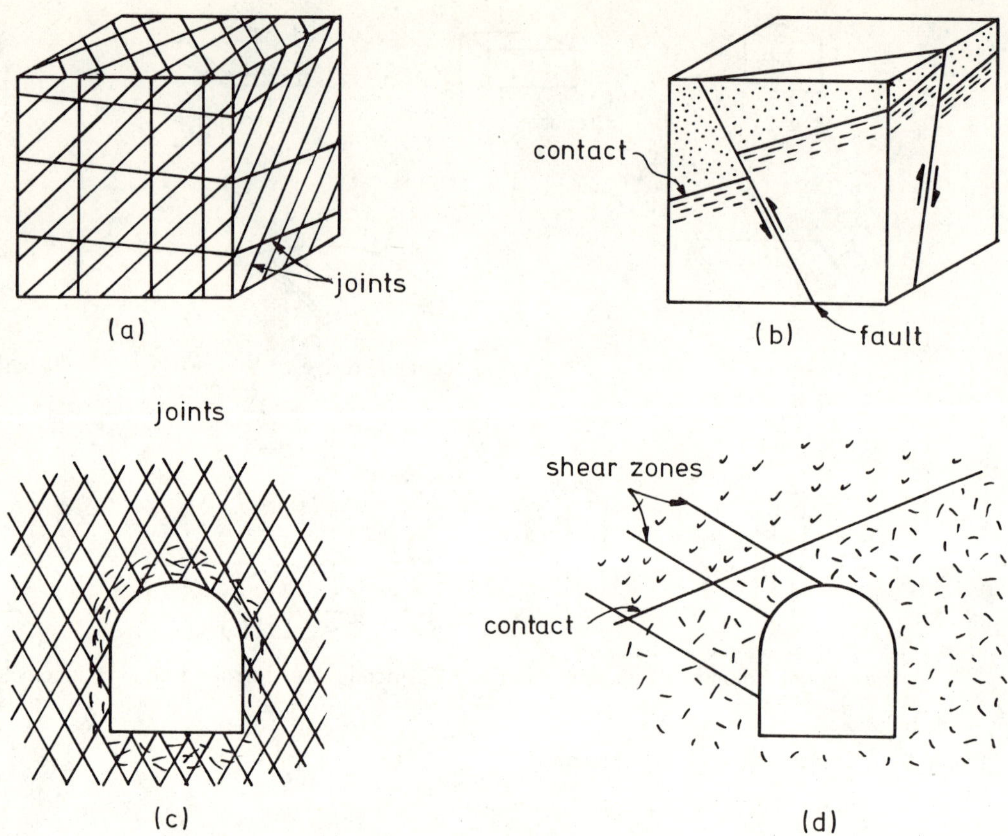

Figure 6-2. Examples of multiple blocks (a and c) and single blocks (b and d).

Lines of intersections between discontinuities give birth to a prevalent failure mode of translation of rock blocks -- two planes remain simultaneously in contact as translation occurs parallel to the line of intersection, as shown in figure 6-3a. This is only possible if the intersection "daylights" into a cutting, that is to say, if it points out of the block into the free space (figure 6-3a). A simple test of this requirement is possible on the stereographic projection. First plot the orientation of the cutting S (figure 6-3b); identify the daylighting portion of the lower hemisphere. Under gravity alone, the slide is possible only if the intersection is pointed downward at such an angle as to overcome friction on the two planes. If the angle between the planes is close to 180°, a slide will ensue whenever the line of intersection plunges more steeply

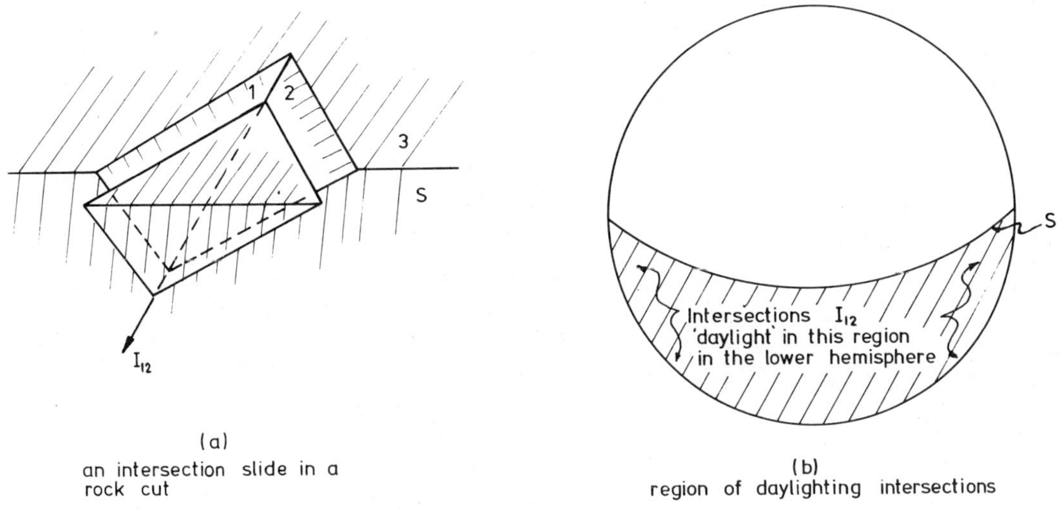

Figure 6-3. Sliding along the intersection of discontinuities.

than the friction angle ϕ (assuming no cohesion). In more acute wedges, the line of intersection must plunge considerably more steeply than ϕ for sliding to be possible (Wittke, 1965). As a simple kinematic test, combine the two requirements, (1) for daylighting, and (2) an intersection plunging steeper than ϕ. A small circle $90 - \phi$ degrees about the vertical contains all lines plunging steeper than ϕ. A daylighting intersection is in the area outside the great circle S. Superposition of these two kinematic tests yields the ruled area shown in figure 6-4a. Figure 6-4b offers an example. Contours of orientations of normals to joints are shown with the preferred orientations, i.e. the central tendencies of three joints as indicated by the symbols $\hat{n}_1$, $\hat{n}_2$ and $\hat{n}_3$ (figure 6-4b).* The line of intersection between two planes with poles $\hat{n}_1$ and $\hat{n}_2$ can be found by constructing the normal to the great circle containing $\hat{n}_1$ and $\hat{n}_2$ ($\hat{I}_{12}$ in figure 6-4b). Of the three intersections formed by $\hat{n}_1$, $\hat{n}_2$ and $\hat{n}_3$, only $\hat{n}_2$ and $\hat{n}_3$ define an intersection in the potentially unsafe region ($\hat{I}_{23}$ in figure 6-4b). This construction was suggested by Markland (1972) and published by Hoek and Bray (1974). It is very useful for a quick analysis in cases where a great many

* The symbol ^ over a letter specifies a unit vector.

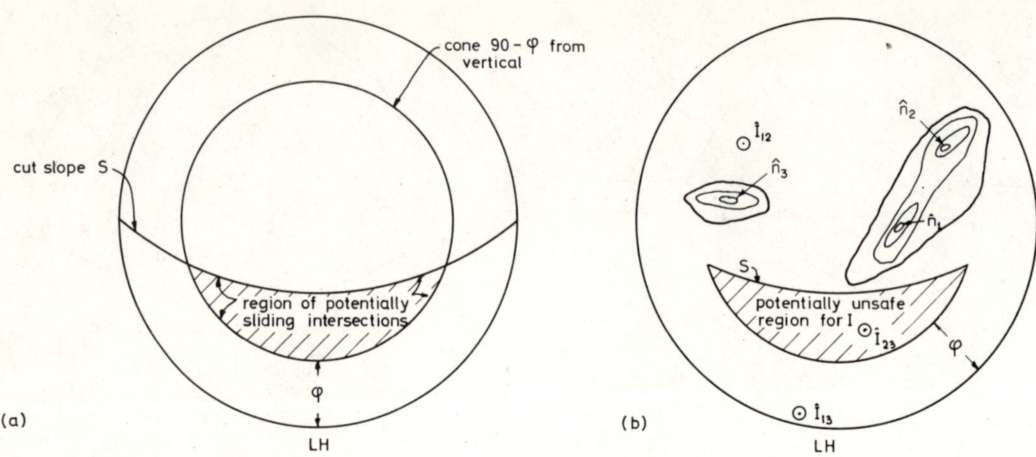

Figure 6-4. Simple tests for sliding along an intersection of discontinuities under gravity loading only; after Hoek and Bray (1974).
(a) superposition of friction and daylighting tests;
(b) application of kinematic tests to a cut (S) in a rock mass.

possibilities exist.

Another construction can test the case of sliding on a single plane, under gravity alone, because in this case the block must translate down the dip of the plane (figure 6-5a). As shown in figure 6-5b, for this mode to exist, the true dip must be contained

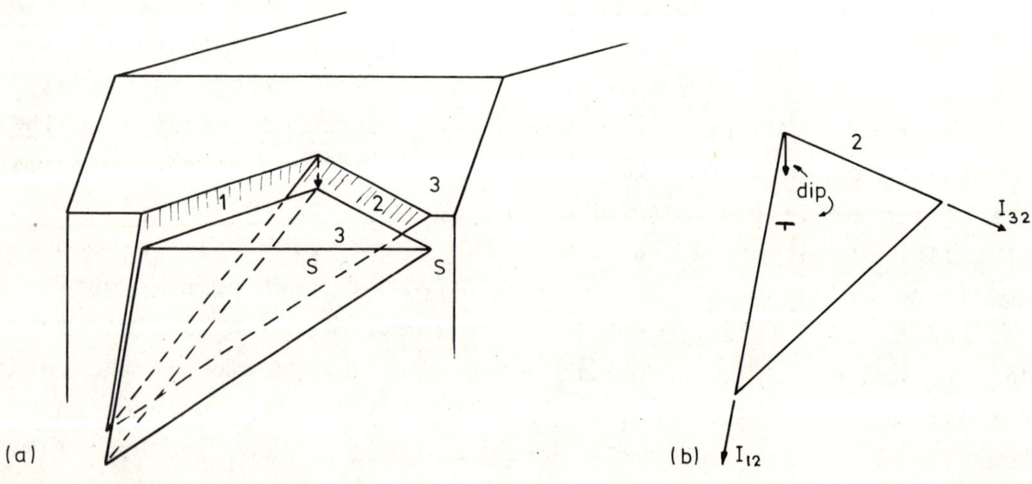

Figure 6-5. Kinematics of a wedge sliding on a single plane under gravity loading only; (a) gravity slide on one plane; (b) kinematic requirement.

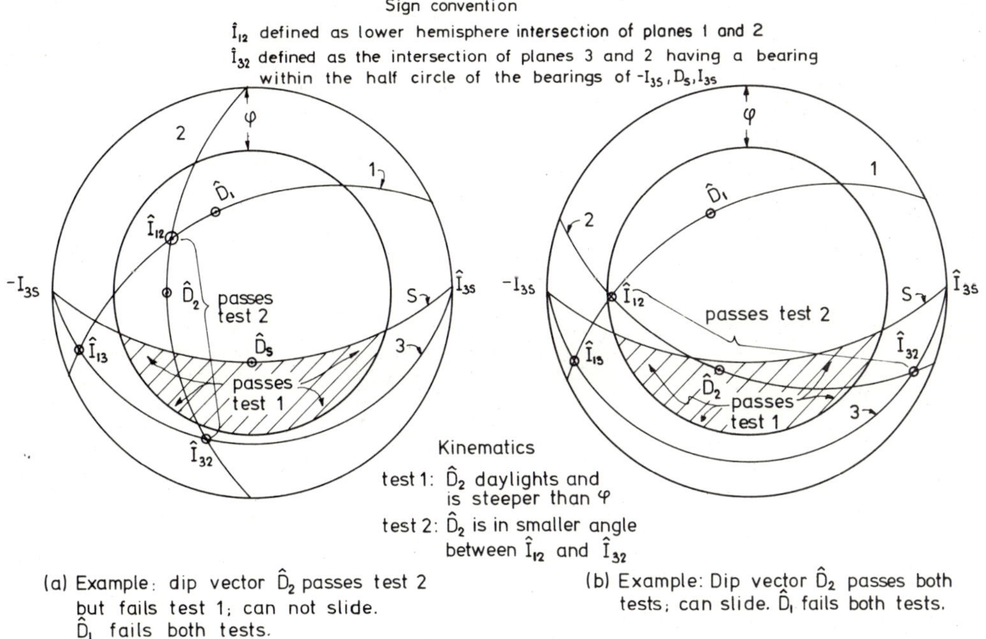

Figure 6-6. Kinematic tests for wedge sliding on a single plane under gravity loading only (see figure 6-5).

between the two intersection lines limiting the extent of the plane of sliding. A line of intersection will be considered positive if it daylights into an excavation; figure 6-6 shows how to identify the positive direction of an intersection for the example of $\hat{I}_{32}$ in figure 6-5a. Form the intersection of plane 3 and the cutting S; the dip of the cutting is denoted $\hat{D}_S$. $\hat{I}_{32}$ is defined then as the intersection of planes 3 and 2 directed to have a bearing within the half circle formed by the lines $\hat{I}_{3S}$, $\hat{D}_S$, $-\hat{I}_{3S}$. The kinematic requirements for sliding down the dip of plane 2, for example, may then be stated in terms of the following two tests, both of which must be passed in order that a slide occur. (1) $\hat{D}_2$ daylights and is steeper than the friction angle ϕ and (2) $\hat{D}_2$ is in the smaller angle between $\hat{I}_{12}$ and $\hat{I}_{32}$. Figure 6-6a shows an example in which the dip vector of plane 2 passes the second test but fails the first test and therefore cannot slide. In the same example the dip vector on plane 1 fails both tests and therefore cannot slide. In figure 6-6b, an example is given in which the dip vector on plane 2 passes both tests

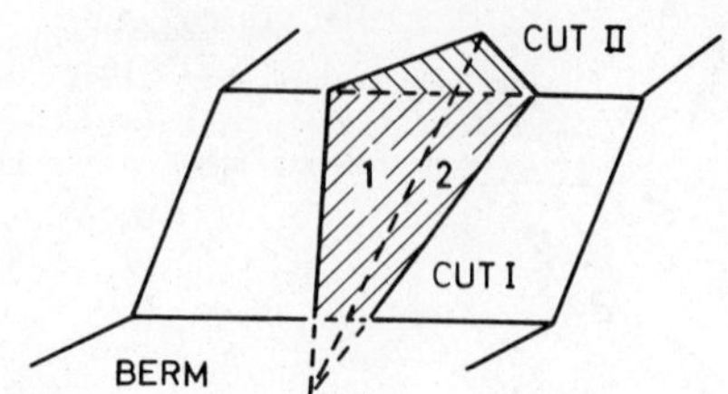

(a) A blocked intersection

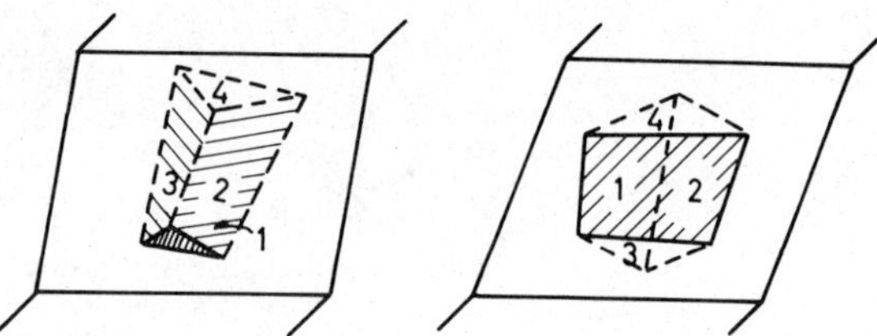

Tapered with respect to faces 1,2,3 Tapered with respect to faces 3,4

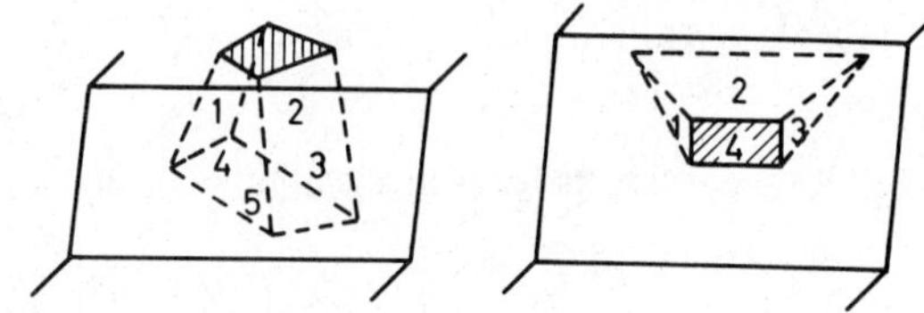

Tapered with respect to faces 1,3 & 2,4 Tapered with respect to faces 1,3

(b) Tapered wedges

Figure 6-7. Tapering and blocking of rock wedges; after Heuze and Goodman (1972).

and therefore can slide. In this same example, $\hat{D}_1$, the dip of plane 1, again fails both tests.

The above kinematic tests are useful for examination of large numbers of blocks. When a single block is under study, further tests can be introduced by tracing the relative locations of the normals to the supporting planes. A block will be called <u>tapered</u> if it grows wider as one traces its extent behind the excavation. Such a block cannot by itself slide forward into an excavation. Figure 6-7b gives four examples of tapered blocks. Figure 8a shows a five-sided block with one free surface denoted S, which is tapered with respect to faces 3 and 4. Such a block is tapered if the normals to opposing faces, in this case 3 and 4, make an angle of less than 180°, as measured in the half space outside of the rock mass. Figure 6-8b shows how this test can be performed on the stereonet. Since $\hat{n}_4$ is

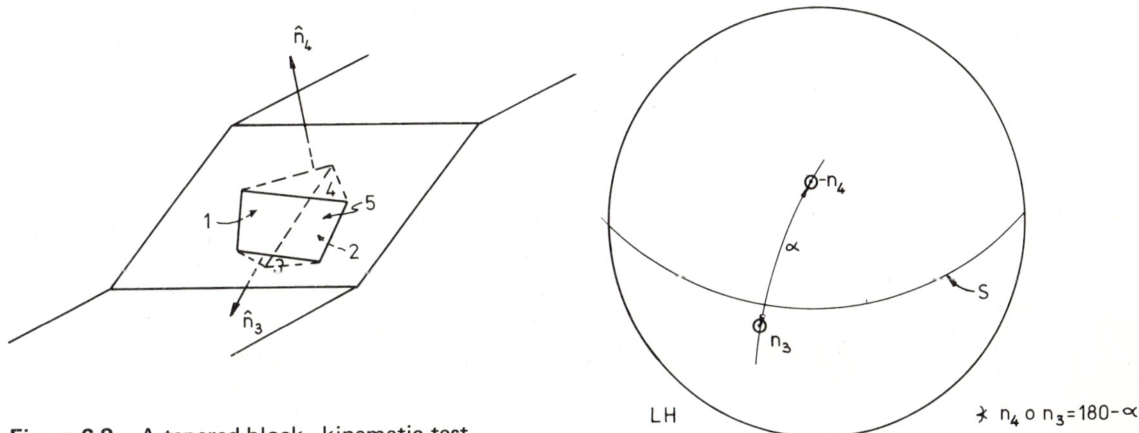

Figure 6-8. A tapered block—kinematic test.

in the upper hemisphere and $\hat{n}_3$ is in the lower hemisphere, the opposite to $\hat{n}_4$ has been shown; therefore the required angle is the supplement of that measured within the lower hemisphere. A further discussion of tapering can be found in Heuze and Goodman (1972).

OPERATIONS WITH VECTORS ON THE STEREONET

The principal contribution of stereographic projection in statics lies in its simplification of operations with vectors. Vector operations -- the cross product, the dot product, addition and subtraction, the triple scalar product and others -- involve an appreciation of the relative orientations of the vectors involved. The usual mathematical method involves decomposing each vector into orthogonal components in three basis directions chosen at will. But the stereographic projection enables the angles between the vectors to be read directly and therefore an artifical decomposition onto an orthogonal basis is no longer necessary.

The Cross Product of Two Vectors

Figure 6-9 shows how the cross product operation is performed on the stereonet. Given vectors $\vec{A}$ and $\vec{B}$ in directions $\hat{a}$ and $\hat{b}$. The directions of unit vectors $\hat{a}$ and $\hat{b}$ are shown on the stereographic projection. Since the direction of the cross product of $\vec{A}$ and $\vec{B}$ is perpendicular to both, the direction of $\vec{A} \times \vec{B}$ lies in the direction of the normal $\hat{n}$, to the plane common to $\hat{a}$ and $\hat{b}$ as shown in figure

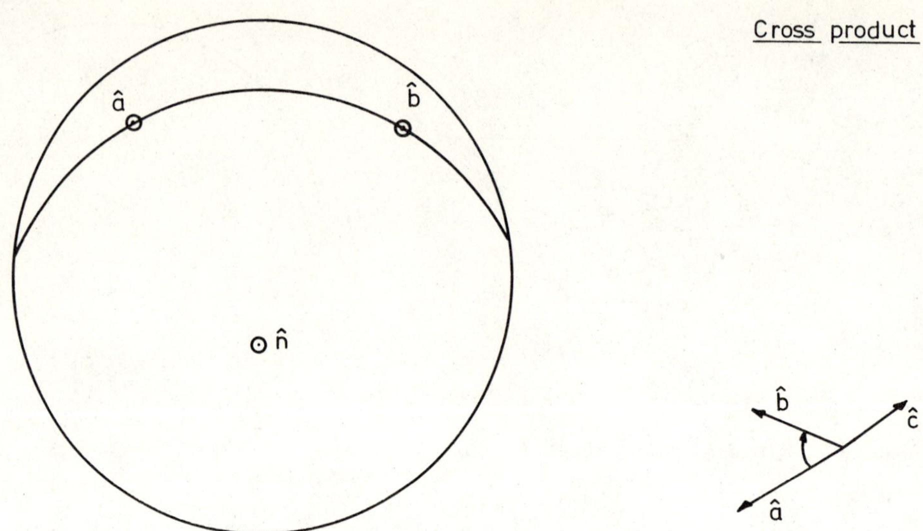

Figure 6-9. The Cross product of two vectors; sign convention—lower hemisphere.

6-9. The sign convention according to the righthand rule is that $\vec{C}$, the cross product of $\vec{A}$ and $\vec{B}$ is in the direction of advance of a right hand screw when $\hat{a}$ is turned into $\hat{b}$; the way this appears in the stereographic projection depends upon the hemisphere convention (figure 6-9).

The magnitude of $\vec{A} \times \vec{B}$ is equal to the magnitude of $\vec{A}$ times the magnitude of $\vec{B}$ times the sine of the angle between $\hat{a}$ and $\hat{b}$; this angle can be read from a stereographic projection along the great circle common to unit vectors $\hat{a}$ and $\hat{b}$. If the vectors involved are in different hemispheres, the operation can be performed with the opposite to either $\hat{a}$ or $\hat{b}$, reversing the order of the cross product, since $\vec{A} \times \vec{B} = \vec{B} \times (-\vec{A})$.

The Dot Product of Two Vectors

The dot product of two vectors $\vec{A}$ and $\vec{B}$ is given by the magnitude of $\vec{A}$ times the magnitude of $\vec{B}$ times the cosine of the angle between them. This angle again can be read in the great circle containing the unit vectors of $\hat{a}$ and $\hat{b}$. Figure 6-10 gives an example.

Addition and Subtraction of Vectors

The addition of two vectors $\vec{A}$ and $\vec{B}$ is formed by the parallelo-

Figure 6-10. Multiplication of vectors—an example.

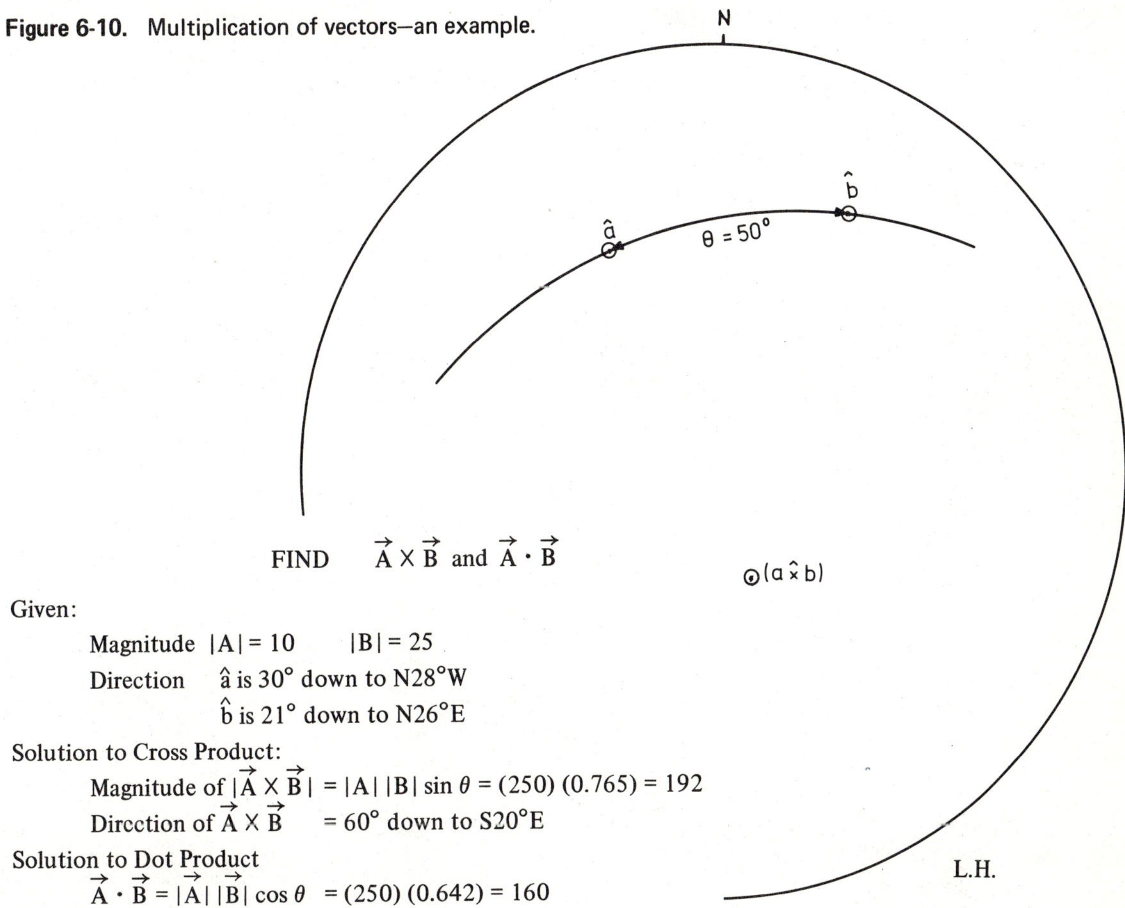

Given:

Magnitude $|A| = 10$ $|B| = 25$

Direction â is 30° down to N28°W

b̂ is 21° down to N26°E

Solution to Cross Product:

Magnitude of $|\vec{A} \times \vec{B}| = |A|\,|B| \sin\theta = (250)\,(0.765) = 192$

Direction of $\vec{A} \times \vec{B}$ = 60° down to S20°E

Solution to Dot Product

$\vec{A} \cdot \vec{B} = |\vec{A}|\,|\vec{B}| \cos\theta = (250)\,(0.642) = 160$

gram rule in the plane common to both vectors. To perform this operation using the stereographic projection, plot the direction of each unit vector and measure the angle between them; then on a supplementary construction, as illustrated in figure 6-11, find the resultant of the two, attaching the known magnitudes to the directions derived from the stereographic projection. It is not necessary to orient the vectors absolutely in the plane of the drawing, but only to measure the angle between the unit vectors $\hat{a}$ and $\hat{b}$. The resultant, $\vec{R}$, of $\vec{A}$ and $\vec{B}$ will lie along the great circle common to $\hat{a}$ and $\hat{b}$, in the smaller of the two angles between $\hat{a}$ and $\hat{b}$. In the case of vector subtraction, the resultant $\vec{A}$ minus $\vec{B}$ will lie in the larger angle between $\hat{a}$ and $\hat{b}$ along the great circle common to $\hat{a}$ and $\hat{b}$ (figure 6-11).

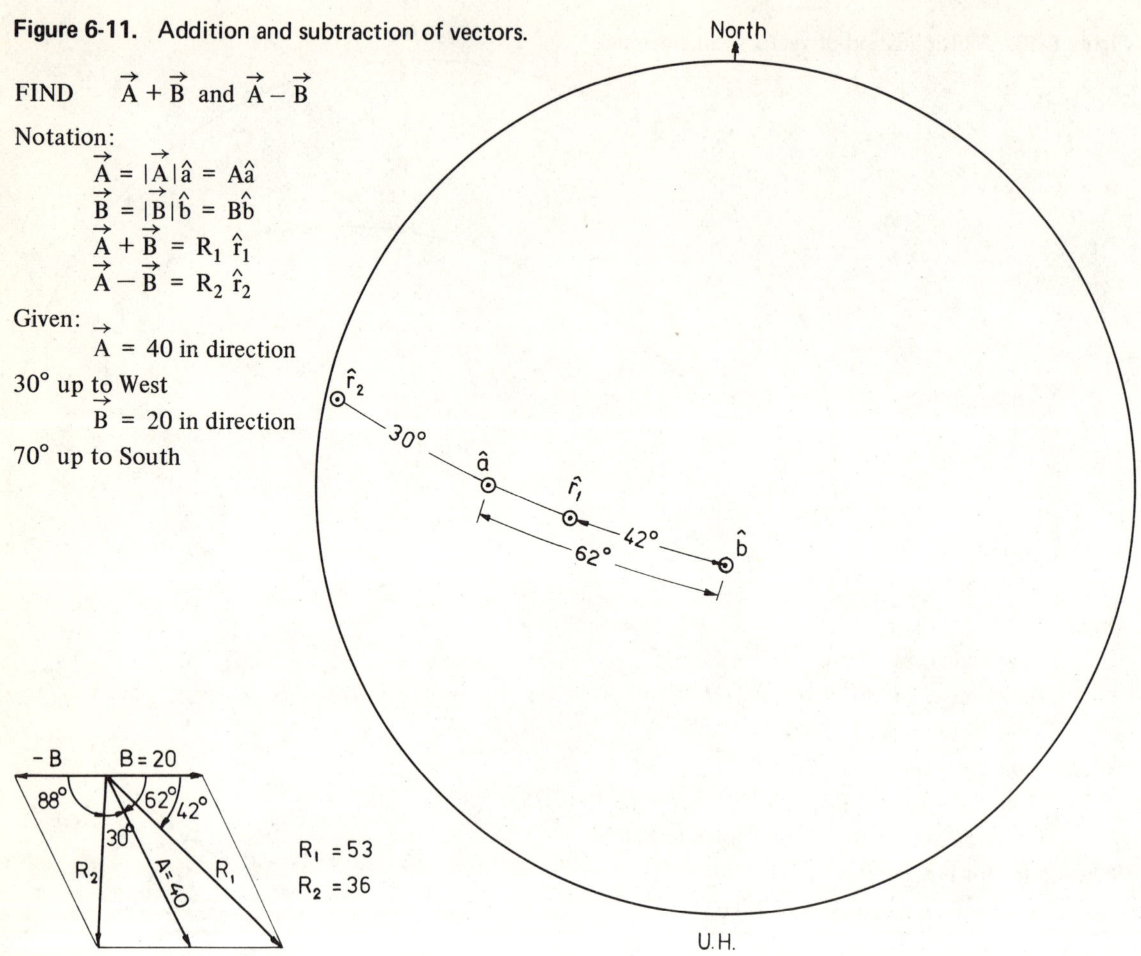

Figure 6-11. Addition and subtraction of vectors.

Addition of Three Vectors by the Intersection of Great Circles

Londe, Vigier and Vormeringer (1970) showed a simple method of summing three vectors on the stereographic projection (figure 6-12). We seek the sum of vectors $\vec{A}$, $\vec{B}$ and $\vec{C}$. Since it is immaterial in which order we add them, we arrive at the same answer if (1) we first add $\vec{A}$ to $\vec{B}$ and then add $\vec{C}$, or (2) we first add $\vec{A}$ to $\vec{C}$ and subsequently add $\vec{B}$. In the first case we will find the resultant $\vec{R}_{ab}$ in direction $\hat{r}_{ab}$ along the great circle between $\hat{a}$ and $\hat{b}$. When $\vec{R}_{ab}$ is added to $\vec{C}$, the resultant $\vec{R}_{abc}$, which we seek, will have direction $\hat{r}_{abc}$ on the great circle between $\hat{r}_{ab}$ and $\hat{c}$, as shown in figure 6-12. However, we can also proceed as in the second case by adding $\vec{A}$ to

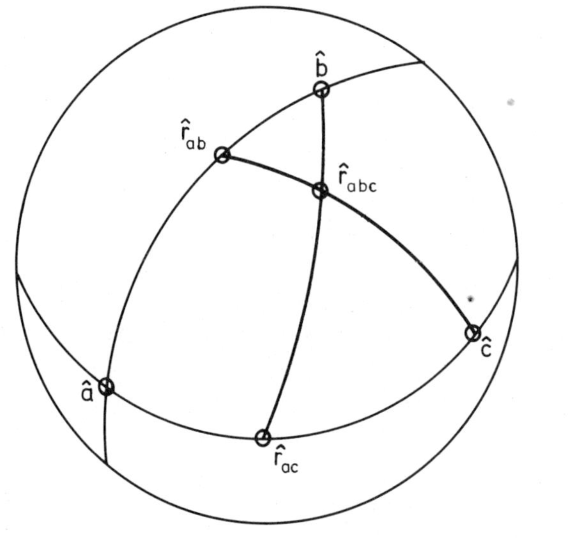

Figure 6-12. Addition of 3 vectors by the intersection of great circles.

$\vec{C}$ with the resultant in the direction $\hat{r}_{ac}$. Adding $\vec{R}_{ac}$ to $\vec{B}$ fixes $\hat{r}_{abc}$ along the great circle common to $\hat{r}_{ac}$ and $\hat{b}$. Thus $\hat{r}_{abc}$ must lie in the intersection of the great circles common to $\hat{r}_{ab}$ and $\hat{c}$, and $\hat{r}_{ac}$ and $\hat{b}$ (figure 6-12). A formula for this construction can be written by introducing notation as follows: let $\overparen{PQ}$ mean the great circle common to lines P and Q; and let $n \wedge m$ signify the line of intersection of planes n and m. Then the unit vector $\hat{r}_{abc}$ defining the direction of $\vec{A} + \vec{B} + \vec{C}$ is

$$\hat{r}_{abc} = \overparen{\hat{r}_{ab}\;\hat{c}} \wedge \overparen{\hat{r}_{ac}\;\hat{b}} \tag{1}$$

The above construction is useful where two of the vectors involved are not known absolutely and are to be left as variables in an analysis. In the example of figure 6-13, vector $\vec{A}$ is considered known while vectors $\vec{B}$ and $\vec{C}$ are left as free parameters. Side construction gives the angle of rotation from $\hat{a}$ corresponding to adding given amounts of $\vec{B}$; thus the great circle $\overparen{\hat{a}\;\hat{b}}$ is "calibrated" with values of $|\vec{B}|$ (figure 6-13). Great circle $\overparen{\hat{a}\;\hat{c}}$ is similarly calibrated. Once these great circles have been labelled, the direction of the resultant of $\vec{A}$ plus any combination of $\vec{B}$ and $\vec{C}$ can be determined

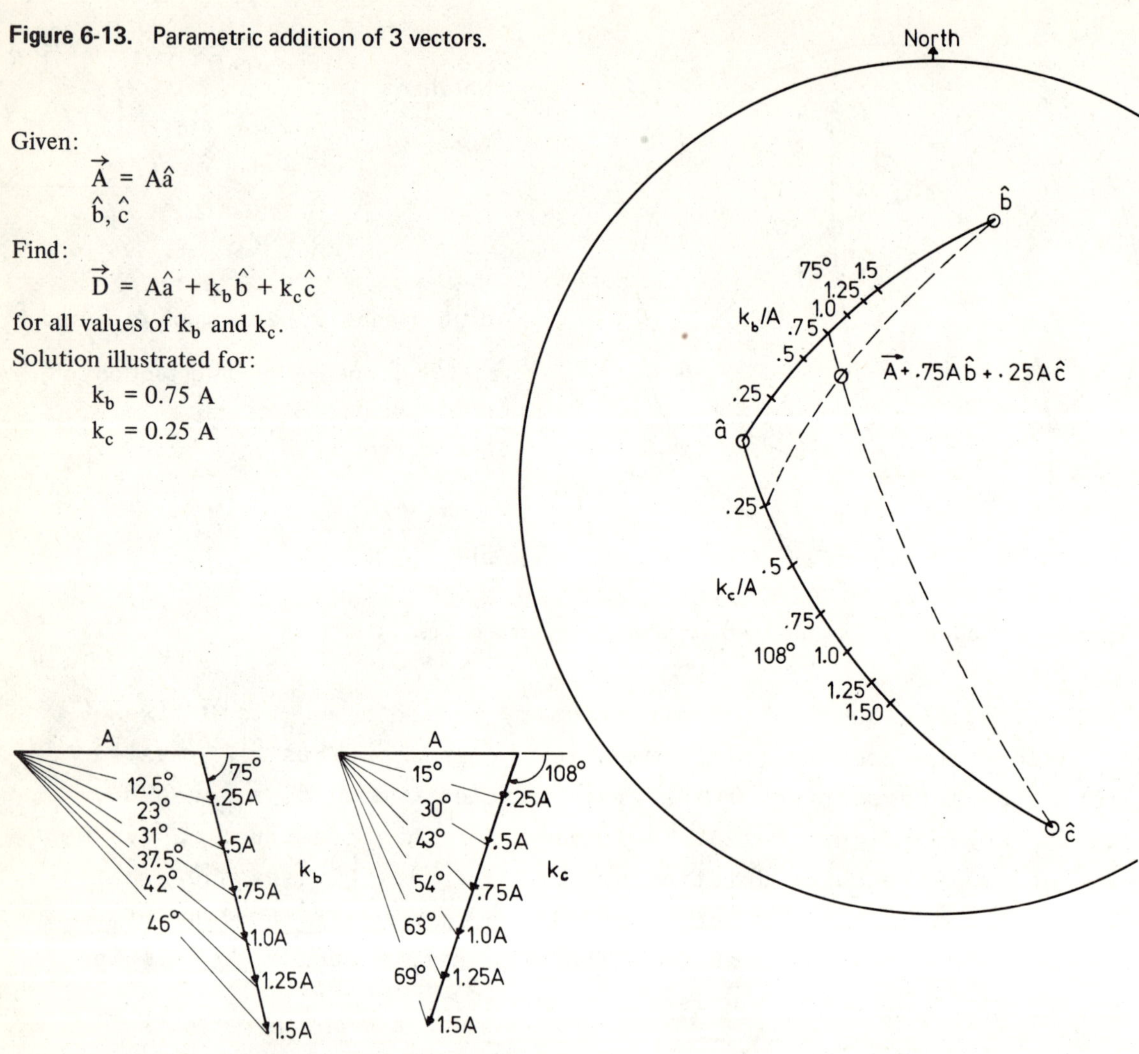

Figure 6-13. Parametric addition of 3 vectors.

by the intersection of great circles, according to the method of figure 6-12, without any further side constructions. This approach is useful in analyzing the stability of blocks of rock which are loaded by water pressure forces on the faces, as for example in the abutments of a dam, in which the magnitudes of water forces will not be known very well during design and will change as the reservoir changes elevation.

The Decomposition of a Vector into Components

Given three basis directions $\hat{b}_1$, $\hat{b}_2$ and $\hat{b}_3$, any vector $\vec{R}$ can be

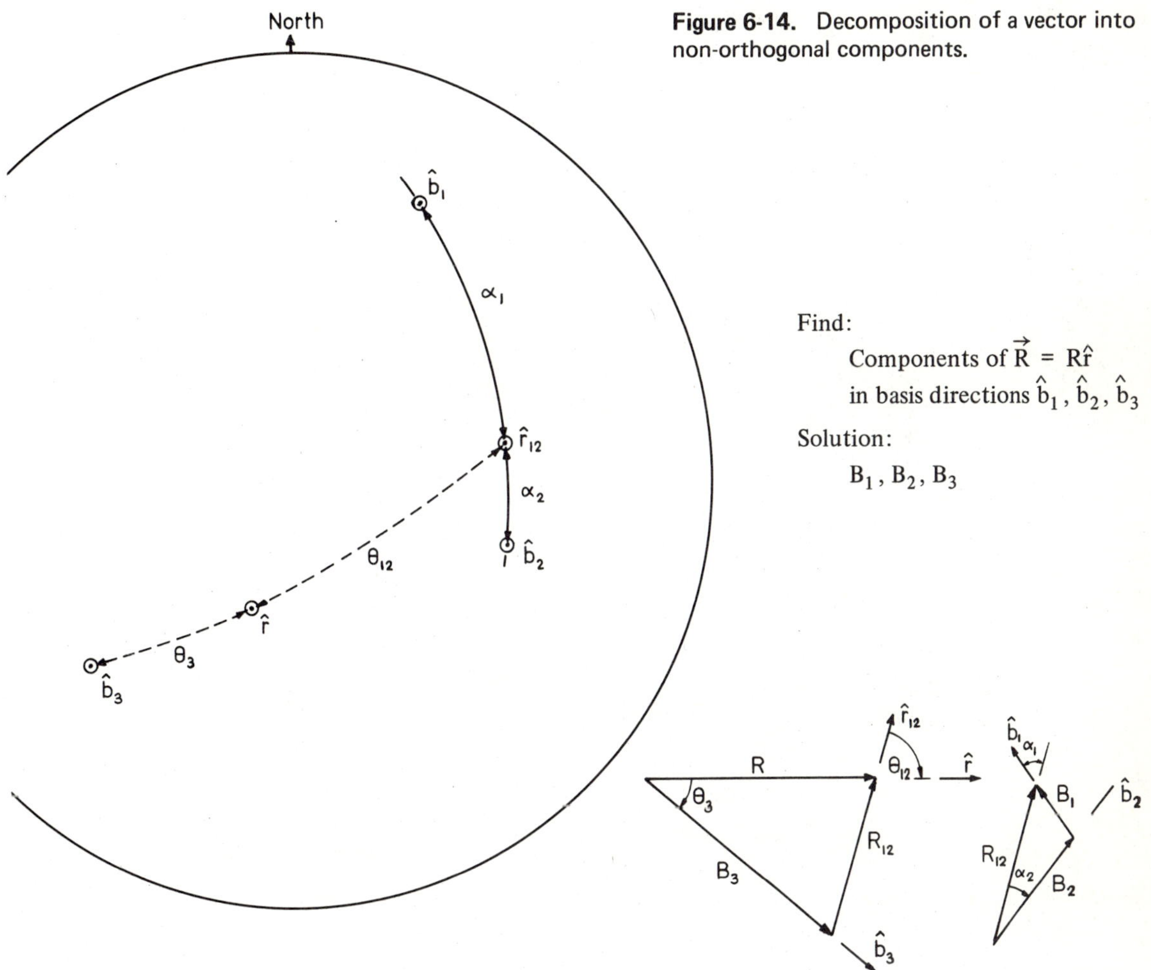

Figure 6-14. Decomposition of a vector into non-orthogonal components.

decomposed into components along them. It is not necessary for the basis to be orthogonal. The method is illustrated in figure 6-14, where $\hat{r}$ is the direction of $\vec{R}$. The angle θ_3 from $\hat{r}$ to $\hat{b}_3$ can be measured in the great circle common to $\hat{b}_3$ and $\hat{r}$. Also construct the great circle common to $\hat{b}_1$ and $\hat{b}_2$. The great circles $\overset{\frown}{\hat{b}_1\ \hat{b}_2}$ and $\overset{\frown}{\hat{b}_3\ \hat{r}}$ intersect at $\hat{r}_{12}$; read angle θ_{12} between $\hat{r}_{12}$ and $\hat{r}$. Now, knowing the angles θ_3 and θ_{12}, construct a triangle of forces to decompose $\vec{R}$ into $\vec{B}_3$, (the component in the direction $\hat{b}_3$) and $\vec{R}_{12}$, (the component in the plane common to $\hat{b}_1$ and $\hat{b}_2$). Then reading the angles α_1 and α_2 from $\hat{r}_{12}$ to $\hat{b}_1$ and $\hat{b}_2$ respectively, decompose $\vec{R}_{12}$ into the components $\vec{B}_1$ and $\vec{B}_2$, as shown in figure 6-14. $\vec{B}_1$, $\vec{B}_2$ and $\vec{B}_3$ are the required components.

Figure 6-15. The largest tetrahedral wedge defined by two sets of discontinuities.

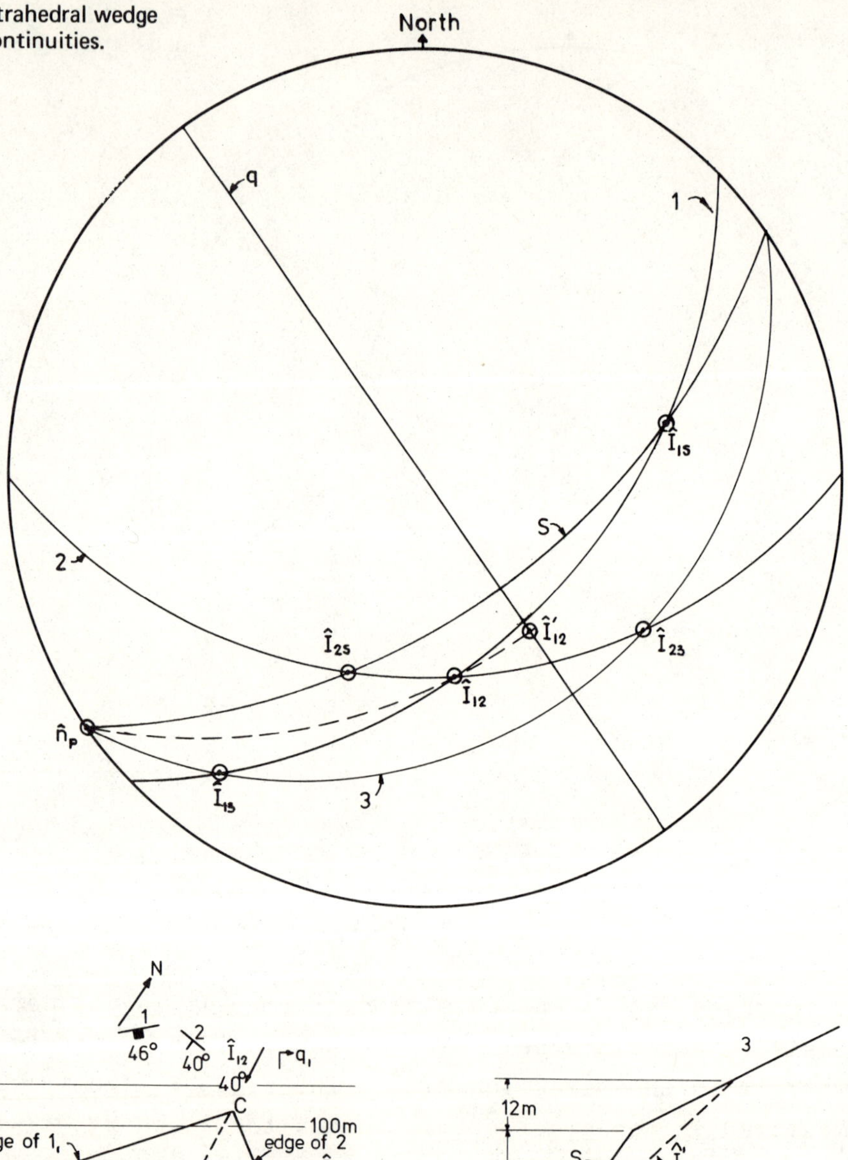

North
q
1
2
S
3
$\hat{I}_{1S}$
$\hat{I}'_{12}$
$\hat{I}_{23}$
$\hat{I}_{2S}$
$\hat{I}_{12}$
$\hat{n}_P$
$\hat{I}_{13}$

(15 a)

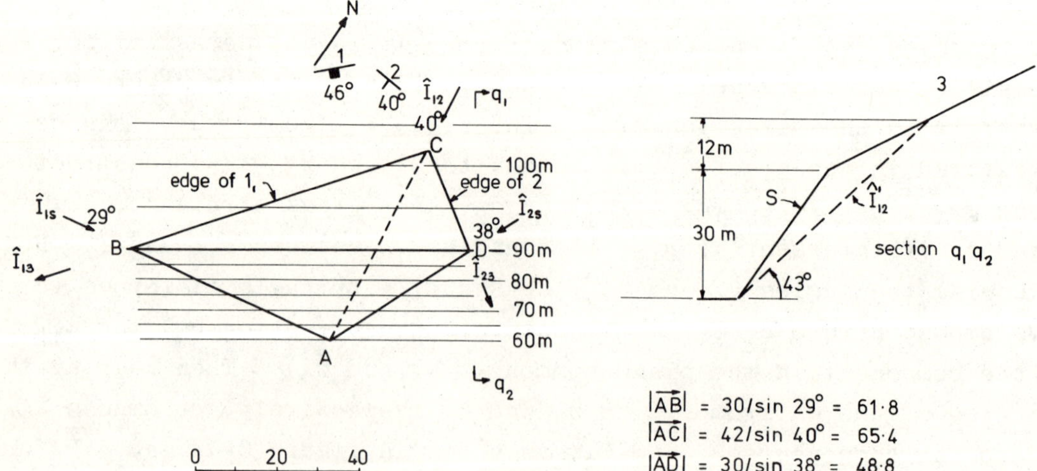

(15b)

APPLICATION OF THE STEREOGRAPHIC PROJECTION IN DEFINING A WEDGE FORMED BY INTERSECTING DISCONTINUITIES

Figure 6-15a presents a hypothetical example of a rock mass containing a system of joints and beds in various directions. A cut to be excavated parallel to plane S will produce daylighting intersections of the joints and beds as shown in figure 6-15b. It can be shown that in the case of a material which has both frictional and cohesive strengths, the most critical member of a set of similar unblocked wedges will be the largest of the set. If the cutting has a specific height, this wedge is uniquely defined; thus even though there are multiple similar blocks formed by the intersections of the joints and beds in this example, for the case of limiting equilibrium only one needs to be analyzed.

In figure 6-15b the surface of the cut is S, and the original slope, which strikes parallel to S, is called plane 3. (The solution would be the same if plane 3 were the underside of another block sitting on top of the wedge.) From the stereographic projection, determine the intersections of the joints and the beds ($\hat{I}_{12}$) and the intersections of each of these discontinuities on the cutting ($\hat{I}_{1S}$) and ($\hat{I}_{2S}$). In the plan view (left half of figure 6-15b) the bearings of $\hat{I}_{1S}$ and $\hat{I}_{2S}$ can be laid off from a vertex A chosen anywhere along the base of the 30 meter high excavation. Also construct the bearing of the line of intersection $\hat{I}_{12}$; then, from the stereographic projection, determine the intersections $\hat{I}_{13}$ and $\hat{I}_{23}$ of planes 1 and 2 with plane 3, the top of the slope surface; laying off the bearings of $\hat{I}_{13}$ and $\hat{I}_{23}$ from points B and D completes the plan view of the wedge. As shown in figure 6-15b, the magnitudes of the vectors $\overrightarrow{AB}$, $\overrightarrow{AC}$ and $\overrightarrow{AD}$ can be computed from the scaled lengths on the drawings and from the known plunge angles listed in figure 6-15b. To check the accuracy of the construction, one can also determine point C by constructing a vertical section in plane q perpendicular to the cut as shown in the right half of figure 6-15b. In this section one does not see the true plunge of the line of intersection $\hat{I}_{12}$ but rather its orthographic projection $\hat{I}_{12}'$ in the plane of the section (the orthographic projection of a line in a plane was shown in figure 3-12).

Figure 6-16. (a) Volume and area of faces for the wedge of figure 6-15. (b) The centroid of the wedge of figure 6-15.

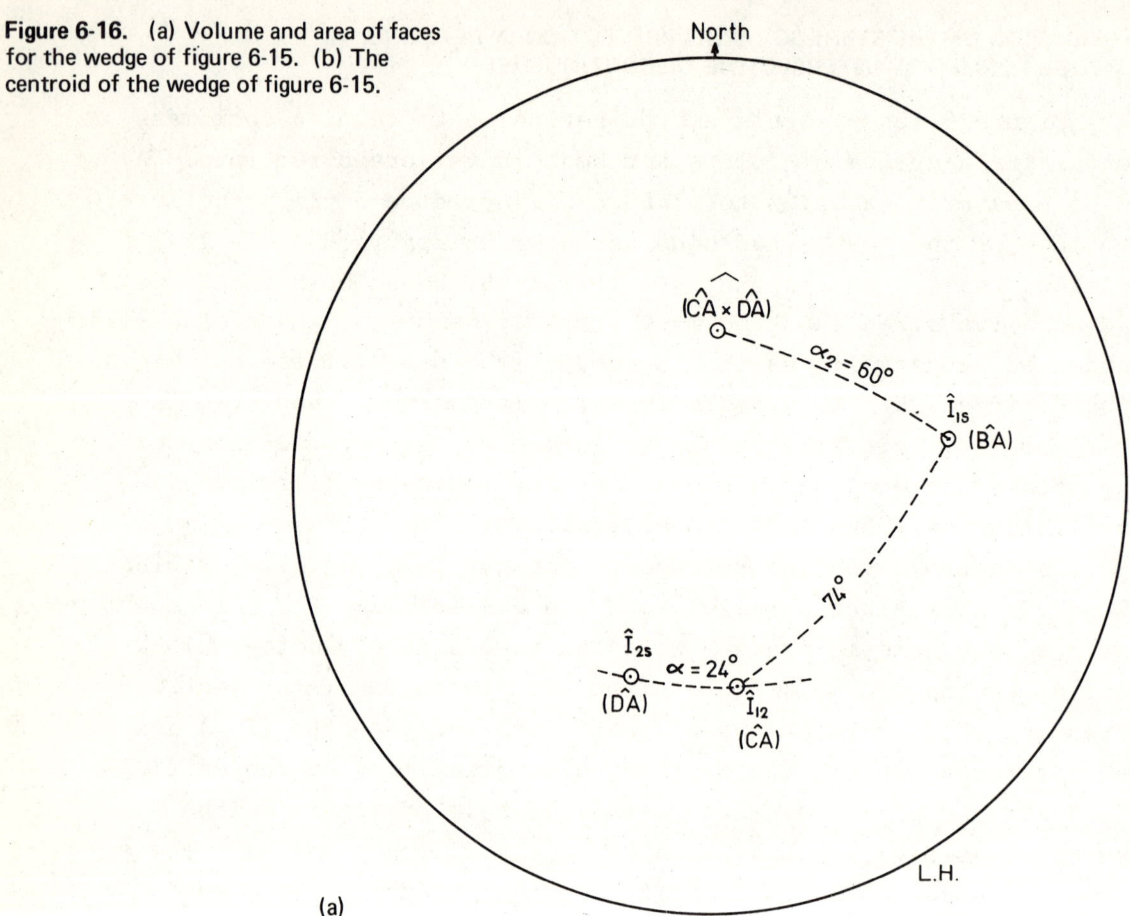

We will need to determine the volume of the wedge ABCD and the area of its contacting faces ABC and ACD. The volume of a tetrahedron can be calculated with the scalar triple product of any three edges radiating from a given corner. Since we know the vectors $\overrightarrow{AB}$, $\overrightarrow{AC}$ and $\overrightarrow{AD}$, we can therefore determine the volume:

$$V = \frac{1}{6} \left| \overrightarrow{AB} \cdot \overrightarrow{AC} \times \overrightarrow{AD} \right| \qquad (2)$$

as shown in figure 6-16a. The operation of the scalar triple product given above is a combination of operations previously defined; it

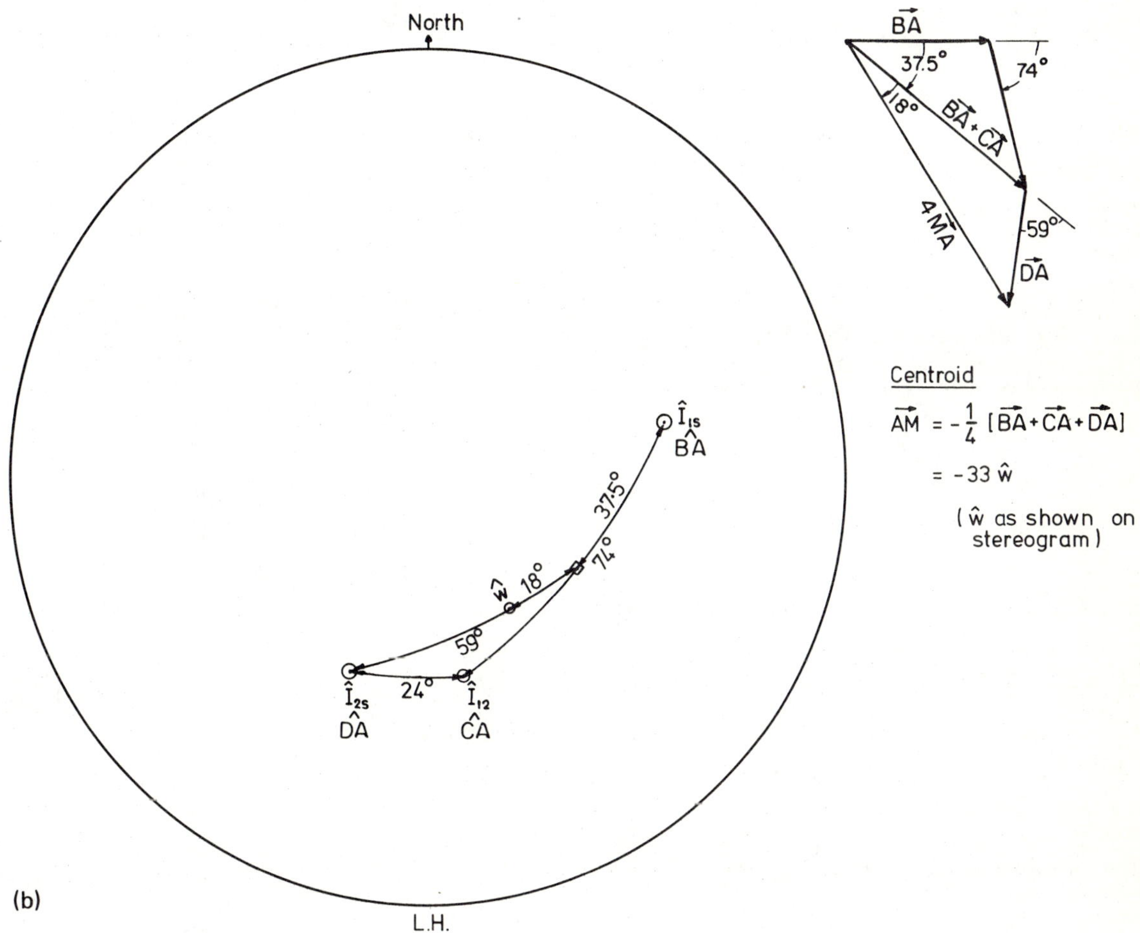

will be necessary only to determine the angles involved for the dot product and the cross product. In this example, the wedge is 6,700 cu. meters in size (knowing the unit weight of the rock, its weight is also determined). The areas of the faces ABC and ADC are calculated by half the magnitude of the cross product of vectors along the edges of the two faces as given in figure 6-16a. For example, the area ABC is $1/2\ |\ \overrightarrow{AB} \times \overrightarrow{AC}\ |$. If the cohesion along the face is known, the maximum cohesive force on each face can be determined from the known areas. The centroid of the tetrahedral wedge can be determined by a vector addition according to the formula

$$\overrightarrow{AM} = \frac{1}{4}\left(\overrightarrow{AB} + \overrightarrow{AC} + \overrightarrow{AD}\right) \tag{3}$$

where $\overrightarrow{AM}$ is the vector from corner A to the centroid of the tetrahedron. The sum of three vectors was discussed previously. In figure 6-16b, the summation is done by a repeated operation of the summation of two vectors; the great circle method discussed earlier for summation of three vectors is not necessary here since all three vectors are known absolutely.

Position Vectors to Fix the Line of Action of a Force

Until now we have considered only free vectors. To fix the position of a vector, one can describe a family of "position vectors" radiating from a known point, whose tips all lie along the desired line of action. In figure 6-17a: force $\vec{F}$ is known in magnitude and direction; point O is a known point near the region of interest; and A is a point known to be on the line of action of force $\vec{F}$. Then the vector $\overrightarrow{OA}$ will have its tip in the line of action of force $\vec{F}$, as will the vector sum $\overrightarrow{OA} + K\vec{F}$ where K is any positive or negative number. The family of vectors $\overrightarrow{OA} + K\vec{F}$ can be considered the vector description of the line of action of $\vec{F}$.

The Vector Equation of a Plane

A family of position vectors can also establish the position of a plane (figure 6-17b). Consider plane p with normal in direction $\hat{n}_p$. For any point C in plane P, the scalar $\overrightarrow{OC} \cdot \hat{n}_p$ is constant. This constant is determined by any known point C_1, in the plane, giving as the equation of the plane

$$\overrightarrow{OC} \cdot \hat{n}_p = \overrightarrow{OC_1} \cdot \hat{n}_p \tag{4}$$

The Piercing Point of a Force on a Plane

By equating the line of action of a force and the equation of a plane, one determines the particular position vector whose tip lies at the piercing point of the force on the plane. Figure 6-18 presents an example and shows how the resulting vector equation can

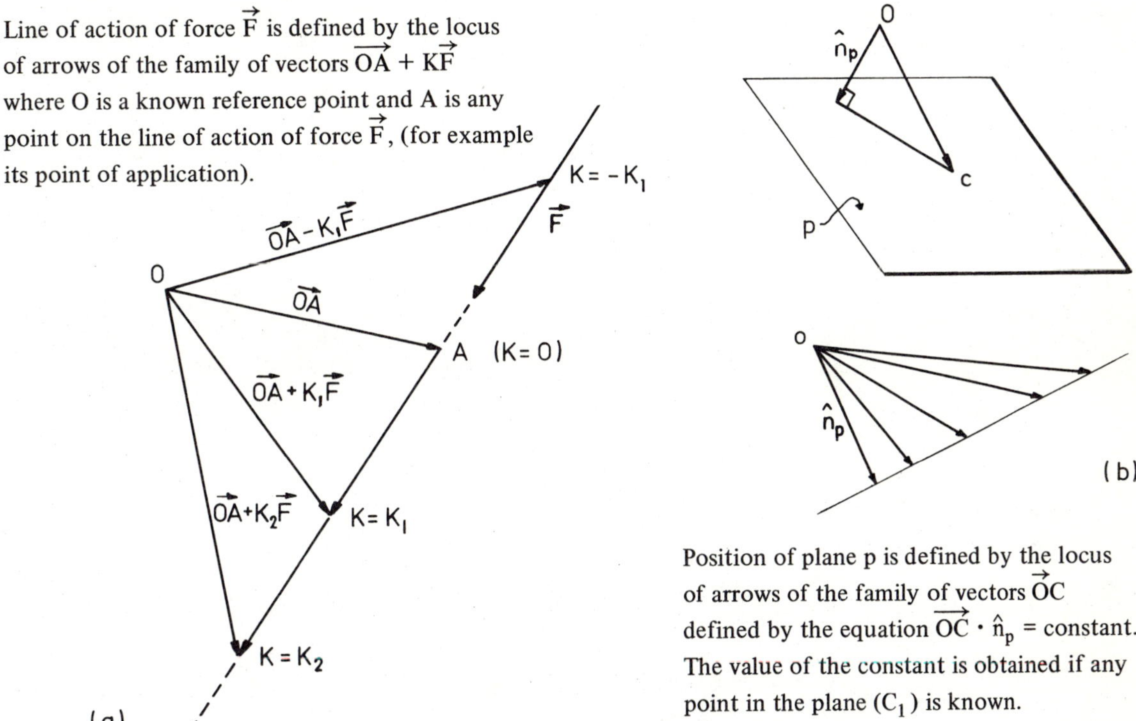

Figure 6-17. Position vectors defining the line of action of a force and the equation of a plane.

be solved iteratively on the stereonet. In this example we wish to find the point where the weight of the wedge of figure 6-15 pierces plane 1. The line of action of the weight force passes through the centroid M whose position is known with respect to point A of the wedge, figure 6-16b. Let point D of figure 6-15b serve as reference. The vector $\overrightarrow{DM}$ can be found by vector subtraction, as shown in figure 6-18b, $\overrightarrow{DM} = \overrightarrow{DA} - \overrightarrow{MA}$. We denote the vertically downward direction as $\hat{Z}$; then the equation of the line of action of the weight force is $\overrightarrow{DM} + K\hat{Z}$. We know the location of point A in the plane so the constant of the equation of plane 1 is determined (constant = $\overrightarrow{DA} \cdot \hat{n}_1$). The piercing point T on the plane is given by a position vector $\overrightarrow{DT}$ found by the value of K giving a solution to

$$(\overrightarrow{DM} + K\hat{Z}) \cdot \hat{n}_1 = \overrightarrow{DA} \cdot \hat{n}_1 \tag{5}$$

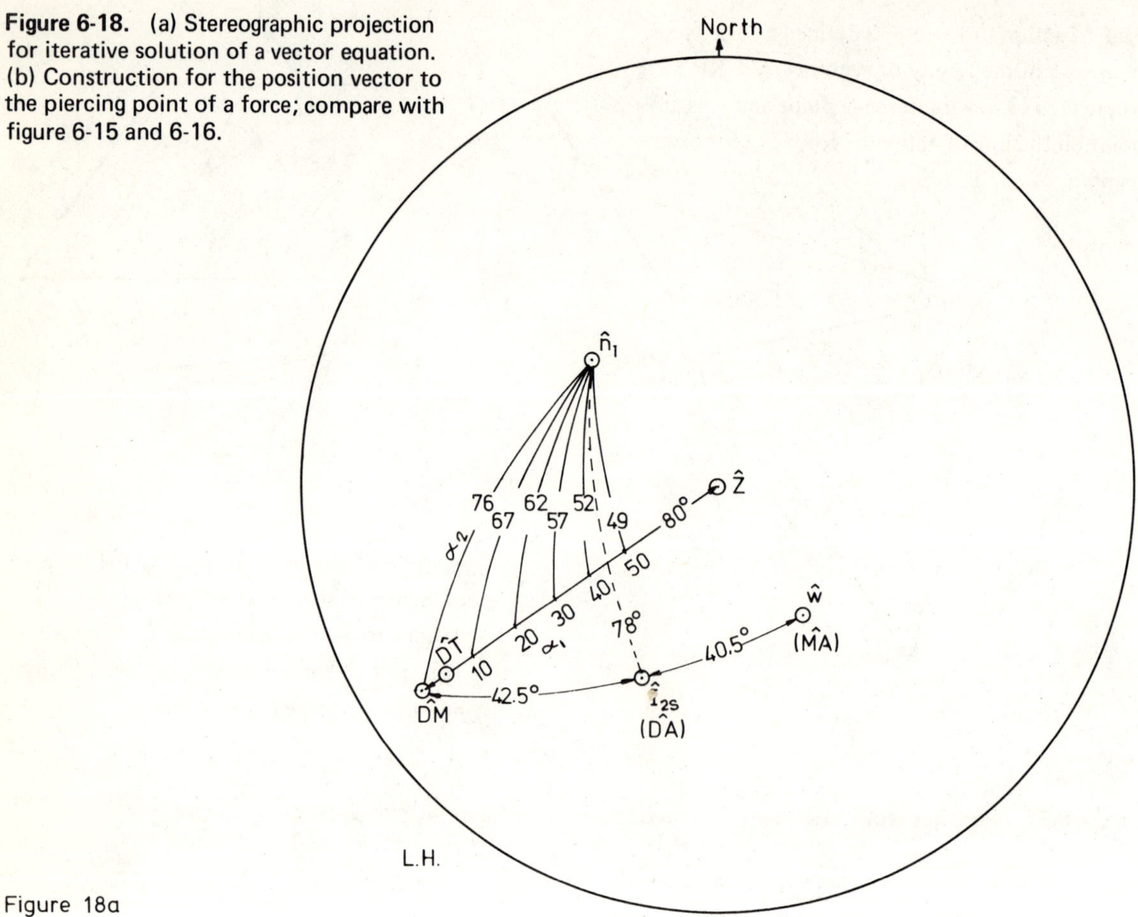

Figure 6-18. (a) Stereographic projection for iterative solution of a vector equation. (b) Construction for the position vector to the piercing point of a force; compare with figure 6-15 and 6-16.

This equation can be solved iteratively on the stereographic projection together with a side construction. For every value K, the resultant $\overrightarrow{DM} + K\hat{Z}$ will lie somewhere along the great circle between the directions of $\overrightarrow{DM}$ and $\hat{Z}$, as shown in figure 6-18a. At each of a number of points along this great circle, the angles to $\overrightarrow{DM}$ and to $\hat{n}_1$ are read from the stereonet and plotted as shown in the upper right of figure 6-18b. The dot product of two vectors is a scalar quantity; it is shown by a circle of appropriate radius about the point D in the lower left of figure 6-18b. In this example, the value $\overrightarrow{DA} \cdot \hat{n}_1$ is equal to 10.2. For a value of K = 0 we can draw the vector $\overrightarrow{DM}$ as shown in the lower left of figure 6-18b and form its dot product so as to produce a value of 10.2 by drawing a tangent to

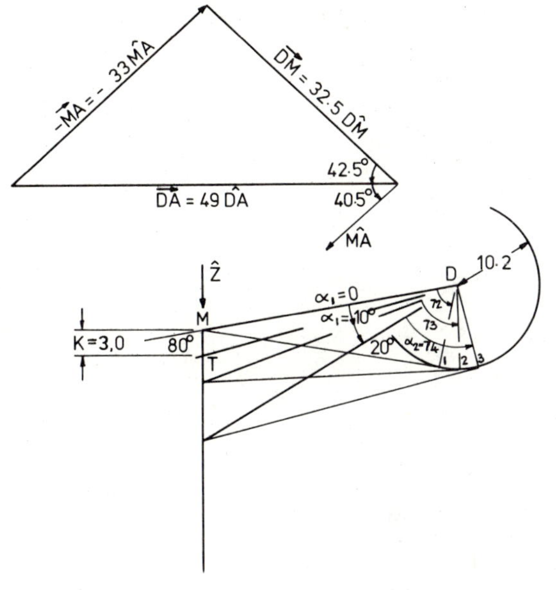

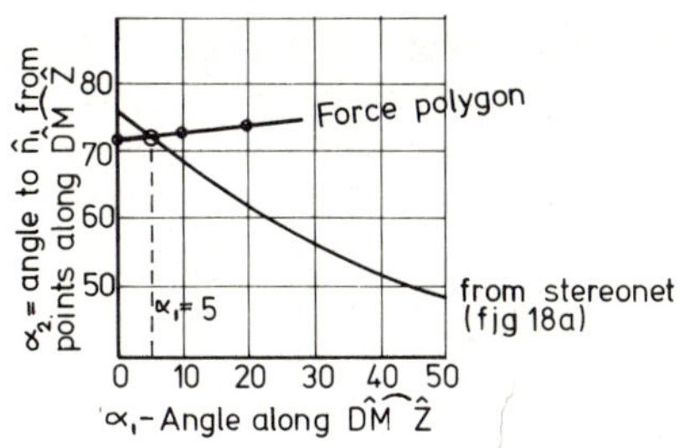

Iterative Solution

Piercing point (T) of weight force on plane

$$\vec{DT} \cdot \hat{n}_1 = (\vec{DM} + K\hat{Z}) \cdot \hat{n}_1 = \vec{DA} \cdot \hat{n}_1$$

$$|\vec{DA}| = 49; \ \angle \text{ DA to } \hat{n}_1 = 78°$$

$$\vec{DA} \cdot \hat{n}_1 = 49 \cos 78 = 10.2$$

From force polygon and stereonet,

$$\alpha_1 = 5° \text{ giving } k = 3.0.$$

$|\vec{DT}| = 33$; $\hat{DT}$ shown on stereogram.

Figure 18b

the scalar circle from point M, meeting the circle at point 1. Corresponding to the choice $\alpha_1 = 0$ that we have made in taking the value of K = 0, we have found the angle MD1 (angle α_2') equal to 72°. Corresponding to each assumption of α_1, there will be another value of α_2' and these can be plotted as in the upper right of figure 6-18b. The intersection of the two curves in the upper right figure of 6-18b defines the value of α_1 and therefore the value of K; K is equal to 3.0 and α_1 equals 5°. The vector $\hat{DT}$ can now be plotted on the stereographic projection.

ANALYSIS OF ROTATION

The previous operations with vectors permit analysis of rotational tendencies in rock wedges. A wedge may be able to topple about its edges or rotate while sliding in a given face, as discussed by Wittke (1965), Goodman and Taylor (1967) and Londe, Vigier and Vormeringer (1969). In this section we will explore the operations necessary to analyze the general problem of rotation about an axis.

Figure 6-19. (a) Rotation in a plane. (b) Analysis of rotation in a plane (see figure 6-19a). (c) Stereographic projection for analysis of rotation.

(a)

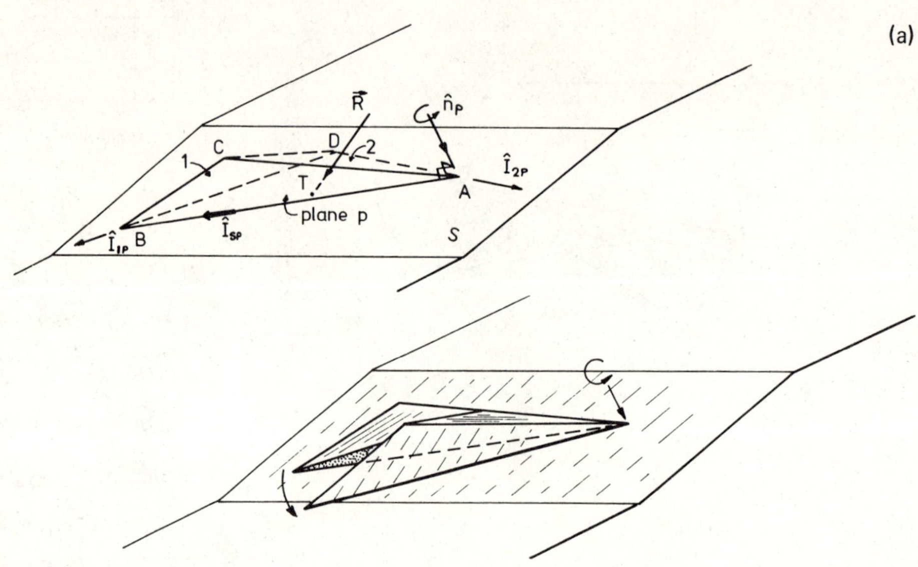

(b)

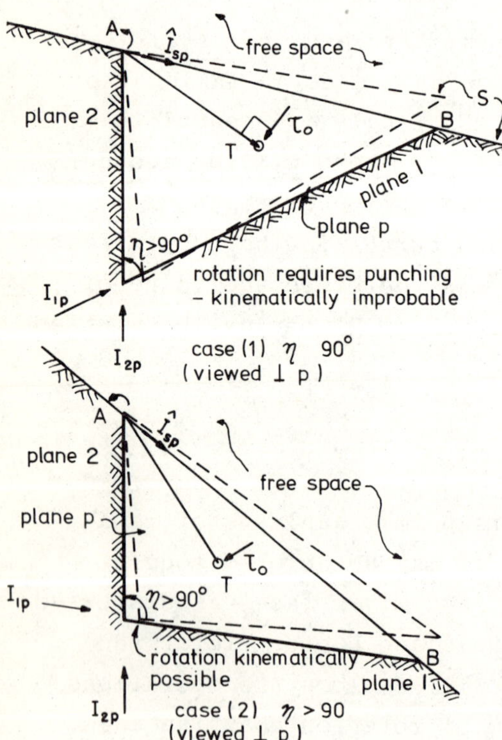

Kinematics of rotation about A

Angle between edges of face p, measured in plane p, must be greater than 90° for rotation (See figure 19c).

$$N > 90^\circ$$

Analysis of rotation about A

Axis through A ⊥ to plane p is in direction $\hat{n}_p$.

T is piercing point of resultant $\vec{R}$ on p. Its location is given by vector $\vec{AT}$.

Overturning moment M_A caused by $\vec{R}$, which acts at T is

$$\vec{AT} \times \vec{R} \cdot \hat{n}_p$$

Maximum resisting moment M_R depends upon the stress distribution. Assume resisting force τ_o is concentrated at T,

$$\tau_{o_{max}} = (\vec{R} \cdot \hat{n}_p) \tan \phi_F$$

$$M_R = \tau_o \cdot |\vec{AT}|$$

Stability requires $M_R > M_A$

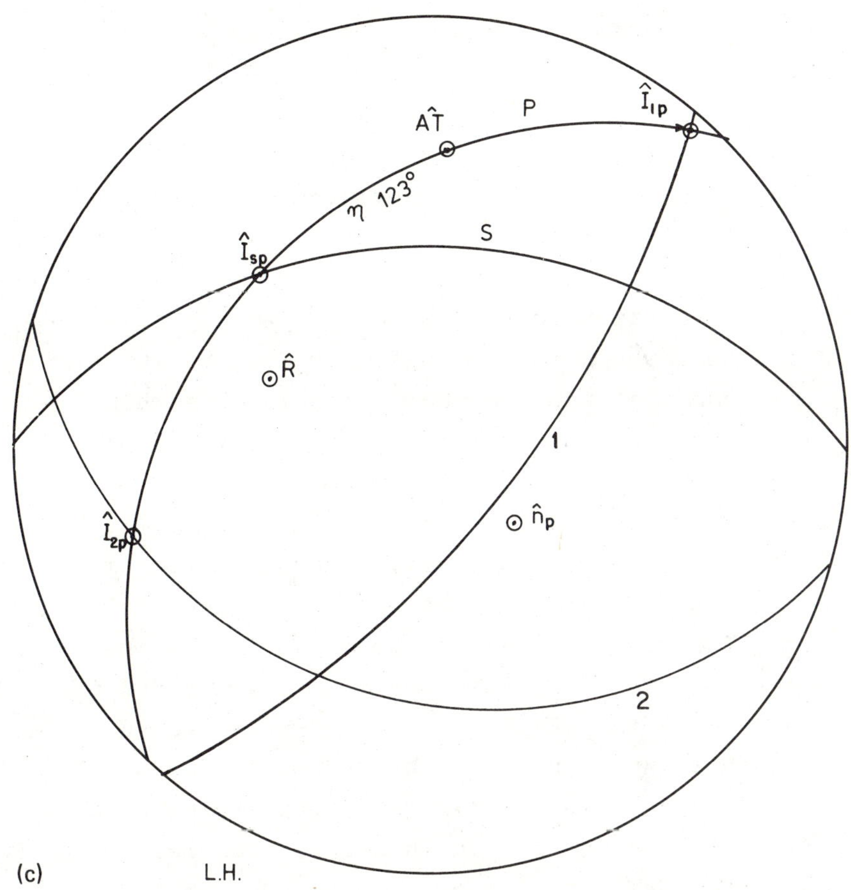

(c) L.H.

In the case of a free block, a translational movement can be viewed as a rotation about an axis directed perpendicular to the plane of sliding and located at infinity; therefore, rotation about any finite lever arm must be less critical than translation. However, if the wedge is not uniform in its properties, for example if it has anisotropic shear strength properties or if it is restricted in its translation in particular directions, then rotational sliding may become critical.

Consider the wedge of figure 6-19a with rotational tendency about the corner A. The equation for the moment of the force $\vec{R}$ about an axis in direction $\hat{d}_A$ through point A is given by $M_A = \overrightarrow{AT} \times \vec{R} \cdot \hat{d}_A$ where T is a point on the line of action of $\vec{R}$. If the moment is positive it means that the sense of rotation is clockwise as viewed from the tail of the vector $\hat{d}_A$. We will examine the moment for the rotational sliding mode of figure 6-19a, that is rotation about an axis ($\hat{n}_p$) perpendicular to plane p through point A. Force $\vec{R}$ pierces plane p at point T. The moment about the axis shown, M_A, is equal to $\overrightarrow{AT} \times \vec{R} \cdot \hat{n}_p$. If $\hat{n}_p$ is pointed into the lower hemisphere, as in this example, then a negative moment corresponds to kinematically possible rotation. If we view the plane of rotational sliding as in figure 6-19b, it will be seen that rotation about A is possible only if the angle η shown is greater than 90°; η is the angle between the intersections $\hat{I}_{1p}$ and $\hat{I}_{2p}$, as shown in figure 6-19c. If η is less than 90°, rotation is possible about an instantaneous center located at the intersection of perpendiculars to plane 2 and plane 1 drawn from the points of contact, as discussed by Wittke, and Londe et al; this case will not be considered here. The driving moment about the axis perpendicular to plane p through A can be calculated from the formula presented above. The resisting moment which must be compared to the overturning moment is more difficult to discuss, as it depends upon the stress distribution in the plane. Dr. John Bray* showed that the most critical condition can be found by assuming that the entire resisting force in plane p is mobilized at the piercing point (T) of the resultant force $\vec{R}$ on plane p. In this case, the resisting moment $M_R = \tau_o$ times distance $|\overrightarrow{AT}|$; τ_o is equal to $\vec{R} \cdot \hat{n}_p \tan \phi_p$. Stability requires that M_R be greater than M_A and therefore the frictional requirements for equilibrium for overturning can be examined.

The method presented above only makes sense if the piercing point of the force $\vec{R}$ intersects the plane of rotational sliding on one of the faces of the wedge. It is convenient to examine this requirement using the stereographic projection. Figures 6-19c and 20a present alternative methods of doing this. With respect to figure 6-19b, the piercing point T will lie in plane p if $\hat{AT}$ is between $\hat{I}_{sp}$ and $-\hat{I}_{2p}$ and the vector $\hat{TB}$ lies between the intersections $\hat{I}_{sp}$ and $\hat{I}_{1p}$. $\overrightarrow{TB}$ can be found by vector subtraction: $\overrightarrow{TB} = \overrightarrow{AB} - \overrightarrow{AT}$. The method of figure 6-20a can also be used to

* Personal communication.

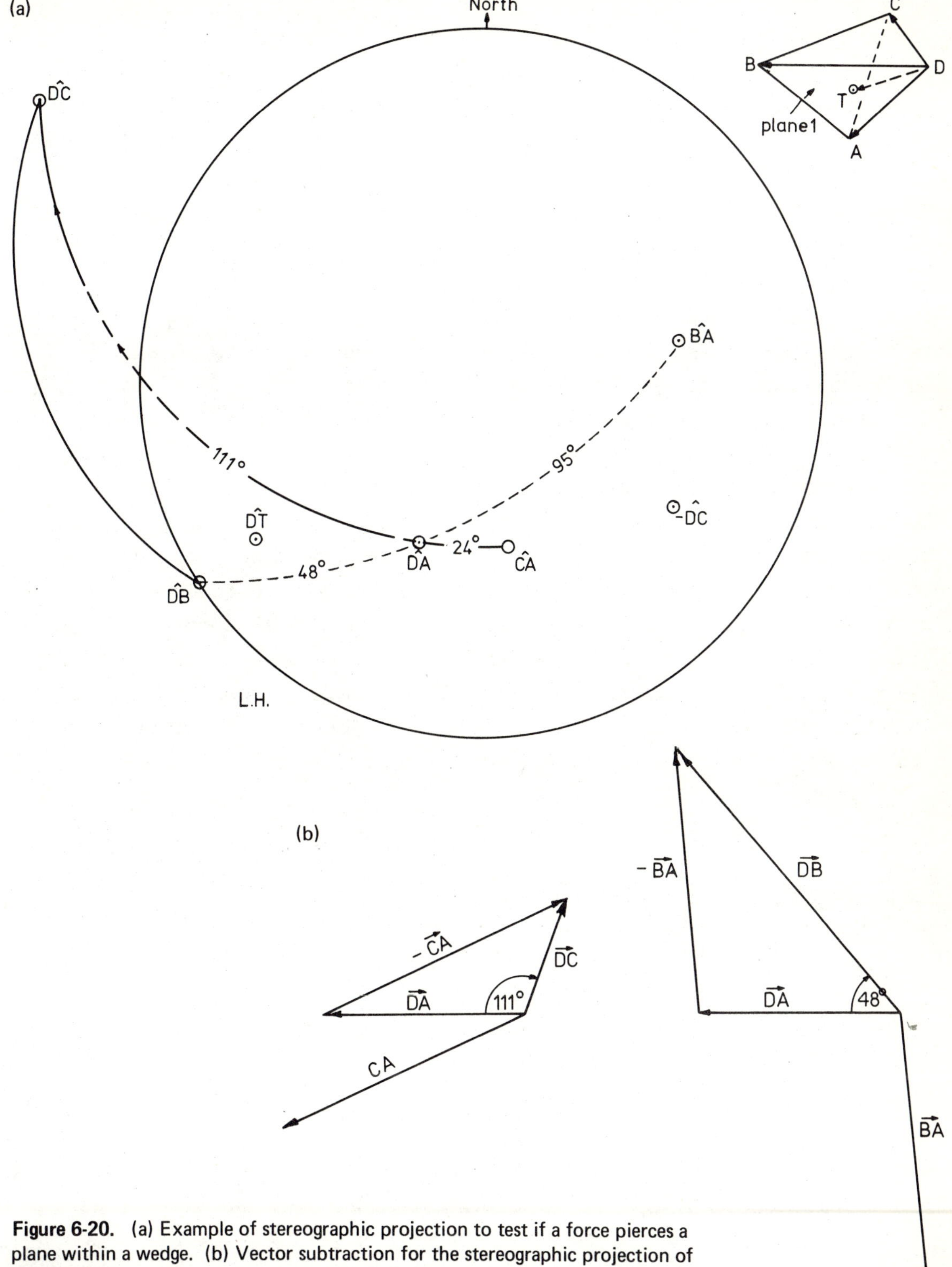

Figure 6-20. (a) Example of stereographic projection to test if a force pierces a plane within a wedge. (b) Vector subtraction for the stereographic projection of figure 5-20a.

satisfy that the piercing point of T on plane 1 is on the face of the wedge. By vector subtraction, form the vectors $\overrightarrow{DA}$, $\overrightarrow{DB}$ and $\overrightarrow{DC}$ as well as the vector $\overrightarrow{DT}$ to the piercing point from D. The unit vectors representing the directions of the edge vectors $\overrightarrow{DA}$, $\overrightarrow{DB}$ and $\overrightarrow{DC}$ form three points on the stereographic projection which, when connected by great circles, define a spherical triangle. If the direction of $\overrightarrow{DT}$ is inside this spherical triangle, the piercing point is on the face of the wedge. In this figure, the piercing point of the weight force has been determined as discussed in figure 6-15, and is found to intersect plane 1 of the wedge.

If one were to add another force (to the problem of figure 6-15) which did not happen to be coplanar with the weight force, it would not be possible to form a single resultant to the applied forces on the wedge. Instead, first determine the piercing point of each force in turn (points P and Q in figure 6-21); then, by taking the dot product of each force with the normal to the plane in question, calculate the contribution of each force to the total normal force on the plane. In this example, the force $\vec{F}_1$ pierces plane 1 at P while force $\vec{F}_2$ pierces plane 1 at Q. The normal forces produced by $\vec{F}_1$ and $\vec{F}_2$ are $\vec{N}_1$ and $\vec{N}_2$ respectively. By taking moments about any point in the plane, for example point A, we will find the distance

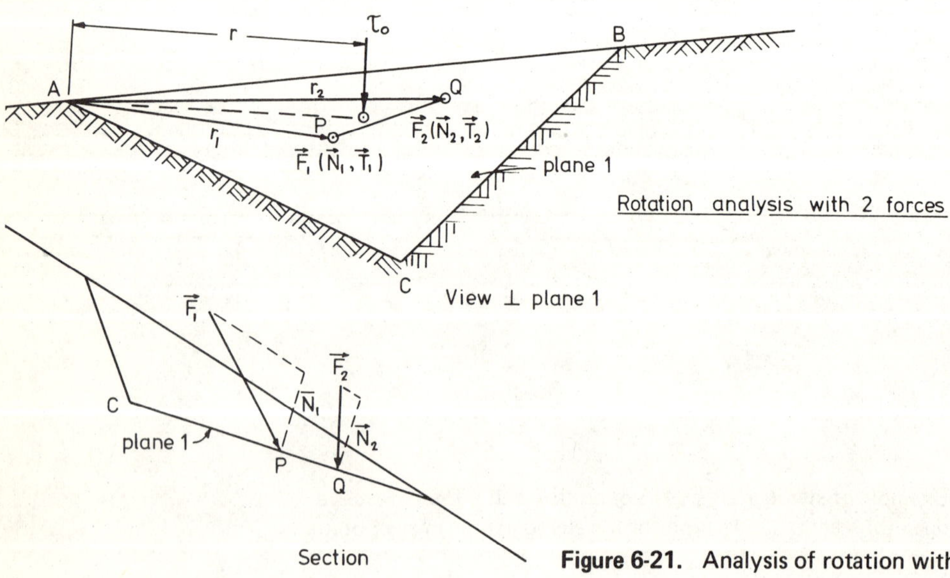

Figure 6-21. Analysis of rotation with multiple forces.

r from A at which $|\vec{N}_1 + \vec{N}_2|$ acts with the same moment as the sum of $\vec{N}_1$ and $\vec{N}_2$.

$$r = \frac{N_1 r_1 + N_2 r_2}{N_1 + N_2} \tag{6}$$

where r_1 and r_2 describe the location of P and Q, figure 6-21. The most critical stress distribution of the resisting force on plane 1 corresponds to the case where the entire frictional resistance τ_o is concentrated at a radial distance r from A. (It does not matter in what direction as long as it is in the face ACB) If M_A is the combined moment of $\vec{F}_1$ and $\vec{F}_2$ about A, determined by repeated application of the moment equation, then the condition for safety is:

$$\tau_o = M_A/r \leq (N_1 + N_2) \tan \phi_{available} \tag{7}$$

where $\phi_{available}$ is the friction angle available on plane 1.

ANALYSIS OF SLIDING OF A BLOCK ON A PLANE–THE FRICTION CIRCLE CONCEPT

Under pure friction, the limit of static equilibrium of a block resting on a plane occurs when the resultant force of the block on the plane becomes inclined at angle ϕ from the normal directed out of the block into the plane. Therefore a block free to move in any direction but which is actually at rest on a plane p, must have a resultant force oriented inside a circular cone of vertex angle 2ϕ centered about the normal to p (figure 6-22). Plotting this "cone of static friction" on the stereographic projection produces a circle about the normal $\hat{n}_p$, as shown in figure 6-22b. The portion of the whole sphere inside the friction circle will be called the "safe region". Outside of this circle it is not possible for a resultant to find an equilibrium reaction (although it may be convenient to discuss the orientations of potential forces outside of the friction circle). The condition of limiting equilibrium is found when the direction of the resultant force of the block on its support lies exactly on

(a)

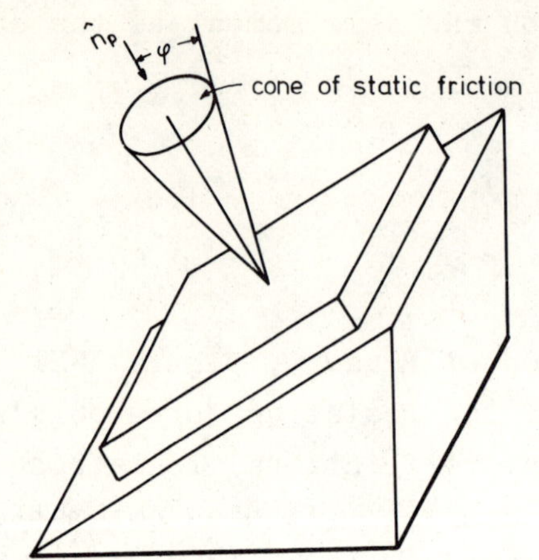

The friction cone about the normal to an inclined plane surface of sliding. If the block is at rest, the resultant force between block and plane is inside the cone. φ is the "friction angle"

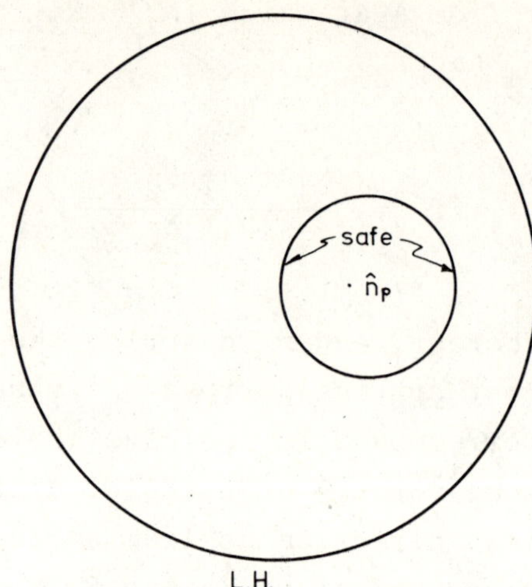

Stereographic projection of friction cone producing a "friction circle"

(b)

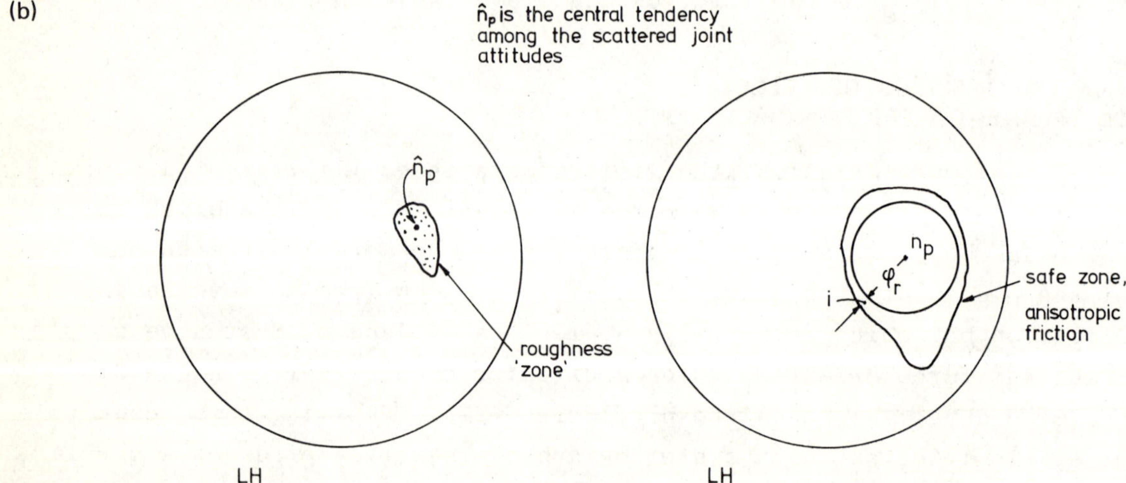

Figure 6-22. (a) The friction circle concept. (b) Generalized friction circle for rough planes.

the circumference of the friction circle. Though simple, this concept is powerful and, as will be shown later, can be generalized for blocks resting on several planes. The basic idea involved is that we sub-divide the whole sphere into a safe region and an unsafe region. The safe region is circular for isotropic friction on a plane. It will be non-circular if the structures in the plane develop anisotropic frictional resistance. As discussed in chapter 5, repeated measurements of the attitude of a rock plane made on a flat plate of given diameter will develop a scatter diagram of poles that can be interpreted to give the roughness angle on the surface in question. If the generally non-circular roughness figure is added to the residual friction value ϕ, it will produce a non-circular safe zone as shown in figure 6-22b. In this illustration, it was necessary to pick a central tendency from within the distribution of poles, presuming it to be a unimodal pole distribution (chapter 3).

While it is convenient to carry out the stability analysis for sliding on a discontinuity having the orientation defined by a single normal ($\hat{n}_p$), discontinuities are actually scattered about this orientation, as discussed in chapter 3. However, it is possible to account for uncertainty of orientation using a reduced friction circle. Suppose that the distribution of normals is such that a probability of occurrence equal to P determines an angle of uncertainty $\psi(P)$ as in figure 3-16. On figure 6-23, a small circle of radius ψ has been

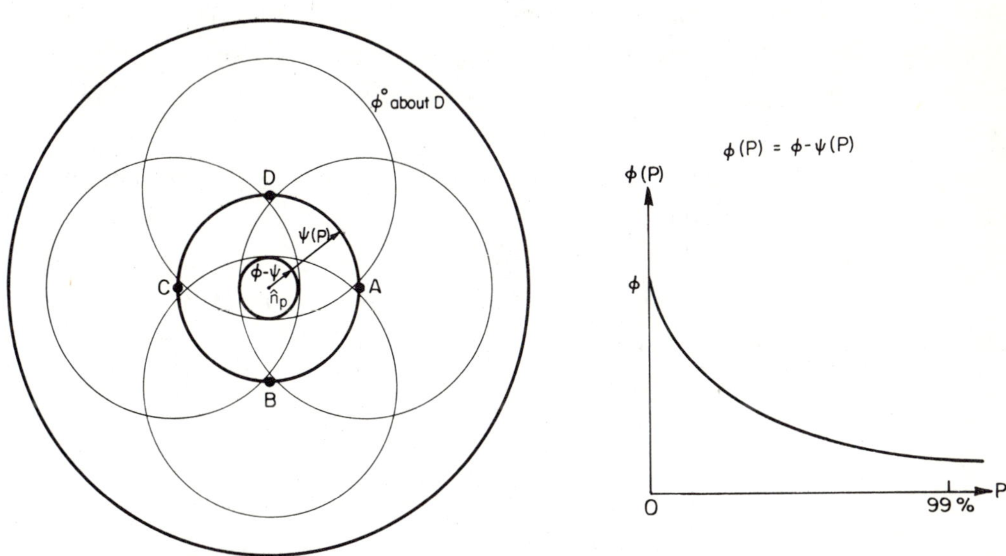

Figure 6-23. The relationship between the friction angle to be used with the mean joint orientation and the probability of occurrence P.

drawn about $\hat{n}_p$, which, for convenience, has been rotated to the center of the lower hemisphere. This small circle, which shall be called the "probability circle" contains (P) x (100%) of the normals to discontinuities of the set being analyzed. We will shrink the safe zone about $\hat{n}_p$ to insure that it is contained in the friction circle of every normal inside the probability circle. Construct a family of friction circles about points on the circumference of the probability circle; the area common to the entire family is a circle of radius $\phi - \psi$. Thus, if we refer the construction of friction circles to the preferred orientation of the set of discontinuities, we must reduce the friction angle to:

$$\phi(P) = \phi_{available} - \psi(P) \tag{8}$$

This conclusion may seem to contradict the construction presented in figure 6-22b, but in that case we were dealing with certain departures from a mean planar orientation for a rigid contacting plate; in figure 6-23, on the other hand, we are concerned with an uncertainty in the actual orientation of the plane itself. It may still be desired to introduce roughness into the probability circle, but then the area common to all the friction circles one can draw within the cone of uncertainty would be rather complex.

What is the relationship between probability of occurrence (P) and the probability of safety (PS)? If the direction of the resultant force on a potentially sliding block plots on the circumference of a friction circle of radius $\phi_{available}$ constructed about $\hat{n}_p$, the probability of safety is one half because half the joints are flatter than the orientation of $\hat{n}_p$. The probability of occurrence (P) that defines such a friction circle about $\hat{n}_p$ is zero. Therefore, the probability of safety and the probability of occurrence are related by $P = 2PS - 1.0$, giving for $PS \geq 0.50$:

$$\phi(PS) = \phi_{available} - \psi(2PS - 1.0) \tag{8a}$$

Applications of the Friction Circle

To illustrate applications of the friction circle concept, various stability calculations will be made with the example of figure

6-24a in which plane p daylights into an excavation, with a potentially sliding block weighing 4,000 tons. The area of the face of the block parallel to p is 200 sq. meters, and the friction angle ($\phi_{available}$) is 30°. Uncertainty in the attitude of plane p can be represented by a normal distribution about the attitude given with K = 120 (see chapter 3). We first examine the condition of stability for the block under its own weight ($\vec{W}$) by constructing a circle of radius 30° about $\hat{n}_p$. Since the weight plots in the center of the lower hemisphere, it lies outside of the friction circle; hence the block is unstable if its weight is the sole contribution to the resultant force between the block and plane p. Rock bolts can be anchored below the slip surface to achieve a desired factor of safety (FS). The factor of safety will be defined as the ratio of the available friction to the friction required for limiting equilibrium under the given forces:

$$FS = \frac{\tan \phi_{available}}{\tan \phi_{required}} \tag{9}$$

A safety factor of 1.0 corresponds to the condition where the resultant force plots on the circumference of the friction circle having radius $\phi = 30^{\circ}$. To achieve this condition, a force can be added to $\vec{W}$ to incline the resultant 20° from vertical. The magnitude of the rock bolt force required to do this depends upon the orientation of the bolts. The required force is smallest if it is directed perpendicularly to the resultant, that is, 20° above the horizontal as measured along the direction of the bearing of $\hat{n}_p$ (figure 6-24b). Since this direction is in the upper hemisphere, its opposite ($-B_{1.0}$) has been shown.

Say a factor of safety of 1.7 is desired. This means that the required friction angle ($\phi_{required}$) is equal to arctan ($\tan \phi_{available}$ / 1.7) or 18.5°. To achieve this the rock bolt force must incline the resultant 31.5° above the vertical along the direction of bearing $\hat{n}_p$. The polygon of forces presented in figure 6-24b, left, determines the required rock bolt forces as 1,400 tons and 2,100 tons respectively for factors of safety of 1.0 and 1.7.

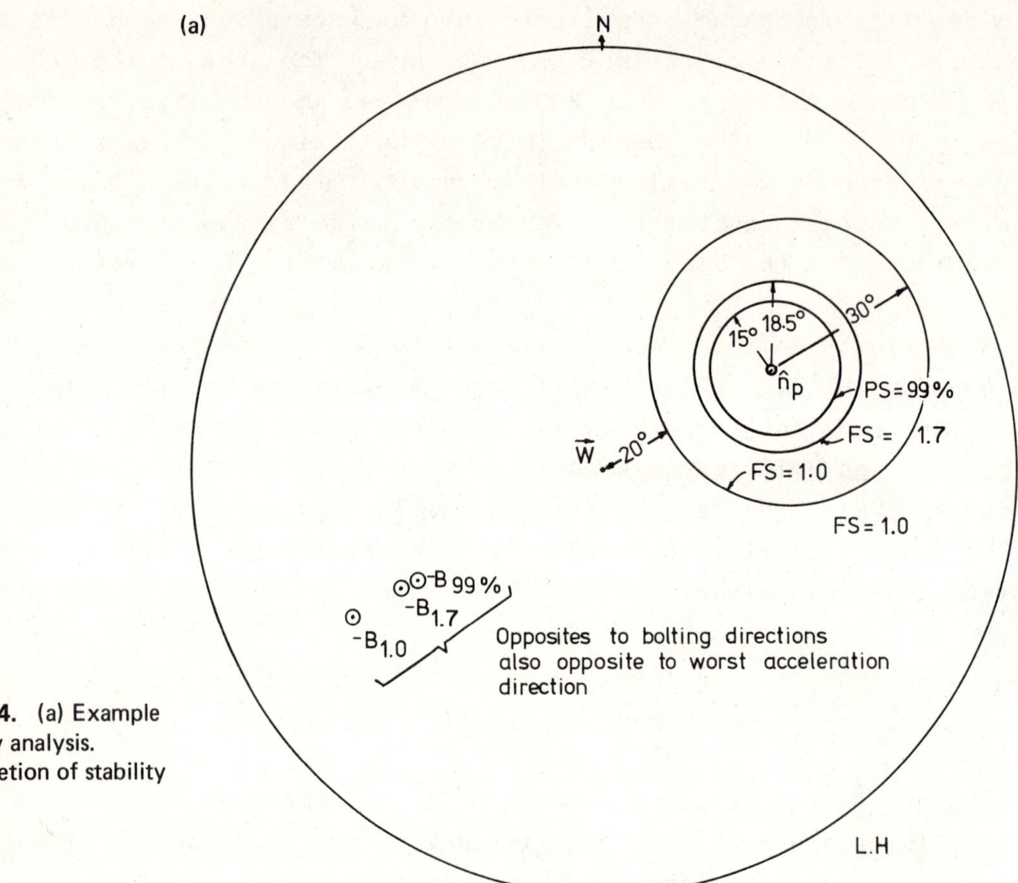

Figure 6-24. (a) Example of stability analysis. (b) Completion of stability analysis.

Though this is the minimum force, and therefore minimizes the required cross-sectional area of steel, it may not be the most economical direction in which to emplace rock bolts in a particular problem; the geometry of the block may dictate other directions for minimum length of drilling necessary to seat the anchorage of the rock bolts well into the support. Every specific case will require the determination of an orientation optimizing the economic factors related to the area and expense of steel on the one hand, and the length and cost of drilling on the other.

Now let us design the rock bolts for a given probability of safety. To design for a probability of safety (PS) equal to 0.99, for example, compute ψ corresponding to P = 0.98. With K = 120, equation 2 in chapter 3 will give ψ (.98) equal to 15°; that is, 98%

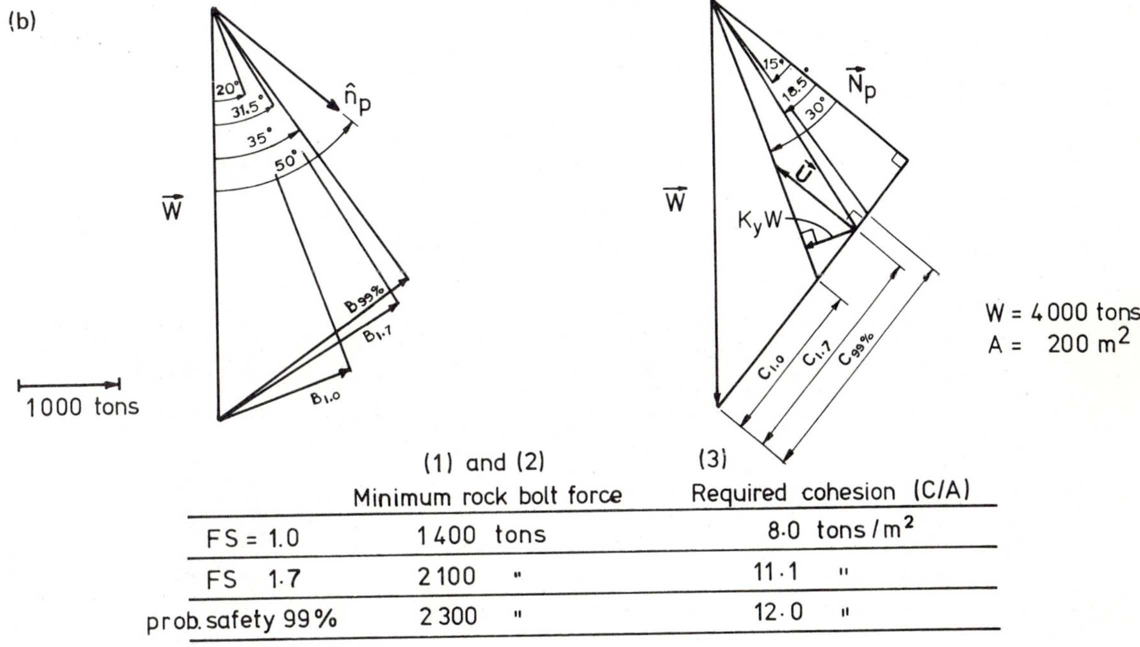

	Minimum rock bolt force	Required cohesion (C/A)
FS = 1.0	1400 tons	8.0 tons/m²
FS 1.7	2100 "	11.1 "
prob. safety 99%	2300 "	12.0 "

(4) the water pressure to initiate slip from an initial factor of safety of 1·7 = U/A = 5 tons/m²

(5) the smallest acceleration K_yg to initiate slip from an initial factor of safety of 1·7 is directed as shown by the vector K_yW; $K_y = 0.11$

of the poles will be oriented within 15° of the direction $\hat{n}_p$. Therefore, to have 99% probability of safety, construct the friction circle with $\phi = \phi_{available} - 15^\circ = 15^\circ$. This leads to a minimum rock bolt force of 2,300 tons in the direction 35° above the horizontal.

When the distribution of normals is known, there is a probability of safety associated with each factor of safety. Even though one might prefer to design for a certain value of FS, he will profit from computing the corresponding value of PS. In the above example, a probability of safety of 0.96 corresponds to FS = 1.7, while the case with PS = 0.99 agrees with FS = 2.15.

We have considered the plane of sliding to be cohesionless; however this limitation can be overcome by calculating the value of cohesive force needed to reach the desired condition of stability (FS or PS). If the available friction angle is 30°, the resultant of the normal force and friction on the plane can be inclined as much as 30° from the normal. If the factor of safety is to be 1.0 without rock bolts, the frictional resistance will have to be augmented by a cohesion of 8.0 tons per square meter, as shown in figure 6-24b. To achieve a factor of safety of 1.7, the resultant of frictional resistance and normal force will be inclined at 18.5° from the normal whereupon the required cohesion is determined respectively as 11.1 tons per square meter (figure 6-24b). For a probability of safety of 99%, the resultant must be inclined 15° from $\hat{n}_p$, demanding a cohesion of 12 tons per square meter.

It may be necessary to consider static water forces on one or another of the faces of a potentially sliding block. What water force on plane p (figure 6-24) will cause slip assuming an initial factor of safety of 1.7? The water pressure acts in the direction opposite to $\hat{n}_p$. As shown in the right side of figure 6-24b, a factor of safety of 1.7 implies that the resultant is inclined 18.5° from the direction of the outward normal ($\hat{n}_p$).* To produce slip, the force $\vec{U}$ shown in the direction opposite to $\hat{n}_p$ must cause the resultant to be inclined 30° from $\hat{n}_p$. The required water pressure ($|\vec{U}|/A$) is 5 tons per square meter.

Similarly, an inertia force produced by a static acceleration equal to K_y times the acceleration of gravity (g) will incline the resultant to produce a factor of safety of 1.0. We can calculate the value of K_y if we know the initial factor of safety. For example, if the initial factor of safety is 1.7, slip occurs when the acceleration becomes equal to 0.11g (i.e., $K_y = 0.11$). The critical direction for the inertia force is shown by the vector $\overrightarrow{K_yW}$ in figure 6-24b (the same direction as the minimum bolting force for a factor of safety of 1.0).

ESTIMATE OF THE DISPLACEMENTS OF A BLOCK UNDER A DYNAMIC IMPULSE

A dynamic load that produces an acceleration greater than K_yg given in the example above will precipitate slip but the ground

* The "outward normal" is directed out of the block (into the supporting plane).

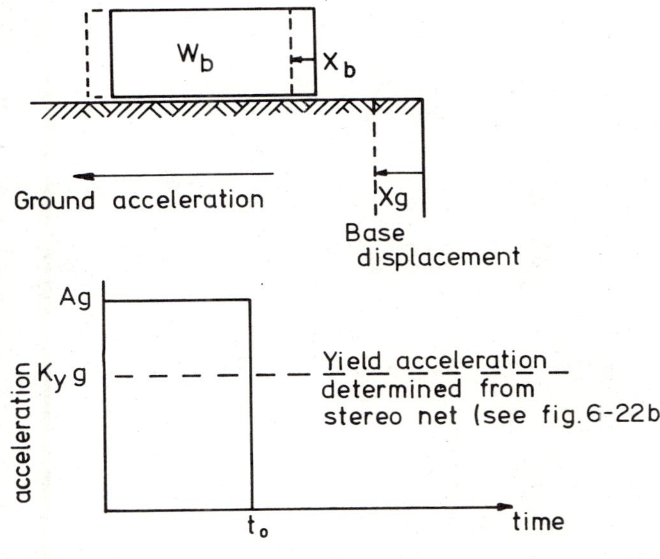

Total displacement of the ground below block

$$X_g = \frac{1}{2} A g t_o^2 + A g t_o (t_m - t_o)$$

Total displacement of block

$$X_b = \frac{1}{2} K_y g t_m^2$$

Slip between block and ground: $\Delta X = X_g - X_b$

$$\Delta X = -\frac{1}{2} A g t_o^2 + A g t_o t_m - \frac{1}{2} K_y g t_m^2$$

Relative motion between block and support ceases at time t_m

$$K_y g t_m = A g t_o$$

$$t_m = \frac{A t_o}{K_y} \quad \text{or} \quad \frac{V}{K_y g}$$

$$A g t_o = V \quad \text{and} \quad t_m = \frac{V}{K_y g}$$

$$\Delta X = \frac{V^2}{2g} \left[\frac{1}{K_y} - \frac{1}{A} \right]$$

Compare ΔX to peak displacement in a shear test

Figure 6-25. Approximate method for estimating the significance of a dynamic load for a potentially sliding block; after Henron et al (1971).

motion may come and go before the block has travelled very far, as discussed by Goodman and Seed (1966). Hendron et al (1971) proposed a simple method of calculating the relative movement of a rock block once the value of its yield acceleration $K_y g$ has been determined. The ground acceleration and its duration, or the ground acceleration and the particle velocity, must be known. Suppose, as in figure 6-25, the ground acceleration rises instantaneously to a value of Ag, remains constant for time t_0 and then drops instantaneously to zero. The block begins to yield when the ground acceleration reaches $K_y g$, and acquires velocity $K_y g t_0$. Relative displacement between the block and the ground (slip) occurs until time t_m when the block acquires a velocity equal to the ground velocity $V = Agt_0$;

$$t_m = \frac{V}{K_y g}$$

The total displacement of the ground at time t_m is

$$x_g = (1/2)\ Agt_0^{\ 2} + Agt_0(t_m - t_0)$$

At time t_m, the total displacement of the block is

$$x_b = (1/2)\ K_y g t_m^{\ 2}$$

The total distance of slip due to the ground motion is:

$$\Delta x = x_g - x_b$$

Substituting $t_0 = V/(Ag)$ and $t_m = V/(K_y g)$, gives

$$\Delta x = \frac{V^2}{2g}\left(\frac{1}{K_y} - \frac{1}{A}\right) \qquad (10)$$

If the yield acceleration $K_y g$ for slip in one direction is greater than the yield acceleration for slip in the opposite direction, as would be the case on an inclined plane, the block will slip unidirectionally with each pulse coming to rest between cycles of acceleration. Equation 10 is not strictly valid for an inclined plane as the direction of relative motion calculated is not the same as the down slope direction of slip; however, it can be used to estimate the severity of a particular impulse on a block having a known yield acceleration $K_y g$. If Δx calculated by equation 10 is greater than the peak displacement in a shear test of the block on the surface, the ground motion might create damage. With each increment of displacement the effective friction angle, and the value of K_y for the next increment, would be reduced as demonstrated experimentally by Goodman and Seed. We saw previously (figure 6-24b) how to determine the yield acceleration ($K_y g$) for unrestrained blocks on a single plane. We will now extend this to single tetrahedral blocks resting on two or more planes.

SLIP OF TETRAHEDRAL WEDGES

Intersecting discontinuity surfaces may liberate tetrahedral wedges as in figure 6-26, at the top of a rock cut. Two of the four faces of the wedge are free surfaces. The wedge can slide in three modes -- on either of the two supporting planes independently, and on both planes simultaneously; in the latter case, sliding is parallel to the line of intersection of the two supporting planes. Conditions of sliding parallel to the line of intersection were discussed

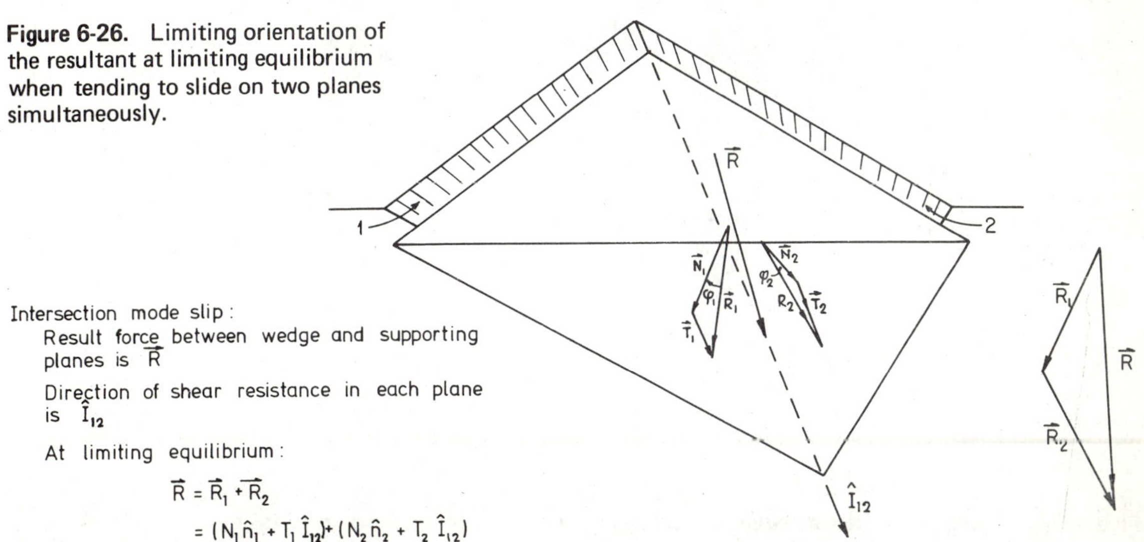

Figure 6-26. Limiting orientation of the resultant at limiting equilibrium when tending to slide on two planes simultaneously.

by John (1968) using the stereographic projection, and previously by Wittke (1965) using vector analysis, and Londe (1965) using other graphical methods. The mechanics of sliding on either plane alone is identical to the examples discussed previously except that there now is a kinematic constraint -- sliding on a single plane is possible only within a restricted set of directions in the plane, as there can be no net displacement towards the adjacent face. If one plane provides the entire reaction to the resultant force initiating sliding, the resultant can be decomposed uniquely into components normal and parallel to that plane. Since the block cannot move into the adjacent plane, the direction of the shear force must lie between the intersection $\hat{I}_{12}$ and its opposite in the half plane containing the outward normal. Thus, on a lower hemisphere stereographic projection, construct the normal to the plane in question and the line of intersection $\hat{I}_{12}$ between the two planes (figure 6-27); the great circle

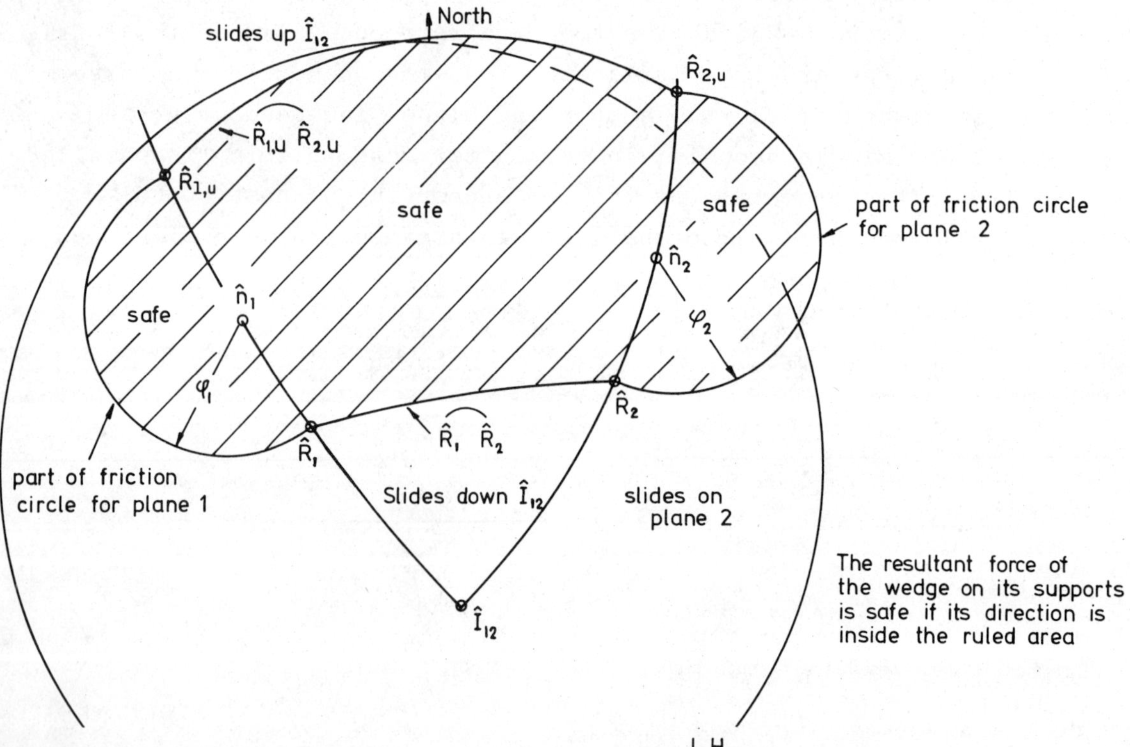

Figure 6-27. Generalized friction circle for sliding of a wedge with two faces in contact.

$\widehat{\hat{I}_{12}\,\hat{n}_1}$ bounds the region of kinematically possible slip on plane 1. Similarly the great circle $\widehat{\hat{I}_{12}\,\hat{n}_2}$ bounds the kinematically meaningful portion of the friction circle for slip on plane 2 alone (figure 6-27). Between these great circles there is a region in which slip is possible along the line of intersection; in this case, the direction of incipient sliding is parallel to the line of intersection and therefore the sum of the shear forces on each plane, at the limit of slip, is parallel to $\hat{I}_{12}$.* In addition there are forces parallel to the two normals. The resultant force of the block on its supports, in the case of sliding down the line of intersection, must therefore be composed of components parallel to $\hat{n}_1$, $\hat{I}_{12}$, and $\hat{n}_2$ and it will accordingly plot in the spherical triangle formed by $\hat{n}_1$, $\hat{I}_{12}$, and $\hat{n}_2$. If it is possible to slide up the line of intersection, the resultant force will have to lie in the spherical triangle between $\hat{n}_1$, $-\hat{I}_{12}$ and $\hat{n}_2$ (figure 6-27).

At the limit of equilibrium in the intersection mode, full friction has been mobilized in both planes simultaneously. The reaction in plane 1 is made up of a normal force and a shear force parallel to $\hat{I}_{12}$ and is $\phi_1{}^{o}$ from $\hat{n}_1$. Therefore the reaction in plane 1 lies at the intersection of the friction circle for plane 1 with the great circle $\widehat{\hat{I}_{12}\,\hat{n}_1}$. Similarly the reaction in plane 2 is made up of a normal force, and a shear force parallel to $\hat{I}_{12}$ and is inclined $\phi_2{}^{o}$ from the normal to plane 2; therefore the reaction in plane 2 will be found at the intersection of the friction circle for plane 2 and the great circle $\widehat{\hat{I}_{12}\,\hat{n}_2}$. Since the resultant force is comprised uniquely of the reactions in plane 1 and plane 2 it must lie along the great circle common to these two reactions, as shown in figure 6-27. Given the friction angles on plane 1 and plane 2, therefore, it is possible to construct a generalized safe zone including all possible modes of translation. Once constructed, this safe zone is used in the same fashion as illustrated previously for the simple friction circle. One examines the orientation of the resultant of applied forces with respect to the safe zone. According

---

* Initial shear stresses that may have existed must change to the limiting values and directions before slip and therefore they are of little interest for the limit equilibrium analysis.

to the position of the resultant not only the degree of safety, but the mode of potential slip is indicated.

Meaning of the Factor of Safety in the Intersection Mode

As discussed by Londe (1965), there is an ambiguity attached to the discussion of the factor of safety when an intersection mode slide is in question. With respect to sliding on a single plane, the factor of safety implied by a particular orientation $\hat{R}$ of the resultant force, inside the safe zone, was found by shrinking the friction circle until its circumference passed through $\hat{R}$. However, for intersection mode sliding, there are an infinite number of ways that the safe zone can be reduced to bring its perimeter exactly through $\hat{R}$. In the illustrative problem, presented in figure 6-28a, we are presented with the normals to two planes and two friction an-

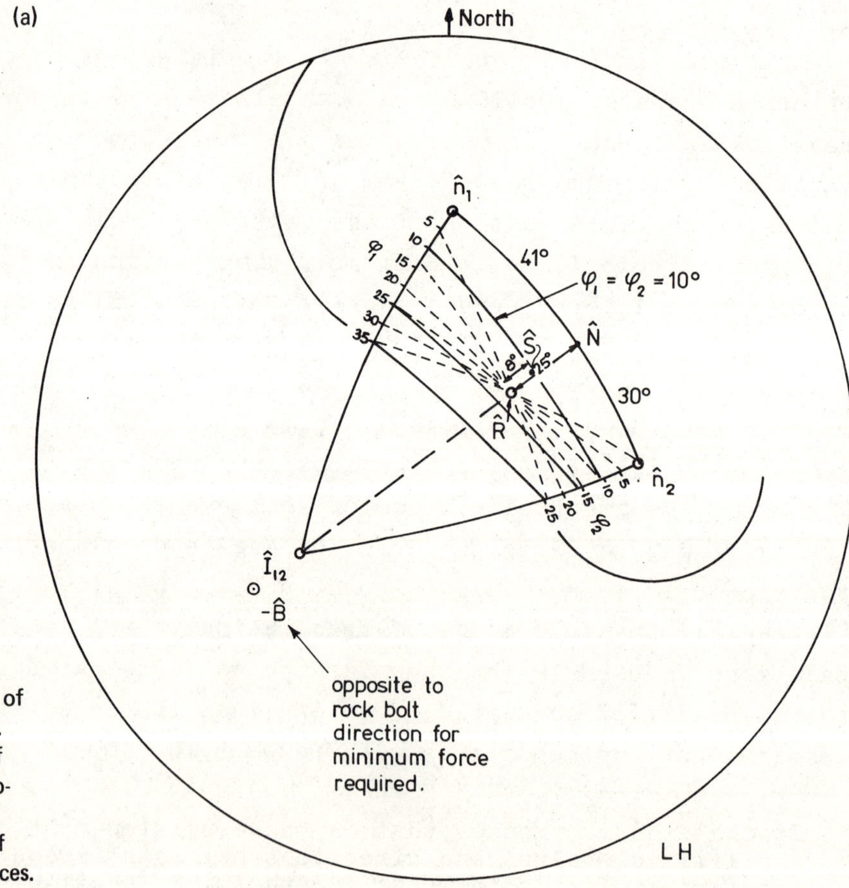

Figure 6-28. (a) Example of intersection mode analysis. (b) infinite combination of friction angles that will provide limiting equilibrium conditons. (c) Inclusion of cohesion and rock bolt forces.

(b)

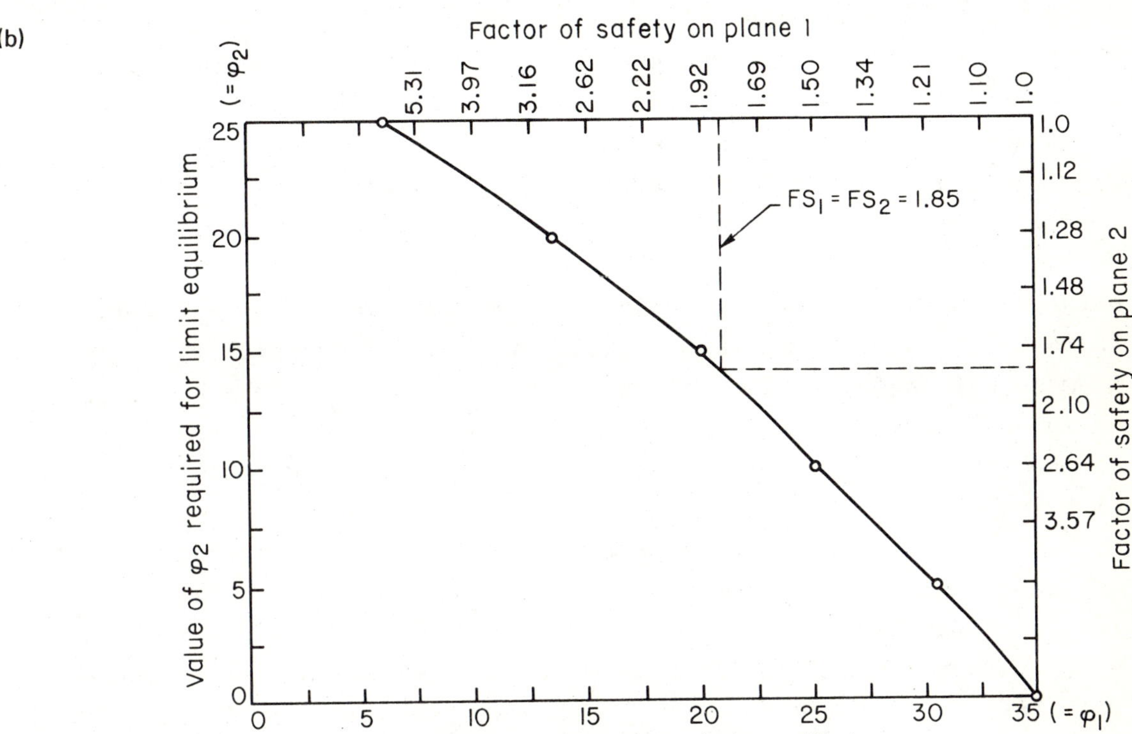

(c)

$\vec{R}$ = 100°

20°

30°

10°

17°

$\vec{N}_2$

$\vec{C}_2 = 9$

$\vec{T}_2$

$\vec{N}$

N_1

10°

25°

41°

$\vec{T}$

$\vec{T}_1$

$\vec{C}_1 = 14$

R = 100

8°

$\vec{B} = 14$

$\varphi_{available} = 10°$ on each plane

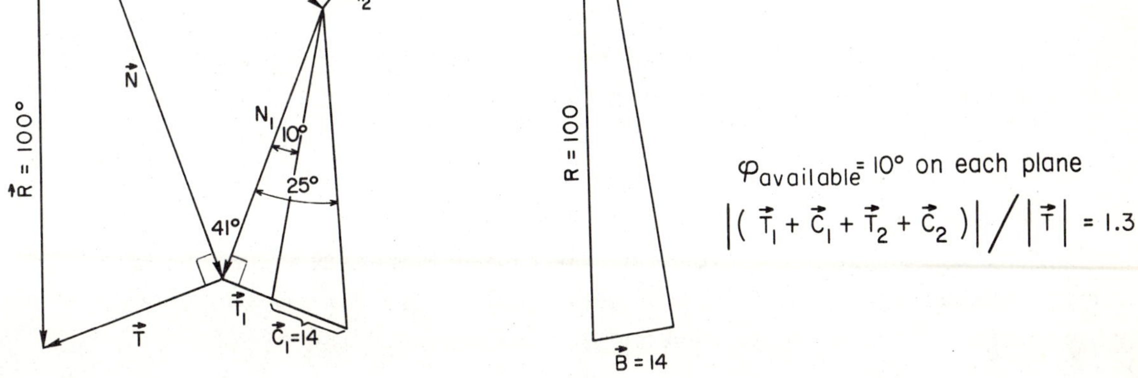

$$\left|\left(\vec{T}_1 + \vec{C}_1 + \vec{T}_2 + \vec{C}_2\right)\right| \Big/ \left|\vec{T}\right| = 1.3$$

gles. The orientation of the resultant given, $\hat{R}$, lies well within the safe zone and therefore the wedge is stable. The dashed lines are great circles obtained by reducing ϕ_1 and ϕ_2 by different amounts to shrink the safe zone through point $\hat{R}$. The values of ϕ_1 and ϕ_2 required for limiting equilibrium ($\phi_{required}$) are then plotted in figure 6-28b. One may adopt a convention that the "factor of safety" is the value which produces the same factor of safety on both planes simultaneously, which in this case is 1.85. However, there is no theoretical justification for such a definition. It is more meaningful to use the curve of $\phi_{1,required}$ versus $\phi_{2,required}$ as the expression of safety and, in particular of the sensitivity of the safety of the wedge towards the change in either of the friction angles.

A probabilistic interpretation of safety as in the example with a single block on a single plane, is more difficult in a case of intersection sliding. This is because a given dispersion of normals about planes 1 and 2 produces a greater dispersion of their intersections (figure 6-29a). This problem was discussed by Ramsay (1967). It will be seen in figure 6-29b that as the angle between the normals becomes small, the dispersion in the position of the line of intersection becomes very large. We can shrink the friction circle on each plane corresponding to the degree of probability required; but it is more difficult to draw the required great circles connecting the normals with the line of intersection, for as the normals wander the line of intersection wanders even more.

The generalized friction circle, i.e. the safe zone is used for a tetrahedral wedge just as for a single block on a plane, as illustrated in figure 6-28a. For example, if the available friction angle is known to be 10° on each plane, we can ask what cohesive forces are required to insure a factor of safety of 1.5 on each plane. The bounds of the safe region in the intersection mode corresponding to ϕ_1 equal ϕ_2 equal 10° are shown in figure 6-28a. In this case $\hat{R}$ is no longer safe and it will be required either to rotate $\hat{R}$ into the safe zone by addition of force or to increase the size of the safe zone by assuming a cohesion. A solution to the latter problem was presented by John (1970). By using the construction of figure 6-14,

Figure 6-29. (a) The dispersion of lines of intersection of planes, each of which is dispersed 10°. (b) Dispersion of intersections as a function of η and ψ.

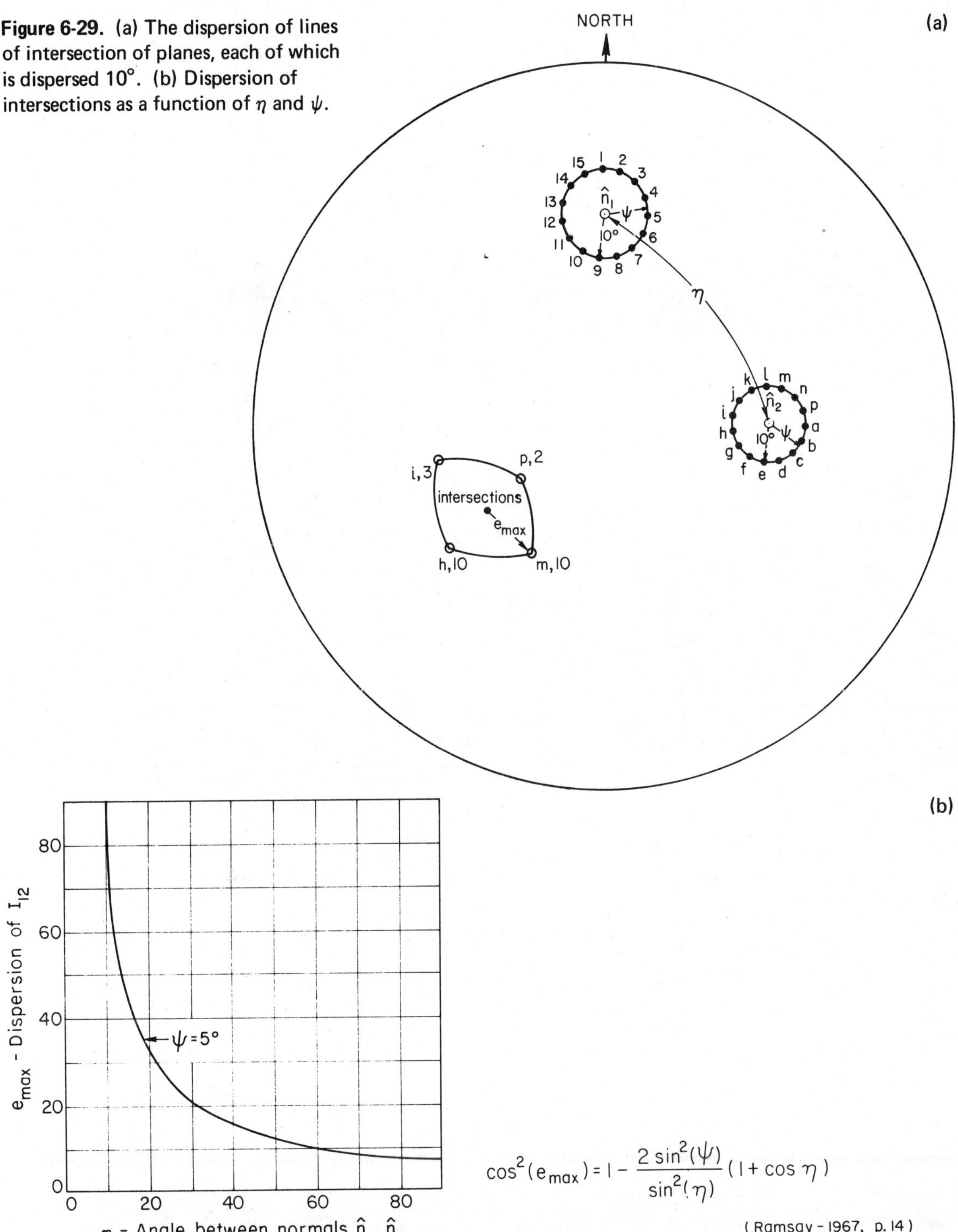

$$\cos^2(e_{max}) = 1 - \frac{2\sin^2(\psi)}{\sin^2(\eta)}(1 + \cos\eta)$$

(Ramsay - 1967, p. 14)

decompose $\hat{R}$ into three components -- one in the direction of sliding, $\hat{I}_{12}$, and the others in the directions of the normals $\hat{n}_1$ and $\hat{n}_2$. The resultant normal force $\vec{N}$ is found from the orthographic projection of $\vec{R}$ on plane $\widehat{\hat{n}_1 \hat{n}_2}$. (Since $\hat{I}_{12}$ is normal to this plane, the intersection of great circle $\widehat{\hat{I}_{12} \hat{R}}$ on plane $\widehat{\hat{n}_1 \hat{n}_2}$ is the same as the orthographic projection of $\vec{R}$ on plane $\widehat{\hat{n}_1 \hat{n}_2}$.) Figure 6-28c shows the polygon of forces for this decomposition. Once the normal $\vec{N}$ has been determined, it is further decomposed into normal forces in each plane, $\vec{N}_1$ and $\vec{N}_2$, making use of the angles 41° and 30° measured in plane $\widehat{\hat{n}_1 \hat{n}_2}$. Figure 6-28c shows how the cohesive forces are added to the force polygon to find effective overall friction angles on each plane. If the friction angles mobilized in these planes are both 10 degrees and the cohesive forces on planes 1 and 2 are respectively 14 and 9, the ratio of resisting to driving forces is 1.3.

Rock bolts can be used to rotate the resultant force into the safe zone so that the mobilized friction does not exceed 10° in either plane. The minimum rotation of force required to produce the desired safety, without cohesion, is 8° (to point $\hat{S}$, figure 6-28a). The direction for bolts to effect this rotation with minimum force is $\hat{B}$, (in the direction opposite to $-\hat{B}$ of figure 6-28a), that is 90° from S along the great circle $\widehat{\hat{R}\,\hat{S}}$. As shown in figure 6-28c, the required bolt force is 14. $\hat{B}$ rises 33° to the N52°E.

Problems with intersection mode sliding in argillite were met by the Corps of Engineers in construction of Libby dam in Montana. Intersections of bedding and jointing plunged 34-38 degrees into excavations on the left abutment (figure 6-30a). Figure 6-30b shows the supporting planes of a wedge approximately 600 feet long after slide debris had been completely cleaned out (see also figures 2-8d and 2-8e). One potential intersection slide involving the upper part of the left supporting plane (bedding) was stabilized with an installation of approximately eighty 200-ton tendons, anchored 65-145 feet deep; these are visible in the upper left of figure 6-30b and in figure 6-30c. Stereographic projection of the wedge shows immediately the potentially dangerous situation, since with only gravity loading, a friction angle of approximately 30° on each plane is required to maintain the wedge.

(a) (b) (c)

Figure 6-30. (a) A daylighting intersection between a bedding fault and jointing at Libby Dam, Montana. Photo by Dennis Lachel; (courtesy of the Corps of Engineers). (b) A portion of the left abutment slide at Libby Dam after removal of the slide debris. The bedding, left, strikes N 26 W and dips 43-5 to the west. The joint surface (right) stikes N 74 E and dips 49 to the NW. Figure 6-30a was taken just to the left of this view. Compare with figure 2-8d and e. Photo by Dennis Lachel; (courtesy of the Corps of Engineers). (c) View from the top of the wedge of 6-30b showing the relationship of the slide to the dam and reservoir. Photo by Dennis Lachel (courtesy of the Corps of Engineers).

SLIDING OF TETRAHEDRAL WEDGES WITH ONLY ONE FREE SURFACE

Londe (1965), and Londe et al (1970) discussed this problem in the context of the abutment stability calculations for an arch dam. Indeed it is believed that the movement of a block resembling that of figure 6-31, permitted the failure of Malpasset Dam (figure 1-3).

One must adopt a sign convention for the directions of the line of intersection and the normal forces. A line of intersection will be considered positive if it points into the free space, i.e. if it daylights. It may be in the upper or in the lower hemisphere according to the orientation of the wedge in space. Figure 6-31 shows a tetrahedral block with one free surface and indicates the positive directions of the lines of intersection and the normals;

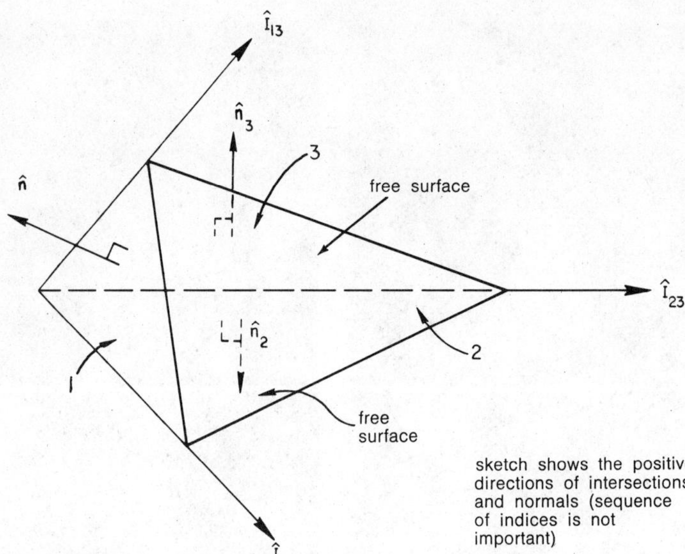

Figure 6-31. Kinematics of a wedge with three planes in contact.

as before, a normal is positive if it points out of the block into the supports. Now there are six translational modes -- three modes of sliding on a single plane, and three modes of sliding along lines of intersection. A mode will be named according to the face which opens to permit the translational movement, according to the terminology of Londe (1965)*. Mode 1 signifies that face 1 opens, and all reaction to sliding is contained on faces 2 and 3. In general, mode i signifies opening from face i with sliding on planes j and k in the direction of the intersection $\hat{I}_{jk}$. A mode of sliding on a single plane requires opening from the other two and therefore will be called mode ij with sliding on plane k; for example mode 1,2 necessitates sliding on plane 3. Each mode of sliding is confined to a particular spherical triangle of the reference sphere, as indicated in Table 6-1. For the mode of sliding in one plane, the resultant force will decompose into components in the directions of the normal to the plane of sliding and the lines of intersection

* In the work presented by Londe (1965) the sign convention for normals and lines of intersections was opposite to that presented here. However the result is the same, for rather than the orientations of resultant forces applied to the block, Londe mapped the orientations of the reaction force of the supports on the wedge.

TABLE 6-1

Mode	Slides on:	Opens from	$\vec{R}$ is inside
0	none	none	$\hat{n}_1\ \hat{n}_2\ \hat{n}_3$
1	2, 3	1	$\hat{n}_2\ \hat{I}_{23}\ \hat{n}_3$
2	1, 3	2	$\hat{n}_1\ \hat{I}_{13}\ \hat{n}_3$
3	1, 2	3	$\hat{n}_1\ \hat{I}_{12}\ \hat{n}_2$
1, 2	3	1, 2	$\hat{I}_{13}\ \hat{n}_3\ \hat{I}_{23}$
2, 3	1	2, 3	$\hat{I}_{12}\ \hat{n}_1\ \hat{I}_{13}$
1, 3	2	1, 3	$\hat{I}_{12}\ \hat{n}_2\ \hat{I}_{23}$
1,2,3	none	all	$\hat{I}_{12}\ \hat{I}_{23}\ \hat{I}_{13}$

along either of its edges. This is because of the kinematic requirement that the direction of motion lies between the two edges of the sliding face. For example, for sliding on plane i the resultant force must be in the spherical triangle $\hat{I}_{ij}$, $\hat{n}_i$, $\hat{I}_{ik}$. The three such modes are identified in Table 6-1. For sliding in two planes simultaneously, as previously discussed for the two plane wedge, the resultant force will decompose into a normal force on each plane and a shear force parallel to the line of intersection. Therefore for mode k with sliding on planes j and i, $\hat{R}$ must lie inside the spherical triangle formed by $\hat{n}_i$, $\hat{I}_{ij}$, $\hat{n}_j$. The three such modes are presented in Table 6-1 Two spherical triangles remain unlabelled: the spherical triangle $\hat{n}_1\ \hat{n}_2\ \hat{n}_3$ which is completely safe (mode 0); and the spherical triangle $\hat{I}_{12}$, $\hat{I}_{23}$, $\hat{I}_{13}$ in which the wedge must lift out (mode 1, 2, 3).

Figure 6-32 presents an example. A rock wedge formed by three intersecting planes lies in the abutment of an arch dam; its weight and the loading of the dam are given, and it is desired to calculate the water force on face 1 which could initiate failure of the wedge. The application of Table 6-1 to this problem is facilitated by recognizing the various normals and lines of intersections (Table 6-2) all of which can be obtained from a subsidiary stereographic

TABLE 6-2

Line	Bearing	Plunge or Rise	Hemisphere
n_1	West	0°	Both
n_2	North	20° up	Upper
n_3	–	Vertical	Lower
I_{12}	South	70° up	Upper
I_{23}	East	0°	Both
I_{13}	South	0°	Both

projection, (or in this simple case by inspection). Figure 6-32b presents the lower hemisphere and figure 6-32c the upper hemisphere

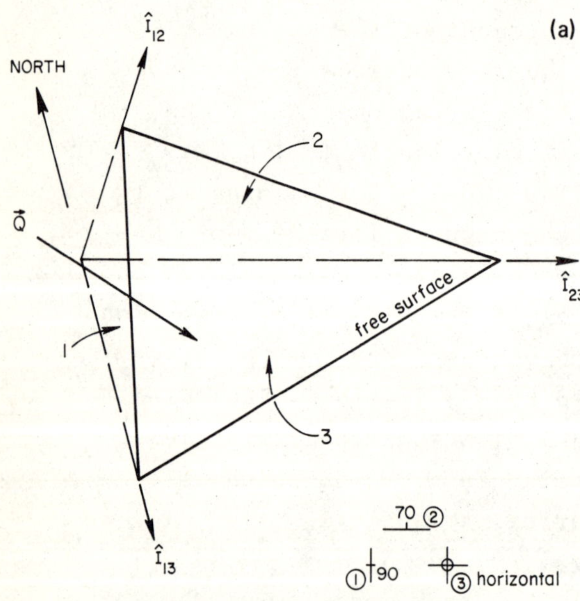

Face	Strike	Dip	Friction angle
1	North	Vertical	30°
2	East	70°	50°
3	—	Horizontal	20°

Figure 6-32. (a) Example problem of a wedge supported on three faces. (b) Lower hemisphere portion of the generalized friction circle (the safe zone) for the example problem.

portion of the reference sphere. The spherical triangles of Table 6-1 are formed by constructing the required great circles. Given the friction angles on each of the planes, we can construct the generalized friction circle for the wedge in question. Suppose $\phi_1 = 30^\circ$, $\phi_2 = 50^\circ$ and $\phi_3 = 20^\circ$. Small circles of the required radii are constructed in the proper spherical triangles. That is, construct: a small circle of radius 20° in the triangle $\hat{I}_{13}$, $\hat{I}_{23}$, $\hat{n}_3$; a small circle of radius 50° about $\hat{n}_2$ in spherical triangle $\hat{I}_{12}$, $\hat{n}_2$, $\hat{I}_{23}$; and a small circle of radius 30° about $\hat{n}_1$ in the spherical triangle $\hat{I}_{13}$, $\hat{n}_1$, $\hat{I}_{12}$. Draw great circles between the intersections of these small circles with the mode boundaries as shown. The only difficulty arises in connection with the great circle from point $\hat{b}$ in

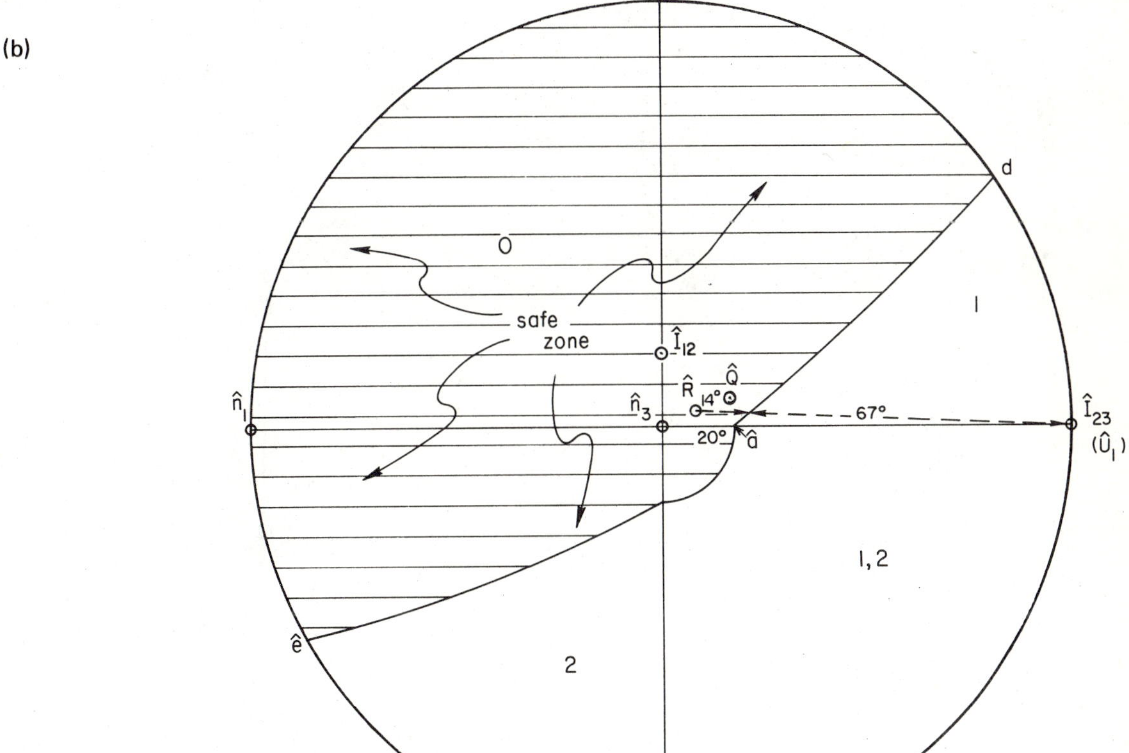

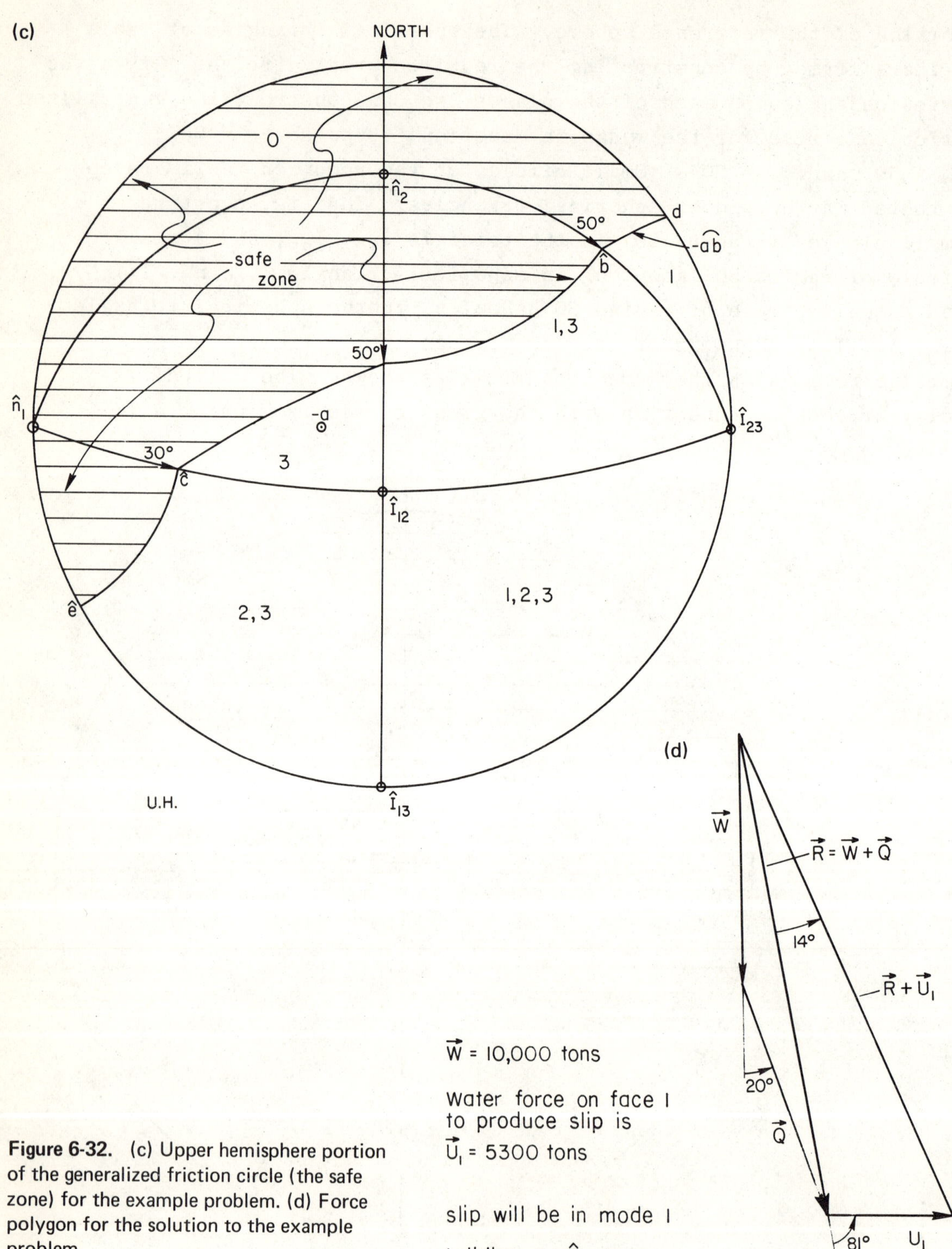

Figure 6-32. (c) Upper hemisphere portion of the generalized friction circle (the safe zone) for the example problem. (d) Force polygon for the solution to the example problem.

the upper hemisphere to point $\hat{a}$ in the lower hemisphere representing the limits of the safe zone in mode 1. In order to construct this great circle plot $-\hat{a}$ (the opposite to $\hat{a}$) in the upper hemisphere projection and draw the great circle $\widehat{-\hat{a}\,\hat{b}}$; this great circle, in the mode 1 region, intersects the primitive circle at point $\hat{d}$, which can be transferred to the lower hemisphere. The required great circle construction is completed by connecting $\hat{d}$ and $\hat{a}$.

The completed construction shows that the orientation of the resultant force, $\hat{R}$, is well within the safe zone. If the water pressure is raised on face 1, a water force on the wedge will point in the direction of the opposite to $\hat{n}_1$, ($-\hat{n}_1 = \hat{I}_{23}$ in this particular problem). Along the great circle $\widehat{\hat{U}_1\,\hat{R}}$ a 14° rotation is required to bring the resultant force to the limit of stability. $\hat{R}$, found by adding the weight force (in direction $\hat{n}_3$) and the thrust of the arch dam ($\hat{Q}$) in the direction given, lies 81° from $\hat{U}_1$. The polygon of forces of figure 6-32d shows that the required water force to initiate slip is 5,300 tons and failure will be by mode 1. Rotations have not been considered; in any practical example one could use the methods discussed previously for detailed analysis of rotations.

SLIDES COMPOSED OF TWO BLOCKS

In general a rock mass consists of numerous mating blocks with little pore space. The movement of one block creates the freedom for its neighbors to move and therefore general problems of rock stability require an appreciation of the interactions of multiple blocks. We will only begin to explore this question here, using the stereographic projection. Figure 6-33 shows a slide consisting of two free blocks, one acting as a passive supporting wedge and the other acting as an active loading wedge. Sliding of the passive wedge is promoted by the weight of the active wedge, which can not be sustained by friction on its base alone. Thus, there is a load transfer from passive to active wedge. Figure 6-33 is a section through the two blocks in the direction of sliding. It need not be parallel to the dip vector in either plane as will be discussed in the later example. A potential two block problem can occur only if the sliding surface of the active wedge daylights into the space

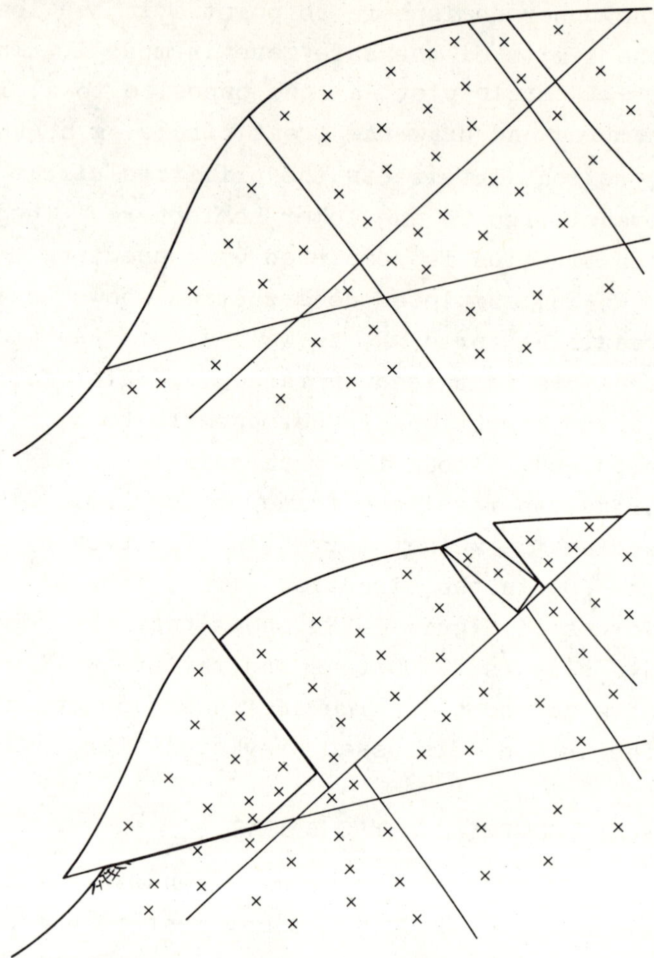

Figure 6-33. Two block slide.

created when the passive wedge moves. Figure 6-34 shows a simple analysis for the mechanics of this case. The principal idea is to separate the blocks along their common plane (plane 3 in this example) which is considered as a free surface in each analysis of each block separately. As opposed to the case in soil mechanics where the division between the active and passive wedge has to be determined by iterations, structural geological information determines the direction of the interface between blocks. In contrast with soil mechanics, there should be no debate about the proper angle of friction for the inclination of the load transferred along the connecting plane, since a very small displacement is enough to mobilize

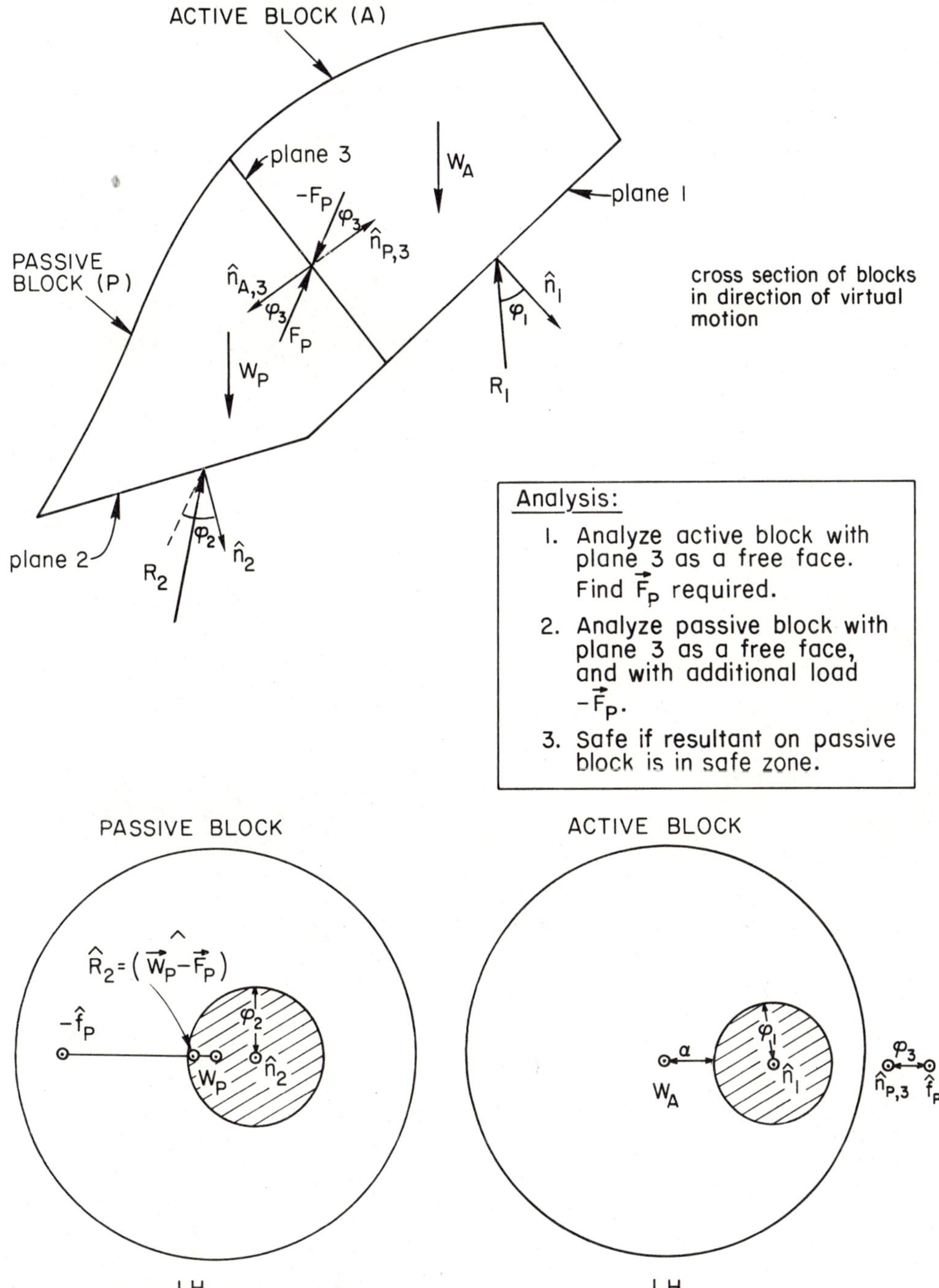

Figure 6-34. Analysis of a two dimensional two block case.

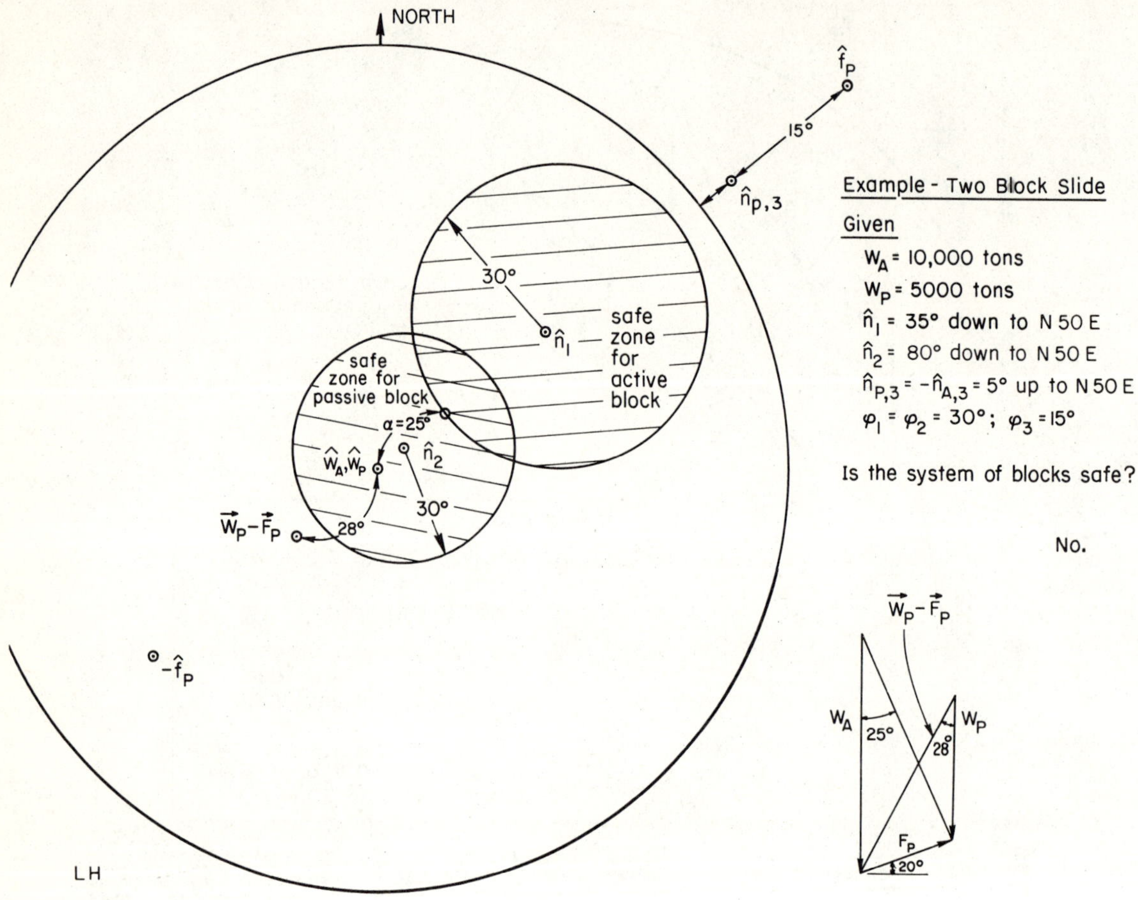

Figure 6-35. Example of a two block analysis; two dimensional case.

the full available friction. Therefore, we will take ϕ_3 as the angle of inclination of the load transferred between blocks, as shown in figure 6-34.

The active block is analyzed in the right half of the stereographic projection in figure 6-34. It is a simple friction circle case with sliding on a single plane. The active wedge is unstable under its own weight; we require a transfer of supporting force from the passive wedge, which is inclined ϕ_3 from $\hat{n}_{p,3}$ -- the direction of the normal pointed into the active wedge. The direction of the transferred load ($\hat{f}_p$) is determined from a force polygon to rotate the resultant through the required α degrees to bring it into the safe zone. Plot the opposite of $\hat{f}_p$ in the stereographic projection for the passive block (its direction is $-\hat{f}_p$). Adding the weight of the passive block to $-\vec{F}_p$ produces a resultant $\vec{R}_2$ inclined with respect to $\vec{W}_p$. If $\hat{R}_2$ is in the safe zone for the passive wedge, then both active and passive blocks are safe. We will not try to balance factors of safety between wedges, but rather proceed to a numerical example (figure 6-35). The active wedge weighs 10,000 tons, the passive wedge weighs 5,000 tons, and the directions of the normals to planes 1, 2 and 3, are given as well as the friction angles. Is the system of blocks safe? Both the active and passive wedge constructions have been superimposed on a single stereogram for this problem. The weight of the active wedge is inclined 25° from the safe zone for the active block and therefore the force $\vec{F}_p$ shown in the force polygon is required. The addition of $\vec{W}_p$ and $-\vec{F}_p$ produces a 28° rotation of the weight force in the passive block, which produces a resultant outside of the safe zone for the passive block. Therefore the system of two blocks is unsafe.

The above example is an illustration of the particular case in which each block slides on a single plane (mode i,j). If either, or both blocks are involved in an intersection type slide, the problem is more difficult. Figure 6-36a illustrates such a case. The active block formed by planes 1, 2 and 3 tends to slide down its line of intersection. Denote by q the point where the line of intersection, $\hat{I}_{12}$ of the active block meets the face of the passive block. If q is inside the face of this block, then movement of the passive block will allow daylighting of the active wedge, a necessary condition for the two block slide. If the passive wedge slides on a single plane, the direction of relative slip on plane 3 will have to be determined by iteration. If on the other hand, both the active and passive wedges move along their line of intersections, there is only one possible direction for the relative slip in plane 3, as will be shown.

Since at first the critical modes are not known, an iterative solution will be illustrated. In figure 6-36b, the active wedge is diagrammed. We will suppose we do not know the direction of slip on plane 3 and will determine a number of solutions for different

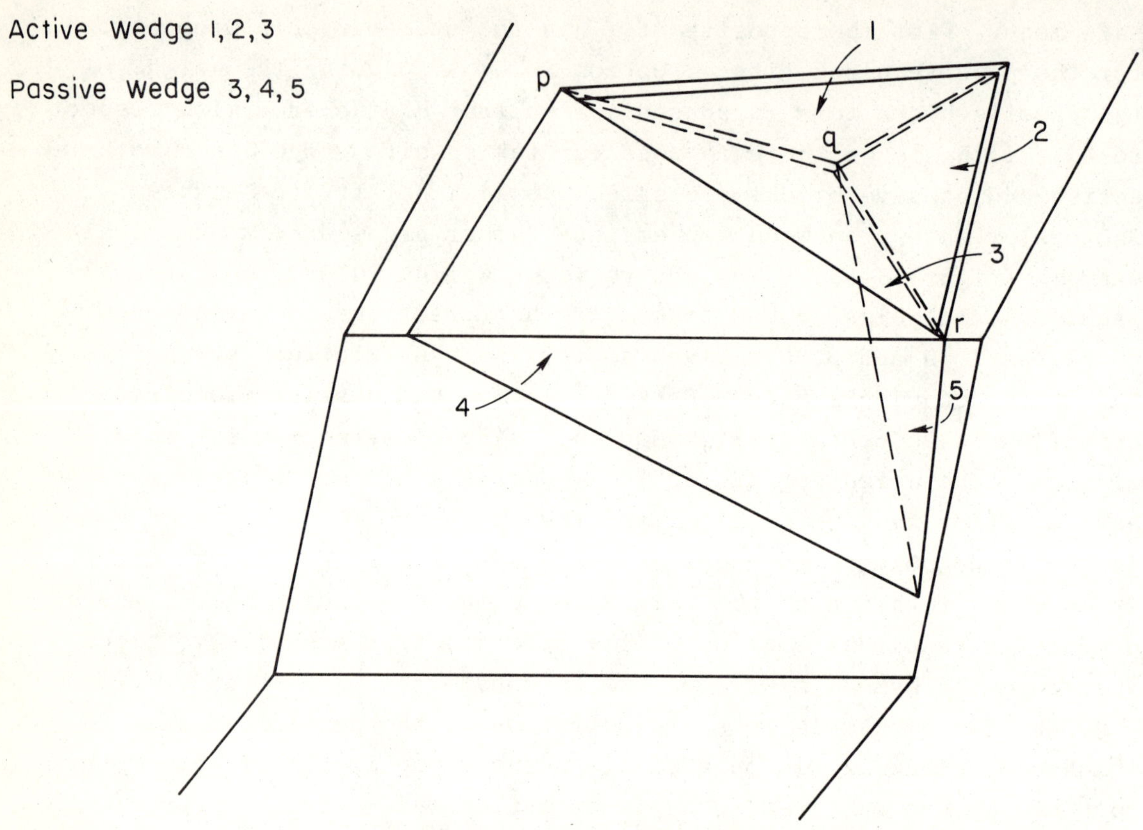

Analysis: Consider plane 3 a free face and determine direction of slip along it as shown in figures 36 b, c.

For intersection sliding of active wedge, $\hat{I}_{12}$ must intersect face pqr of passive wedge.

(a)

Figure 6-36. (a) A two wedge slide. (b) Analysis of the active wedge. (c) Analysis of the passive wedge. (d) Force polygons for analysis of the two wedge case. (e) Construction for compatible velocities of neighboring blocks, each of which slides on its line of intersection.

possibilities. First construct the safe zone for the active wedge; then around the normal $\hat{n}_{p,3}$, construct a small circle of radius equal to ϕ_3. Points $\hat{a}$, $\hat{b}$, $\hat{c}$, $\hat{d}$, $\hat{e}$ and $\hat{f}$ are various possibilities for the direction of the load transference between the active and the passive wedge. Corresponding to each we can read the required rotation α of the weight force to produce a resultant in the safe zone of the active wedge, as tabled in figure 6-36b. Suppose ϕ

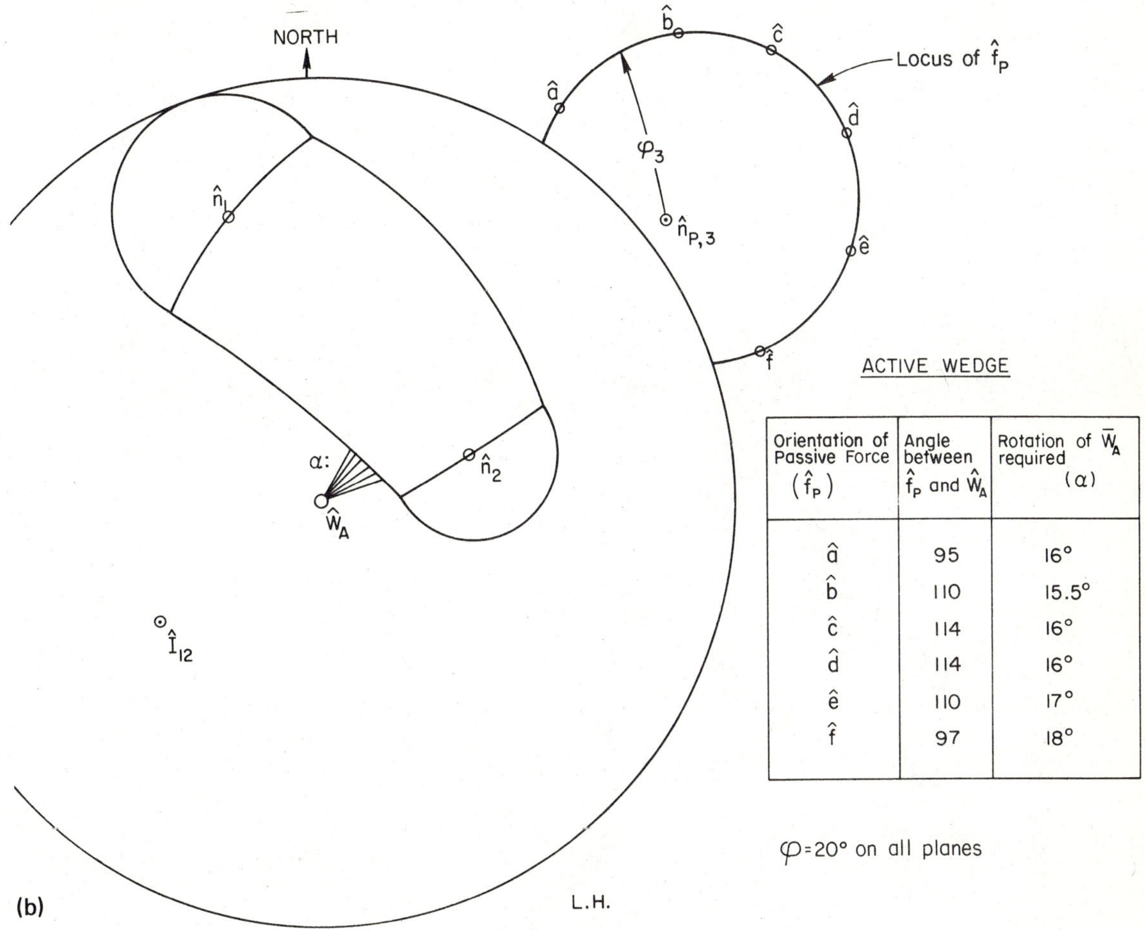

Orientation of Passive Force ($\hat{f}_P$)	Angle between $\hat{f}_P$ and $\hat{W}_A$	Rotation of $\overline{W}_A$ required (α)
$\hat{a}$	95	16°
$\hat{b}$	110	15.5°
$\hat{c}$	114	16°
$\hat{d}$	114	16°
$\hat{e}$	110	17°
$\hat{f}$	97	18°

equals 20° on all planes; what angle of friction is necessary for stability in the passive wedge? Since the angle between the direction of $\hat{f}_p$ and the weight force of the active wedge is known for each assumption of direction $\hat{f}_p$, a force polygon can be constructed, as shown in figure 6-36d, defining the magnitude of $\vec{F}_p$ corresponding to each assumption for $\hat{f}_p$ ($\hat{a}$ through $\hat{f}$). Transfer the opposites to points $\hat{a}$ through $\hat{f}$ onto the stereographic projection for the passive wedge (figure 6-36c) as well as the opposite to $\hat{n}_{p,3}$ (point $\hat{n}_{a,3}$). Read the angle to $\hat{W}_p$ from each point just plotted and construct a series of force polygons for each point to determine the rotation of $\vec{W}_p$ produced by each assumption of slip direction in plane 3. This has been done (figure 36d) for two assumptions of the weight

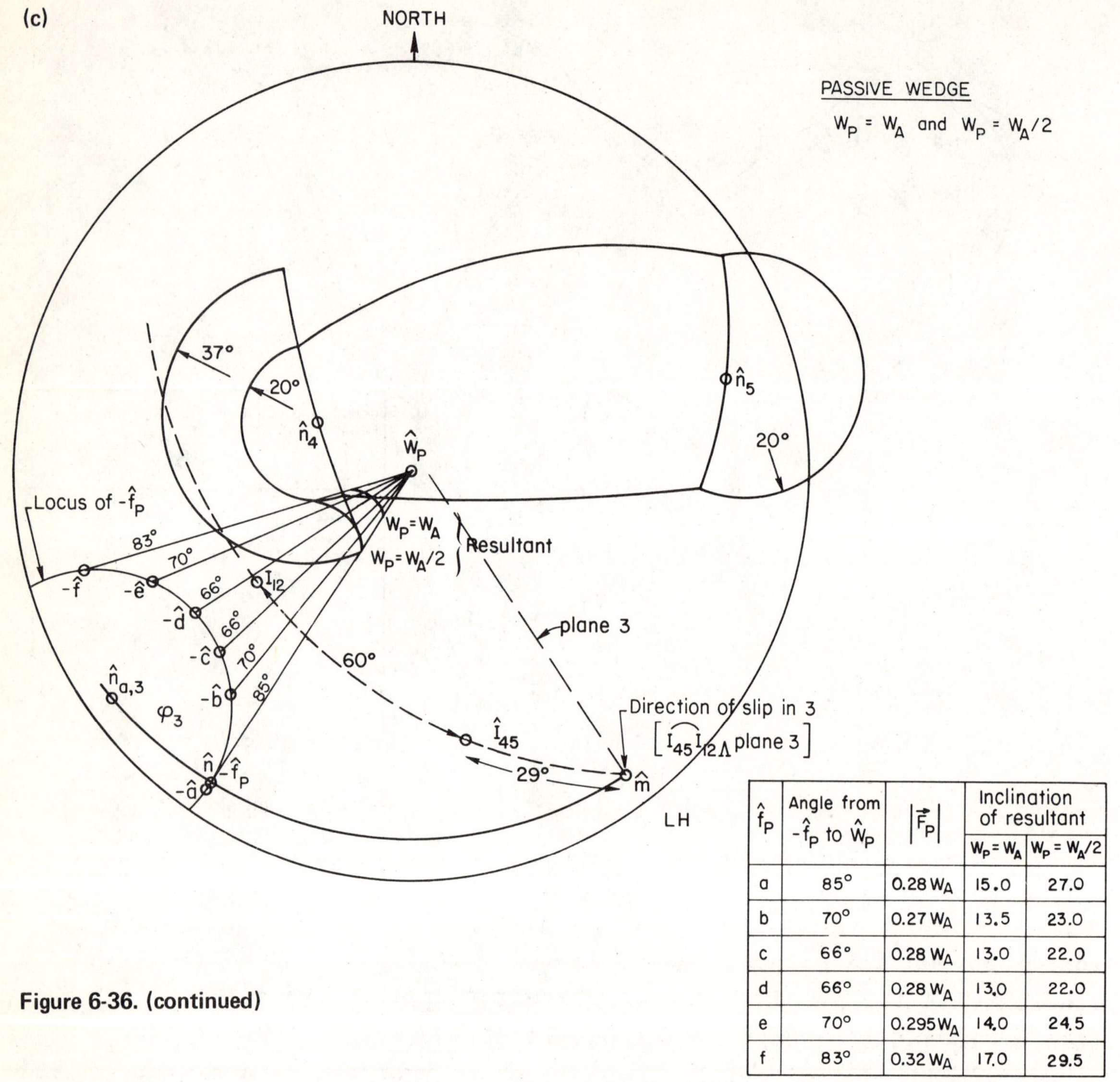

$\hat{f}_P$	Angle from $-\hat{f}_P$ to $\hat{W}_P$	$\lvert\vec{F}_P\rvert$	Inclination of resultant	
			$W_P = W_A$	$W_P = W_A/2$
a	85°	0.28 W_A	15.0	27.0
b	70°	0.27 W_A	13.5	23.0
c	66°	0.28 W_A	13.0	22.0
d	66°	0.28 W_A	13.0	22.0
e	70°	0.295 W_A	14.0	24.5
f	83°	0.32 W_A	17.0	29.5

Figure 6-36. (continued)

of the passive wedge: $W_p = W_a$, and $W_p = (1/2)W_a$.

In a more general problem, other forces might be included, e.g. the thrust of a dam, or water pressure forces on any given face. It is seen that the resultant is unstable and the passive wedge must fail by sliding in the direction of its line of intersection. The direction of slip on plane 3 is therefore determinable as follows (figure 6-36c). Construct the great circle $\widehat{\hat{I}_{12}\,\hat{I}_{45}}$, and mark its intersection, $\hat{m}$, with plane 3. Then construct great circle $\widehat{\hat{m}\,\hat{n}}_{a,3}$.

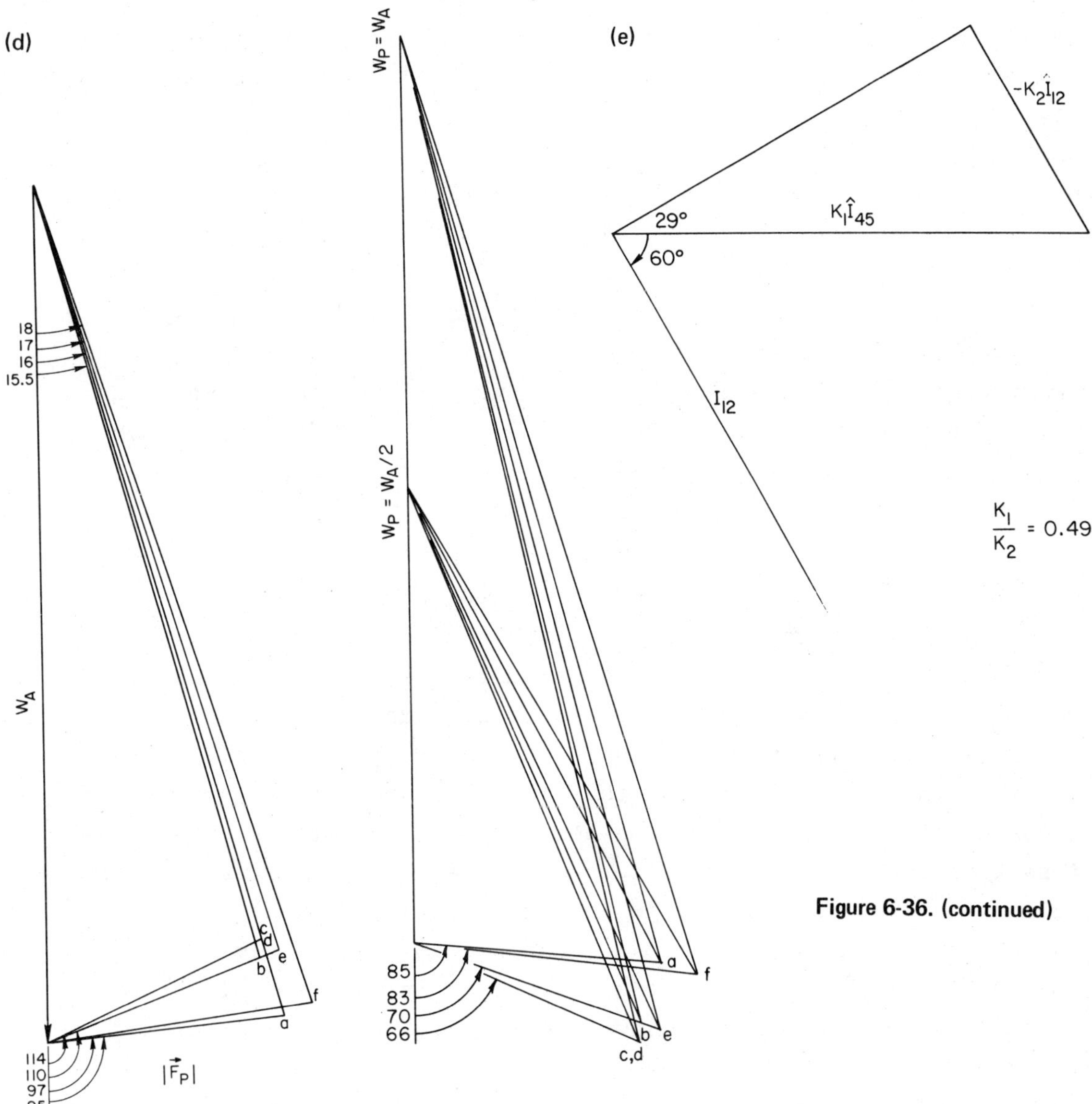

Figure 6-36. (continued)

The intersection of the latter with the friction circle about $\hat{n}_{a,3}$ defines point $\hat{n}$ which is the direction of the resultant load transfer from the active to the passive wedge. (The direction $\hat{m}$ corresponds to the direction of slip in plane 3.) The wedge is unsafe. An expansion of the friction circle about $\hat{n}_4$ to 37 degrees will produce limiting equilibrium.

Figure 6-36e shows the vector triangle, used to construct the vector subtraction to determine point $\hat{m}$. The relative slip in plane 3 was obtained by the difference of vectors pointed in the direction of slip on each wedge, since it is the relative movement between the wedges which provokes the load transference. The orientation of the vector difference $\hat{I}_{45} - \hat{I}_{12}$ is known to be in plane 3 since the blocks maintain contact during sliding (assuming there is no rotation). The relative magnitudes of the vectors parallel to $\hat{I}_{12}$ and $\hat{I}_{45}$ indicated by the vector triangle, figure 6-36e, indicates the relative velocity of each wedge down its line of intersection. Observation of the displacement of one wedge, therefore, can be used to infer the compatible movement of its neighbor. In this instance the active wedge moves 0.49 as far as the passive wedge in each increment of slip.

THE STATE OF STRESS IN ROCKS

Rocks differ from most other materials in that they may already possess significant stresses before additional loadings or unloadings are constructed. Excavation of surface or underground space leads to stress rearrangement with concentration or spreading of the lines of force, which, to the extent that the rock properties vary with stress, render the rock mass non-homogeneous. If stress concentrations cause extreme stress differences, the rock may break, sometimes explosively as in deep mines in South Africa and Canada. On the other hand, if stress reductions leave blocks of rock almost free from their neighbors, they may fall out, as in the roof of a tunnel at shallow depth and in the exposed corners of excavated rock walls in complex underground openings like underground power plants (see figure 1-5b). Since properties of joints are highly stress dependent, a jointed rock mass with a non-homogeneous stress field will change in character from point to point. In such rocks, we are particularly concerned with the real possibility that a state of stress may imply sliding or opening of individual planes. Whether or not such movements will provoke rock falls depends upon the direction of the sliding or opening tendency with respect to the kinematic freedoms of the rock mass. Given a state of stress, we will show how to calculate the normal and shear stresses local to a given discontinuity, and how to depict their directions with respect to the plane. The stereographic projection is useful here, particularly as regards the directions of shear stresses.

Stress

The state of stress at a point in a solid is described by the set of three vectors representing the force per unit of area transmitted across three perpendicular faces radiating from the point in question. Figure 6-37a shows the naming convention for these vectors and their components.

Choosing axes arbitrarily in orthogonal directions $\hat{x}$, $\hat{y}$, and $\hat{z}$, the plane perpendicular to $\hat{x}$ is called the x plane and the vector of force per unit area across it is $\vec{P}_x$. It is usual to represent $\vec{P}_x$ by the following three components: one normal to the x plane, σ_x; one in the x plane parallel to y, τ_{xy}; and one in the x plane parallel to z, τ_{xz}. The state of stress is thus represented by nine components; but consideration of moment equilibrium about each axis in turn leads to three results of the form $\tau_{yx} = \tau_{xy}$ and thus the complete state of stress at a point is designated by six independent components.*

$$\{\sigma_{xyz}\} = \begin{pmatrix} \sigma_x & \tau_{yx} & \tau_{zx} \\ \tau_{yx} & \sigma_y & \tau_{yz} \\ \tau_{zx} & \tau_{yz} & \sigma_z \end{pmatrix} \qquad (11)$$

It the state of stress is known with respect to axes $\hat{x}$, $\hat{y}$, and $\hat{z}$, the vector of force per unit area may be computed for a plane in any other orientation, e.g. with normal parallel to $\hat{x}'$ and containing perpendicular lines $\hat{y}'$ and $\hat{z}'$. It will be necessary to determine the angles: from $\hat{x}'$ to $\hat{x}$, $\hat{y}$, and $\hat{z}$; from $\hat{y}'$ to $\hat{x}$, $\hat{y}$, and $\hat{z}$; and from $\hat{z}'$ to $\hat{x}$, $\hat{y}$, and $\hat{z}$. The table of direction angles is most easily determined using stereographic projection, as in figure 6-37b. Denote the cosines of the first three angles as l_1, m_1, and n_1, of the

---

* The brackets { } denote a column matrix; arrays are enclosed by ( ).

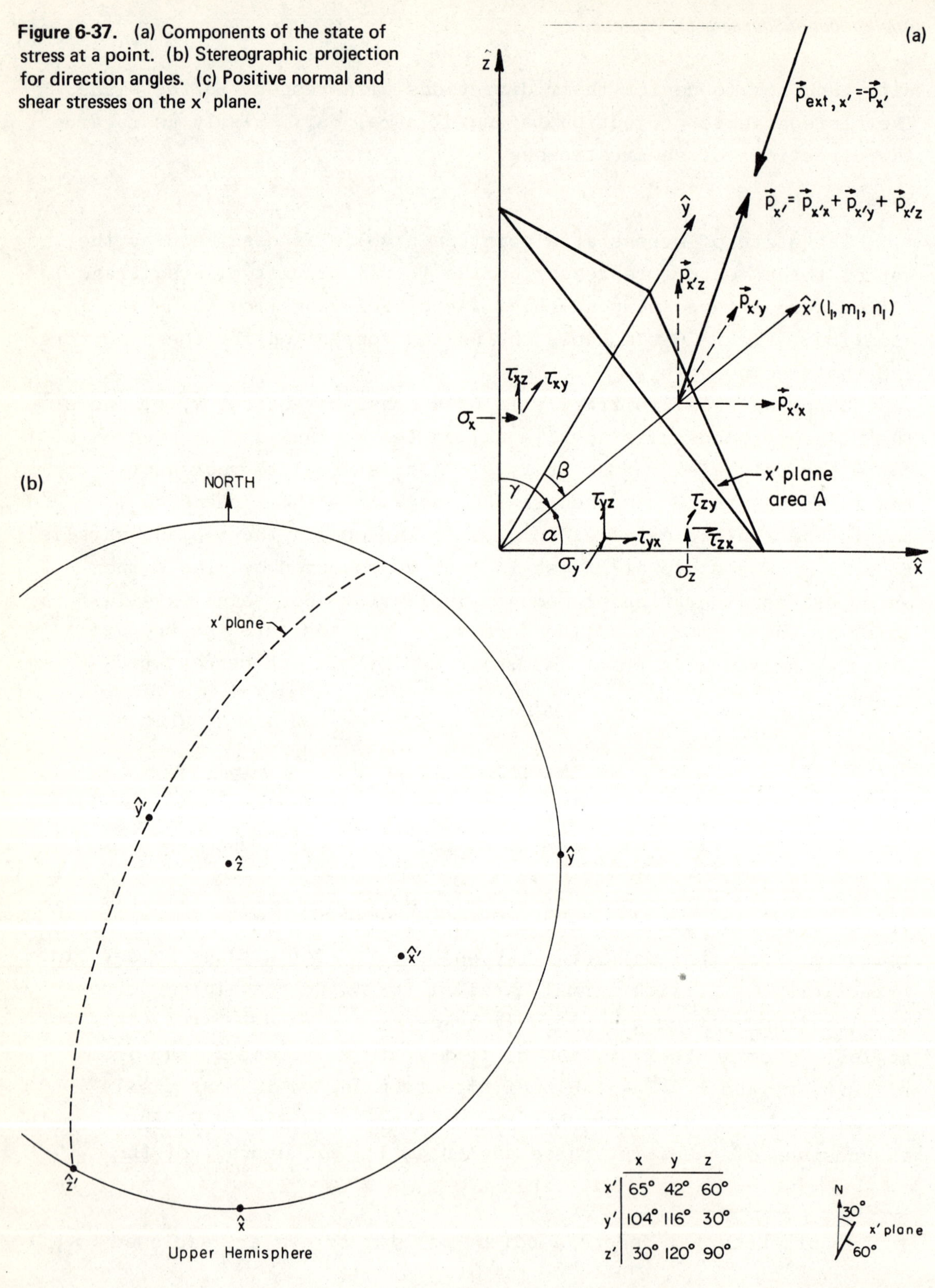

	x	y	z
x′	65°	42°	60°
y′	104°	116°	30°
z′	30°	120°	90°

Figure 6-37. (a) Components of the state of stress at a point. (b) Stereographic projection for direction angles. (c) Positive normal and shear stresses on the x′ plane.

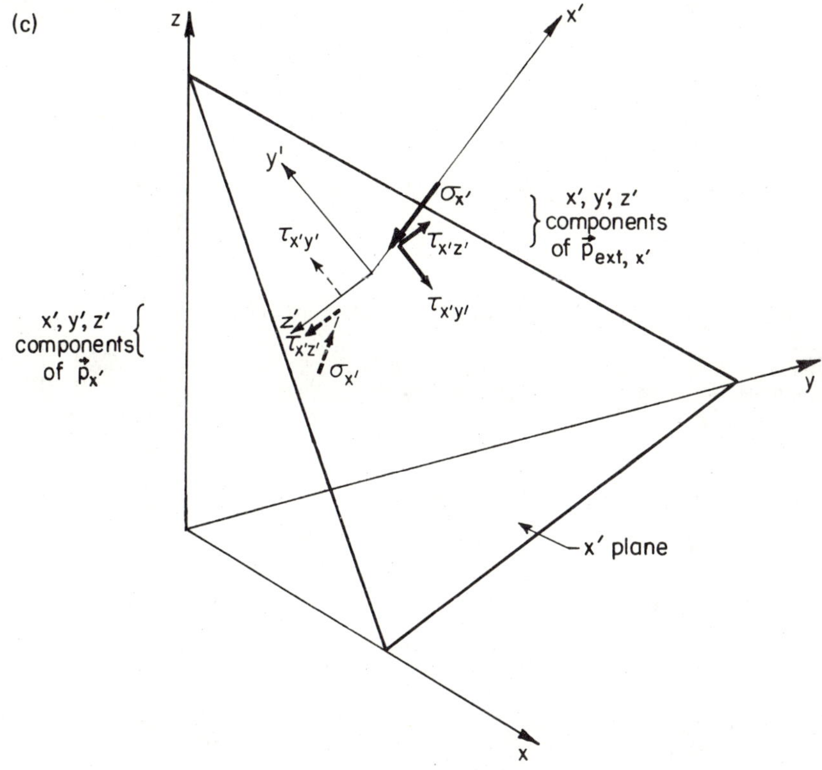

second three angles as l_2, m_2, and n_2, and of the third set as l_3, m_3, and n_3. The x, y, and z components of the force per unit area across the x' plane ($P_{x'}$) are:

$$(P_{x'x}\ P_{x'y}\ P_{x'z}) = (l_1\ m_1\ n_1)\ (\sigma)_{xyz} \tag{12}$$

The equilibrium of the wedge shown in figure 6-37a demands that the force per unit area of the x' plane transmitted from the neighboring material into the wedge across the x' plane is the opposite of $\vec{P}_{x'}$, the vector resultant of the components calculated by equation (12). It is useful to decompose $\vec{P}_{x'}$ into components in the direction x' normal to the x' plane, and in directions y' and z' chosen conveniently in the x' plane; this defines stress components on the x' plane:

$$(\sigma_{x'}\ \tau_{x'y'}\ \tau_{x'z'}) = (l_1\ m_1\ n_1)\ (\sigma)_{xyz}\ (T)^t \tag{13}$$

where

$$(T) = \begin{pmatrix} l_1 & m_1 & n_1 \\ l_2 & m_2 & n_2 \\ l_3 & m_3 & n_3 \end{pmatrix}$$

We may take positive $\sigma_{x'}$ to signify compressive stress. Then if $\sigma_{x'}$ is directed in the positive coordinate direction, the direction of positive x', $\tau_{x'y'}$ and $\tau_{x'z'}$ are positive in the positive y' and z' directions respectively.

As an example, given stresses referred to x, y, z axes (figure 6-37b)

$$(\sigma)_{xyz} = \begin{pmatrix} 10. & 1. & 0 \\ 1. & 4. & 2. \\ 0 & 2. & 0 \end{pmatrix}$$

first we find

$$(T) = \begin{pmatrix} .422 & .743 & .500 \\ -.242 & -.438 & .866 \\ .866 & -.500 & 0 \end{pmatrix}$$

By equation (13)

$$(\sigma_{x'}\ \tau_{x'y'}\ \tau_{x'z'}) = (6.2, \ -1.8, \ 2.1)$$

A complete stress transformation to new coordinates x', y', z' requires repetition of this procedure for each of the three orthogonal coordinate planes. The result is

$$(\sigma)_{x'y'z'} = (T)\,(\sigma)_{xyz}\,(T)^t \qquad (14)$$

Further Use of Stereographic Projection

The above procedure completely solves the stress transformation problem but gives a mathematical rather than physical result.

The stereographic projection shows how the stresses are oriented on the plane of interest.

Consider the x' plane of figure 6-37b, whose direction cosines were previously obtained. The x, y, and z components of the traction $\vec{P}_{x'}$ can be computed from equation (12), giving for the example computation

$$(P_{x'x}, P_{x'y}, P_{x'z}) = (4.96, 4.39, 1.49)$$

Using the methods previously considered for addition of three force components (figure 6-13) we can obtain the magnitude and direction of the resultant $\vec{P}_{x'}$. On the stereographic projection (figure 6-38), $\hat{A}$ is the orientation of $(\vec{P}_{x'x} + \vec{P}_{x'y})$ and $\hat{B}$ is the orientation of $(\vec{P}_{x'x} + \vec{P}_{x'z})$; they can be plotted using the angles $\delta_{xy} = \tan^{-1}(P_{x'y}/P_{x'x})$ and $\delta_{xz} = \tan^{-1}(P_{x'z}/P_{x'x})$. Then the direction of $\vec{P}_{x'}$ is $\hat{P}_{x'}$ located at the intersection of great circles $\widehat{\hat{A}\,\hat{z}}$ and $\widehat{\hat{B}\,\hat{y}}$ while the magnitude of $\vec{P}_{x'}$ is $|\vec{P}_{x'}| = (P_{x'x}^2 + P_{x'y}^2 + P_{x'z}^2)^{\frac{1}{2}}$ In the example previously considered, $\delta_{xy} = 41.5^{\circ}$, $\delta_{xz} = 17.5^{\circ}$ and $|\vec{P}_{x'}| = 6.8$.

Finally, the orientation of the maximum shear stress in the x' plane is $\hat{\tau}_{x',max}$ found as the nearest intersection of the x' plane and the plane common to $\hat{x}'$ and $\hat{P}_{x'}$. As shown in figure 6-38, $\tau_{x',max}$ is in the lower hemisphere pitching 39° below horizontal from the southwest. In this upper hemisphere representation, its opposite is the direction of the <u>external</u> shear force on the plane, as shown in the inset to figure 6-38. The magnitudes of $\tau_{x',max}$ and $\sigma_{x'}$ are obtained by reading the angle between $\hat{P}_{x'}$ and $\hat{x}'$, which is here 24°. Then $|\sigma_{x'}| = 6.8 \cos 24^{\circ} = 6.2$ and $|\tau_{x',max}| = 6.8 \sin 24^{\circ} = 2.8$. The signs on the stereonet can be interpreted as follows. $\hat{P}_{x'}$ produces <u>compression</u> if it makes an angle <u>less</u> than 90° with $\hat{x}'$, which must be visualized as the outward normal to the wedge. Sliding of the contiguous block will tend to occur <u>down</u> the face of the wedge if $\tau_{x',max}$ is in the <u>upper</u> hemisphere.

CONCLUSION

The constructions utilized in this chapter are basic vector

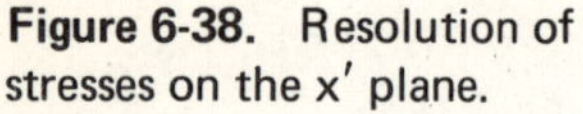

Figure 6-38. Resolution of stresses on the x′ plane.

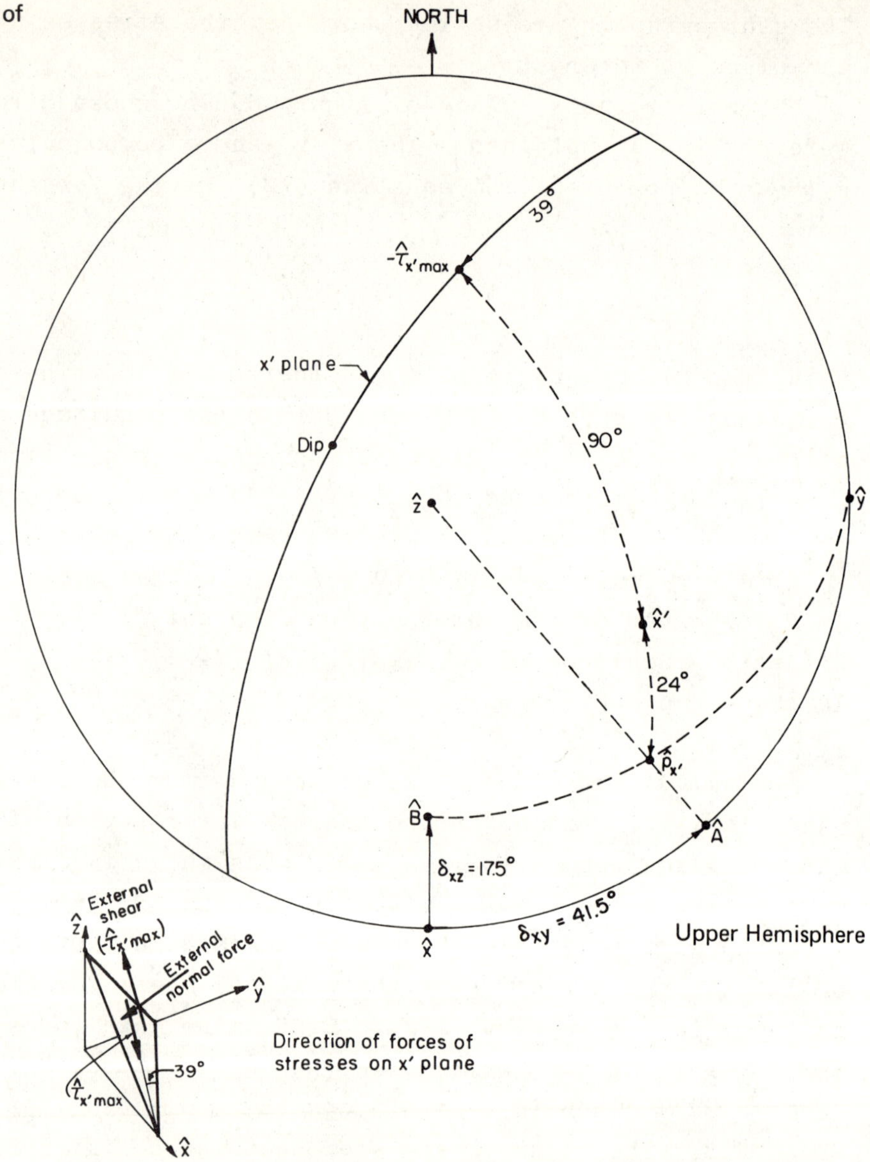

operations. The illustrations are pertinent to stability of blocks, and resolution of stresses; but the methodology can be followed for other problems of rock masses involving vectors. Among these are water flow through jointed rock systems, analysis of measured rock movements, interpretation of geophysical measurements, and other fields yet to be identified.

7
physical models

Since structures in discontinuous rock masses usually provoke opening or shearing of some discontinuities, almost every real rock engineering problem is too complicated for closed form, mathematically based calculations. As noted in the introduction to chapter 6, rock behavior can be assessed if the most likely modes of failure can be identified correctly; simple conceptual models are useful for this purpose. Once identified, the modes of failure can be weighed experimentally in scaled physical models, or computationally, using numerical models. Numerical methods are the subject of Chapter 8. In this chapter we will consider physical model methods, by means of which the behavior of discontinuous rock masses may be explored and extrapolated to prototype dimensions and conditions.

KINEMATIC MODELS

The word prototype refers to an idealization of the field problem in which only those factors considered essential and relevant have been retained. In the physical model study, the prototype will be duplicated at a convenient scale with a minimum of distortion with respect to the more important properties. It is good practice to experiment first with simpler, distorted models to determine the essentials which must be duplicated. Such a preliminary kinematic model study, for example, may observe the changing modes of behavior

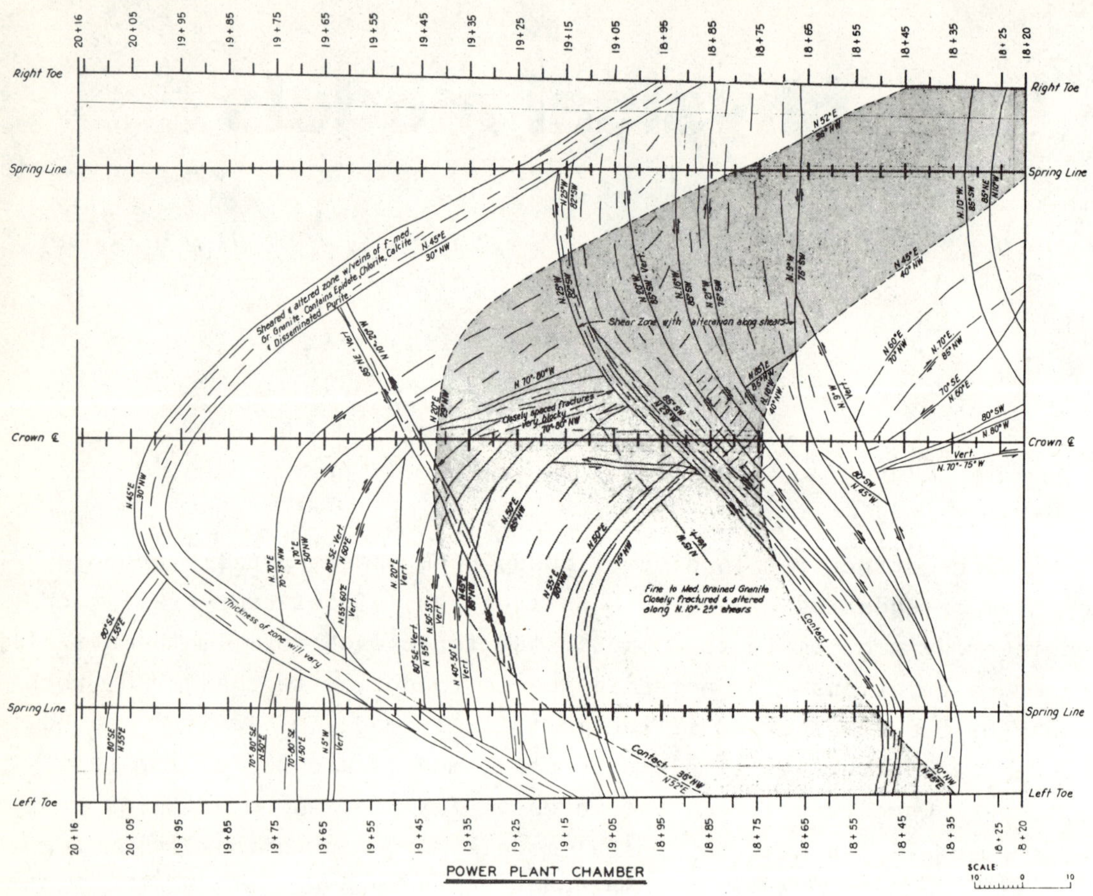

Figure 7-1. A geological log of an underground excavation prepared by Corps of Engineers geologist James Zeltinger (Omaha District) in such a way that it transforms into a geological model of the discontinuities when rolled into the proper shape. The method of preparing such logs is described by Jack (1969). Courtesy of the Corps of Engineers.

corresponding to different jointing styles as determined by geological mapping and study of geological models (figure 7-1).

A useful material for simple two-dimensional kinematic models of excavations in discontinuous rocks is a mixture of flour, cooking oil, and sand, similar to children's "Play Dough." Such a material, mixed to below its plastic limit, combines cuttability and rigidity. When smoothed out into a sheet 1/4" to 1/2" thick, lines of discontinuities may be pressed or cut. Each new cut compresses the model by the width of the blade so that it is possible to produce a variety of jointing styles -- open or closed, planar or imbricated -- by programming the cutting sequence. According to the oil content, the mixture may be made relatively plastic or brittle, and heterogeneous structures can be modelled readily.

The main requirements for such a study concern cost and facility, since it is by repetitions and parameter variations that the full

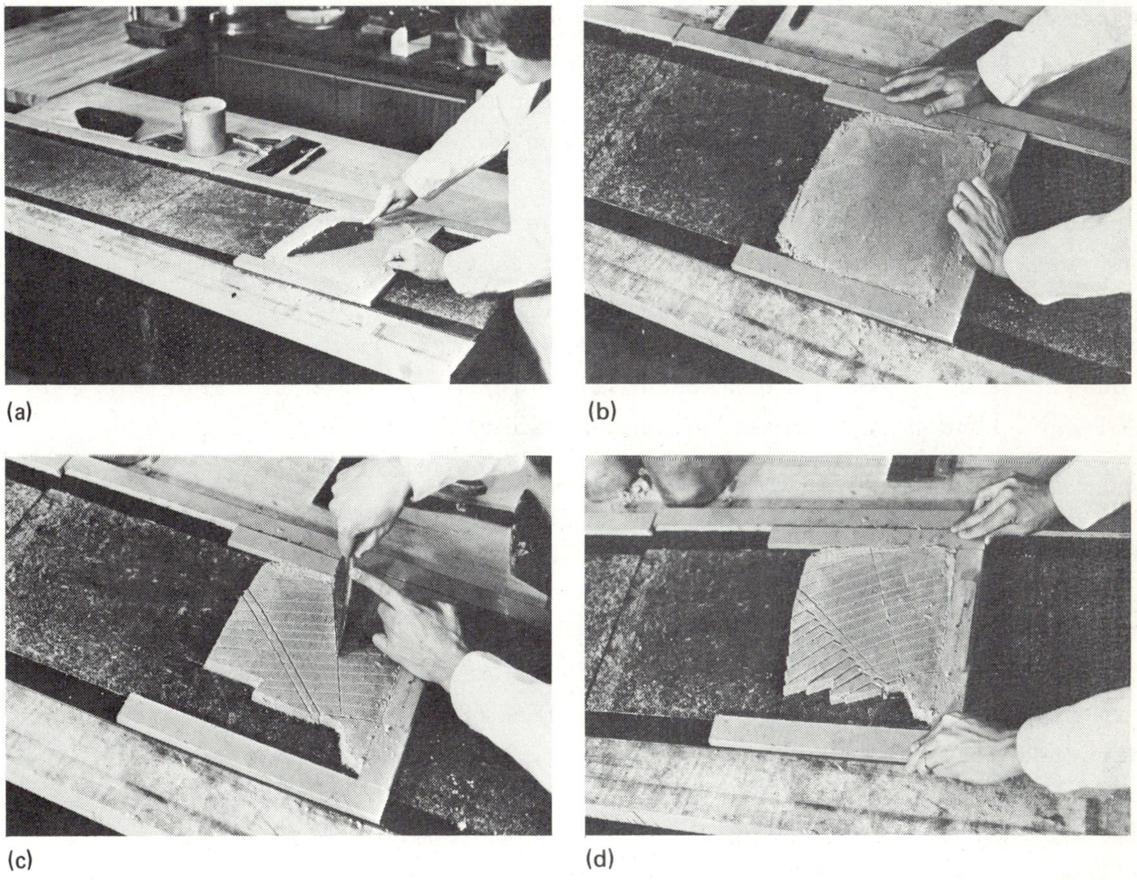

(a) (b) (c) (d)

Figure 7-2. Procedure for conducting a base shear kinematic model study of a gravity loaded rock structure.
(a) Using a trowel, smooth out a sheet of model material.
(b) Give the sheet a push to break the bond along its base.
(c) Cut the outline of the excavation (in this example, a rock cut with benches) and the system of discontinuities.
(d) Push the model against the direction of gravity. The slope is failing by toppling, with the lower limit of toppling defined by the discontinuity inclined towards the free surface.

spectrum of deformational modes can be studied. A mixture of oil, sand, and flour is cheap, universally available, and immediately reusable. However, any mixture of ingredients which is pressed into its final consistency rather than hardened, e.g., by cooling or cementing, can also be used easily in such experiments. Precut blocks of plastic (Trollope, 1966), sugar cubes, wood, and cork, have been used advantageously.

Figure 7-2 shows the sequence of preparing and running one model experiment, in this case under simulated gravity loading. The experiment is conducted by pushing the prepared model along a roughened surface. Each element of the model feels a shear force along its base which "follows" the deformation. The lower edge of the model represents a line of constant elevation, i.e., stationary points on the base surface are effectively "moving downward" as the model is pushed upward. Alternatively, the base may be moved while the model is restrained, as in demonstrations conceived by Dr. Evert Hoek* conducted on the surface of an overhead projector and displayed on a projection screen. Figure 7-3 shows a large simple and inexpensive "base friction" modelling machine similar to machines at Imperial College, London, and at Golder and Brawner's offices in Vancouver. A continuous sand paper belt is driven at constant velocity by a small motor, creating the gravity effect in the model which is restrained from following the sand paper by a fixed barrier.

For experiments with excavations loaded by boundary pressures where the loading does not pursue the deformations, the model may be loaded by applying suitable thrusting provisions in the style of the prototype. It is also possible to combine gravity and non-following directed loads, although the complexity of the required set-up may defeat the basic premise of the supposedly simple kinematic analysis. Release of initial stresses cannot easily be modelled, but the relative effects and behavior styles following excavation in media of different ratios of horizontal to vertical initial stress may be considered by pushing in a direction opposite to the vector sum of the vertical and horizontal initial normal stresses (Goodman, 1973).

---

* Innaugural Lecture, Imperial College of Science and Technology, London, Feb. 1971.

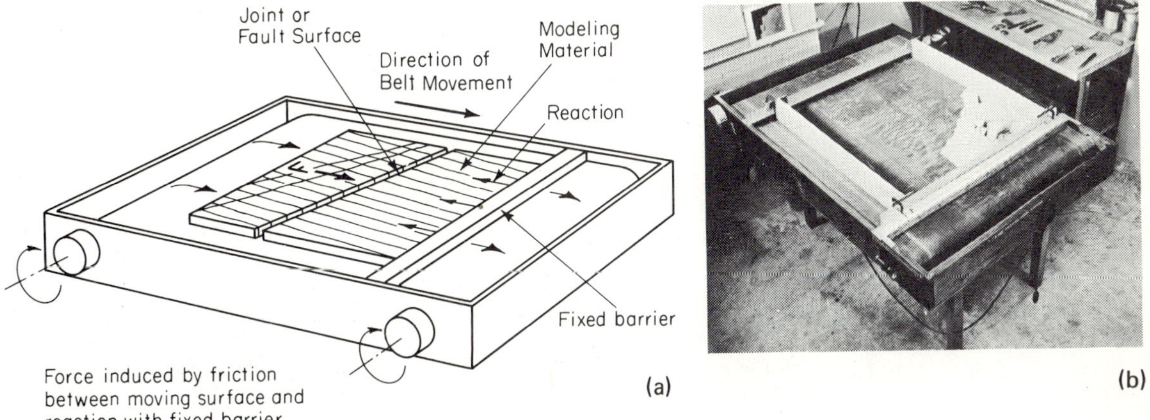

Figure 7-3. (a) Kinematic model machine used by Golder-Brawner Association to study strata movement in underground mining ventures. (b) Kinematic modelling machine in the rock mechanics laboratory at the University of California, Berkeley.

Suppose the belt of the apparatus in figure 7-3 is driven at a constant velocity v, small enough that inertia forces are negligible. In a model subjected to true gravity, g, a falling block moves a distance $s = \frac{1}{2}gt^2$ in time t. In the base friction model in time t_b an unsupported block "falls" a distance $s = vt_b$. The displacement at time t in the prototype is, therefore, correctly indicated by the base friction model at time t_b given by

$$t_b = \frac{gt^2}{2v} \tag{1}$$

Dr. John Bray* found this similitude requirement also to be correct for the case of a block sliding in a given direction on a uniform discontinuity. If the discontinuity is inclined at an angle $i > \phi$, the gravity loaded model accelerates downslope, acquiring displacement $s(t) = \frac{1}{2}g\,(\sin i - \cos i \tan \phi)t^2$. In the base friction model, the block and its support move differentially with the relative velocity vector $\vec{v}'$ directed parallel to the resultant force $\vec{R}$ across the discontinuity (figure 7-4).

---

* Unpublished notes on similitude in the base friction model, March 1973, Imperial College, London.

Figure 7-4. Mechanics of base friction model of a block on an inclined plane.

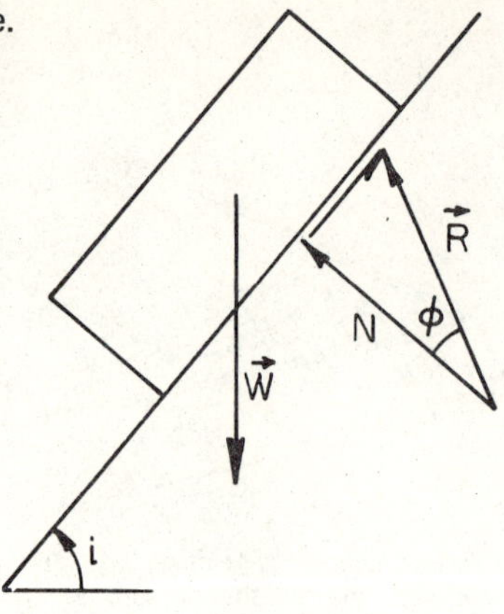

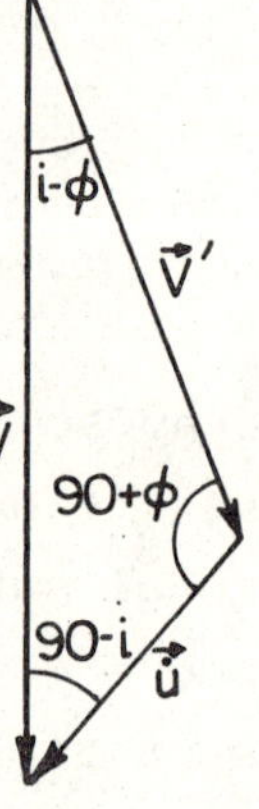

$\vec{u}$ = downslope velocity of block
$\vec{V}$ = sandpaper belt velocity
$\vec{V}'$ = relative velocity of block and support

The velocity vector triangle (figure 7-4) gives

$$\frac{v}{\sin(90 + \phi)} = \frac{\dot{u}}{\sin(i - \phi)}$$

or

$$\dot{u} = \frac{v \sin(i - \phi)}{\cos \phi} = v(\sin i - \cos i \tan \phi) \qquad (2)$$

In time t_b, the block subjected to base shear moves downslope a distance $s(t_b) = v(\sin i - \cos i \tan \phi)t_b$. Equating displacement in base shear and gravity models, we again find the result of (1). We lack such a simple result for similarity in cases of overturning, although qualitatively the model results appear to be defensible. The base shear method cannot duplicate the correct response when the moving body acquires horizontal momentum since there is no mechanism

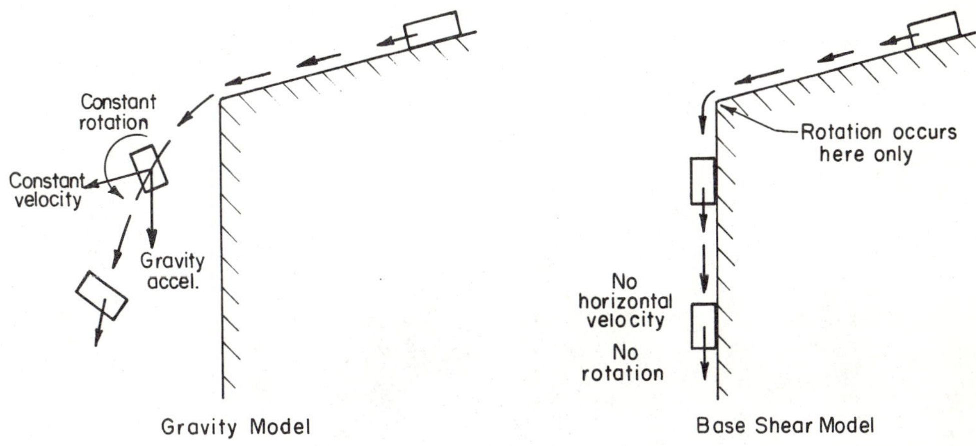

Figure 7-5. Dissimilarity of base friction and gravity loaded models where momentum is not negligible; after Dr. John Bray.

for persistence of translation and rotation after "impact," or change in friction of the surface of sliding (figure 7-5).

Figures 7-6 through 7-9 present examples of simple kinematic models in which deformational modes of discontinuous rock masses were examined. Figure 7-6 shows base shear models of rock slopes while figure 7-7 presents underground excavations loaded by the base shear technique. The latter figure illustrates potential usefulness of this modelling technique in comparing shapes for underground excavations. Figures 7-8 and 7-9 illustrate kinematic models loaded by pushing the structure in the direction of the vector of net load on the structure. The embankment dam models in figure 7-8 display foundation cracking, opening of joints, rotation and flexure in the foundation, and cracking inside the embankment. The arch dam case, figure 9, is crudely simulated by pushing the whole arch directly downstream, with a concentrated push at its center; it demonstrates wedge sliding and flexural failure modes. In the case of an actual dam, cracks once formed may fill with water at some high percentage of the reservoir head, further damaging the rock mass. This is hard to model kinematically.

In making use of the results of kinematic model exercises, it must be understood that the behavior modes realized in the models may be quite unlikely in the actual structure. The business of assessing and weighing the various possible behavior modes requires attention to dynamic similitude. Whereas, kinematic models are made in an attempt to understand the problem better, the physical models will be studied in an attempt to solve it.

Figure 7-6. Kinematic base shear models of rock cuts with two systems of discontinuities.
(a) A cut with horizontal benches, before pushing.
(b) The same cut as (a) after pushing. Sliding of a wedge occurs, causing overturning of loose blocks under the benches. Note the caves between joints inside the loosened rock mass. If the wedge were to stop moving, for some reason, these caves would be preserved as evidence of the previous slide movement.
(c) Another wedge slide in development, causing caves to from between joint blocks.

PHYSICALLY SCALED MODELS

If dynamic similitude can be closely approached, a model can reproduce deformation and failure of a discontinuous rock mass. With regard to excavations and foundations in rock masses, we are interest-

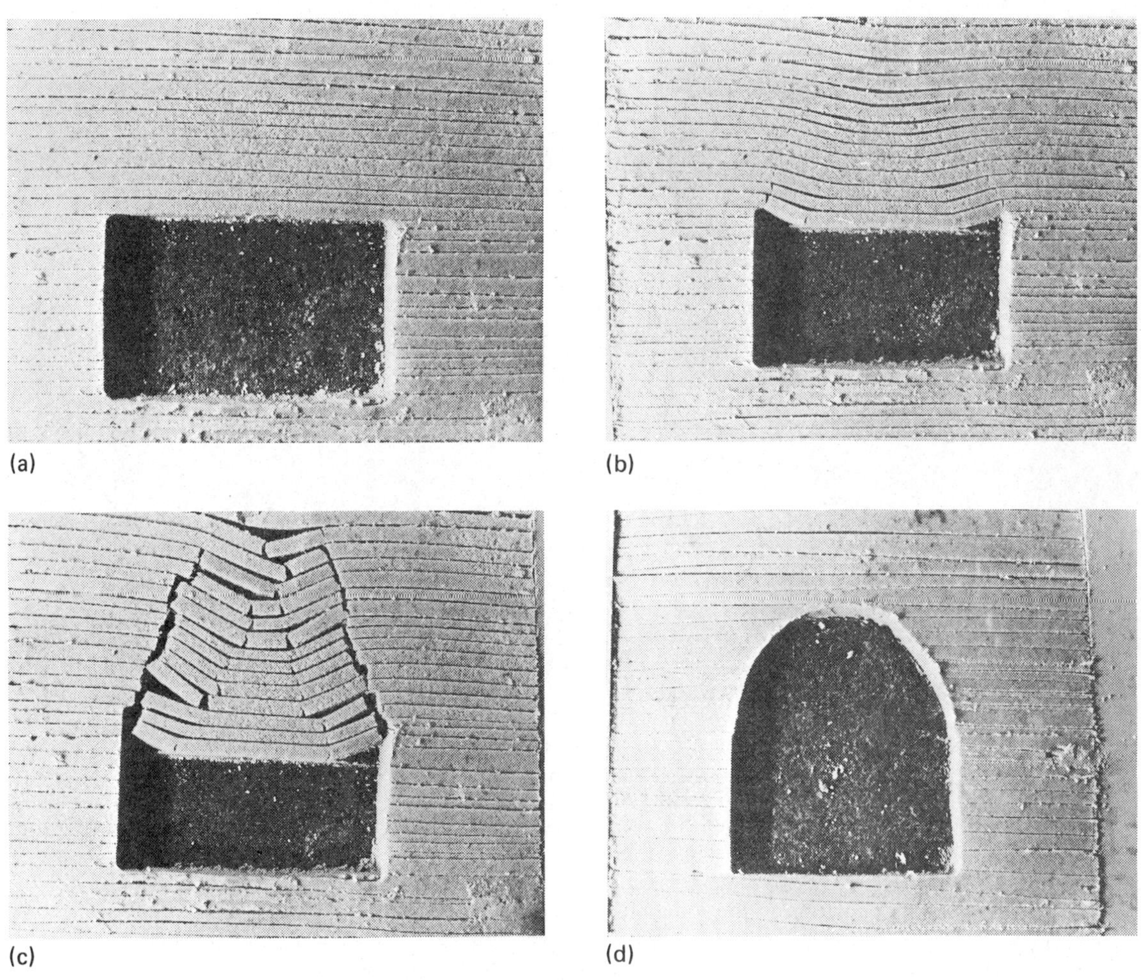

(a) (b) (c) (d)

Figure 7-7. Kinematic base shear model showing the effect of excavation shape on the deformation mode.
(a) Flat roofed opening with horizontal bedding before "turning on gravity" by pushing the model.
(b) Classical symmetrical flexure pattern; the upper lines of flexural cracks extend through eight layers, with tolerable bending in the layers above.
(c) Further propagation of cracks and the beginning of a fall from the roof.
(d) An arch shaped roof of the same span; stable.

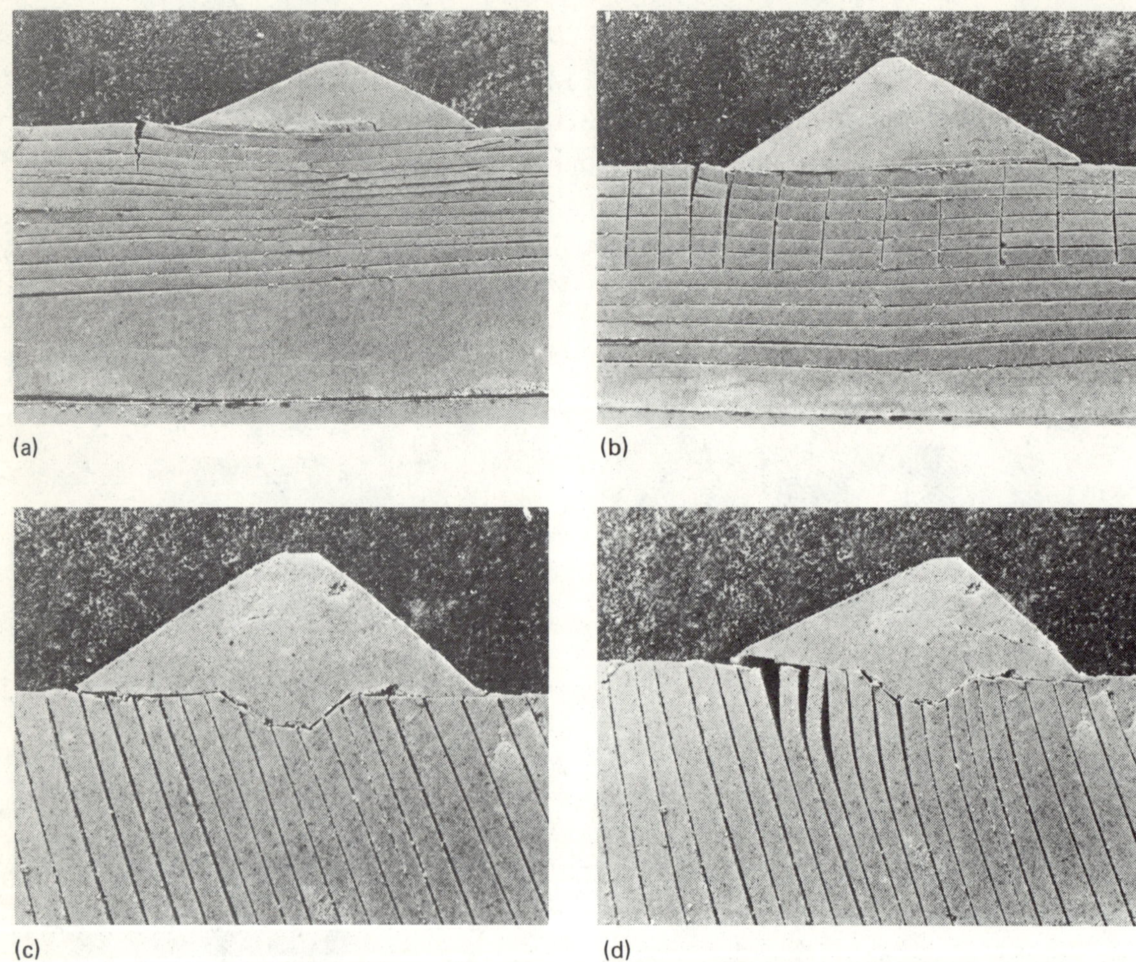

Figure 7-8. Kinematic models of earth dams on bedded and jointed foundations.

(a) Horizontal beds in the foundation close as the dam is pushed in the direction of the load transmitted to the foundation by the dam and reservoir. Tension cracks form in the upstream portion of the dam foundation and the differential deformation of the foundation causes cracking in the embankment.

(b) A similar case, except that vertical joints in the upper part of the foundation accommodate the downstream deformation without new tension cracking. The dam does not crack, perhaps because the additional joints permit shear deformation in the foundation which reduces the shear stress at the base of the embankment. Some rotation of blocks occurs in the upstream part of the foundation.

(c) A rock-fill dam with a shear key on a roack mass with beds dipping steeply downstream.

(d) Case (c) after pushing the dam in the direction of the load transmitted to the foundation by the dam and reservoir. Flexure of the layers in the foundation has occurred, lifting the dam in the downstream part of the toe trench and in the downstream shell as the embankment rotates about its lower middle region.

Figure 7-9. (a) A kinematic model of an arch dam in a canyon with vertical beds striking into the left abutment and slightly downstream. Initial condition; the arch is made of material stiffer than the model material. (b) First stage of deformation as the arch is pushed downstream. Shear failure of the edge of one layer has taken place on the left abutment and flexure of the downstream layers has initiated. On the right side, a tension crack has severed the layer under the thrust of the arch. (c) Continued pushing of the dam downstream has increased the flexure and initiated flexural cracking on the left abutment, while sliding of a wedge defined by the bedding and tension cracks occurs on the right abutment. (d) The dam swings around an axis on the right side as the flexural cracks propagate on the left side.

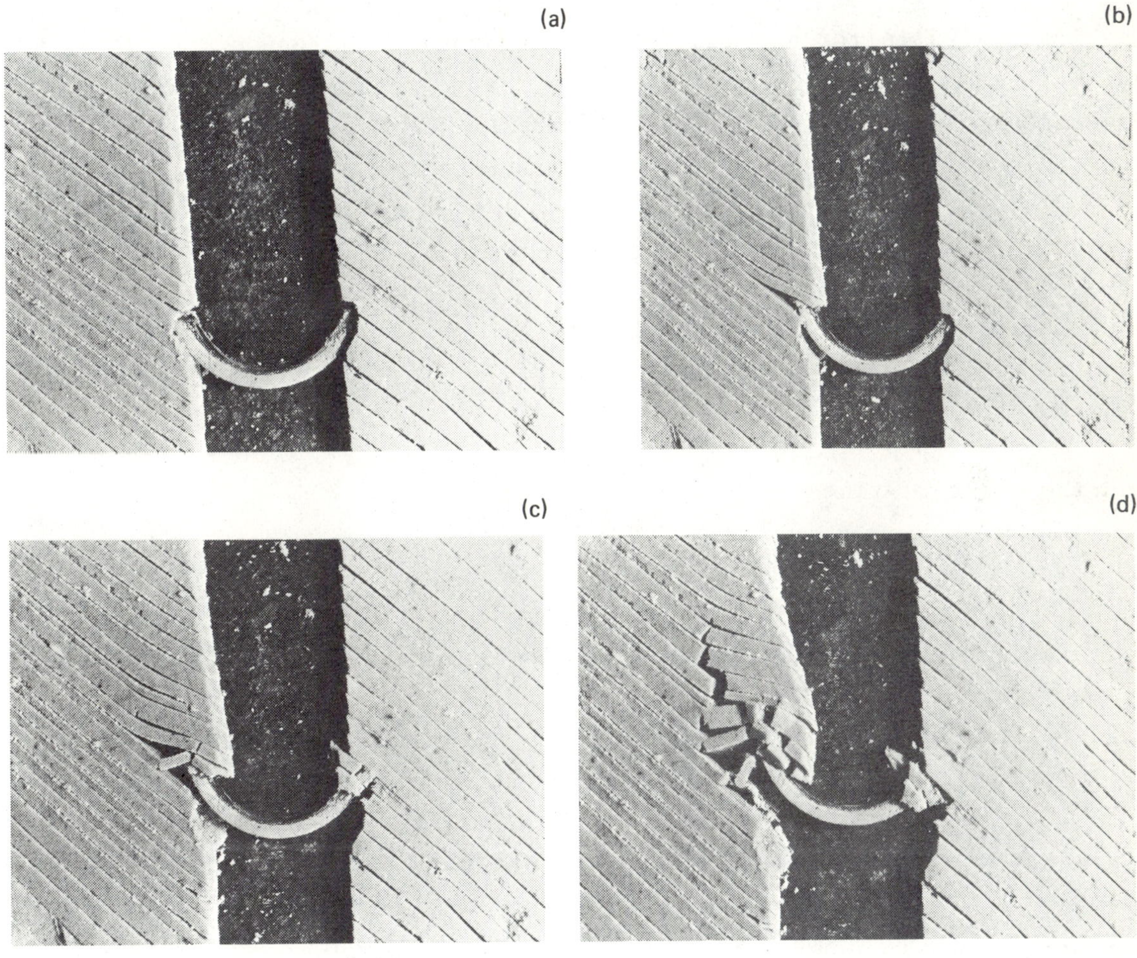

ed in similitude with respect to strengths, deformabilities, lengths, and time. Similarity requirements can be expressed in terms of two independent dimensionless ratios (Fumagalli, 1968) as follows:

$$\xi = \frac{\text{strength of prototype}}{\text{strength of model}} \tag{3}$$

and

$$\lambda = \frac{\text{length of prototype}}{\text{length of model}} \tag{4}$$

The scaling ratio for any quantity having the dimensions of force per unit area is ξ. The scaling ratio for any quantity having dimensions of length is λ. It follows that the scale ratio ρ for quantities with dimensions force/volume (density) will be

$$\rho = \frac{\xi}{\lambda} \tag{5}$$

the scaling ratio ψ for forces will be

$$\psi = \lambda^3 \rho \tag{6}$$

and the scale of time

$$\tau = \sqrt{\lambda/a} \tag{7}$$

where a is the scaling ratio of accelerations; a will be unity unless a centrifuge is used. Dimensionless quantities like Poisson's ratio, angle of friction, and strain should be the same in model and prototype.

It is not possible with a given material to achieve the required scaling ratio for all quantities. For example, the length scale is fixed once the model size is selected. It may be necessary to establish this scale mainly to use existing equipment. Once fixed, the length scale governs other lengths such as the scale of shear displacement at failure, and this may or may not be obtainable with material which has been chosen to meet other requirements. If two quantities of the prototype which have the same dimensions are in fact not equally scaled, the model is said to be distorted. A certain amount of distortion is inevitable. As another example, both the unconfined compressive strength and the modulus of elasticity have dimensions of force per unit area. The ratio of modulus to strength for many rocks is of the order of 500 (Deere et al 1967); in the case of the flour-oil-sand mixture mentioned earlier, this ratio is of the order of 4, so either the deformations or the strength will be distorted. The choice of materials must represent a compromise in which all factors bearing on the results are considered and weighted in importance and the necessary distortions relegated to the least important quantities. For this reason there cannot be one modelling material for all purposes.

In many cases, rock deformations are of first interest, for example in the problem of finding the stresses in a dam resulting from displacement and rotation of its rock abutments. If all rock deformation is in the elastic region, primary attention in selecting a material will be addressed to scaling its modulus of elasticity and Poisson's ratio. Scaling the deformation of the discontinuities necessitates choosing the width of aperture to give the correct value of maximum closure V_{mc} and the correct shear stiffness (see Chapter 5). However, as discussed by Barton (1972), the peak shear displacement in direct shear tests increases with sample size, while the peak strength is independent of sample size. Therefore, the shear stiffness displays a strong scale effect, and the model will give distorted scaling for all but one scale ratio.

Where plastic behavior of the discontinuities or the rock is predictable, as in study of rock failure processes, the material will be selected to scale ϕ and c and the compressive and tensile strengths, while for the discontinuities the peak and residual friction angles and the dilatancy will become the principal objects.

Stimpson (1970) reviewed modelling materials used in rock mechanics. Granular materials such as cement and plaster are dilatant while plastics tend to be non-dilatant. Tables give the unconfined compressive and tensile strength, elastic properties, and ultimate strain for many materials. Materials used to represent rock in model tests in the non-elastic domain include: cement, sand and water; sand, wax, and mica; sand and clay; and plaster alone or mixed with sand, clay, mica, barite, lead oxide, diatomite, sawdust, or lime. Sand and wax mixtures tend to be plastic while plaster or cement sand mixtures are brittle. Strength, deformability, and unit weight can be varied over wide ranges by controlling curing and mixing time, and mixing additional materials. For example, sand plaster mixes are weakened by curing at 90 degrees C, or by adding crushed mica; they are made more brittle with chalk or clay admixture; and adding powdered lead or barite raises their specific gravity.

Joints and other discontinuities have been simulated in many ways -- as rough extension fractures caused by breaking the solid modelling material, as untreated or varnished saw cuts in the solid, and as thin partings of wax, grease, talc, limestone dust, wax paper, oiled tracing paper, lens tissue, graphite, and clay. It is easy to duplicate the low friction and ready parting of important seams and faults, e.g., by grease partings, but it is harder to raise the friction angle above 40° and to control dilatancy. Krsmanovic and Associates developed the technique of imbricating joint blocks, i.e., offsetting blocks slightly to produce block interlocking. The resulting joints show high strength and dilatancy at low normal pressures but they are highly directional in their shear behavior. Barton (1972) developed an indirect tension splitting device (termed a "guillotine") to introduce rough extension fractures in cast plates of model material. In contrast to cut joints, the oldest set of such split discontinuities is the only one which is planar, since all younger discontinuities are offset wherever they cross the earlier sets.

A simple, plane stress model study of a hypothetical underground power station has been used for several years as a class exercise in the rock mechanics course at Berkeley. The model, at 1:150 length scale, is prepared in about one half day by placing hot mixtures of sand, crushed mica, and paraffin (2-4%) according to the desired lithologic section. The model apparatus consists of a wooden box of inside dimensions 30 inches wide by 25 inches high by 4 inches thick (figure 7-10). To construct the model, the cross section is drawn to scale and attached to the plywood front piece (figure 7-11a).

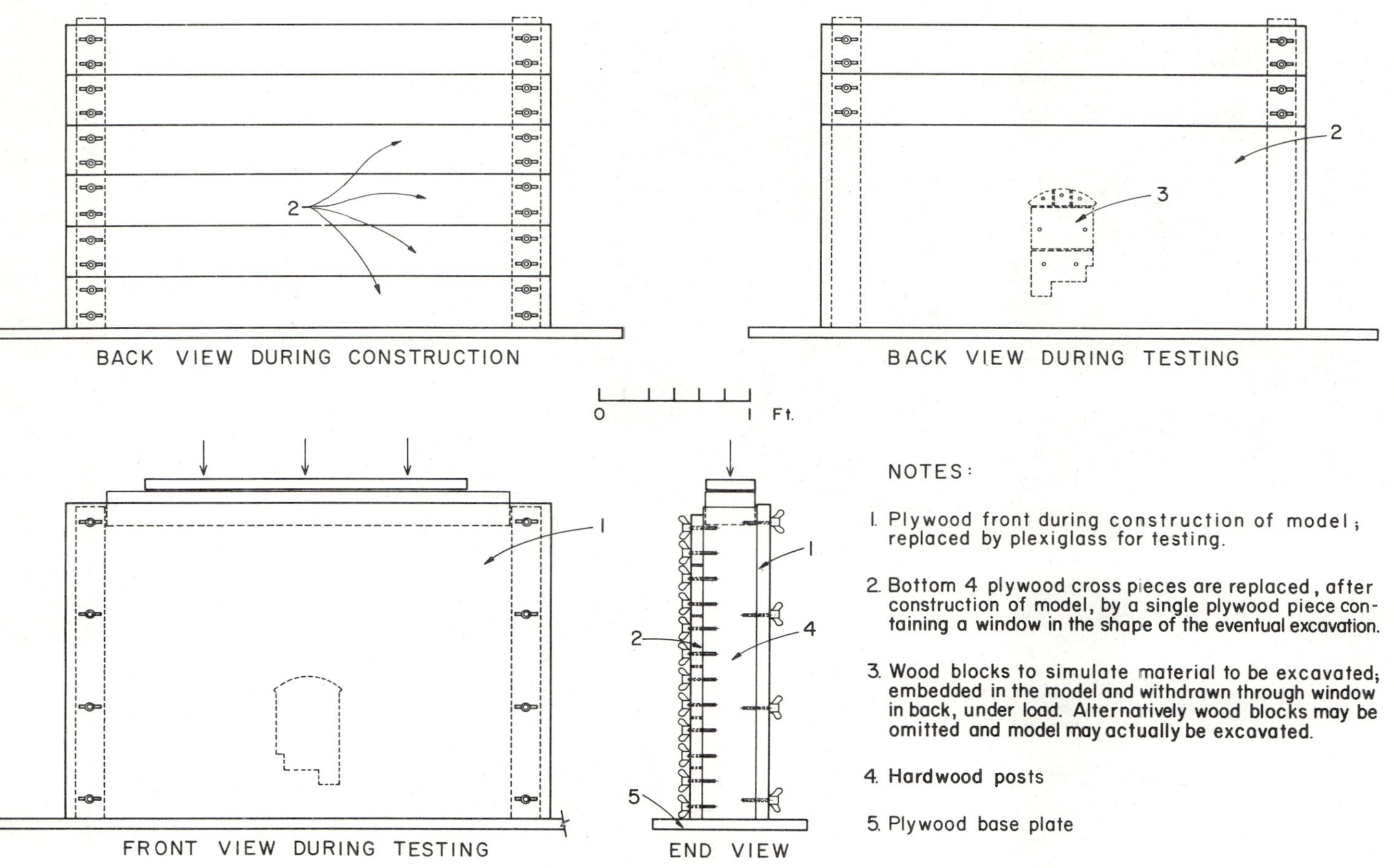

Figure 7-10. An apparatus for two dimensional scale model studies.

(a)

(b)

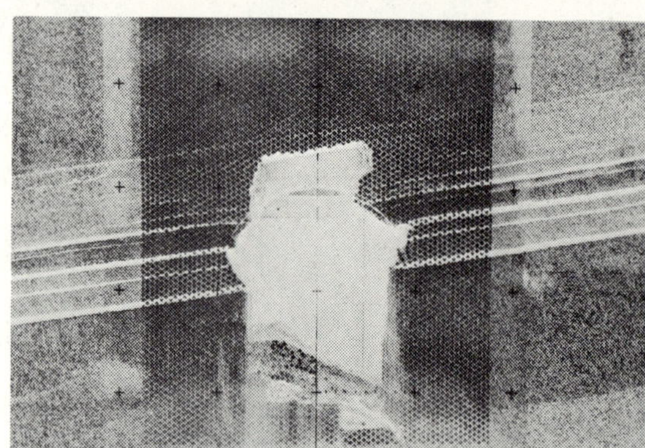

(c)

Figure 7-11. Scale model study using the apparatus of figure 7-10.
(a) Model during construction, viewed from the back.
(b) Before testing, viewed from the front.
(c) After testing to failure.

The back of the model is installed as the model is built up; then the plywood front is removed, a grid of reference points is sprayed over the surface, and the clear plastic front is attached. To simulate the excavation sequence, hardwood blocks having the shape of each excavation stage are buried in the model (figure 7-11b) and later withdrawn under load, through a hole in the plywood back.

Simple studies such as this are useful for many facets of engineering for underground and surface excavations, for example: identifying the critical points of an excavation; studying variations of shape and location; developing rational bases for designing supports; and putting in context the deformation readings of individual instruments. Much more elaborate studies are possible, of course, and in fact may even serve as the basis for quantitative design decisions. Figure 7-12 is a typical result from many coal mining

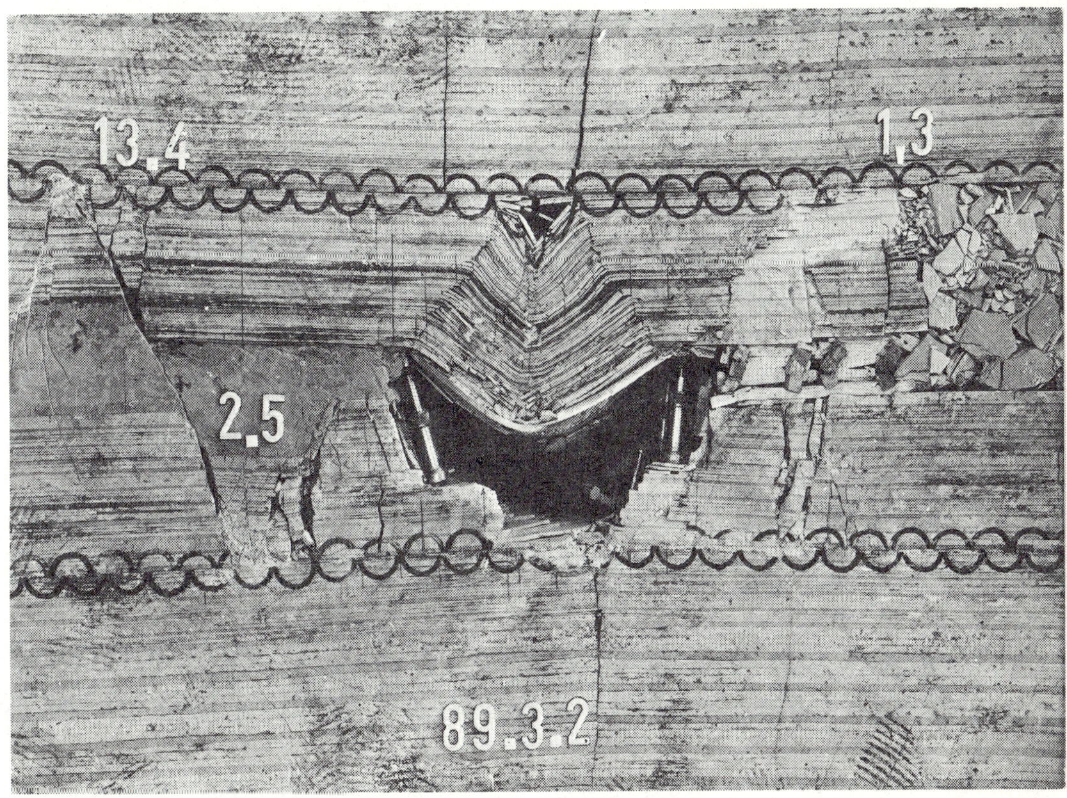

Figure 7-12. Flexural failure of a laminated mine roof; a model studied by Dr. Everling at the Bergbau-Forschung, Essen, West Germany. Courtesy of Dr. Everling.

studies in the Ruhr district of Germany.

Two dimensional models of long structures such as tunnels, shafts, road cuts, and long-wall mine faces will usually be more closely correct in a plane strain configuration than in plane stress. There is no fundamental difference between them in elastic problems, but in models carried into the rupture region there may be great differences between plane strain and plane stress. Failure will tend to occur in and along the discontinuities in either case, but new fractures in the rock will propagate in the plane of least principal stress. In the plane stress model, if there is no tension stress, the material will crack longitudinally parallel to the model plane. To prevent such cracks, plane strain conditions can be approached by providing stiff, constrained front and back plates. Unfortunately, this introduces significant friction as well as limiting access and observability.

Three dimensional models, more difficult to build and to instrument, are justifiable for many problems which are truly three dimensional since no other engineering method of analysis can yet duplicate complex kinematic, geologic, and structural requirements in three dimensions. Examples of three dimensional problems are caving and subsidence above mines, roof conditions in underground intersections, conditions at the face of a tunnel, an arch dam and its foundation and abutments, and an open pit mine or spillway excavation. Further, almost any anisotropic rock mass necessitates a three dimensional analysis as only rarely will the planes of elastic and shape symmetry coincide. The following illustrates a careful three dimensional model study of a problem in discontinuous rock.

Example—Grancarevo Dam, Yugoslavia

A model of this 120 meter high arch dam and its bedded and jointed foundation was made by Prof. Krsmanovic and co-workers Langof and Tufo of the faculty of Civil Engineering, Sarajevo. The limestone beds, 0.2 to one meter thick, separated frequently by clay partings, dip gently upstream and into the right bank (figure 7-13). The rock is also extensively jointed with smooth and rough unfilled discontinuities and has calcite and clay-filled minor faults. Statistical analysis led to a simplified picture with two joint sets and additional important individuals in other orientations. Joint and rock properties were measured in the foundation excavation using flat jack and field shear tests ($5m^2$ area) and geophysical measurements. Eleven homogeneous zones were defined with the modulus of deformability in the range 7,600 to 21,500 MN/m^2. Samples of beds, joints and minor faults of the rock were also tested in the laboratory, resulting in classification of the discontinuities into several behavioral types with friction angles from 15 to 25 degrees for clay-filled seams and considerably higher for rough, unfilled joints. In the latter case, ϕ varied continuously with normal stress.* Properties

* ϕ is defined at any normal stress (σ) as $\tan^{-1}(\tau_p/\sigma)$. A "c - ϕ material" by this approach becomes a material in which ϕ varies with σ.

Figure 7-13. Bedded limestone in the left abutment of Grancarevo Dam, Yugoslavia; courtesy of Dr. M. Popovic, Institute for Geotechnics and Foundation Engin., Faculty of Civil Engin., Sarajevo Yugoslavia.

of the intact rock were established as follows: E = 80,000 to 100,000 MN/m^2, q_u = 100 to 150 MN/m^2, and specific gravity = 2.5.

The scale ratios selected were:

λ = 80 (the ratio of lengths)

ρ = 1 (the ratio of unit weights)

giving ξ = 80 (the ratio of stresses).

The material selected for the dam actually had a lower density than required; additional weight was added externally at distributed points. The material selected for the rock blocks - - a clay, plaster mixture - - had values of $E = 1100\ MN/m^2$, $q_u = 1.4\ MN/m^2$, and specific gravity 2.5. The latter high value was achieved by admixing lead grains.

Thirty three of the more important individual discontinuities in varying attitudes were represented with friction angles assigned either 25 or 40 degrees. The recurrent joints and bedding were approximated by an imbricated pattern of blocks with an average of 2mm of offset per 9cm of length. Choosing one size of blocks fixes the joint spacing and would determine the deformability identically throughout if the joints were identical. However, by varying the initial normal compression and fit during model building, the initial normal deformability could be adjusted within the required range. 111,000 cubes 4cm on edge were built into the model. Clay layers were duplicated by grease partings, giving ϕ of about 14 degrees. Continuous variation in bedding plane attitudes was not realized, but the volume of the model was subdivided into seven zones of constant dip; partial cubes were placed to complete the space at zone boundaries.

For construction, points were located in space using plumb lines suspended from a plan placed above the model (figure 7-14). The dam was poured in place and loaded by hydraulic cylinders on the upstream face. Body forces due to seepage in the abutment were not duplicated in the model but to simulate uplift in the foundation, a rubber pillow was inflated inside the rock mass under the dam upstream of the grout curtain. Instrumentation consisted of pressure sensing blocks emplaced in the rock mass and strain gages and displacement points on the rock and dam surface.

The investigators learned that the rock mass took less time to adjust to load or unload increments than did the dam itself, presumably due to the dam's higher stresses. First the reservoir load was applied and removed quickly; then, before proceeding with the

(a)

(b)

Figure 7-14. (a) View of the model during construction; note the use of plumb lines to locate points in plan; courtesy of Dr. M. Popovic, Sarajevo. (b) A general view of the model; courtesy Dr. M. Popovic, Sarajevo.

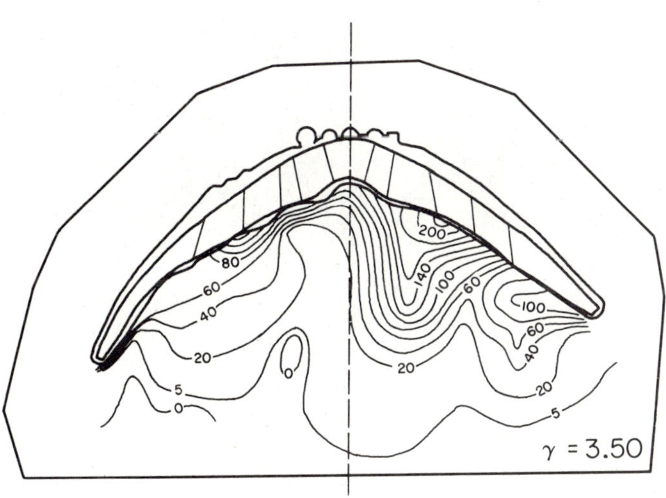

Figure 7-15. Lines of equal deformation on the rock surface under 3.5 times the reservoir load. Displacements in mm. at the prototype scale.

next level of loading, the load was sustained for 5 to 10 days. Figure 7-15 shows lines of equal displacement of points on the rock surface at a loading corresponding to three and a half times that of the full reservoir ($\gamma = 3.5$). The foundation deformations in the lower left abutment were the object of a reinforcement scheme using prestressed cables as shown in figure 7-16. That part of the

Figure 7-16. Reinforcement scheme to stiffen and strengthen the foundation in the left abutment. It was designed on the basis of the results of the model study.

foundation contains unfavorably oriented strata and clay seams.

The results of this comprehensive model study evaluated for the designers what the reserve of strength was likely to be and showed where the weaker points lay in the dam and in the foundation. The studies also determined the size of the volume of rock participating in the structural behavior of the dam under full reservoir load. The model indicated where improvement could be made in design and where the rock might be strengthened. And it provided a framework for evaluating the response of instruments in the actual structure. Only a physical model, presently, can provide this degree of knowledge about a structure on discontinuous rocks. The principal difficulty in such a study is that after tests to rupture, it cannot be repeated readily with different parameters. Numerical models, which have this capability, may eventually be able to duplicate the three dimensional rock details as faithfully as the physical model.

8
the finite element method

INTRODUCTION

Finite element analysis is a digital computer method for stress analysis and other tensor field problems of large size. It is especially powerful for non-linear rock problems introduced by heterogeneities and discontinuities for which closed form solution methods are difficult and special.

Not all problems require a finite element solution. Problems of elastic continua in two dimensions, and some three dimensional cases, can be solved for many boundary conditions by direct application or superposition of published results of the theory of elasticity. Limiting equilibrium of blocks with simple geometry in slopes for which intermediate stress states are of little interest, are more readily solved by methods explored in chapter 6. Finite element analysis is immediately useful, on the other hand, in situations where displacements need to be known, particularly in heterogeneous or discontinuous rock masses. This need arises when interpreting deformation readings in-situ, when scaling up model experiments or extrapolating prototype studies, and when studying the action of rock foundation movements on the behaviour of structures. A number of references illustrating various applications of finite element analysis are given in the bibliography.

THE METHOD

The finite element method is an application of the direct stiffness method of structural analysis. In this approach, displacements of representative points within the structure, termed nodal points, are the variables of a set of simultaneous equations. The coefficient matrix *, describing the geometrical and physical properties of the structure, is termed the structural stiffness matrix.

"Stiffness" measures the amount of force necessary to produce equilibrium in a body undergoing differential displacements. It is a system property, since it reflects both the shape and size of the body and its physical properties. Consider for example, the stiffness (k) of a coil spring, defined by $W = k(\Delta u)$, where W is the force stretching the spring and Δu is the resulting stretch; k depends not only on the kind of steel used, but on the cross sectional area of the wire, and on its coil configuration (figure 8-1a). Similarly, as shown in figure 8-1b, the compressional stiffness of a cylindrical specimen of elastic rock depends not only on its modulus of elasticity (E) but on its cross sectional area (a) and length (L).

$$k_1 = \frac{aE}{L} \tag{1}$$

When a body has multiple freedoms to deform in various modes, one can divide the whole body into sub-elements and develop a structural stiffness matrix from the individual stiffness components of each element. In the direct stiffness method, after the structural stiffness matrix is assembled, specific forces and constraints are introduced to obtain a specific "solution", that is, the set of displacement vectors at each nodal point. Then the forces or stresses in each component are determined by multiplying the known displacements by the known stiffness terms for each element.

---

*The notations and operations of matrix algebra simplify the material of this Chapter. Matrix algebra is discussed by Wylie (1960).

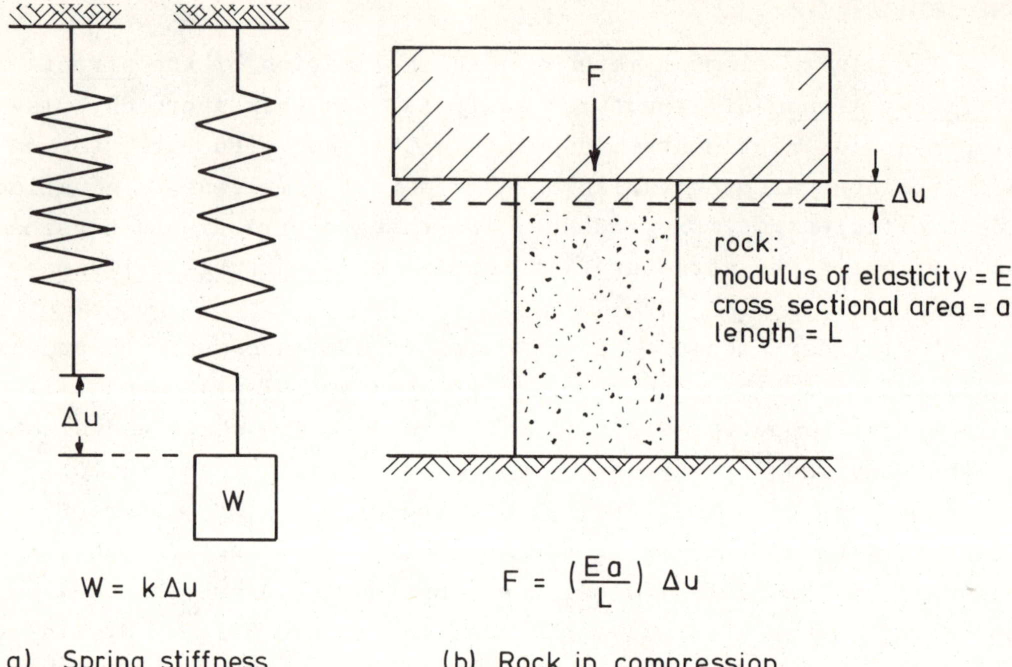

Figure 8-1. Concept of "stiffness".

A simple one dimensional example will illustrate the essential steps followed in the finite element scheme.

Example 1

The structure in figure 8-2a, consisting of a column of two different rock cylinders is loaded by a force P. Find the displacements and stresses throughout the column when the bottom is moved upward by an amount y and held in that place.

<u>Step 1</u>. - subdivide the structure into idealized elements with a finite number of nodal points as shown in figure 8-2b.

<u>Step 2</u>. - form the element stiffness matrix for each element. We rewrite Δu as $U_I - U_J$. Then the external forces at Nodes I and J of element A required by the relative displacement $U_I - U_J$ are

$$F_I = -F_J = k_A (U_I - U_J) \tag{2}$$

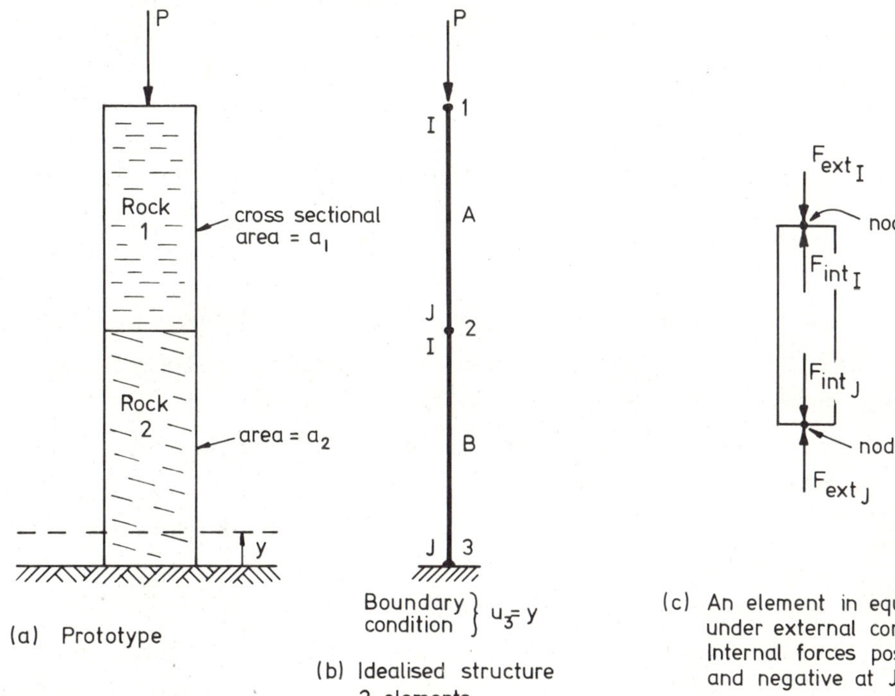

Figure 8-2. Simple illustrative problem—example 1.

The sign convention used recognizes forces and displacements positive when pointed upward. The forces F_I and F_J are external to the element. Their existence implies an equal and opposite reaction inside the element. A downward applied external force P produces a compression in each element, which necessitates internal forces $-F_I$, and $-F_J$ tending to stretch the body back to its original length if P is removed, as shown in figure 8-2c.

The element stiffness matrix for element A is obtained by rewriting equation (2) in matrix notation.*

$$\begin{Bmatrix} F_I \\ F_J \end{Bmatrix}_A = \begin{pmatrix} k_A & -k_A \\ -k_A & k_A \end{pmatrix} \begin{Bmatrix} U_I \\ U_J \end{Bmatrix}_A \tag{3a}$$

*{A} means A is a column matrix while <A> means A is a row matrix.

Similarly, for element B:

$$\begin{Bmatrix} F_I \\ \\ F_J \end{Bmatrix}_B = \begin{pmatrix} k_B & -k_B \\ \\ -k_B & k_B \end{pmatrix} \begin{Bmatrix} U_I \\ \\ U_J \end{Bmatrix}_B \tag{3b}$$

We may term the column matrices on the lefthand side "nodal point external force vectors" $\{F_e\}$ while the righthand column matrices may be termed "nodal point displacement vectors" $\{u\}$.

Step 3. - Assemble the structural stiffness matrix. The external forces for all nodal points of the structure can now be expressed as the products of the nodal point displacements (unknowns) and appropriate stiffness terms. The stiffness term k_{pq} gives the contribution to the force at nodal point p caused by the displacement at nodal point q; it is the sum of all element stiffness terms linking p and q and therefore is zero if p and q have no elements in common. For example (figure 8-2b), the increment of external force F_2 at nodal point 2 produced by a displacement u_2 at nodal point 2 for the structure considered is $k_{22}u_2$ where $k_{22} = k^A_{JJ} + k^B_{II} = k_A + k_B$.*
Considering each nodal point in turn we can write

$$\begin{Bmatrix} F_1 \\ F_2 \\ F_3 \end{Bmatrix} = \begin{pmatrix} k_a & -k_a & 0 \\ -k_a & (k_a + k_b) & -k_b \\ 0 & -k_b & k_b \end{pmatrix} \begin{Bmatrix} u_1 \\ u_2 \\ u_3 \end{Bmatrix} \tag{4}$$

or since the internal forces are the negatives of the external forces (figure 8-2c)

$$\{F_i\} = -(K)\{u\} \tag{4a}$$

where: $\{F_i\}$ is the nodal point internal force vector for the whole

---

* The double subscript IJ indicates the row and column respectively of the term in the appropriate element stiffness matrix.

structure. In a two-dimensional problem $\{F_i\}$ has rank $2n \times 1$, where n is the number of nodal points; $\{u\}$ is the nodal point displacement vector for the structure, with rank $2n \times 1$ in a two dimensional problem; and (K) is the structural matrix, (defined to be positive), with rank $2n \times 2n$ in a two-dimensional problem.

Step 4. - Introduce the external applied forces and write the equations of equilibrium. The vector of external forces applied to each nodal point may include contributions from external pressures, water, accelerations, temperature effects, or other sources. Initial stresses may be included as external forces, as discussed later. In example 1, the only applied force is a downward load (-P) at nodal point 1. The net external forces (applied minus initial) induce deformations until the resulting internal forces come into equilibrium with them. This is equivalent to specifying, for a static problem, that the sum of external and internal forces equals zero.

$$\{F_e\} + \{F_i\} = 0 \tag{5}$$

In the example being discussed, $\{F_e\} = \langle -P, 0, X\rangle^T$ where X is the reaction to the displaced and fixed boundary condition at nodal point 3. Introducing this value for $\{F_e\}$ and also substituting for $\{F_i\}$ from (4a) yields the simultaneous equations of the structure

$$\begin{Bmatrix} -P \\ 0 \\ X \end{Bmatrix} = \begin{pmatrix} k_a & -k_a & 0 \\ -k_a & k_a + k_b & -k_b \\ 0 & -k_b & k_b \end{pmatrix} \begin{Bmatrix} u_1 \\ u_2 \\ u_3 \end{Bmatrix} \tag{6}$$

Step 5. - Introduce displacement boundary conditions. When any nodal point displacement is specified, the list of unknowns is shortened. Thus we can set aside (partition) the row of the structural stiffness matrix corresponding to the row of the known displacement. Also since the stiffness terms are known, we can transfer to the known external applied load vector (on the left side), the products of known stiffnesses and known displacements. This deletes, from the structural stiffness matrix, the column having the same number as the row number of the known displacement.

Whenever a displacement is specified, an external force at that nodal point is implied. It is not necessary to solve for this force, to solve the system, since the row involved is removed from the simultaneous equations. The unknown reaction can be found later, if desired, from the known nodal point displacements of the elements containing the specified nodal point.

Returning to the example, remove the third equition

$$X = \langle 0 \; -k_b \; k_b \rangle \begin{Bmatrix} u_1 \\ u_2 \\ y \end{Bmatrix} \tag{7a}$$

Then since the displacement of nodal point 3 was specified equal to y as a boundary condition, move the products of y to the left-hand side of what remains of the system of equations (6) giving:

$$\begin{Bmatrix} -P \\ k_b y \end{Bmatrix} = \begin{pmatrix} k_a & -k_a \\ -k_a & k_a + k_b \end{pmatrix} \begin{Bmatrix} u_1 \\ u_2 \end{Bmatrix} \tag{7b}$$

Step 6. - Solve for the unknown displacements. A solution to (7b) is possible if the determinant of the coefficient matrix is not zero. The solution is

$$u_1 = y - \frac{P}{k_a} - \frac{P}{k_b}$$

and

$$u_2 = y - \frac{P}{k_b}$$

All nodal points displacements are now known.

$$\{u\} = \begin{Bmatrix} y - \frac{P}{k_a} - \frac{P}{k_b} \\ y - \frac{P}{k_b} \\ y \end{Bmatrix} \tag{8}$$

Step 7. - Find the stresses in each element and the external reactions at the supports. Knowing the individual element stiffness matrices, multiplication by the nodal point displacements, (3a) and (3b), yields the external forces. The stresses are related to the forces in this simple case by $\sigma = F/a$ so we can modify the element stiffness matrices to yield stresses directly; taking tension stress positive*

$$\sigma_A = \frac{F_I}{a_1} = \frac{F_J}{a_1} = \frac{-P}{a_1}$$

where F_I and F_J are external forces in element A.

Similarly, in element B,

$$\sigma_B = \frac{F_I}{a_2} = \frac{-F_J}{a_2} = \frac{-P}{a_2}$$

This may seem a long process for such an obvious result, but it becomes attractive with only a few elements since the matrix operations are efficient on a digital computer.

FORMULATION OF ELEMENT STIFFNESS MATRICES AND EXTERNAL LOADS

The steps outlined above are the same when a more general two dimensional structure is solved; only the element stiffness matrix

*For finite element work it is a good idea to suffer the temporary inconvenience associated with a tension positive sign criterion. Even though rock is usually in compression, commercial programs, usually based on computer programs written by structural engineers, take tension positive. Errors, particularly in shear stress signs, are likely to creep in from partial mixing of sign conventions.

requires further elaboration. Two kinds of elements are used in the computer program presented in Appendix 1 -- orthotropic constant strain triangular elements to represent the rock material, and linear linkage elements to represent individual discontinuities. Many other types of elements are used in finite element programs -- e.g. bars, shells, quadrilaterals, axisymmetric solids, tetrahedra. But we will discuss only these two to simplify the presentation and the program.

A general formula for the stiffness matrix of a finite element is as follows (Zienkiewicz, 1971):

$$K = \int_{area} (L_o)^T (C)(L_o)da \tag{9}$$

where L_o is the matrix relating strains to nodal point displacements, and C is the matrix relating stresses to strains. We will use this formula to develop the stiffness of a constant strain triangle.

THE CONSTANT STRAIN TRIANGLE

The constant strain triangle, introduced by Clough (1960), develops a constant state of strain throughout the triangular area between three neighbouring nodal points as an approximation to the varying field of strain actually occurring. In practice, where strains vary over short distances, the triangles must be made small. Figure 8-3 shows a triangular element in a "mesh" of contiguous triangles representing a continuous part of a rock mass.

Strain-Displacement Relationship (L_o)

As a result of a deformation of the triangle (figure 8-3) an interior point at co-ordinates x, y is displaced by amounts u_x, u_y given by:

$$u_x(x,y) = a + a_1x + a_2y$$

and

$$u_y(x,y) = b + b_1x + b_2y$$

this can be written

$$\begin{Bmatrix} u_x(x,y) \\ u_y(x,y) \end{Bmatrix} = \begin{pmatrix} 1 & 0 & x & 0 & y & 0 \\ 0 & 1 & 0 & x & 0 & y \end{pmatrix} \begin{Bmatrix} a_o \\ b_o \\ a_1 \\ b_1 \\ a_2 \\ b_2 \end{Bmatrix} \qquad (10)$$

or $\{u(x,y)\} = (\Phi)\ \{\alpha\}$ (10a)

This variation of displacement with position preserves the connection

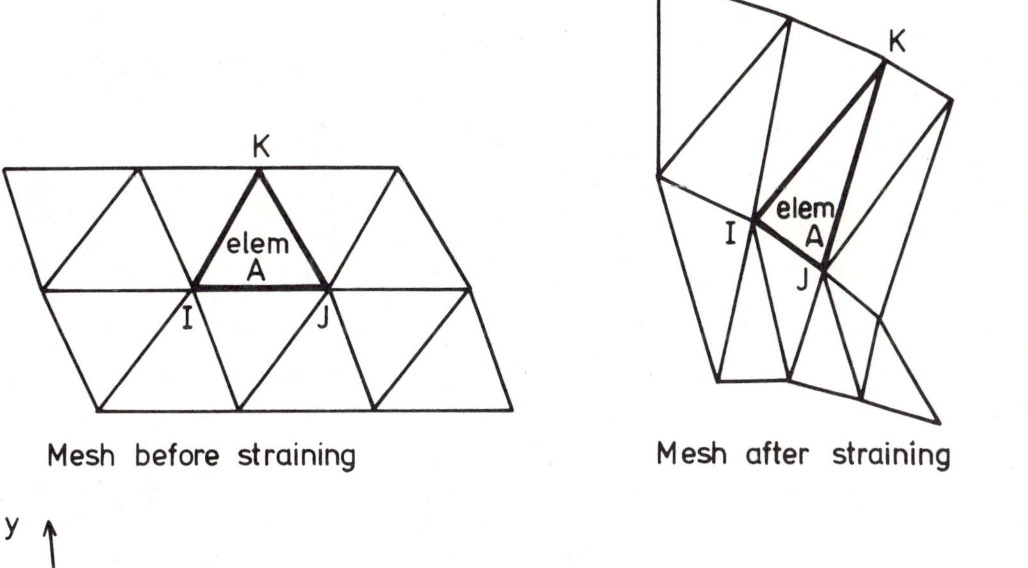

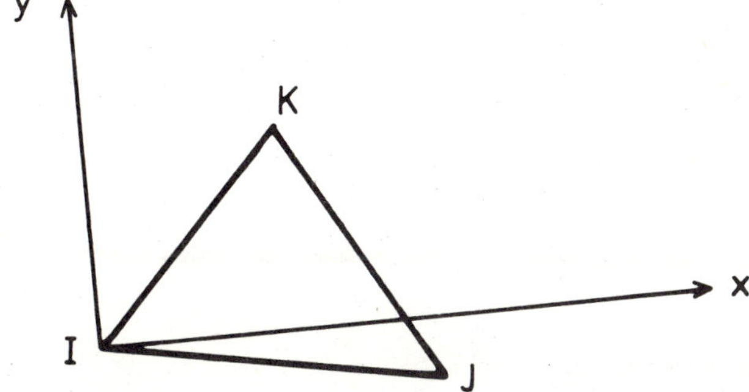

Figure 8-3. Constant strain triangles.

between adjacent triangles no matter how large the strain, (figure 8-3). Since we are particularly interested in displacement at the nodal points, apply (10) to each of these points in turn. The result is simplified by taking a local origin at nodal point I, an allowable step since the stiffness matrix of an element must be independent of its position in the structure. This yields:

$$\begin{Bmatrix} u_{x,I} \\ u_{y,I} \\ u_{x,J} \\ u_{y,J} \\ u_{x,K} \\ u_{y,K} \end{Bmatrix} = \begin{pmatrix} 1 & 0 & 0 & 0 & 0 & 0 \\ 0 & 1 & 0 & 0 & 0 & 0 \\ 1 & 0 & x_J & 0 & y_J & 0 \\ 0 & 1 & 0 & x_J & 0 & y_J \\ 1 & 0 & x_K & 0 & y_K & 0 \\ 0 & 1 & 0 & x_K & 0 & y_K \end{pmatrix} \begin{Bmatrix} a_o \\ b_o \\ a_1 \\ b_1 \\ a_2 \\ b_2 \end{Bmatrix} \tag{11}$$

$$\text{or } \{u\} = (\Phi_o)\,\{\alpha\} \tag{11a}$$

We now can obtain an expression for $\{\alpha\}$ in terms of the nodal point displacements.

$$\{\alpha\} = (\Phi_o)^{-1}\{u\} \tag{12}$$

Performing the required matrix inversion gives:

$$(\Phi_o)^{-1} = \frac{1}{x_J y_K - x_K y_J} \begin{pmatrix} x_J y_K - x_K y_J & 0 & 0 & 0 & 0 & 0 \\ 0 & x_J y_K - x_K y_J & 0 & 0 & 0 & 0 \\ y_J - y_K & 0 & y_K & 0 & -y_J & 0 \\ 0 & y_J - y_K & 0 & y_K & 0 & -y_J \\ x_K - x_J & 0 & -x_K & 0 & x_J & 0 \\ 0 & x_K - x_J & 0 & -x_K & 0 & x_J \end{pmatrix} \tag{12a}$$

which can be confirmed by multiplying by Φ_o.

The determinant $x_J y_K - x_K y_J$ equals 2a, where a is the area of the element. The strain components in the triangle are:

$$\varepsilon_x = \frac{\partial u_x}{\partial x}$$

$$\varepsilon_y = \frac{\partial u_y}{\partial y}$$

and

$$\gamma_{xy} = \frac{\partial u_x}{\partial y} + \frac{\partial u_y}{\partial x}$$

this may be written:

$$\left\{\begin{matrix} \varepsilon_x \\ \varepsilon_y \\ \gamma_{xy} \end{matrix}\right\} = \begin{pmatrix} \frac{\partial}{\partial x} & 0 \\ 0 & \frac{\partial}{\partial y} \\ \frac{\partial}{\partial y} & \frac{\partial}{\partial x} \end{pmatrix} \begin{pmatrix} u_x(x,y) \\ u_y(x,y) \end{pmatrix} \tag{13}$$

Substituting for $\{u(x,y)\}$ from (10a) and performing the differentiations yields:

$$\{\varepsilon\} = (\psi)\{\alpha\} \tag{14}$$

where

$$(\Psi) = \begin{pmatrix} 0 & 0 & 1 & 0 & 0 & 0 \\ 0 & 0 & 0 & 0 & 0 & 1 \\ 0 & 0 & 0 & 1 & 1 & 0 \end{pmatrix} \tag{14a}$$

Finally substituting for $\{\alpha\}$ from (12) in equation (14) gives:

$$\{\varepsilon\} = (\Psi)(\phi_o^{-1})\{u\} = (L_o)\{u\} \tag{15}$$

Since (ϕ_o^{-1}) was determined (equation 12a) the strain-displacement matrix (L_o) of equation (9) is now determined.

$$(L_o) = (\Psi)(\phi_o^{-1}) = \frac{1}{x_J y_K - x_K y_J} \begin{pmatrix} y_J - y_K & 0 & y_K & 0 & -y_J & 0 \\ 0 & x_K - x_J & 0 & -x_K & 0 & x_J \\ x_K - x_J & y_J - y_K & -x_K & y_K & X_J & -y_J \end{pmatrix} \tag{16}$$

$$\text{or } (L_o) = \frac{1}{2a}(L_1) \tag{16a}$$

where (L_1) is the matrix on the right hand side of (16).

Stress-Strain Relationship for the Rock

The stresses inside the triangle can be expressed in terms of the strains and the deformability properties of the material. Assume the rock is bedded or banded with the direction of the normal, (n) in the plane of the mesh, (figure 8-4). The trace of the layers in the element is direction s, and the direction perpendicular to the mesh is called t. The counterclockwise angle from x to s, in the triangular element, is α. First we will develop the stress-strain relationships in directions s, n in which the properties are most

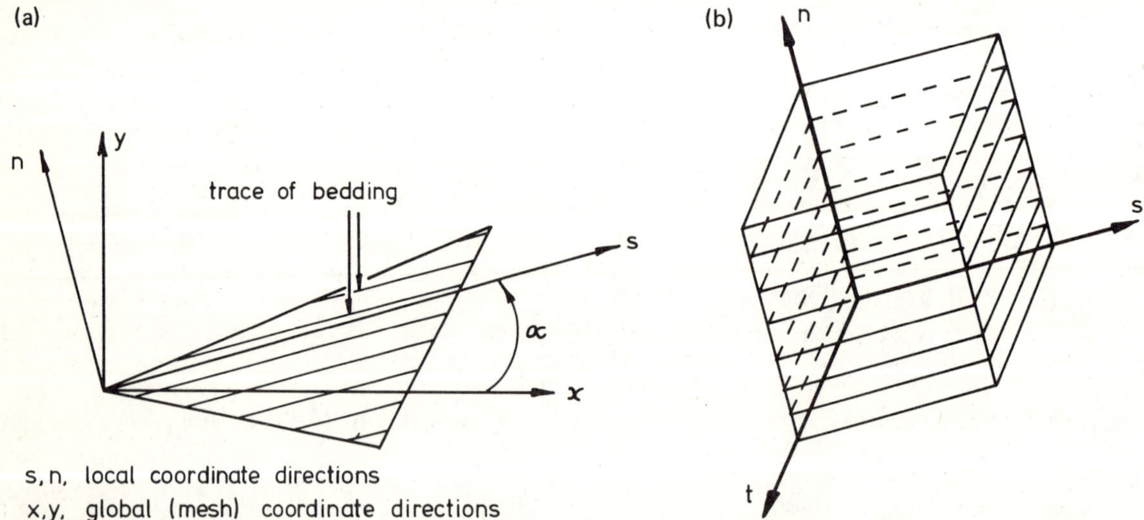

Figure 8-4. (a) Definition of local and global coordinate directions; (α) positive. (b) Relationship of local coordinates and elastic symmetry directions—transversely isotropic rock.

naturally expressed; then we will rotate them to the mesh co-ordinates x, y.

The relationship between increments of strain and stress is presumed to be elastic. Denote by the term ν_{ij} the Poisson's ratio that tells how much strain occurs in direction j due to a stress in direction i. Then the strain increments are related to the stress increments by:

$$\varepsilon_s = \frac{1}{E_s}\Delta\sigma_s - \frac{\nu_{ns}}{E_n}\Delta\sigma_n - \frac{\nu_{ts}}{E_t}\Delta\sigma_t$$

$$\varepsilon_n = \frac{-\nu_{sn}}{E_s}\Delta\sigma_s + \frac{1}{E_n}\Delta\sigma_n - \frac{\nu_{tn}}{E_t}\Delta\sigma_t \qquad (17)$$

and

$$\varepsilon_t = \frac{-\nu_{st}}{E_s}\Delta\sigma_s - \frac{\nu_{nt}}{E_n}\Delta\sigma_n + \frac{1}{E_t}\Delta\sigma_t$$

The stress-strain relationship is symmetric. If the material is bedded parallel to s (figure 8-4) it is isotropic in the st plane. The following relationships then hold:

$$E_s = E_t$$

$$\nu_{st} = \nu_{ts}$$

$$\nu_{sn} = \frac{E_s}{E_n}\nu_{ns}$$

$$\nu_{nt} = \frac{E_n}{E_s}\nu_{sn} \qquad (18)$$

$$\nu_{tn} = \nu_{sn}$$

$$G_{st} = \frac{E_t}{2(1 + \nu_{ts})}$$

For plane strain, $\varepsilon_t = 0$, and (17) and (18) yield:

$$\Delta\sigma_t = \nu_{st}\Delta\sigma_s + \nu_{sn}\Delta\sigma_n \tag{19}$$

and the strain-stress relationship simplifies to

$$\{\varepsilon\} = (D)\{\Delta\sigma\} \tag{20}$$

where $\{\varepsilon\} = \langle\varepsilon_s \varepsilon_n \gamma_{sn}\rangle^T$

and:

$$(D) = \begin{pmatrix} \frac{1}{E_S}(1 - \nu_{st}^2) & \frac{-\nu_{sn}}{E_s}(1 + \nu_{st}) & 0 \\ \frac{-\nu_{sn}}{E_s}(1 + \nu_{st}) & \frac{1}{E_n} - \frac{\nu_{sn}^2}{E_s} & 0 \\ 0 & 0 & \frac{1}{G_{sn}} \end{pmatrix} \tag{20a}$$

The stress-strain relationship is $\{\sigma\} = (C)\{\varepsilon\}$ where $(C) = (D)^{-1}$. Performing the matrix inversion yields:

$$(C) = \frac{1}{m} \cdot \begin{pmatrix} E_s - E_n\nu_{sn}^2 & E_n\nu_{sn}(1 + \nu_{st}) & 0 \\ E_n\nu_{sn}(1 + \nu_{st}) & E_n(1 - \nu_{st}^2) & 0 \\ 0 & 0 & G_{sn}m \end{pmatrix} \tag{20b}$$

and

$$m = (1 + \nu_{st})[1 - \nu_{st} - (2E_n/E_s)\nu_{sn}^2]$$

Anisotropic stress-strain properties may originate from inherently anisotropic rock; or they may be introduced to describe the overall behaviour of rhythmically bedded or banded rock. In this case the terms ν_{st}, ν_{sn}, E_s, E_n, and G_{sn} can be calculated from the deformability properties of the rock and the discontinuities (Goodman and Duncan, 1971).

For the rock mass depicted in figure 8-4b, with joint spacing h, and isotropic rock:

$$E_n = \frac{1}{\frac{1}{E} + \frac{1}{k_n h}}$$

$$E_s = E$$

$$G_{sn} = \frac{1}{\frac{2(1 + \nu)}{E} + \frac{1}{k_s h}} \qquad (21)$$

$$\nu_{sn} = \frac{E_n}{E_s} \nu$$

and $\nu_{st} = \nu$

where k_n and k_s are the normal and shear stiffnesses of the joints (discussed in chapter 5) and E and ν are the elastic modulus and Poisson's ratio of the rock.

The stress-strain relationship in the local co-ordinate system sn can now be written:

$$\{\Delta\sigma\}_{s,n} = (C)\{\varepsilon\}_{s,n} \qquad (22)$$

We can generalize this relationship to the global co-ordinates of the mesh, x, y, by expressing the strain and stress increments in the system s, n in terms of the strains and stress increments in the system x, y as follows:

$$\{\varepsilon\}_{s,n} = (T_\varepsilon)\{\varepsilon\}_{x,y}$$

and $\{\Delta\sigma\}_{s,n} = (T_\sigma)\{\Delta\sigma\}_{x,y}$

substituting these expressions in (22) and using

$$(T_\sigma)^{-1} = (T_\varepsilon)^T$$

gives $\{\Delta\sigma\}_{x,y} = (T_\varepsilon)^T(C)(T_\varepsilon)\{\varepsilon\}_{x,y}$ (22a)

where (T_ε) is defined by (Jaeger & Cook, 1969):

$$(T_\varepsilon) = \begin{pmatrix} \cos^2\alpha & \sin^2\alpha & \frac{1}{2}\sin 2\alpha \\ \sin^2\alpha & \cos^2\alpha & -\frac{1}{2}\sin 2\alpha \\ -\sin 2\alpha & \sin 2\alpha & \cos 2\alpha \end{pmatrix} \quad (23)$$

For an isotropic body $E_s = E_n = E_t = E$, all Poisson's ratios = ν, and $G_{sn} = \frac{E}{2(1+\nu)}$. The stress-strain relationship for plain strain simplifies to:

$$(C) = \frac{E(1-\nu)}{(1+\nu)(1-2\nu)} \begin{pmatrix} 1 & \frac{\nu}{1-\nu} & 0 \\ \frac{\nu}{1-\nu} & 1 & 0 \\ 0 & 0 & \frac{1-2\nu}{2(1-\nu)} \end{pmatrix} \quad (20c)$$

Element Stiffness Matrix

We may now substitute the strain-displacement relationship (L_o) (16) into the stress-strain relationship (22a) and introduce both into (9). Since all the terms of (L_o) and (C) are constants, the integration yields only the area of the triangle, which as noted is: $a = \frac{1}{2}(x_j y_k - x_k y_j)$. Then

$$(K) = \frac{1}{4a}(L_1)^T(T_\varepsilon)^T(C)(T_\varepsilon)(L_1) \quad (24a)$$

where L_1 is the quantity defined by (16a), (defined in global co-ordinates x, y). For an elastically isotropic rock

$$(K) = \frac{1}{4a}\,(L_1)^T(C)(L_1) \tag{24b}$$

in which (C) is given by (20c).

The computer program performs the matrix operations directly rather than substituting in an explicit formula obtainable by expanding (24a). $(L_1)(3 \times 6)$ is defined by (16) and (16a); (T_ε) (3×3) is defined by (23); and $(C)(3 \times 3)$ is defined by (20b). Since (K) relates 6 forces to 6 displacements, it is 6×6 in size.

INITIAL STRESSES IN THE ROCK

In many practical problems with rock the initial stresses are of comparable order of magnitude to the stress changes anticipated and furnish important contributions to the vector of applied loads. Equilibrium under an initial stress state implies initial internal forces throughout each element resisting the pressure of adjacent elements. When new free boundaries are excavated, the equilibrium is disturbed and the previous set of internal forces deforms the rock producing stress changes sufficient to create a new equilibrium. The total stress at any time, (σ), is defined by:

$$\{\sigma\} = \{\sigma_o\} + \{\Delta\sigma\}_{x,y} \tag{25}$$

where $\{\sigma_o\}$ is the vector of initial stresses referred to x, y.

$$\{\sigma_o\} = \begin{Bmatrix} \sigma_{x,o} \\ \sigma_{y,o} \\ \tau_{xy,o} \end{Bmatrix}$$

The external force vector F_o in equilibrium with the initial stress is:

$$\{F_o\} = \int_{area} (L_o)^T \{\sigma_o\}\, da \tag{26}$$

Substituting (L_o) from (16a) in equation (26), and observing that the integrand is constant, the external forces equilibrating the initial stresses are:

$$\{F_o\} = \frac{1}{2} (L_1)^T \{\sigma_o\} \tag{27}$$

The net load vector on the structure due to applied forces $\{F_e\}$ with initial forces $\{F_o\}$ is $\{\Delta F\} = \{F_e\} - \{F_o\}$. The equilibrium equations of the structure then are:

$$\{F_e\} - \{F_o\} = (K)\{u\} \tag{27a}$$

Example 2

A triangular finite element (figure 8-5) has vertices I, J and K at co-ordinates (x, y) = (0, 0), (2, 0), and (0, 1) respectively. The material is isotropic with E = 1000 and $\nu = 0.2$* and has initial compressive stresses $\sigma_x = -10$, $\sigma_y = -5$, $\tau_{xy} = -1$. Form the element stiffness matrix and load vector.

Figure 8-5. Example 2.

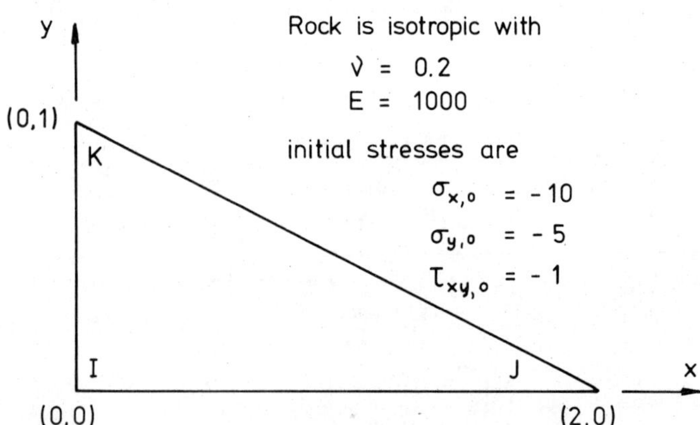

Find the stiffness matrix and internal forces for the triangular finite element.

*The dimensions for the problem are of no interest here. Any consistent set of units may be inferred, e.g. σ, E in MN/m^2 and co-ordinates in meters; or σ, E in p.s.f., and co-ordinates in feet. Avoid mixed units, e.g. E in p.s.i. and co-ordinates in feet.

We first form the matrix (L_1), (16) and (16a).

$$(L_1) = \begin{pmatrix} -1 & 0 & 1 & 0 & 0 & 0 \\ 0 & -2 & 0 & 0 & 0 & 2 \\ -2 & -1 & 0 & 1 & 2 & 0 \end{pmatrix}$$

The stress-strain matrix is formed from (20c)

$$(C) = \begin{pmatrix} 1111 & 278 & 0 \\ 278 & 1111 & 0 \\ 0 & 0 & 417 \end{pmatrix}$$

Thus, the stiffness matrix (K) is, by (24b)

$$(K) = \frac{1}{4} \begin{pmatrix} -1 & 0 & -2 \\ 0 & -2 & -1 \\ 1 & 0 & 0 \\ 0 & 0 & 1 \\ 0 & 0 & 2 \\ 0 & 2 & 0 \end{pmatrix} \begin{pmatrix} 1111 & 278 & 0 \\ 278 & 1111 & 0 \\ 0 & 0 & 417 \end{pmatrix} \begin{pmatrix} -1 & 0 & 1 & 0 & 0 & 0 \\ 0 & -2 & 0 & 0 & 0 & 2 \\ -2 & -1 & 0 & 1 & 2 & 0 \end{pmatrix}$$

$$\text{giving } (K) = \begin{pmatrix} 695 & 347 & -278 & -208 & -417 & -139 \\ 347 & 1215 & -139 & -104 & -208 & -1111 \\ -278 & -139 & 278 & 0 & 0 & 139 \\ -208 & -104 & 0 & 104 & 208 & 0 \\ -417 & -208 & 0 & 208 & 417 & 0 \\ -139 & -1111 & 139 & 0 & 0 & 1111 \end{pmatrix}$$

The external forces in equilibrium with the initial compressive stresses are, by (27)

$$\begin{Bmatrix} F_{x,I} \\ F_{y,I} \\ F_{x,J} \\ F_{y,J} \\ F_{x,K} \\ F_{y,K} \end{Bmatrix} = \frac{1}{2} \begin{pmatrix} -1 & 0 & -2 \\ 0 & -2 & -1 \\ 1 & 0 & 0 \\ 0 & 0 & 1 \\ 0 & 0 & 2 \\ 0 & 2 & 0 \end{pmatrix} \begin{Bmatrix} -10 \\ -5 \\ -1 \end{Bmatrix} = \begin{Bmatrix} 6 \\ 5.5 \\ -5 \\ -0.5 \\ -1 \\ -5 \end{Bmatrix}$$

The nodal point displacements {u} will therefore be obtained by solving the simultaneous equations

$$\{F_e\} + \begin{Bmatrix} -6. \\ -5.5 \\ 5 \\ 0.5 \\ 1. \\ 5. \end{Bmatrix} = (K)\{u\}$$

CONSTANT STRAIN JOINT ELEMENT

A discontinuity can be considered as a special kind of link between faces of blocks -- one that parts in response to tension, slides in response to shear, and transmits any force in response to compression. Each of these modes of deformation contributes primarily non-elastic displacements to the rock mass. We will develop an elastic linkage element and then, by an iterative solution procedure, constrain the element to obey the non-elastic, non linear deformation laws appropriate for a discontinuity in a rock mass.

Figure 8-6 shows a four nodal point joint element as an idealization of an actual joint. It has a small thickness (e) simulating the irregular and variable region between the joint walls. For simplicity, we will consider it essentially as a linear feature. We first form the joint element stiffness in the local co-ordinates s, n.

Joint Deformation Modes

The strain displacement relationship (L_o) describes the relative

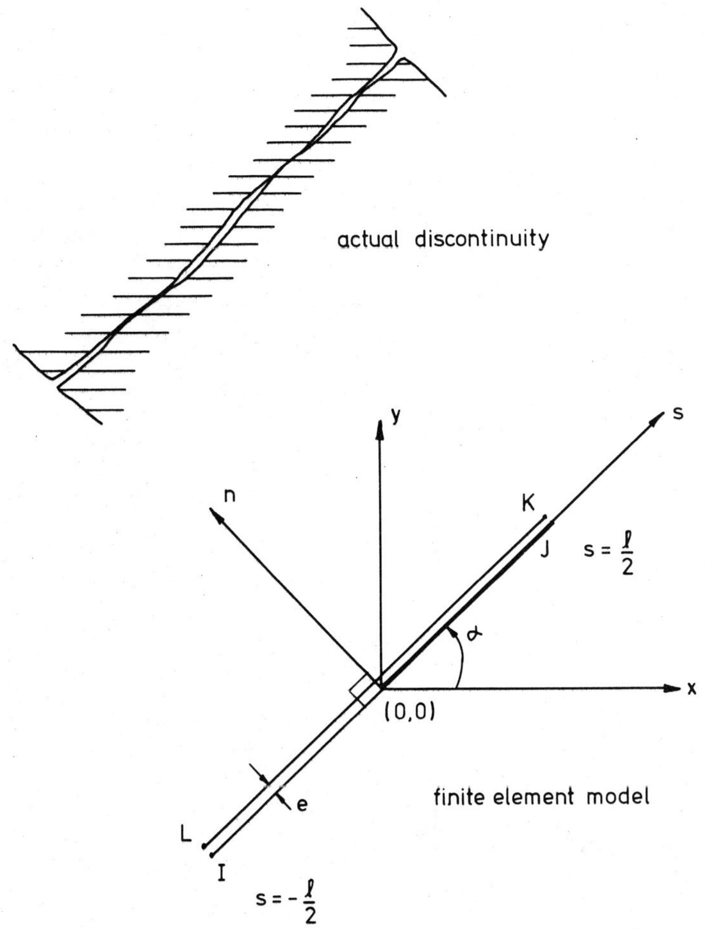

Figure 8-6. "Joint" element.

displacement between the rock walls K, L and I, J, as a function of nodal point displacements

$$\{u\}_{s,n} = \langle u_I, v_I, u_J, v_J, u_K, v_K, u_L, v_L \rangle^T$$

where u_I and v_I are the total displacements of node I in the local directions s, n (respectively parallel and perpendicular to the original orientation of the joint walls), etc. Analogous to a dimensionless strain quantity in the rock, we introduce joint deformations defined at the center (s = o).

$$\{\varepsilon_j\} = \begin{Bmatrix} u_o \\ v_o \\ \omega \end{Bmatrix} = \begin{pmatrix} \frac{u_K + u_L}{2} - \frac{u_I + u_J}{2} \\ \frac{v_K + v_L}{2} - \frac{v_I + v_J}{2} \\ \frac{v_K - v_L}{\ell} - \frac{v_J - v_I}{\ell} \end{pmatrix} \tag{28}$$

It is tempting to make $\{\varepsilon_j\}$ uniformly dimensionless by dividing u_o and v_o by the joint length ℓ; however, the definition given for $\{\varepsilon_j\}$ seems preferable on physical grounds. Figure 8-7 shows how the various modes of deformation are expressed in the joint element and indicates the sign convention of (28) consistent with tension positive in the rock. Relative counterclockwise rotation of the joint wall KL is positive.

We can now identify the strain displacement matrix (L_o).

$$\begin{Bmatrix} u_o \\ v_o \\ \omega \end{Bmatrix} = \begin{pmatrix} -\frac{1}{2} & 0 & -\frac{1}{2} & 0 & \frac{1}{2} & 0 & \frac{1}{2} & 0 \\ 0 & -\frac{1}{2} & 0 & -\frac{1}{2} & 0 & \frac{1}{2} & 0 & \frac{1}{2} \\ 0 & \frac{1}{\ell} & 0 & -\frac{1}{\ell} & 0 & \frac{1}{\ell} & 0 & -\frac{1}{\ell} \end{pmatrix} \begin{Bmatrix} u_I \\ v_I \\ u_J \\ v_J \\ u_K \\ v_K \\ u_L \\ v_L \end{Bmatrix} \tag{29}$$

or $\{\varepsilon\}_j = (L_o)\,\{u\}_{s,n}$ (29a)

"Stress-Strain" Relationship for the Joints

Joint element "stresses" need to be defined, since the actual load transfer across a rough joint may occur at point contacts. The normal and shear stresses on the joint wall describe the total normal and shear forces per unit of area (the thickness of the element is unity).

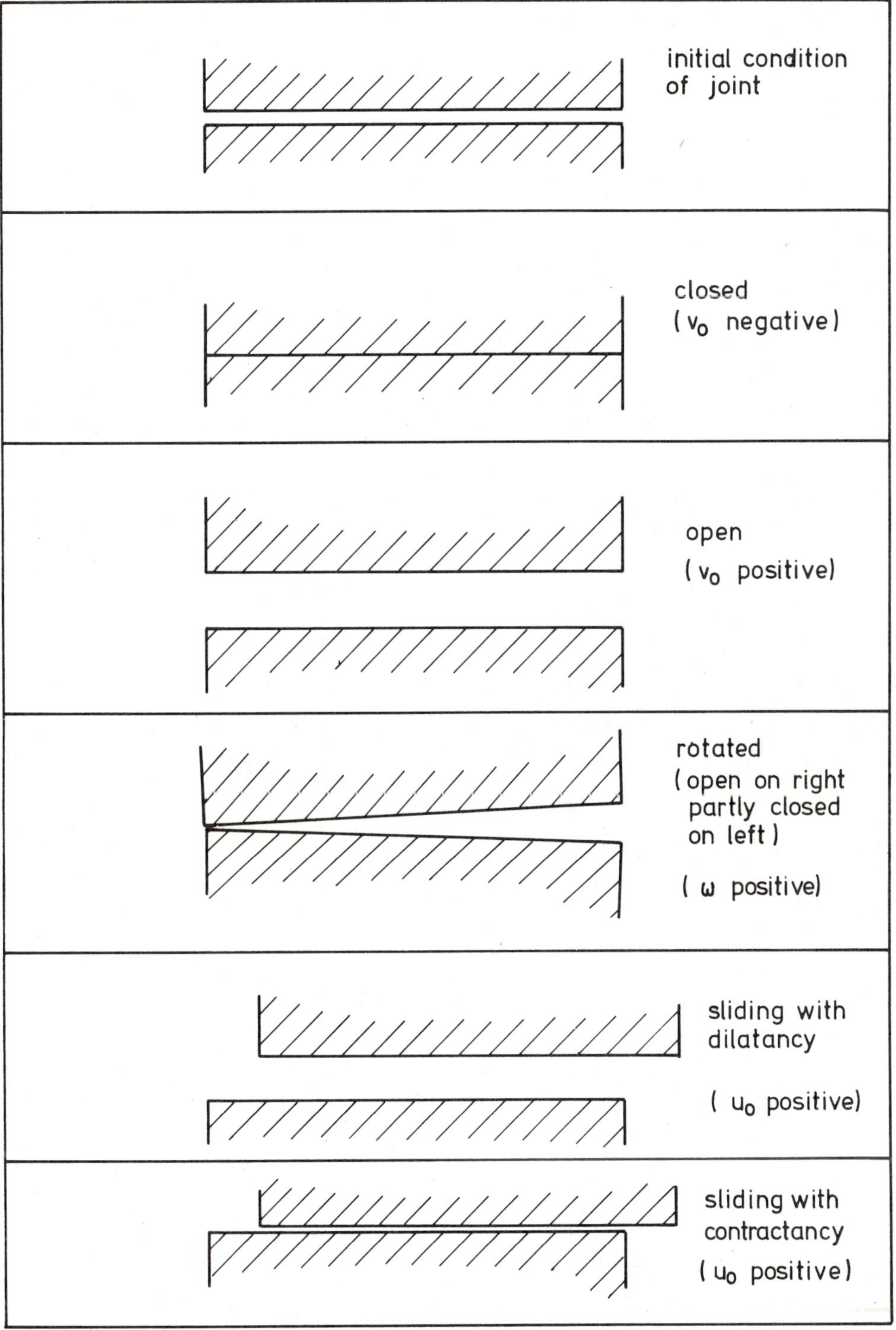

Figure 8-7. Modes of deformation of a joint element.

$$\Delta\sigma = \Delta\sigma_n = \frac{1}{\ell}(\Delta F_{n,K} + \Delta F_{n,L}) \tag{30}$$

and $$\Delta\tau = \Delta\tau_{ns} = \frac{1}{\ell}(\Delta F_{s,K} + \Delta F_{s,L}) \tag{30a}$$

The "delta" symbol (Δ) is attached to the forces and stresses to indicate that it is the increments of force and stress caused by a deformation $\{u\}_{s,n}$ which we calculate. The total stress is the initial stress (equation 25) plus the stress increment. The moment (M_o) of the nodal point forces about the center of the joint expresses the distribution of stress in the joint. (Counterclockwise external moment on face KL is positive).

$$M_o = \Delta F_{n,K}\frac{\ell}{2} - \Delta F_{n,L}\frac{\ell}{2} \tag{30b}$$

Equations (30) and (30b) require a particular distribution of the normal forces in the joint element. Inverting equations (30), (30a), and (30b) and stipulating that $F_I = -F_L$ and $F_J = -F_K$, we can write the nodal point forces in terms of the "stresses":

$$\begin{Bmatrix} \Delta F_{s,I} \\ \Delta F_{n,I} \\ \Delta F_{s,J} \\ \Delta F_{n,J} \\ \Delta F_{s,K} \\ \Delta F_{n,K} \\ \Delta F_{s,L} \\ \Delta F_{n,L} \end{Bmatrix} = \begin{pmatrix} -\frac{\ell}{2} & 0 & 0 \\ 0 & -\frac{\ell}{2} & \frac{1}{\ell} \\ -\frac{\ell}{2} & 0 & 0 \\ 0 & -\frac{\ell}{2} & -\frac{1}{\ell} \\ \frac{\ell}{2} & 0 & 0 \\ 0 & \frac{\ell}{2} & \frac{1}{\ell} \\ \frac{\ell}{2} & 0 & 0 \\ 0 & \frac{\ell}{2} & -\frac{1}{\ell} \end{pmatrix} \begin{Bmatrix} \Delta\tau_{ns} \\ \Delta\sigma_n \\ M_o \end{Bmatrix} \tag{31}$$

or

$$\{\Delta F\}_{s,n} = (B)\{\Delta\sigma_J\}_{s,n} \tag{31a}$$

Joint "stresses" and "strains" are linked through joint system stiffness properties as follows:

$$\begin{Bmatrix} \Delta\tau_{ns} \\ \Delta\sigma_n \\ M_o \end{Bmatrix} = \begin{pmatrix} k_s & 0 & 0 \\ 0 & k_n & 0 \\ 0 & 0 & k_\omega \end{pmatrix} \begin{Bmatrix} u_o \\ v_o \\ \omega \end{Bmatrix} \tag{32}$$

or

$$\{\Delta\sigma_J\} = (C_J)\{\varepsilon_J\} \tag{32a}$$

We observed, in Chapter 5, that k_s and k_n change with changing stress. Therefore, the linear formulation of (32) must be coupled with an iterative solution technique, to yield a physically acceptable solution. One can introduce a term k_{ns} to describe the contribution to normal stress because of dilatancy (Goodman and Dubois, 1972). As discussed in Chapter 5, in initially closed joints a shear displacement in either direction will develop a joint thickening: i.e. $\Delta v\ (\tau)$ is positive when τ is either negative or positive. Therefore k_{ns} must be non-linear. Non linearities, including dilatancy, will be discussed in a later section. By omitting k_{ns} from the stress-strain relationship at this stage, we simply place the representation of dilatancy effects outside of the stiffness matrix.

The coefficient k_ω, coupling joint rotation to the stress dis-

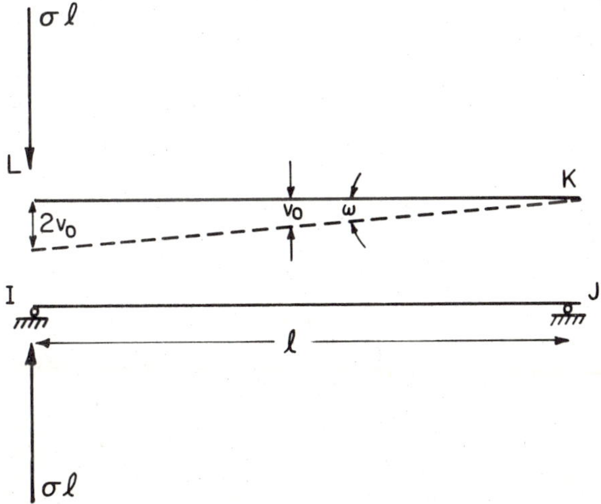

Figure 8-8. Rotational stiffness.

tribution, is a new, unknown parameter. However, it can be expressed as a function of k_n if we make an assumption about the mode of behavior. Assume that the closure at any nodal point pair across a joint is proportional to the increment of normal force at the corresponding nodal points. Consider the case when all the normal force is at pair I, L (figure 8-8). The assumption demands a rotation $\omega = 2v_o/\ell$ where v_o, as defined by (28), is the normal displacement at the center of the joint. The moment, M_o, corresponding to this case is $\frac{1}{2}\Delta\sigma\ \ell^2$. Since, by equation (32), $v_o = \Delta\sigma/k_n$, we conclude that $k_\omega = k_n(\ell^3)/4$. Any other case with a different stress distribution will produce the same result. Thus we can write.

$$(C_J) = \begin{pmatrix} k_s & 0 & 0 \\ 0 & k_n & 0 \\ 0 & 0 & \frac{1}{4}\ell^3 k_n \end{pmatrix} \tag{33}$$

The reason to account for moment and rotation in the linkage element is simply that the stresses and deformations in a rock mass subjected to rotation are quite different than in a rock mass which is not. For example, the two block systems in figure 8-9 are both in equilibrium under the applied loading but in very different conditions. Without rotation, all the rock blocks are in compression, and are confined in intimate face to face contacts (figure 8-9a). In contrast as a result of a virtual rotation, exaggerated in figure 8-9b, each block becomes loaded by eccentric edge to face (line) contacts and may suffer indirect tension as in a point load test.

Joint Element Stiffness Matrix

We can now write the relationship between displacements at nodes of the joint element and the corresponding increments of external force.

$$\{\Delta F\}_{s,n} = (B)\{\Delta\sigma_J\} = (B)(C_J)\{\varepsilon_J\}$$

or finally,

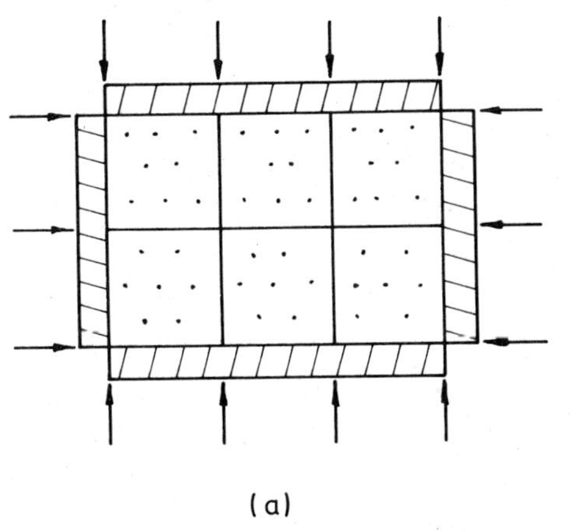

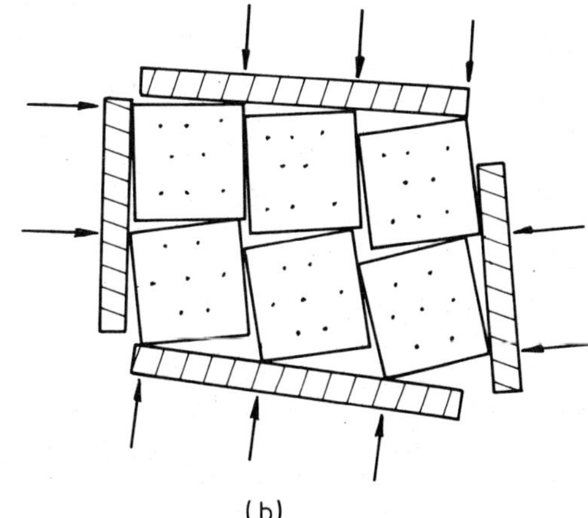

Figure 8-9. Effect of block rotation on stress state in jointed rock masses.

$$\{\Delta F\}_{s,n} = (B)(C_J)(L_o)\{u\}_{s,n} \tag{34}$$

where

(B) is defined by (31) and (31a),

(C_J) is defined by (33),

and

(L_o) is defined by (29) and (29a).

Performing the multiplication demanded by (34), the joint element stiffness is:

$$(K)_{s,n} = \frac{1}{4}\ell \begin{pmatrix} k_s & 0 & k_s & 0 & -k_s & 0 & -k_s & 0 \\ 0 & 2k_n & 0 & 0 & 0 & 0 & 0 & -2k_n \\ k_s & 0 & k_s & 0 & -k_s & 0 & -k_s & 0 \\ 0 & 0 & 0 & 2k_n & 0 & -2k_n & 0 & 0 \\ -k_s & 0 & -k_s & 0 & k_s & 0 & k_s & 0 \\ 0 & 0 & 0 & -2k_n & 0 & 2k_n & 0 & 0 \\ -k_s & 0 & -k_s & 0 & k_s & 0 & k_s & 0 \\ 0 & -2k_n & 0 & 0 & 0 & 0 & 0 & 2k_n \end{pmatrix} \tag{35}$$

It remains to rotate the stiffness matrix to the global mesh coordinate system x, y; with reference to figure 8-6, at any nodal point:

$$\begin{Bmatrix} \Delta F_s \\ \Delta F_n \end{Bmatrix} = \begin{pmatrix} \cos\alpha & \sin\alpha \\ -\sin\alpha & \cos\alpha \end{pmatrix} \begin{Bmatrix} F_x \\ F_y \end{Bmatrix} \qquad (36)$$

$$\alpha = \arctan \frac{y_J - y_I}{x_J - x_I}$$

Thus we can write:

$$\{\Delta F\}_{s,n} = (T)\{\Delta F\}_{x,y}$$

and (37)

$$\{u\}_{s,n} = (T)\{u\}_{x,y}$$

where

$$(T) = \begin{pmatrix} \cos\alpha & \sin\alpha & 0 & 0 & 0 & 0 & 0 & 0 \\ -\sin\alpha & \cos\alpha & 0 & 0 & 0 & 0 & 0 & 0 \\ 0 & 0 & \cos\alpha & \sin\alpha & 0 & 0 & 0 & 0 \\ 0 & 0 & -\sin\alpha & \cos\alpha & 0 & 0 & 0 & 0 \\ 0 & 0 & 0 & 0 & \cos\alpha & \sin\alpha & 0 & 0 \\ 0 & 0 & 0 & 0 & -\sin\alpha & \cos\alpha & 0 & 0 \\ 0 & 0 & 0 & 0 & 0 & 0 & \cos\alpha & \sin\alpha \\ 0 & 0 & 0 & 0 & 0 & 0 & -\sin\alpha & \cos\alpha \end{pmatrix} \qquad (37a)$$

and since $(T)^{-1} = (T)^T$, we can rewrite (34) as:

$$\{\Delta F\}_{x,y} = (T)^T (K_{s,n})(T)\{u\}_{x,y} \qquad (38)$$

or

$$\{\Delta F\}_{x,y} = (K_{x,y})\{u\}_{x,y} \tag{38a}$$

Initial Stresses in Joints

A joint element under initial compression or shear releases stored energy when the equilibrating external forces are removed. Therefore, the finite element program must associate initial stresses with the joints as well as with the solids. Since the joint element is very thin, its normal and shear stresses are the same as the stress components in the adjacent elements, referred to the joint axis s,n.

$$\{\sigma_{o,J}\}_{s,n} = \begin{Bmatrix} \tau_{sn,o} \\ \sigma_{n,o} \\ 0 \end{Bmatrix} = \begin{pmatrix} -\frac{1}{2}\sin 2\alpha & \frac{1}{2}\sin 2\alpha & \cos 2\alpha \\ \sin^2\alpha & \cos^2\alpha & -\sin 2\alpha \\ 0 & 0 & 0 \end{pmatrix} \begin{Bmatrix} \sigma_{x,o} \\ \sigma_{y,o} \\ \tau_{xy,o} \end{Bmatrix} \tag{39}$$

or

$$\{\sigma_{o,J}\}_{s,n} = (T_{\sigma,J})\{\sigma_o\}_{x,y} \tag{39a}$$

The joint stresses are given by

$$\{\sigma_J\}_{s,n} = \{\sigma_{o,J}\}_{s,n} + \{\Delta\sigma_J\}_{s,n} \tag{39b}$$

Initial forces $\{F_{o,J}\}$ must be subtracted from the structural external load vector $\{F_e\}$ to account for initial stresses in the joints.

$$\{F_{o,J}\} = \{F_{o,J}\}_{x,y} = (T)^T\{F_{o,J}\}_{s,n} = (T)^T(B)(T_{\sigma,J})\{\sigma_o\}_{x,y} \tag{40}$$

where (T) is defined by (37a).

(B) is defined by (31) and (31a)

and $(T_{\sigma,J})$ is defined by (39) and (39a).

ASSEMBLY OF THE STRUCTURAL EQUATIONS

A structure with n nodal points produces 2 n simultaneous equations. Denoting initial forces in the rock by $F_{o,R}$:

$$\{F_e\} - \{F_{o,J}\} - (F_{o,R}) = (K)\{u\} \tag{41}$$

or

$$\underset{(2n \times 1)}{\{F\}} = \underset{(2n \times 2n)}{(K)} \underset{(2n \times 1)}{\{u\}} \tag{41a}$$

Assembly of the structural stiffness matrix (K) is exactly as in step 3 of example 1 and is best examined through another illustration.

Example 3

Assemble the stiffness matrix and load vector to establish the simultaneous equations for displacements in the structure of figure 8-10 (6 nodal points, 3 elements). The properties and initial stresses

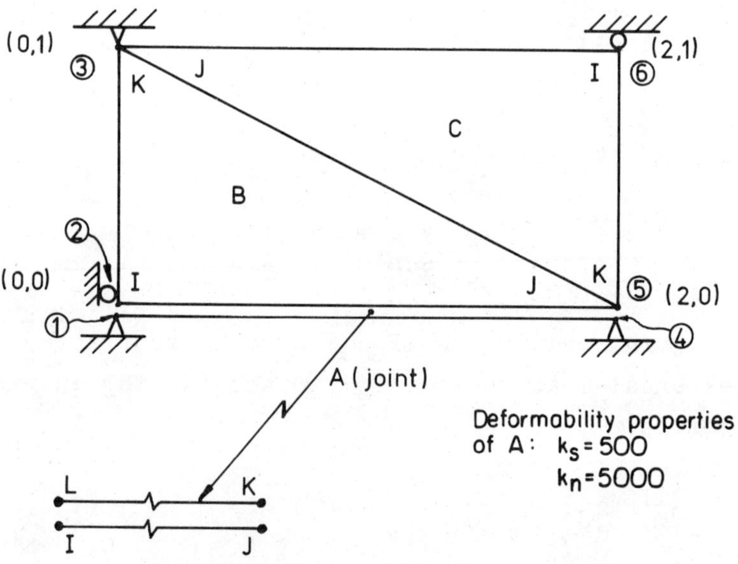

Figure 8-10. Example 3.

of the solid elements are the same as in example 2 (figure 8-5).

Since elements B and C are identical, the rock element stiffness has to be formulated only once if we name the corners identically in each element. The result was given previously (see example 2).

The element stiffness for the joint, element A, is given by (35). (Since $\alpha = 0$ in the example, (T) (equation 37a) is just the identity matrix and $(K)^A{}_{x,y} = (K)^A{}_{s,n}$). Let $(K)^N_{IJ}$ be a 2 × 2 matrix of stiffness terms for an element N giving the increment of force at nodal point I, $(\vec{F}_I)^N$, caused by x and y displacements at nodal point J, $(\vec{u}_J)^N$. Then, the 12 × 12 structural stiffness matrix for the problem of figure 8-10 is:

$$\begin{pmatrix} k^A_{II} & k^A_{IL} & 0 & k^A_{IJ} & k^A_{IK} & 0 \\ k^A_{LI} & (k^A_{LL} + k^B_{II}) & k^B_{IK} & k^A_{LJ} & (k^A_{LK} + k^B_{IJ}) & 0 \\ 0 & k^B_{KI} & (k^B_{KK} + k^C_{JJ}) & 0 & (k^B_{KJ} + k^C_{JK}) & k^C_{JI} \\ k^A_{JI} & k^A_{JL} & 0 & k^A_{JJ} & k^A_{JK} & 0 \\ k^A_{KI} & (k^A_{KL} + k^B_{JI}) & (k^B_{JK} + k^C_{KJ}) & k^A_{KJ} & (k^A_{KK} + k^B_{JJ} + k^C_{KK}) & k^C_{KI} \\ 0 & 0 & k^C_{IJ} & 0 & k^C_{IK} & k^C_{II} \end{pmatrix}$$

Substituting the required stiffness terms gives (K) =

$$\begin{pmatrix} 250 & 0 & -250 & 0 & 0 & 0 & 250 & 0 & -250 & 0 & 0 & 0 \\ 0 & 5000 & 0 & -5000 & 0 & 0 & 0 & 0 & 0 & 0 & 0 & 0 \\ -250 & 0 & 945 & 347 & -417 & -139 & -250 & 0 & -28 & -208 & 0 & 0 \\ 0 & -5000 & 347 & 6215 & -208 & -1111 & 0 & 0 & -139 & -104 & 0 & 0 \\ 0 & 0 & -417 & -208 & 695 & 0 & 0 & 0 & 0 & 347 & -278 & -139 \\ 0 & 0 & -139 & -1111 & 0 & 1215 & 0 & 0 & 347 & 0 & -208 & -104 \\ 250 & 0 & -250 & 0 & 0 & 0 & 250 & 0 & -250 & 0 & 0 & 0 \\ 0 & 0 & 0 & 0 & 0 & 0 & 0 & 5000 & 0 & -5000 & 0 & 0 \\ -250 & 0 & -28 & -139 & 0 & 347 & -250 & 0 & 945 & 0 & -417 & -208 \\ 0 & 0 & -208 & -104 & 347 & 0 & 0 & -5000 & 0 & 6215 & -139 & -1111 \\ 0 & 0 & 0 & 0 & -278 & -208 & 0 & 0 & -417 & -139 & 695 & 347 \\ 0 & 0 & 0 & 0 & -139 & -104 & 0 & 0 & -208 & -1111 & 347 & 1213 \end{pmatrix}$$

Note that K is symmetric. Since each element stiffness matrix is symmetric, the structural matrix will always be also.
Let F_I^N represent the 2 × 1 matrix of external force components at nodal point I of element N. The initial stress contributions to the external load vector {F} (equations 41 and 41a) are

$$\underset{(12\times 1)}{\{F\}} = \begin{Bmatrix} \vec{F}_1 \\ \vec{F}_2 \\ \vec{F}_3 \\ \vec{F}_4 \\ \vec{F}_5 \\ \vec{F}_6 \end{Bmatrix} = \begin{Bmatrix} F_I^A \\ F_L^A + F_I^B \\ F_K^B + F_J^C \\ F_J^A \\ F_K^A + F_J^B + F_K^C \\ F_I^C \end{Bmatrix}$$

The external force contributions from initial stress in element B were calculated in example 2. Because traangle C is obtainable from triangle B by rotation through 180°, its external load vector is the negative of the load vector of triangle B. The contributions to the external load vector for the joint element (A) are calculated using (40). (T) is the identity matrix (I), so that

$$\{F_{o,J}\} = (I) \begin{pmatrix} -1 & 0 & 0 \\ 0 & -1 & +\frac{1}{2} \\ -1 & 0 & 0 \\ 0 & -1 & -\frac{1}{2} \\ 1 & 0 & 0 \\ 0 & 1 & +\frac{1}{2} \\ 1 & 0 & 0 \\ 0 & 1 & -\frac{1}{2} \end{pmatrix} \begin{pmatrix} 0 & 0 & 1 \\ 0 & 1 & 0 \\ 0 & 0 & 0 \end{pmatrix} \begin{Bmatrix} -10 \\ -5 \\ -1 \end{Bmatrix} = \begin{Bmatrix} 1 \\ 5 \\ 1 \\ 5 \\ -1 \\ -5 \\ -1 \\ -5 \end{Bmatrix}$$

Introducing these results and changing the sign, as required by (41) the load vector is:

$$\begin{Bmatrix} F_{x,1} \\ F_{y,1} \\ F_{x,2} \\ F_{y,2} \\ F_{x,3} \\ F_{y,3} \\ F_{x,4} \\ F_{y,4} \\ F_{x,5} \\ F_{y,5} \\ F_{x,6} \\ F_{y,6} \end{Bmatrix} = \begin{Bmatrix} -1 \\ -5 \\ -5 \\ -0.5 \\ -4 \\ 4.5 \\ -1 \\ -5 \\ 5 \\ 0.5 \\ 6 \\ 5.5 \end{Bmatrix}$$

Equilibrium has been disturbed by the removal of support from the right side and the body will stretch until equilibrium is restored. In this case, $F_{x,6}$ and $F_{x,5}$ must become zero. The displacements will be found by solving the set of simultaneous equations (as in example 1), which after introduction of the fixed boundary conditions will reduce to four equations in four unknowns.

ITERATIVE SOLUTION TO SIMULATE REAL PROPERTIES OF JOINTS

The linear equations developed for the displacements of nodal points provide a first solution. However the indicated displacements may presume tension in some joints, or excessive shear in others; and in any case shear displacement will produce a dilatancy tendency which has not yet been accounted for. Therefore the first solution must be examined at each nodal point along the joints and compared with real behaviour. This comparison yields information as to how the problem can be restarted or continued in order to produce an entirely acceptable answer. Through an iterative process, one

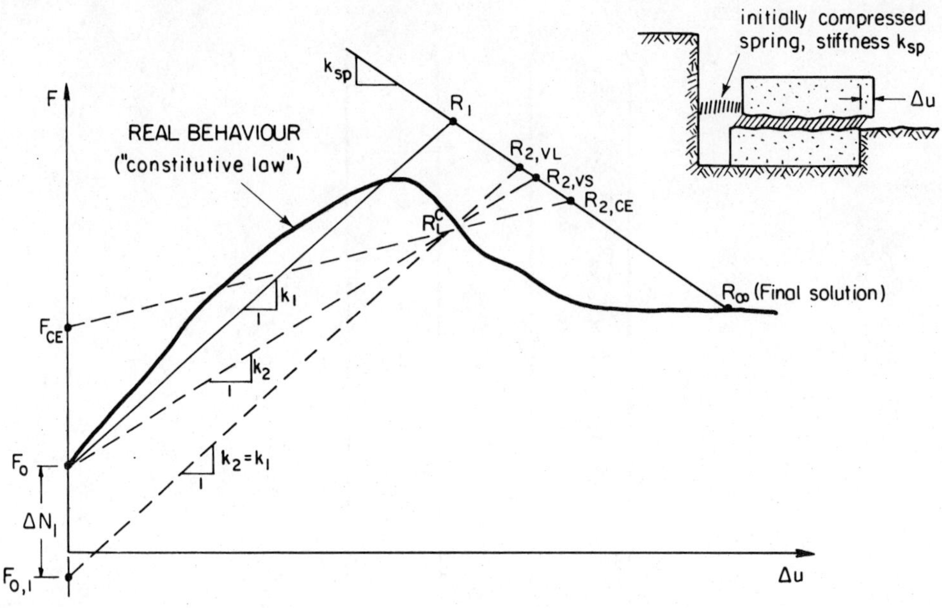

Figure 8-11. Alternative iterative schemes to constrain the solution to obey the real material behavior.

will try to compute entirely acceptable deformations and stresses throughout the structure. These methods are similar to "predictor-corrector" methods for numerical solution of differential equations (Dorn and McCracken, 1972).

Load Transfer and Variable Stiffness Methods

Various iterative processes are illustrated in figure 8-11, which considers a direct shear specimen with an initial shear load (F_o) being forced by an initially compressed spring. To begin the calculation we choose the initial stiffness k, representative of the elastic portion of the load-deformation curve for the joint. Because of the initial load in the spring, the spring and joint come to equilibrium at point R_1, which is unacceptable according to our estimate of the joint's real load-deformation behaviour (its "constitutive law"). We can restart the problem with new information by "aiming" the solution towards a point on the constitutive curve near R_1, for example point R_1^C. For the second run, we may re-define the stiffness to k_2 (<u>variable stiffness method</u>) calculated to pass through R_1^C; if we do this, the new solution will be point $R_{2,VS}$, still unacceptable but closer to the correct answer (R_∞).

To save recalculating the stiffness matrix, Zienkiewicz (1971) suggested restarting with the same stiffness ($k_2 = k_1$) but a new initial load, $F_{0,1}$ (the load transfer method); in this example, $F_{0,1}$ is simply $F_{x,0} + \Delta N_1$ where $\Delta N_1 = R_1^C - R_1$. This yields as a solution point $R_{2,VL}$. The load transfer method generally requires more iterations than the variable stiffness method to reach convergence but requires fewer calculations per iteration. It is the method used in the computer program presented in the Appendix. Other modification paths are possible, and may be preferable in certain instances. For example (Goodman and Dubois, 1972), if the stiffness matrix is to be altered at each run anyway, it will be only slightly more expensive to modify the load vector as well; then one can choose a path such as in figure 8-11 from F_{CE} with a slope such that the area under $F_{CE}R_1^C$ is the same as the area under the actual load-deformation curve up to R_1^C. Such an approach would usually speed convergence; however, it would not converge in all cases as shown by Dubois (1972). All of these methods can be viewed as modifications of the Newton-Raphson scheme for non-linear functions, (see for example Dorn and McCracken, 1972) in which the stiffness is updated to the value of the slope of the constitutive surface in the neighbourhood of the current solution.

Interlaced Joint Elements

The stiffness matrix, representing the coupling from one nodal point to another, was constructed from elements filling the space between the nodal points. The "solution" gives the forces and displacements at nodal points and it is here where modifications to the load vector must be made to restart the problem on the next iteration. Thus it is natural to re-define the set of joint elements at this stage, as shown in figure 8-12. In practice it is only necessary to store a list of lengths (ℓ') and orientations (α)* to be associated with each modified joint element. In the

---

*If the orientation changes along a line of joint elements, the scheme must be handled differently.

Joints for formation of stiffness matrix.

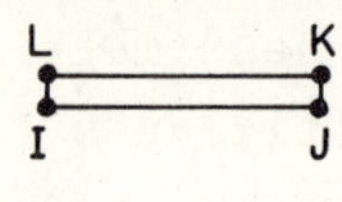

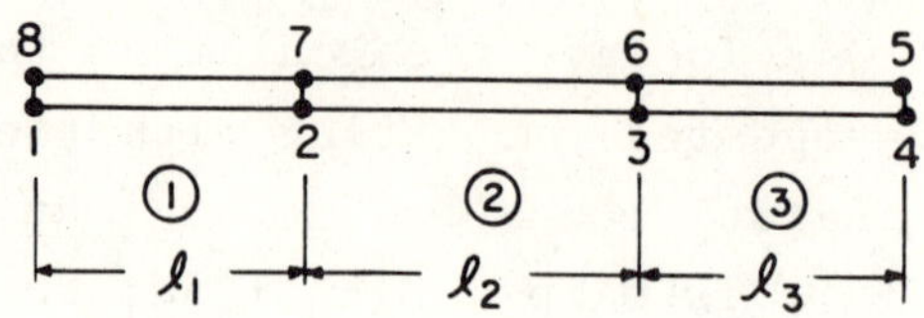

Modified joints for load transfer.

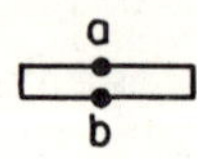

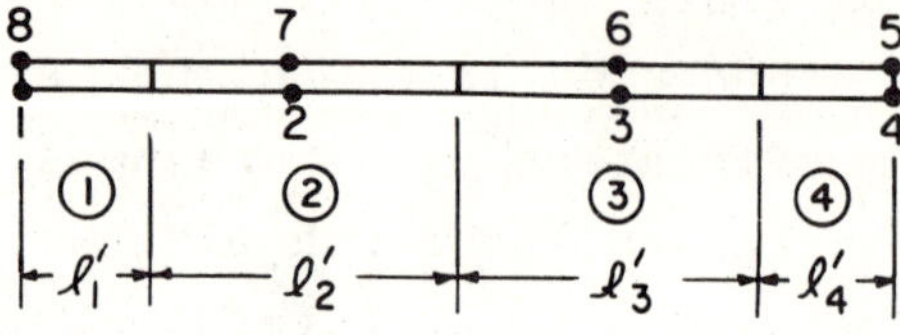

Figure 8-12. Modified joint elements.

example of figure 12, $\ell_1' = \ell_1/2$, $\ell_2' = (\ell_1/2 + \ell_2/2)$ etc.

For computing forces from known displacements, the joint stiffness properties k_s', and k_n' to be associated with a modified joint element are simply the weighted average of stiffness properties from the two joint element halves of which it is composed. For example:

$$(k_s')_1 = (k_s)_1$$

$$(k_s')_2 = \frac{1}{\ell_2'}\left((k_s)_1 \frac{\ell_1}{2} + (k_s)_2 \frac{\ell_2}{2}\right)$$

$$(k_s')_3 = \frac{1}{\ell_3'}\left((k_s)_2 \frac{\ell_2}{2} + (k_s)_3 \frac{\ell_3}{2}\right) \qquad (42)$$

$$(k_n')_1 = (k_n)_1$$

$$(k_n')_2 = \frac{1}{\ell_2'}\left((k_n)_1 \frac{\ell_1}{2} + (k_n)_2 \frac{\ell_2}{2}\right)$$

Joint Opening and Closing

In chapter 5 it was presumed a joint can not sustain tension and furthermore, that it becomes thinner under a compressive normal load, until a maximum closure V_{mc} (figure 8-13). The following discussion illustrates how the law of normal deformation (equation 8 of chapter 5) can be obeyed by means of a load transfer process. Substituting $t = 0$, $A = 1$, and $\xi = F_{n,o}/\ell'$ in equation 8 of chapter 5 gives

$$F_n = \left(\frac{\Delta v}{V_m - \Delta v} + 1\right) F_{n,o} \tag{43}$$

where: $F_{n,o}$ is the initial external force at a nodal point; Δv is the difference of normal displacements between the individuals of a nodal point pair caused by an increment of normal force $(F_n - F_{n,o})$ and V_m is the maximum closure beginning from initial load $F_{n,o}$. For the load transfer operation assume (43) applies both for loading and for unloading.

Let Δv be the joint opening (positive displacement) on releasing the initial stress ($\sigma_n = \sigma_o$) to the seating pressure

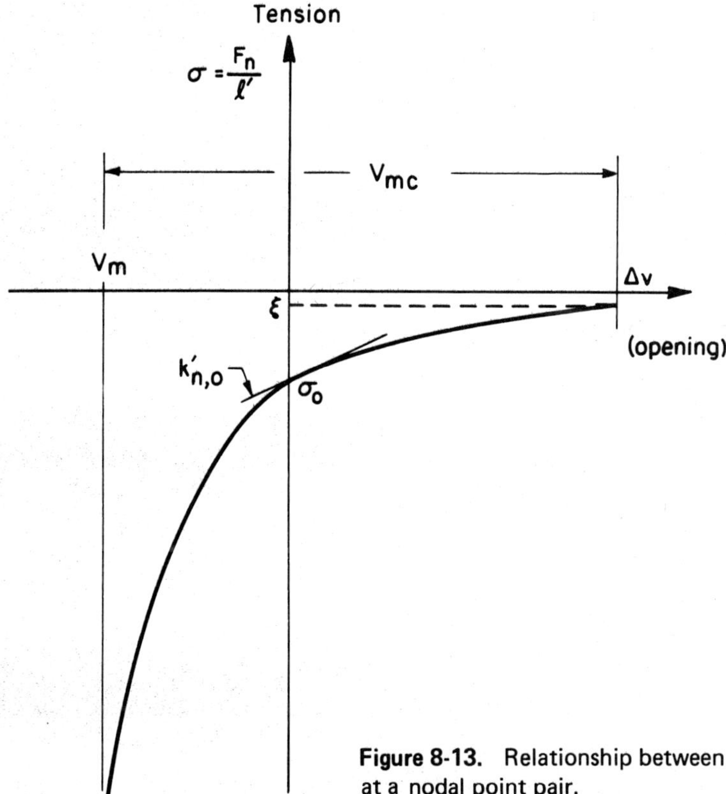

Figure 8-13. Relationship between normal stress and normal displacement at a nodal point pair.

$\sigma_n = \xi$ used to measure V_{mc}. From (43) $\Delta v = V_m - (F_{n,o}/F_n)V_m = V_m - (\sigma_o/\xi)V_m$. Since $V_{mc} = \Delta v(\xi) - V_m$ (figure 8-13):*

$$V_m = \frac{-V_{mc}}{\sigma_o}\,\xi \tag{44}$$

The computation is initiated with a unit normal stiffness $k_{n,o}$ (force/length^3) derived by differentiating (43) with respect to Δv:

$$k_{n,o} = \frac{\sigma_o}{V_m} = \frac{-\sigma_o^2}{\xi V_{mc}} \tag{44a}$$

where σ_o is the initial normal stress in the joint element; this value of normal stiffness produces a solution with displacements V_a and V_b at nodal points a and b along a modified joint element**; then:

$$\Delta v_1 = v_a - v_b \tag{45}$$

(in general, there will also be a tendency for dilatancy $\Delta v(\tau)$, as will be discussed later). From the initial stress σ_o, reference pressure ξ, and maximum closure V_{mc} of the two joint element halves forming the modified element, we find the normal stiffness (k'_n) (stress/length) using (44a) and (42). If both joints comprising a modified element have the same initial stress and maximum closure,

$$k'_n = \frac{\sigma_o}{V_m} = \frac{F_{n,o}}{V_m \ell'} \tag{46}$$

---

*Assuming unit joint thickness. ξ and σ_o are negative since they are compressive. V_m is a negative displacement. k_n and V_{mc} are defined to be positive.

**In the modified element, a is K or L of a joint element while b is I or J of a joint element (figure 8-12).

where ℓ' is the length of the modified element and V_m is given by (44).

Then, the external force at node a of the pair a, b is:

$$F_{n,1} = \ell' k_n'(\Delta v_1) + F_{n,o} \tag{46a}$$

Thus, as in figure 8-14 the first iteration produces point $R_1 = (F_{n,1}, \Delta v_1)$. Only in a rare instance will R_1 be precisely on the constitutive curve (43).

<u>Joint opening</u> (Δv_1 positive) will be considered first. In figure 8-14, Δv_1 is positive, and R_1 is above the curve. A point on the constitutive curve with $\Delta v = \Delta v_1$ defines point R_1^C (compare with figure 8-11) and the distance $R_1^C - R_1$ determines the initial load for the second iteration, $(F_{n,o})_2$ as follows (figure 8-14):

$$(F_{n,o})_2 = (F_{n,o})_1 + \Delta N_1$$

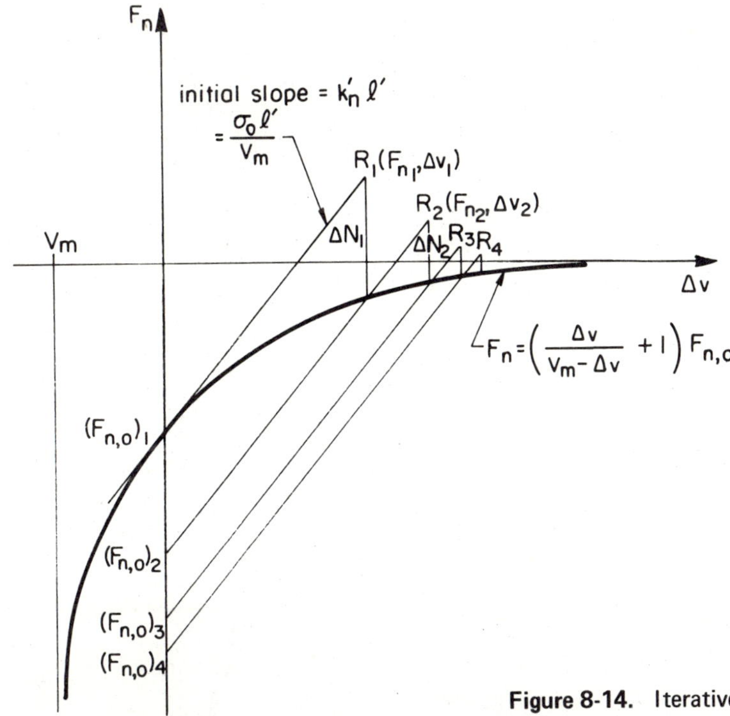

Figure 8-14. Iterative process for joint opening (Δv positive).

where (47)

$$\Delta N_1 = \left(\frac{\Delta v_1}{V_m - \Delta v_1} + 1 \right) F_{n,o} - F_{n,1}$$

For the (i + 1)th iteration

$$F_{n,i} = \ell' k_n' \Delta v_i + (F_{n,o})_i$$

and (47a)

$$(F_{n,o})_{i+1} = (F_{n,o})_i + (\Delta N_i)$$

where

$$\Delta N_i = \left(\frac{\Delta v_i}{V_m - \Delta v_i} + 1 \right) F_{n,o} - F_{n,i}$$

and (47b)

$$\Delta v_i = V_a - V_b + (\Delta v(\tau))_i$$

The process must be repeated until ΔN_i is smaller than a satisfactorily small number ε in all elements, and other non-linear constraints to be discussed have also been satisfied.

Combining equations (47a), and (47b) and introducing (45) gives a load transfer formula for joint opening:

$$(F_{n,o})_{i+1} = \left(\frac{\Delta v_i V_m}{(V_m - \Delta v_i)} + (V_m - \Delta v_i) \right) k_n' \ell' \qquad (48)$$

The iterations are the same whether R_i (figure 8-14) is above or below the constitutive curve; in the latter case, application of (48) will produce a positive value of ΔN_i. Occasionally, iteration will yield an oscillating convergent series of ΔN_i values.

Joint closing is handled similarly, except that the point R_i^C must be guided by the force rather than the displacement computed in the previous iteration (figure 8-15). For this purpose we rewrite the constitutive law (43) as

$$\Delta v = \frac{v_m(F_n - F_{n,o})}{F_n} \tag{49}$$

The results of the first iteration now show Δv_1 negative, with point $R_1 = \Delta v_1, F_{n,1}$. The initial load for the second iteration is:

$$(F_{n,o})_2 = (F_{n,o})_1 + \Delta N_1$$

where

$$\Delta N_1 = \left(\Delta v_1 - \frac{v_m(F_{n,i} - F_{n,o})}{F_{n,i}}\right) k_n' \ell' \tag{50}$$

For the $(i + 1)^{th}$ iteration

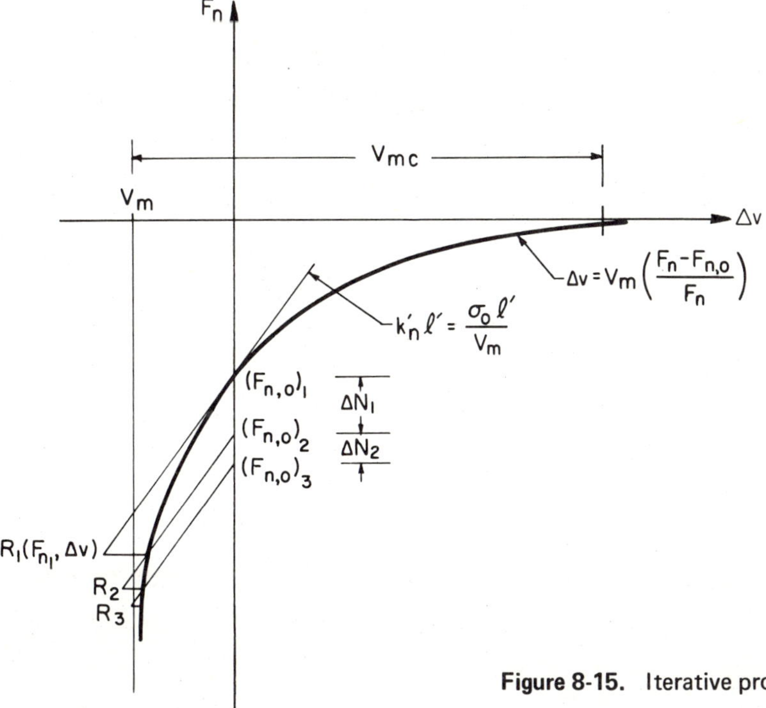

Figure 8-15. Iterative process for joint closing (Δv negative).

$$F_{n,i} = \ell' k_n' \Delta v_i + (F_{n,o})_i$$

and

$$(F_{n,o})_{i+1} = (F_{n,o})_i + (\Delta N_i) \tag{50a}$$

where

$$\Delta N_i = \left(\Delta v_i - \frac{v_m(F_{n,i} - F_{n,o})}{F_{n,i}} \right) k_n' \ell'$$

Combining equations (50a) and introducing (45) gives a load transfer recursion formula for joint closing:

$$(F_{n,o})_{i+1} = k_n' \ell' \left(2\Delta v_i - V_m + \frac{V_m^2}{\Delta v_i + (F_{n,o})_i/(k_n' \ell')} \right) \tag{51}$$

As in the case of joint opening, iterations are required whether R_i is to the left or to the right of the constitutive curve; in the case depicted in figure 8-15 with R_i to the left, ΔN is negative, whereas if R_i is to the right ΔN_i will be positive. Several examples will be discussed later.

Joint Shearing and Dilatancy

Joint Shearing can be treated in the same fashion as joint opening; the limiting shear stress criterion to be imposed on the shear-deformation between blocks is analogous to the "no tension" criterion imposed on the normal stress-opening curve. Since we lack a universal model describing the shear deformation-shear stress behaviour of joints, a simple constitutive law will be assumed (figure 8-16). The initial stress, τ_o, and initial shear stiffness, k_s define the elastic region, whose limiting stress τ_p depends upon σ according to a criterion of peak shear strength:

$$\tau_p = f_1(\sigma) \tag{52}$$

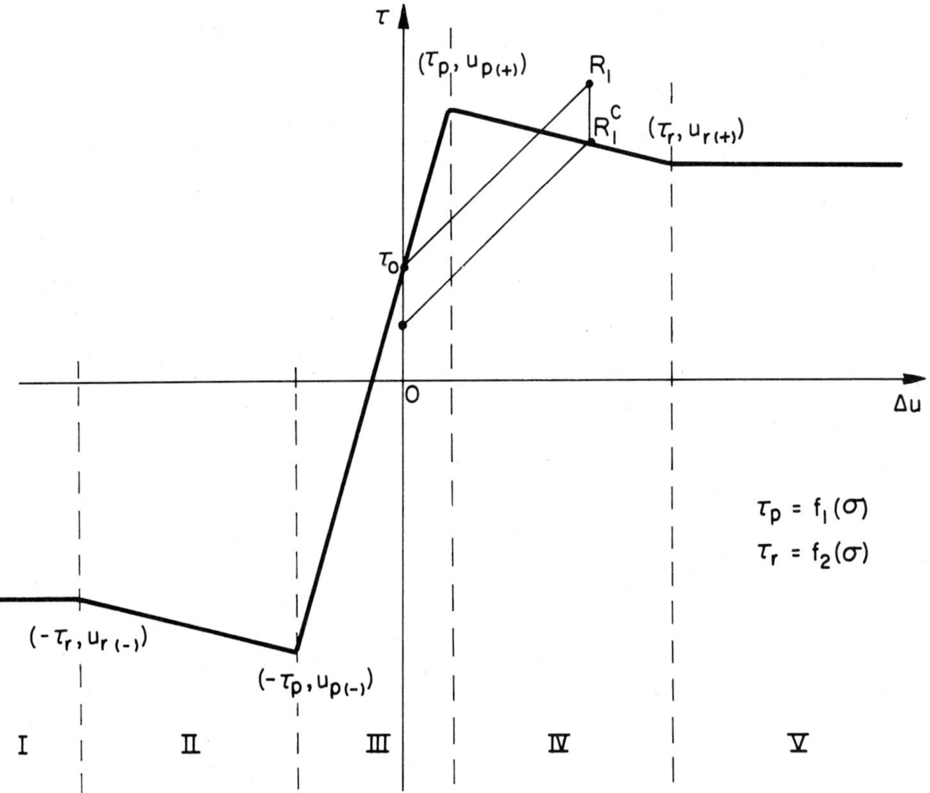

Figure 8-16. Constitutive law for shear deformation.

If τ_p is exceeded, the strength falls, attaining a residual value, τ_r, when the displacement u_r has been attained.

$$\tau_r = f_2(\sigma) \tag{53}$$

Unfortunately, we do not yet know enough about the variation of peak and residual displacements (u_p, u_r) with normal stress σ. As new information becomes accepted, it can supplant the model assumed here, which is: for τ positive (right lateral shear)*

* τ_p, τ_r, u_p, u_r, are defined as positive quantities.

$$u_{p(+)} = u_p = \frac{\tau_p - \tau_o}{k_s}$$

$$u_{r(+)} = u_r = \frac{M(\tau_p - \tau_o)}{k_s}, \quad M > 1. \tag{54}$$

For τ negative,

$$u_{p(-)} = -u_p - \frac{2\tau_o}{k_s}$$

$$u_{r(-)} = -u_r - \frac{2\tau_o}{k_s} \tag{54a}$$

Let

$$\Delta u_i = u_a - u_b \tag{54b}$$

be the relative shear displacement in a modified joint element; then the constitutive relationship (figure 8-16) may be expressed formally as follows:

$$\text{I} \qquad \tau = -\tau_r; \qquad \Delta u \leq u_{r(-)}$$

$$\text{II} \qquad \tau = -\tau_p + \frac{\tau_p - \tau_r}{u_p - u_r}(\Delta u - u_{p(-)}); \quad u_{r(-)} \leq \Delta u \leq u_{p(-)}$$

$$\text{III} \qquad \tau = k_s \Delta u + \tau_o; \quad u_{p(-)} \leq \Delta u \leq u_{p(+)}$$

$$\text{IV} \qquad \tau = \tau_p + \frac{\tau_p - \tau_r}{u_p - u_r}(\Delta u - u_{p(+)}); \quad u_{p(+)} \leq \Delta u \leq u_{r(+)}$$

$$\text{V} \qquad \tau = \tau_r; \qquad \Delta u \geq u_{r(+)} \tag{55}$$

As before, the load transfer is in terms of forces in the modified joint elements. The computation, begun with a stiffness k_s in each joint element, gives $\Delta u_1 = u_a - u_b$ (figure 8-12). The corresponding shear sorce at nodal point a is

$$F_{s,1} = \ell' k_s' \Delta u_1 + (F_{s,o})_1 \tag{56}$$

where k_s' is given by (42) and $(F_{s,o})_1 = \tau_o \ell'$ (compare with (46a)). Then, as for joint opening, the initial force at node a for iteration 2 is

$$(F_{s,o})_2 = (F_{s,o})_1 + \Delta S_1$$

with (57)

$$\Delta S_1 = \tau_1 \ell' - F_{s,1}$$

where τ_1 is given by (55) with $\Delta u = \Delta u_1$.
Similarly, to begin the $i + 1^{th}$ iteration,

$$F_{s,i} = \ell' k_s' \Delta u_i + (F_{s,o})_i$$

and

$$(F_{s,o})_{i+1} = (F_{s,o})_i + \Delta S_i \tag{58}$$

with

$$\Delta S_i = \tau_i \ell' - F_{s,i}$$

Combining equations gives:

$$(F_{s,o})_{i+1} = \tau_i \ell' - k_s' \ell' \Delta u_i \tag{59}$$

where τ_i is given by (55) which in turn depends upon the choice of f_1 and f_2 (52) and (53). Any consistent specific experimental or

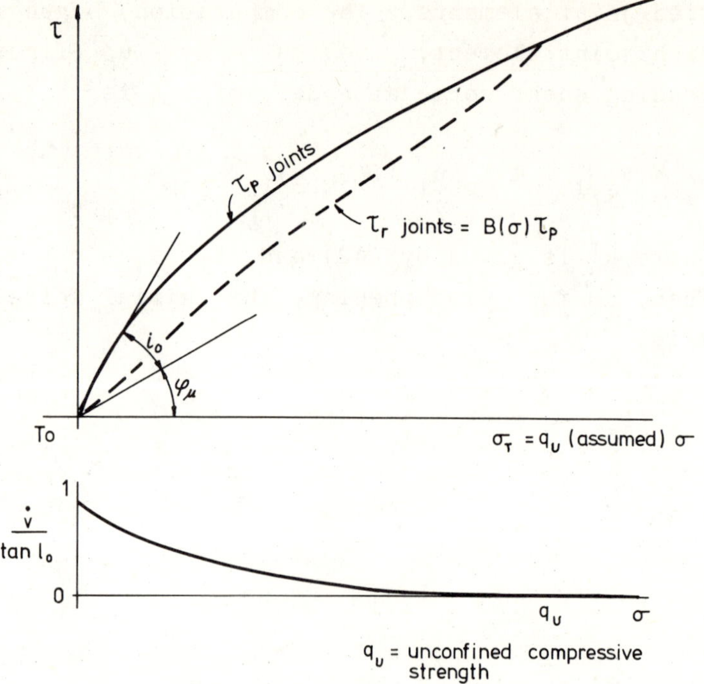

Figure 8-17. Assumed variation of peak and residual shear strength and peak dilatancy ($\dot{v}$) with normal stress; based on Ladanyi and Archambault (1970).

empirical results can be used to define (f_1) and (f_2). As an illustration we will input the formulas given by Ladanyi and Archambault (1970), chapter 5, equations (14), (15), and (16), (figure 8-17).

Unfortunately we know little about the variation of residual shear strength, τ_r, with σ. We will input f_2 in a consistent manner as follows. At high normal stresses, rock becomes plastic, i.e. exhibits a ratio $\tau_r/\tau_p = 1$. We will presume therefore that the ratio τ_r/τ_p increases from B_o, ($0 \leq B_o \leq 1$) at $\sigma = 0$* to 1 at $\sigma = q_u$.

$$\tau_r = \tau_p \left(B_o + \frac{1 - B_o}{q_u}\sigma\right) ; \qquad \sigma \leq q_u$$

and (60)

$$\tau_r = \tau_p ; \quad \sigma \geq q_u$$

*This value will have to be obtained in general by extrapolating data to zero.

B_o can be left as an input parameter for a computer program to permit one to contrast the behaviour of brittle and plastic joints.

Dilatancy must also be introduced into the analysis. Dilatancy describes the normal displacement (joint thickening) $\partial \Delta v(\tau)/\partial u$ caused by shear. Let $\dot{v} = \Delta v(\tau_p)/u_p$ (the secant dilatancy rate). Then as an approximation, the normal displacement caused by dilatancy at a shear displacement Δu is:

$$\Delta v_i(\tau) = -\dot{v}(|\Delta u| + |\frac{\tau_o}{k_s}|); \qquad u_{r(-)} \leq \Delta u \leq u_{r(+)}$$

and (61)

$$\Delta v_i(\tau) = -\dot{v}(u_{r(+)} + |\frac{\tau_o}{k_s}|); \qquad \Delta u \geq u_{r(+)} \text{ or } \Delta u \leq u_{r(-)}$$

The absolute value function and the minus sign insure that the joint thickens regardless of the sign of the shear stress, as discussed in chapter 5.

The variation of $\dot{v}$ with σ was given by equation 16b of chapter 5. Substituting this equation in (61) with $\sigma_T = q_u$* and considering stresses in the modified joint element gives for the ith iteration:

$$\Delta v_i(\tau) = \left(\frac{(F_n)_{i-1}}{\ell' q_u} - 1\right)^4 \tan i_o \left(|\Delta u_i| + \frac{|F_{s,o}|}{k_s \ell}\right)$$

$$\text{for } u_{r(-)} \leq \Delta u \leq u_{r(+)}$$

---

*As noted previously, q_u is here a negative quantity, since tension is positive.

and (62)

$$\Delta v_i(\tau) = \left(\frac{|F_n|_{i-1}}{\ell' q_u}\right)^4 \tan i_o \left(u_{r(+)} + \frac{|F_{s,o}|}{k_s' \ell'}\right)$$

$$\text{for } \Delta u \geq u_{r(+)}$$
$$\text{or } \Delta u \leq u_{r(-)}$$

At the end of the ith iteration, we will know the shear displacement Δu_i at each nodal point pair. Inserting Δu_i in (62) for each modified joint element, we determine $\Delta v_i(\tau)$.

If all dilatancy is prevented by the adjacent elements, there must be external compressive forces applied to joint nodes a and b equal in magnitude to the dilatancy multiplied by the joint stiffness. Redefining the joint stiffness as the slope of the compression curve (43) evaluated at the previous normal stress $(F_{n,i-1}/\ell')$, the increment in initial normal stress in the joint due to dilatancy calculated for the ith iteration is

$$(\Delta\sigma_{n,o})_i = \frac{\Delta v_i(\tau)}{\xi V_{mc}} \left(\frac{F_{n,i-1}}{\ell'}\right)^2 \qquad (62a)$$

To start the $i + 1^{st}$ iteration, the initial stress incremented by $(\Delta\sigma_{n,o})_i$ produces external forces on the neighboring elements, which in fact deform, thereby automatically relaxing the initial assumption that dilatancy is prevented.

For the first iteration we must use $(F_n)_{i-1} = (F_n)_o = F_{n,o}$.

Updating the Loads

The final 'step' in the load transfer procedure is to rotate $(F_{n,o})_{i+1}$ and $(F_{s,o})_{i+1}$ to global coordinates and update the load vector. With the sign convention used in the modified joint elements, $(F_n)_i$ and $(F_s)_i$ are external forces at nodal point a (figure 8-12). We will update the load vector as follows. If a nodal point is not in a joint, the load at that nodal point does not change. At every nodal point pair a, b belonging to a modified joint element, the new load vector terms are:

$$\begin{Bmatrix} F_{x,a} \\ F_{y,a} \\ F_{x,b} \\ F_{y,b} \end{Bmatrix}_{i+1} = \begin{pmatrix} \cos\alpha & \sin\alpha & 0 & 0 \\ -\sin\alpha & \cos\alpha & 0 & 0 \\ 0 & 0 & \cos\alpha & \sin\alpha \\ 0 & 0 & -\sin\alpha & \cos\alpha \end{pmatrix} \begin{Bmatrix} (F_{s,o})_{i+1} \\ (F_{n,o})_{i+1} \\ -(F_{s,o})_{i+1} \\ -(F_{n,o})_{i+1} \end{Bmatrix} \quad (63)$$

$\{F\}_{i+1}$ is then multiplied by the inverted stiffness matrix $(k)^{-1}$ to yield the new estimate of displacement $\{u\}_{i+1}$. Table 8-1 summarizes the steps in the load transfer scheme, as discussed here.*

SOURCES OF EXTERNAL LOAD

The contributions of residual stress to the load vector have been explored. Here we will consider additional forces from water pressure distributions, gravity, pseudo-static accelerations (inertia forces), and active or passive supports.

Water Forces

Water forces tend to change as a result of the joint deformations they cause. Therefore, any complete analysis of the action of water becomes a problem in coupled stress and flow, as discussed by Rodatz and Wittke (1972), Noorishad et al. (1972), Gale (1975), and others. This class of problems, which will not be treated here, can be set up as follows. One assumes an initial water pressure distribution, which generates force contribution at the nodal points. Iterative solution by load transfer as discussed in this chapter

---

*Certain economies are realizable by following a somewhat modified scheme for load transfer. Store the displacements $\{u_i\}$. Use terms ΔN_i and ΔS_i, rotated to global co-ordinates, for each modified joint element to build a force increment vector $\{\Delta F_i\}$, which when multiplied by $(k)^{-1}$ gives displacement increments for all nodal points. (The force increments at nodal points inside solid elements are zero). The incremental displacements are cumulated and added to the displacements of the first increment. This modified procedure replaces many products by zero but requires summing displacements.

TABLE 8-1

Summary of Steps in Finite Element Analysis of Jointed Rock Masses by the Load Transfer Method

$i = 1$ (Parentheses identify relevant equation numbers)

1. Read input: geometric and material properties; initial stresses; accelerations; pore pressures; and support loads.
2. Form solid element stiffness matrix for each type, orientation, and shape of solid element. (24a)
3. Form joint element stiffness matrix for each type length and orientation of joint element. (35) (38)
4. Assemble structural stiffness matrix (K) (see example 3). (Actually this step is performed simultaneously with 2 and 3.)
5. Assemble residual stress contributions to the load vector (27) and (40). (This step is also done simultaneously with 2 and 3). Change sign and store in load vector (41).
6. * Add external forces from other sources: water pressures; weight; active and passive supports. Total load vector = $\{F\}_i$.
7. Invert the structural stiffness matrix (K). For small computers this may be done outside the rest of the program as it only needs to be done once. Store $(K)^{-1}$
8. * Determine displacements by matrix multiplication. $\{u\} = (K)^{-1}\{F\}_i$
9. Form modified joint element stiffnesses and initial forces (42) (46) and relative displacements Δv_i and Δu_i (45), (62) and (54b).
10. Determine normal force $\{F_n\}_i$ (47a) and shear force $\{F_s\}_i$ (58) in each modified joint element.
11. Find ΔN and ΔS in each element (47b) or (50a), and (58). If $\Sigma\Delta N + \Sigma\Delta S > \varepsilon$, where ε is some small number, update $F_{n,o}$ and $F_{s,o}$ by (51) and (59). Rotate to global coordinates and update $\{F\}_i$.
12. $i = i + 1$. Go to step 8.

---

*For incremental loading, divide total load vector into small increments and cumulate their effect.

defines the change of volume of each modified joint element.

$$\theta = \frac{\Delta vol}{vol} = \frac{\Delta vol}{\ell' e} \qquad (64)$$

where e is the initial thickness of the joint. The change in water pressure in the modified element is therefore:

$$p_i = \frac{\Delta vol}{C_w \ell' e} \qquad (64a)$$

where C_w is the compressibility of water.

The result of the first iteration will thus produce a new water pressure distribution which can be expected to initiate flow through the network of joints. We allow flow to occur until a new pressure distribution has been established, calculate the new water pressures and begin the stress problem anew. The problem is unfortunately quite path dependent.

In each triangular rock element, water forces are input as applied external forces at the nodal points in the direction opposite to the hydraulic gradient. If as a simplification we assume the pressure gradient is constant over the element, it follows mathematically (Zienkiewicz (1971) that the water forces distribute equally to each node. Taking the global y axis positive upwards gives*:

$$\left\{F_{ext}\right\}_{water} = \begin{Bmatrix} F_{x,I} \\ F_{y,I} \\ F_{x,J} \\ F_{y,J} \\ F_{x,K} \\ F_{y,K} \end{Bmatrix} = \frac{a}{3} \begin{Bmatrix} \frac{\partial p}{\partial x} \\ \frac{\partial p}{\partial y} \\ \frac{\partial p}{\partial x} \\ \frac{\partial p}{\partial y} \\ \frac{\partial p}{\partial x} \\ \frac{\partial p}{\partial y} \end{Bmatrix} = \frac{\gamma_w a}{3} \begin{Bmatrix} -\frac{\partial h}{\partial x} \\ 1-\frac{\partial h}{\partial y} \\ -\frac{\partial h}{\partial x} \\ 1-\frac{\partial h}{\partial y} \\ -\frac{\partial h}{\partial y} \\ 1-\frac{\partial h}{\partial y} \end{Bmatrix} \qquad (65)$$

---

*$\partial h/\partial y$ is positive for downward flow; $\partial h/\partial x$ is positive for flow to the left.

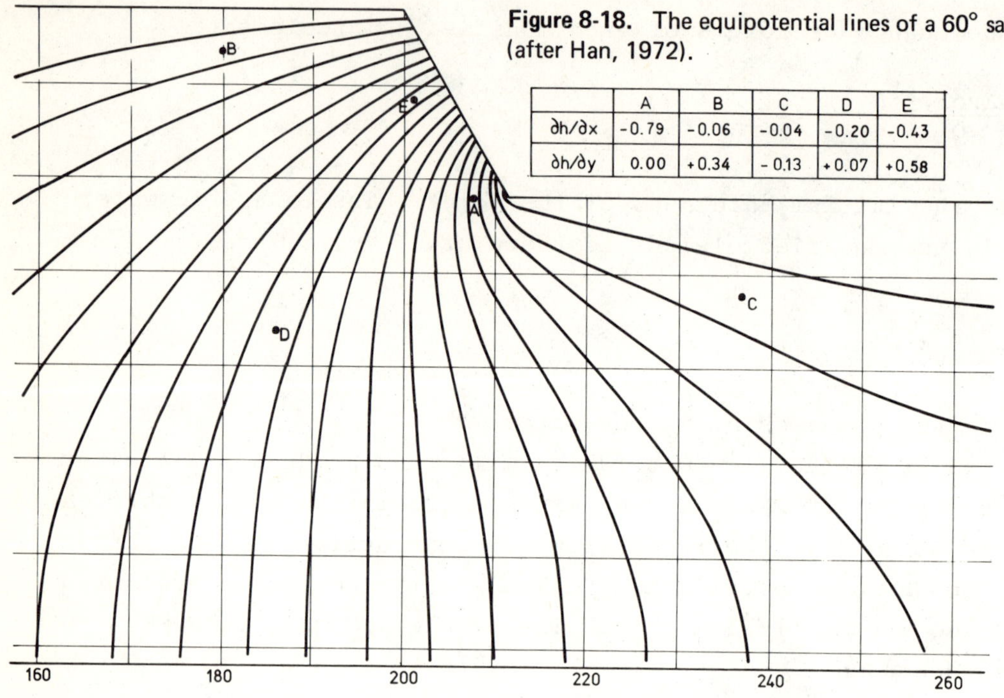

Figure 8-18. The equipotential lines of a 60° saturated slope (after Han, 1972).

	A	B	C	D	E
$\partial h/\partial x$	-0.79	-0.06	-0.04	-0.20	-0.43
$\partial h/\partial y$	0.00	+0.34	- 0.13	+0.07	+0.58

For example figure 8-18 shows lines of equal potential for a 60° saturated slope, obtained by Han (1972) using a conducting paper method. The gradients for elements centered at five points have been calculated from the diagram and are tabulated on the figure. These values will be input in the respective elements and the nodal point forces computed internally by (65). (Above the water table, we assign $\gamma_w = 0$.)

Weight Forces

The initial stresses input in the analysis may reflect the action of self weight forces on the region under investigation. But the residual stress input can not duplicate the action of gravity because "loading" by an initial stress and the loading of gravity are fundamentally different. The residual stress load is like that of a precompressed spring; it can not continuously follow the deformations it causes in its neighbours. Gravity, on the contrary, produces an inertia force that pursues the deforming body. Therefore, to simulate the gravity action in a beam we require additional forces input as external forces at the nodal points; (we should also reduce the initial stresses accordingly so that gravity is not counted twice). Since gravity produces a force distributed over each unit of mass, it can be treated like the seepage forces just considered. Denoting by γ the weight per unit volume of the rock (total, not buoyant), the required applied external forces in an element are:

$$\begin{Bmatrix} F_{x,I} \\ F_{y,I} \\ F_{x,J} \\ F_{y,J} \\ F_{x,K} \\ F_{y,K} \end{Bmatrix} = \frac{a\gamma}{3} \begin{Bmatrix} 0 \\ -1 \\ 0 \\ -1 \\ 0 \\ -1 \end{Bmatrix} \tag{66}$$

(As before, we assume the global y axis is positive upwards).

Pseudo Static Acceleration

Pseudo static acceleration can be included for purposes of a limit equilibrium analysis. If a body is accelerated kg in a direction 180 + α degrees from the positive x axis (measured counter-clockwise from x), it will experience inertia forces in the direction α; these forces can be input with gravity by rewriting (66) as follows:

$$\begin{Bmatrix} F_{x,I} \\ F_{y,I} \\ F_{x,J} \\ F_{y,J} \\ F_{x,K} \\ F_{y,K} \end{Bmatrix} = \frac{a\gamma}{3} \begin{Bmatrix} k \cos \alpha \\ -1 + k \sin \alpha \\ k \cos \alpha \\ -1 + k \sin \alpha \\ k \cos \alpha \\ -1 + k \sin \alpha \end{Bmatrix} \tag{66a}$$

Rock Bolts

Tensioned rock bolts provide a pair of forces compressing the rock at the bearing plate and anchor ends. One must layout the configuration of nodal points with the locations of rock bolts in mind. The forces are simply added as external loads parallel to the bolt axes at the nodal points in question. Rock bolts produce two additional effects, however. First, the steel stiffens the rock. This can be included by increasing the appropriate stiffness terms for the nodal points along the line of the bolt. It may be more satisfactory to input one-dimensional bar elements along the line of the bolt (see example 1); these should be constrained to yield at the appropriate load through load transfer. Secondly, the bolts act as shear keys in crossing each joint; since the bolt is confined by the rock, the rock must crush around the bolt to permit it to shear. The bolt action can be simulated by supplementing the peak and residual shear strength, of the modified joints containing the rock bolt, by an amount equal to the shear strength of the bolt, plus an increase in parameter a_s in Ladanyi's peak shear strength criterion.

EXAMPLE PROBLEMS

Example 4—Joint Closing

In figure 8-19, a block under an initial stress of 5 MN/m^2 compression is next to a joint with initial compression 1 MN/m^2, held by a constraint (not shown). When the constraint is removed, the momentary disequilibrium destresses the block and compresses the joint to restore equilibrium. The speed of convergence depends upon the initial stress in the joint.

Putting eq (44) in (43) and substituting $F_{n,o} = \ell'\xi$ and $\sigma_o = \xi$, $\sigma_n = \left(\frac{\Delta v}{-V_{mc} - \Delta v}\right)\xi + \xi$. The normal deformation curve corresponding to $V_{mc} = 0.05$ and $\xi = -0.1$ passes through the points $(F_n, \Delta v)$ in Table 8-2.

As shown in the graphical solution (figure 8-19) about 5 iterations are required for convergence for the given data, with $\sigma_o =$ -1 MN/m^2 in the joint, whereas when σ_o is -0.5 and -0.2 MN/m^2 con-

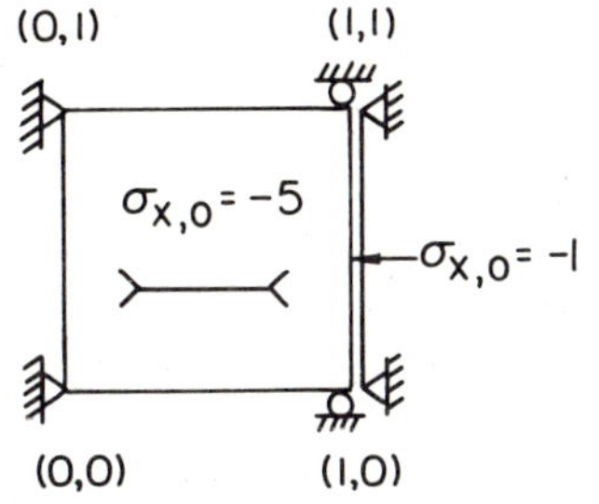

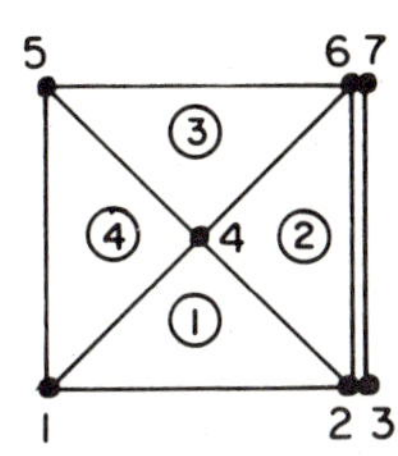

Figure 8-19. Example 4—joint closing.

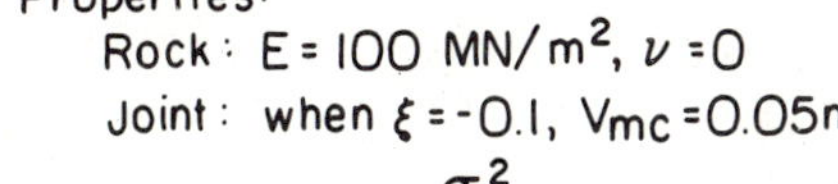

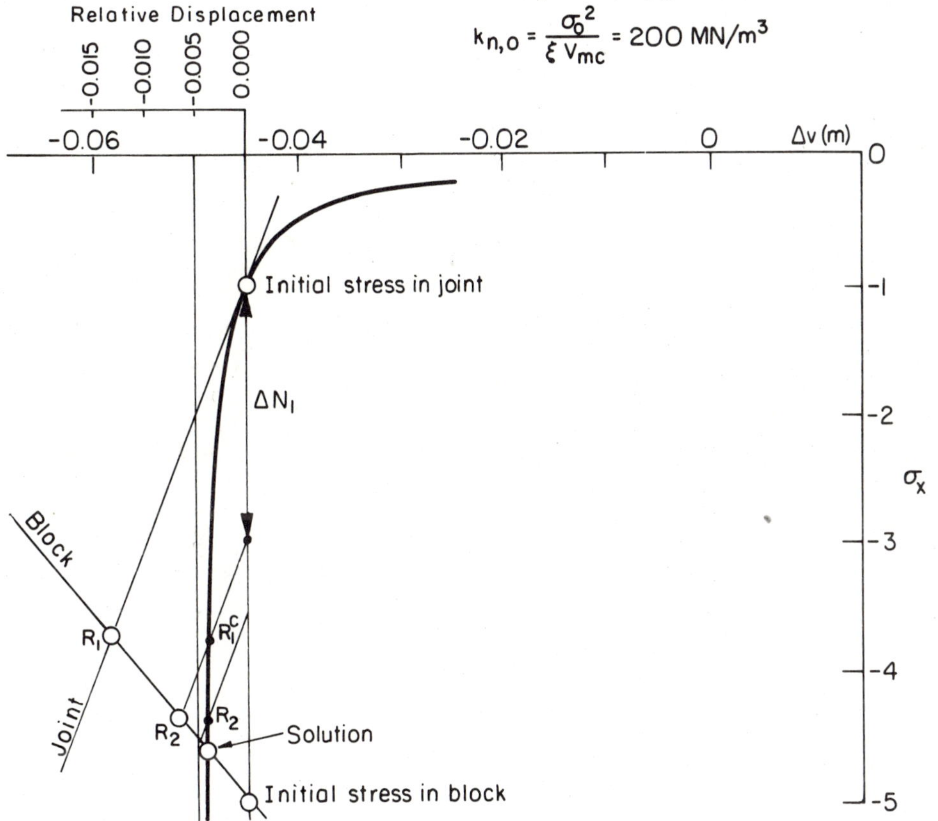

vergence requires 12 and 35 iterations, respectively. In such a case, convergence can be accelerated.

Example 4a—Accelerating the Convergence

Had we carried out example 4 with an initial stress of $-0.2\ \text{MN/m}^2$,

TABLE 8-2

Δv	$\sigma_x = F_n$
0	-.100
-.010	-.125
-.020	-.167
-.030	-.25
-.035	-.333
-.040	-.500
-.042	-.625
-.045	-1.0
-.046	-1.25
-.048	-2.5
-.049	-5.0

35 iterations would have been required for convergence (figure 8-20). It is likely that the requested number of iterations would fall short. On restarting to continue the computations, several ways to accelerate the convergence are possible, as in figure 8-20. Say the first computations with initial joint stress at A terminated after 2 iterations, yielding point R_2, and associated points C ($=R_2^C$) and E ($=F_{o,3}$). Three alternative accelerating schemes are shown.

(a) A Newton Raphson correction restarts from C with stiffness redefined to the tangent to the compression curve at C. The initial stress in the rock must then be changed, for restart, from B to D or the 'solution' will shift. The displacements will have to be stored from the first run and added cumulatively on each restart. Convergence will be achieved in 6 more iterations.

(b) Another approach, the dashed lines, restarts at E with the stiffness redefined to the slope of the compression curve at C. This produces an over correction but convergence is reached in 7 more iterations.

(c) A third approach is given by the dotted lines. The restart point is the initial starting point A, and the stiffness is redefined to that corresponding to line AC. Convergence

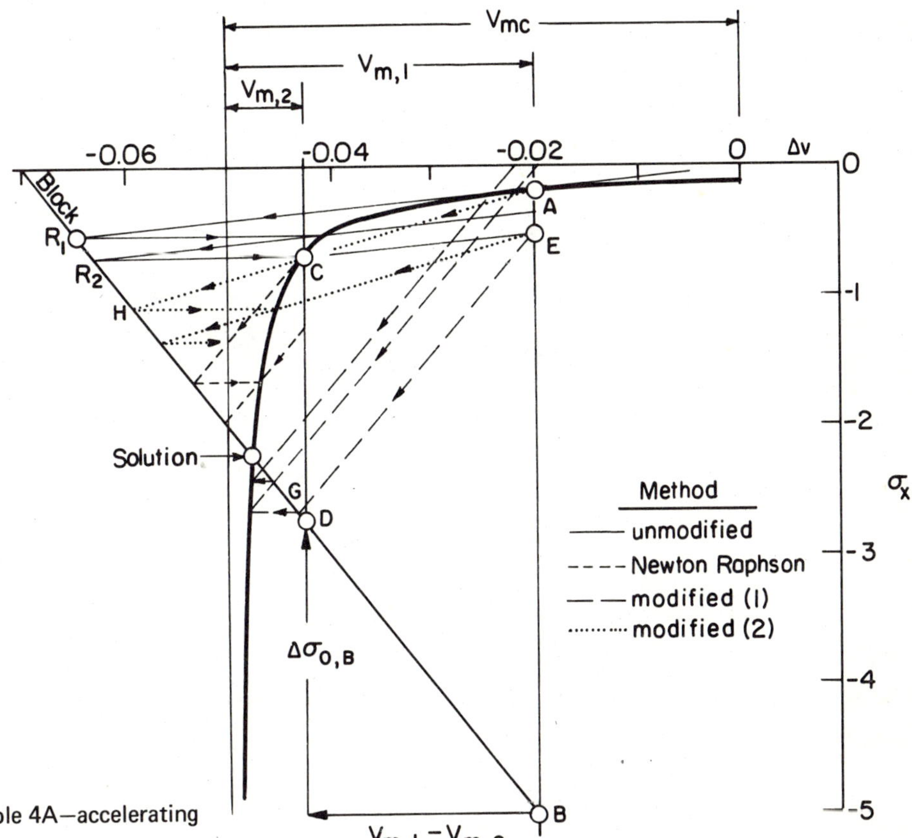

Figure 8-20. Example 4A—accelerating the convergence.

is achieved in an additional 16 iterations. The first run writes the new joint stiffnesses on tape which is read on restarting.

With reference to figure 8-15, the unit normal stiffness in a closing joint upon restart after i iterations is

$$k_{n,i} = \frac{1}{\ell} \frac{F_{n,i} - F_{n,o}}{\Delta v_i} = \frac{F_{n,i}}{\ell V_m} \tag{2}$$

Example 5—Joint Opening

Example 5 is a case of joint decompression (opening) which eliminates an initial block tension. Figure 8-21 shows a graphical solution with convergence in 22 iterations. Restarting after the second iteration with redefined normal stiffness according to modified methods 1 and 2 accelerates the convergence to 9 and 18 iterations respectively. The computer output for the first 5 iterations gives the results of Table 8-3.

TABLE 8-3

iteration	σ_x in solid	ΔN at Nodes 2 and 6	u_x at nodes 2 and 6
Start	1.0	0	0
1	0.667	-1.52	-0.00333
2	0.497	-1.16	-0.00503
3	0.368	-0.93	-0.00631
4	0.264	-0.76	-0.00736
5	0.179	-0.64	-0.00821

The modified unit normal stiffness corresponding to modified method 2 (figure 8-21, dotted lines) will be calculated for the case of joint opening. With reference to figure 8-14, the stiffness (stress/displacement) for accelerated restart after i iterations in an opening joint is:

$$k_{n,i} = \frac{1}{\ell}\,\frac{(F_{n,i} - F_{n,o})}{\Delta v} = \frac{F_{n,o}}{\ell(V_m - \Delta v)} \tag{68}$$

Examples 6 and 7—Sliding

These examples show how a shear failure is indicated by diverging output. (Unfortunately, numerical difficulties can sometimes produce divergence in a stable case so that 'failure' can only be suggested by the output, not confirmed by it). A block under an initial vertical stress is pushed by a 'following' force past two joints, whose peak strength is insufficiently large. In example 6A (figure 8-22) the joints exhibit peak-residual behaviour (B_o = 0.333) while in example 6B (figure 8-23) the joints are plastic (B_o = 1.0). The addition of a 15° dilatancy angle to the joints, example 7, (figure 8-24) brings stability in both cases. The dilatancy raises the peak shear strength for the first iteration (from 0.66 to 0.83) but this is still short of the applied stress (1.0). However, the shear

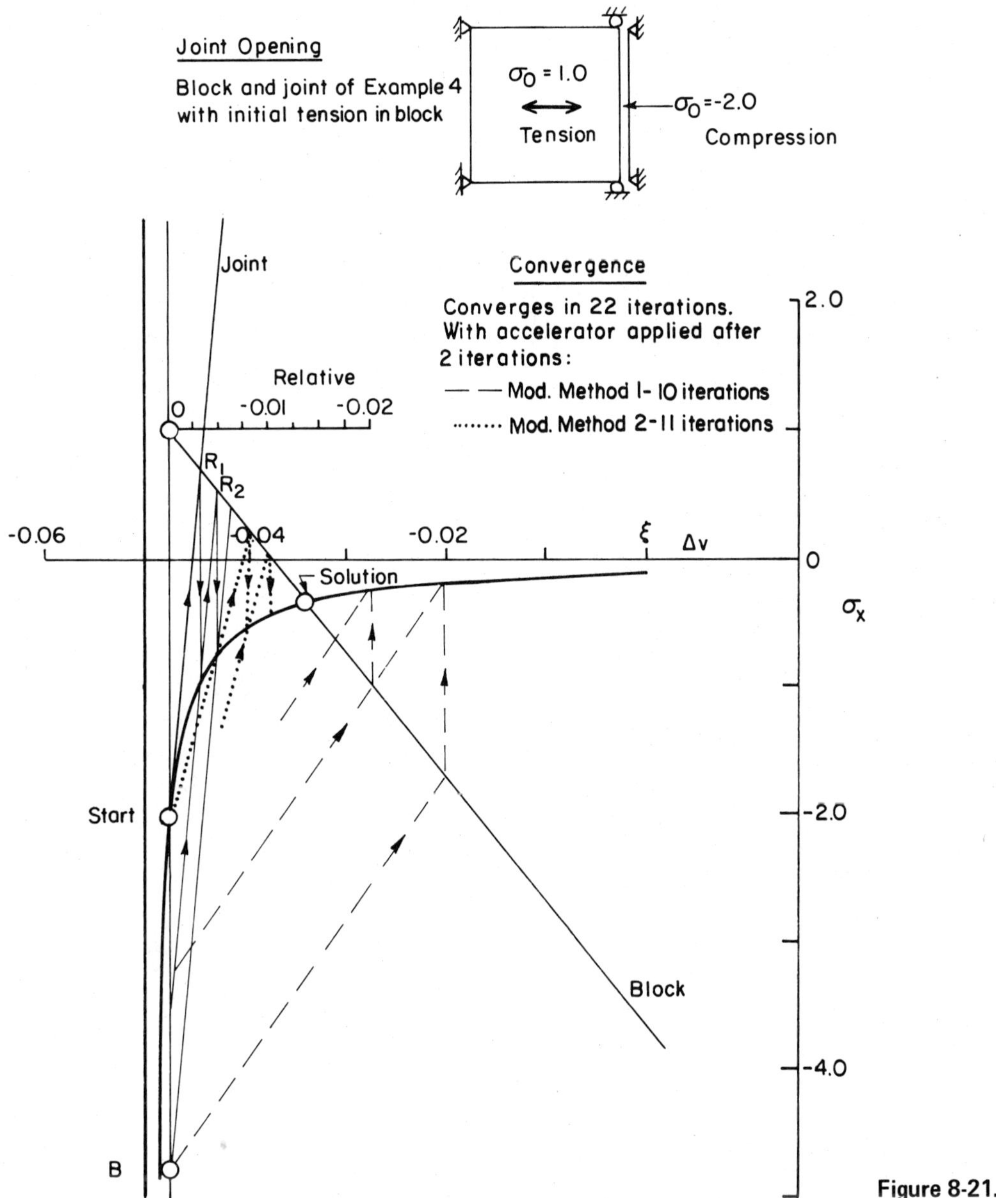

Figure 8-21. Example 5.

displacement on the first iteration raises the normal stress which further increases τ_p above 1.0. Table 8-4 contrasts the convergent solution of example 7, which is stable, with the divergence seen in example 6A, in which failure occurs.

Figure 8-22. Example 6A.

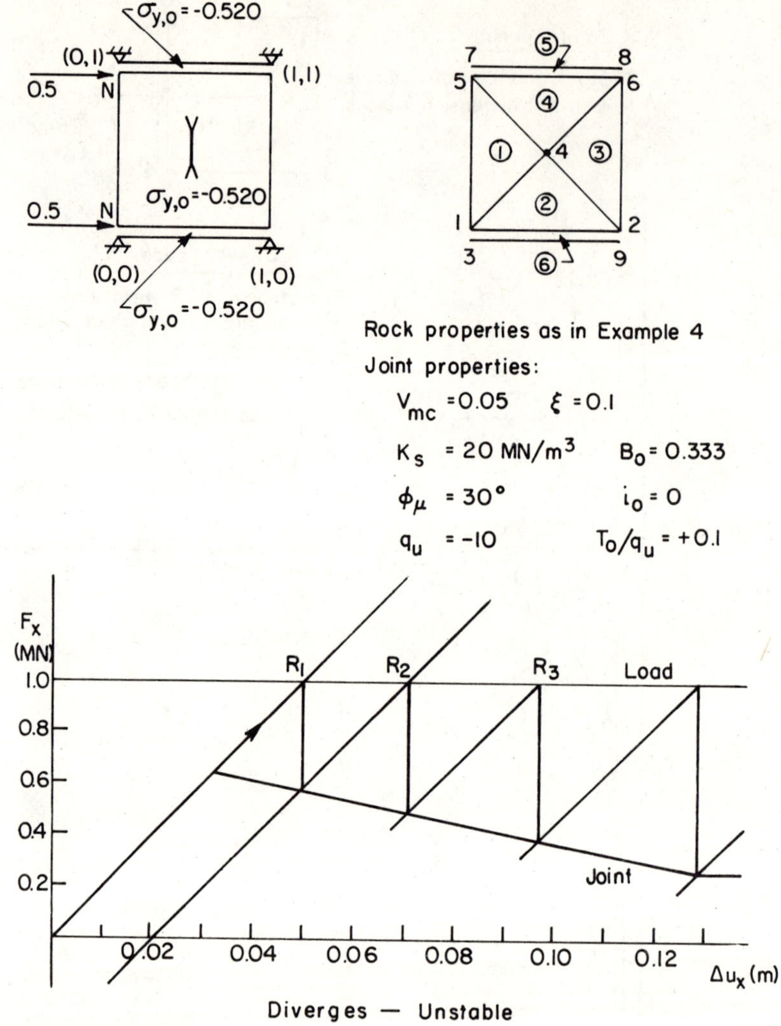

TABLE 8-4

Output Displacements in Examples 6A and 7

iteration	u_x at node 4 (m)	
	example 6A	example 7
1	0.0500	0.0500
2	0.0710	0.0587
3	0.0963	0.0500
4	0.1271	0.0500

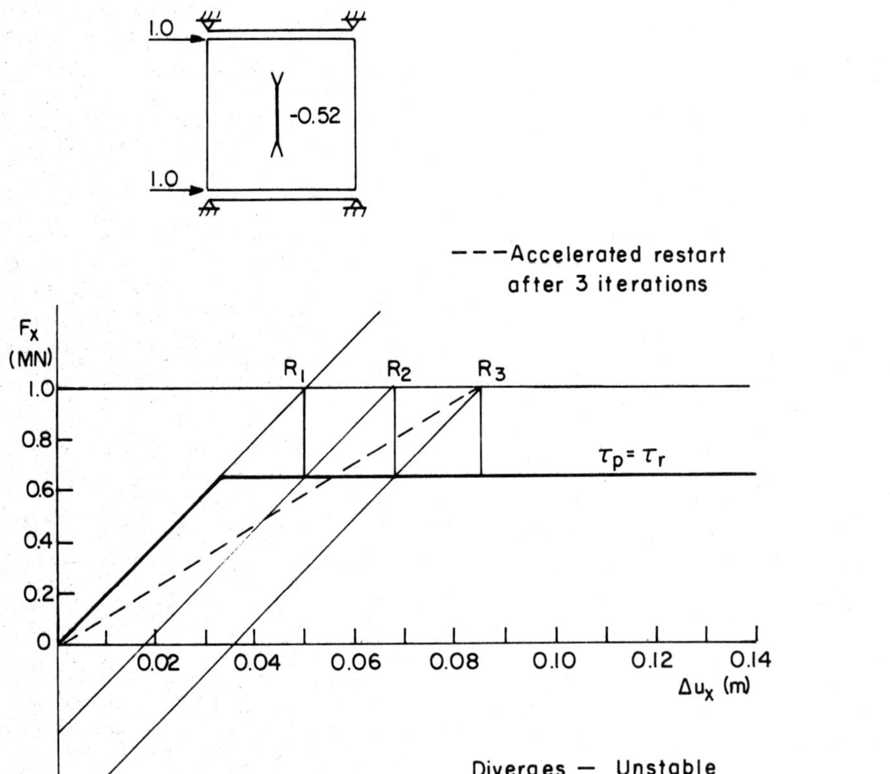

Figure 8-23. Example 6B—shear failure on non-dilatant joints.

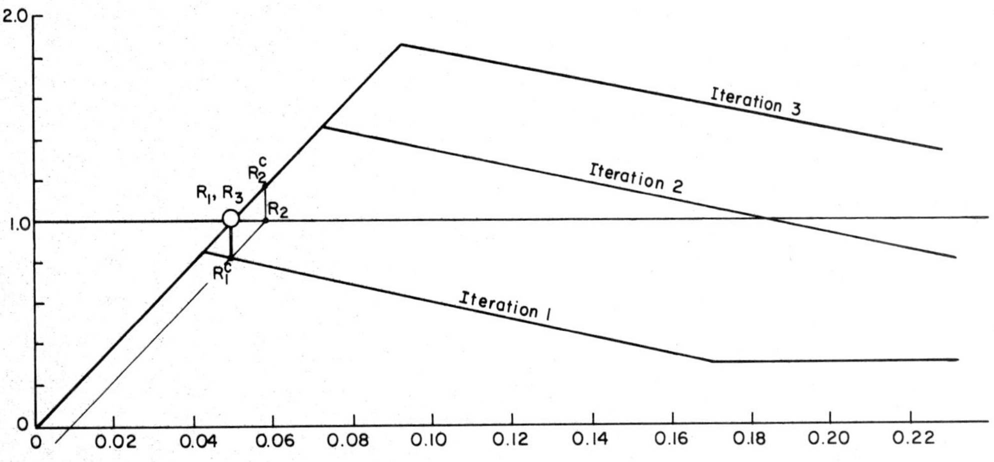

Figure 8-24. Example 7—shear on dilatant joints.

Accelerated restart is performed as shown by the dotted path in figure 8-23. The revised unit shear stiffness for restarting after i iterations is

$$k_{s,i} = \frac{\tau_i - \tau_o}{\Delta u} \tag{69}$$

Example 8—Rotation of a Block

A square block between two horizontal joints rotates in response to a clockwise moment (figure 8-25). The first iteration produces an overclosing of the joints. The solution after 10 iterations shows normal displacements within the limit; as the block deforms to meet the maximum closure restraint, the stress state changes from biaxial to approximately uniaxial, directed along the block diagonal. A larger loading of the same style would produce tension in the solid.

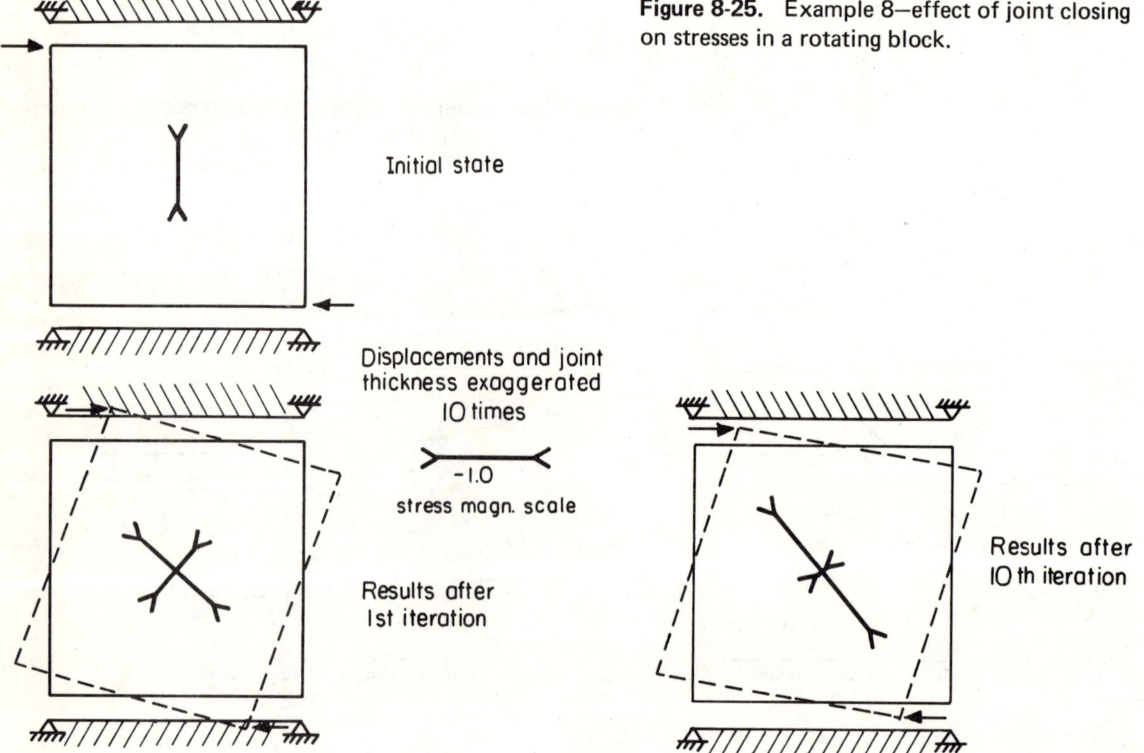

Figure 8-25. Example 8—effect of joint closing on stresses in a rotating block.

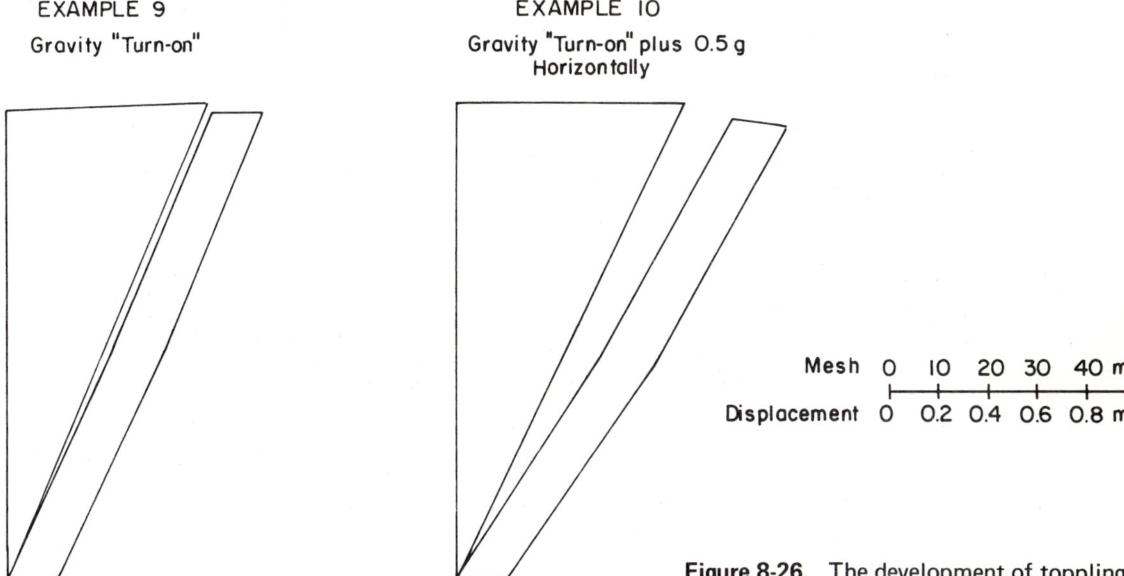

Figure 8-26. The development of toppling.

Shearing and Overturning of Rock Slopes—Examples 9, 10, 11, and 12

Figure 8-26 shows intermediate states in an unstable case involving overturning of a slab on an overhanging slope. The "turn-on" of gravity, example 9, produces an opening af the discontinuity at the top of the slope, and a closing at the toe; however the rate of overturning is very slow without an accelerated restart. In example 10, the rate of divergence is much faster because a horizontal acceleration has been superimposed on gravity. Figure 8-27 shows the importance of accurately reproducing the path of actual loading in these problems involving gravity. In example 11A, gravity has been "turned-on" and the downslope movement of the wedge is confused with the lowering of the surface as gravity is applied. A better way to perform this analysis (example 11), is to restrain the free slope for the first run in which gravity is applied; then introduce the stresses corresponding to the results of the first run as initial stresses while a second run is made with the slope restraints removed. The only deformations remaining will be those associated with the downslope sliding of the block (presuming the stresses indicated in the output to the first run are in equilibrium and acceptable in all elements). In example 12 (figure 8-28) an unstable case is indicated

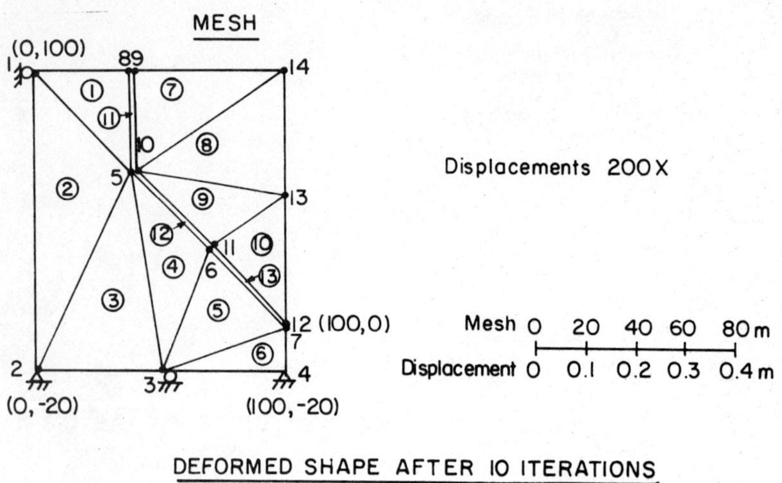

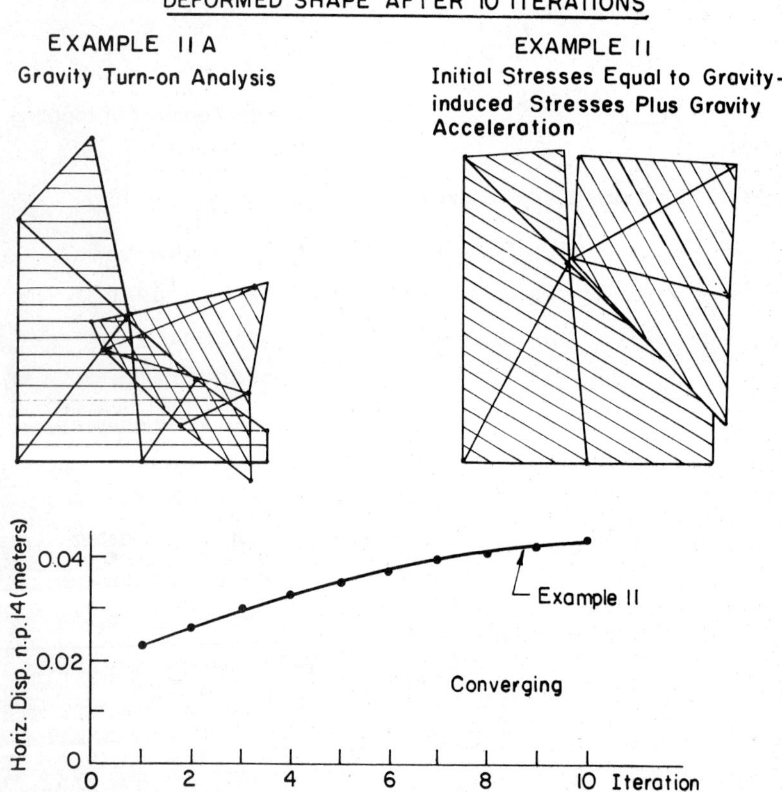

Figure 8-27. Effect of loading path in a gravity-loaded wedge.

Figure 8-28. Failure of a slope indicated by diverging output.

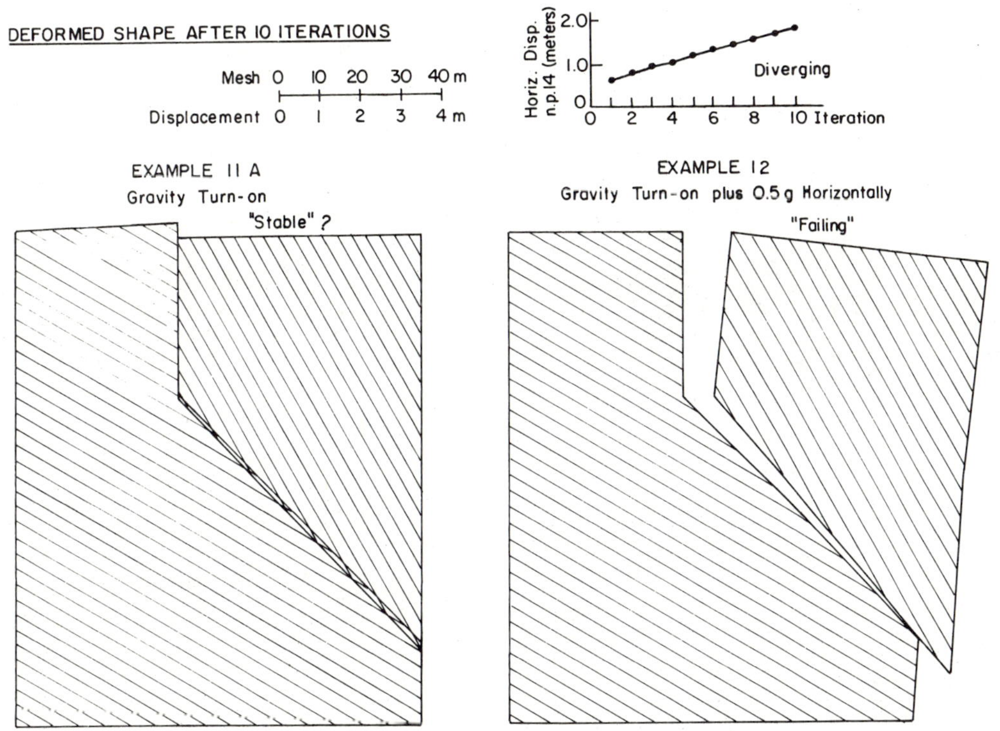

by the output divergence when a horizontal acceleration of 0.5g is superimposed on the gravity acceleration.

Example 13—Study of Underground Excavation

Figure 8-29 shows a mesh used to investigate progressive relaxation and failure of rock around an underground power station. To model the layers of shale, coal, sandstone, and siltstone, 288 elements and 329 nodal points were used in a load transfer program.* Since every bedding plane could not be represented, the stratigraphy was simplified by division into a manageable number of units. Rows of joint elements were placed between each of these units. A calculation provided anisotropic properties for each layer which took

---

*This analysis was performed in collaboration with J. Dubois and T. A. Lang.

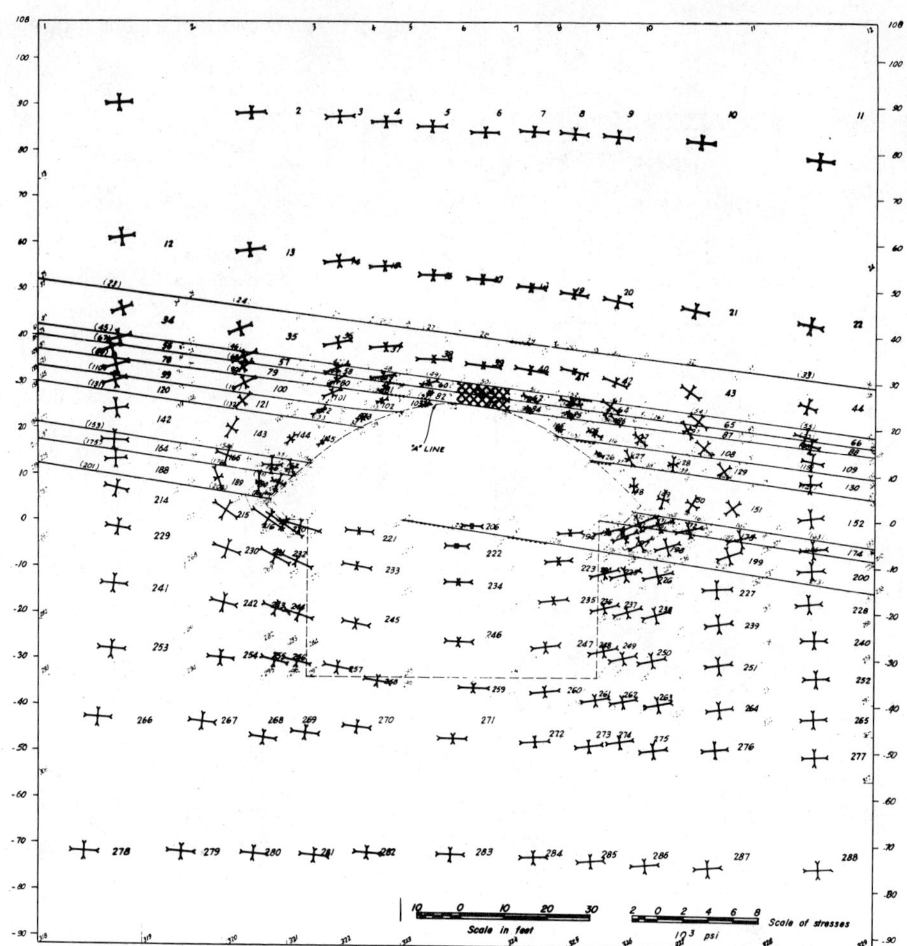

Figure 8-29. A finite element study of an underground excavation.

into account the number of individual bedding planes it might contain. The mesh at first included the first stage excavation as an initial opening. When the initial stress believed to exist in the rock was input, the walls and roof of the stage I excavation deformed and the stresses changed; some elements were over-stressed or placed in significant tension. Judgement was applied as to whether or not such elements might "fall out" and if it was thought they could, additional computations were run with lower stiffness properties prescribed for these elements. Eventually the second stage of excavation was effected, elements were "removed" or "modified" in properties according to the severity of the stresses and subsequent excavation stages applied. Figure 8-29 shows the results at an intermediate stage of excavation.

INCREMENTAL LOADING

The opening and closing, shear and dilatancy of discontinuities introduce a large degree of non-linearity into structural calculations. If there are numerous joints, and if loading creates large changes from a stable initial condition, the iterative process may not converge. It is a good practice, in such cases, to use an incremental solution, in which the loading is divided into small steps and applied cumulatively. For a gravity loaded problem, for example, an initial increment of perhaps ten percent of the gravity load vector is applied and iterations pursued by load transfer or by variable stiffness methods until convergence is approached. Then, the next increment of load is applied by increasing the acceleration of gravity by another ten percent, and the convergent displacements and stresses are accumulated. The process is repeated until the whole load has been applied.

For an initial stress problem, where the initial equilibrium state of stress is to be disturbed by an excavation, the normal and shear loads on the excavated boundary must be zeroed gradually. This can be done by adding normal and shear forces to these boundaries which are directed opposite to the initial forces and are of say ten percent of their magnitude. The excavation will then slowly approach the final configuration as the unload increments are accumulated. Thus, incremental loading can be used for both loading and unloading problems to improve convergence.

In initial stress problems, one must be careful to insure that the input stresses are in fact in equilibrium. With heterogeneous or anisotropic properties, this may not be easy. It would be fortuitous if measured residual stresses were all compatible with the geological structure and input properties. Thus, an initial cycle of iterations may be warranted, using the assumed initial stresses in a mesh without a free boundary, and without any load or unload increments. When this problem is computed, there will be displacements inside the mesh and changes in the initial stresses. These adjusted stresses can be used as improved estimates of the initial stress state, since they will be in equilibrium.

RIGID BLOCK ANALYSIS

When there are a large number of closely spaced joints in hard rock, the joint deformations so overshadow the block deformations that the blocks can be considered rigid. Economies in computer storage are thus possible in which case larger systems of blocks can be accommodated. Calculation of the deformations of a rigid block system begins with input of incremental forces in the joints. These are transformed to increments of force and moment at the centroid of each block. For a short time step, the resulting accelerations of each block are integrated to give new positions and orientations for the block centroids. As a result of these block motions, block sides and corners are displaced, deforming the joints and causing new increments of joint forces. This starts the second cycle of computation, etc. By integrating with small time steps, large deformations can be accumulated. Cundall (1971 and 1974), Burman (1972), and Byrne (1974) have contributed to the development of rigid block analysis. Cundall programmed the solution with interactive computer graphics with simplified input and output. The subject holds great potential, especially for kinematic study of large block systems under either static or dynamic loads.

appendix ONE
an illustrative finite element program

PURPOSE AND SCOPE OF THE COMPUTER PROGRAM

This appendix presents a computer program based upon the material of Chapter 8. The program was written by Dr. Christopher St. John* in collaboration with the author. The sole motivation for preparing this program was to enable an interested reader to follow the coding of equations for analysis of jointed rocks. Thus, the program has been kept small, mainly by omitting obvious refinements and generalizations. However, even a more sophisticated and more general program would soon become obsolete in this rapidly moving field. If you have an acquaintance with the Fortran language, the processes and algorithms presented will help you to understand the principles better.

As presently structured and dimensioned, the program JETTY, listed at the end of this appendix, calls for a digital computer having a 21,000 word core memory and will solve problems with up to 25 nodal points, 23 elements, and 20 different types of materials or joints.

A number of improvements and enlargements beyond the scope of this program will be desirable when using the finite element method in practise. Constant strain triangles were used here to simplify the representation of rock. Better and more convenient elements are

---

* Lecturer, Royal School of Mines, Imperial College, London

now available. In addition, special elements can be introduced to represent beams, tunnel liners (shells), rock bolts, and cables. A Berkeley computer library matrix inversion routine was called to solve the structural equations. For use elsewhere, this can be replaced with the CDC routine MATRIX, or other direct solution schemes. However, the stiffness matrices used here are symmetric; furthermore they can be narrow banded, i.e. numbered so that the relatively few non-zero terms cluster about the main diagonal of the structural stiffness matrix. Therefore, when solving larger problems, one will be able to economize greatly by using symmetric, narrow banded equation solvers such as BANSOL of Professor E. Wilson.* Also, to extend the size of a program beyond that which can be stored simultaneously in core, it is possible to utilize supplementary storage with a block solution procedure, as done in BANSOL. Plotting routines prove invaluable in debugging large meshes and in interpreting the output displacements and stresses.

With respect to problems with discontinuities, it may sometimes be desirable to introduce algorithms other than the load transfer procedures discussed in Chapter 8 and incorporated in the subroutine JSTR. In particular, when the initial normal stress in joint elements is either very low or very high, the initial normal stiffness may prescribe a structural stiffness which requires too many iterations to converge. For such problems, a variable stiffness approach is more appropriate. The load transfer method was used because the stiffness matrix has to be inverted only once; therefore, joint non-linear behavior is solvable with only matrix multiplications after the first iteration and, consequently, problems of considerable complexity can be finished on small computers readily available to design engineers and geologists. Also, the load transfer method allows computation of shear deformations after the peak load. As discussed in Chapter 8, incremental loading and unloading is a good procedure which will improve convergence in many cases.

---

* Professor Edward L. Wilson, "Analysis of Axisymmetric Solids", University of California, Berkeley, Department of Civil Engineering, SESM Computer Programming Series (February, 1967) -- Subroutine BANSOL.

The list of input parameters describing properties of discontinuities has been kept short by coding much of the "model" in the body of the program. These assumptions, discussed in Chapter 8, are reviewed in Table A-1. Minor programming will be necessary to change the details of the discontinuity model. Major programming may be desirable to enlarge the scope in other ways. For example, a program like JETTY can serve as the static elasticity "module" with other subroutines to solve problems in coupled fluid flow, coupled heat flow, dynamics, non-linear rock behavior, etc. One simple way to enlarge the scope so as to permit study of construction sequence or incremental loading type problems is to introduce modifications manually when restarting continuing problems.

We now will consider the structure of the program in comparison with the material presented in Chapter 8. To facilitate reading the listing, equations are identified in numerous comment cards.

PROGRAM STRUCTURE

Table A-2 shows the sequence of computations and subroutine calls for JETTY. The main program reads the input data and calls subroutine STIFF to assemble the structural stiffness (KS) and load vector {R}; this is done with the aid of subroutines TRIA and JSTIF for triangles and joints respectively. The structural stiffness matrix is inverted by a library matrix inversion routine (not listed) which uses Gauss Elimination. The inverse of the stiffness matrix, including the boundary conditions, is stored in the array (KS). The main iterative loop of the program is entered after initializing the displacements {u}. The total displacement at this point becomes {u} = (KS)*{R}. From these displacements, solid stresses are calculated in STRESS while joint deformations, and initial load corrections, are computed in JSTR. Initial force increments for the next iteration are returned by JSTR in {R}, and the iterative loop is re-entered with updated displacements {u} = (KS)*{R} + {u}.

The sequence of computations and the equations programmed are almost identical to those presented in Chapter 8 except for the following points. Rather than compute an updated initial load vector to begin each new iteration, an incremental load vector is used and the displacements computed from each iteration are added to those previously computed. The incremental initial load vector will approach zero if convergence is neared. Another point of difference is the way in which the displacement constraints are introduced. Rather than partition the stiffness matrix by removing the rows and columns of known displacements, the stiffness term corresponding to the row and column of a fixed nodal point displacement is set to the arbitrarily high value of 10^{20}. A non-zero displacement condition, i.e. a moved and then fixed nodal point can be input by reading into the load vector {R}, at the constrained node, the value of the given displacement multiplied by 10^{20}. Modified joint elements (figure 8-12) are not explicitly identified in the computation. Instead, the left and right halves of each joint element are considered separately in JSTR. Thus, each nodal point pair (a,b) of a modified joint element receives its initial force increment computations first from the half element on one side and then from the half element on the other side. Limits have been set on k_n: $0.01\ q_u/V_{mc} \leq k_n \leq 100.\ q_u/V_{mc}$.

No test for convergence has been programmed. Problems with diverging or unstable results can arise when stepping out very far from an initial equilibrium, in which case loading or unloading will have to be done in increments, with stress output from any increment introduced as residual stress to begin the next increment. JETTY has no provision to do this automatically. Numerical instability can result when individual triangles are connected to more than one joint, since incompatible load corrections may be required.

The program includes a number of WRITE statements intended for debugging purposes. These will be executed if the word INTERMEDIATE is introduced as an optional control card. Thus, a complete printout of intermediate computations can be obtained whenever desired. Samples of such output for examples 3 and 4 are presented later. The intermediate results make the output long but will enable you to follow the program logic.

TABLE A-1

Summary of Assumed Material Property Relationships

ROCK

Linear, transversely isotropic solid. The rock elements may represent uniformly bedded or slabby rock. There is no provision for failure of the rock elements.

JOINTS

Asperities obey Fairhurst's failure criterion, consisting of a parabolic envelope fitted to Mohr circles for the unconfined compression and tension tests (Fairhurst, 1964).

Peak shear strength τ_p is given by Ladanyi and Archambault's equation with the transition pressure $\sigma_T = q_u$ of the wall rock (equations 14 to 16 of Chapter 5).

Dilatancy $\dot{v}$ and area of contact a_s vary with normal pressure as given by Ladanyi and Archambault, with $\sigma_T = q_u$.

Residual shear strength $\tau_r = B(\sigma)\tau_p$ where $B(\sigma)$ decreases linearly from B_o at $\sigma = 0$ to 1 at $\sigma = q_u$ (60)*

The ratio of peak to residual displacements, M, measured in a test beginning from $\tau = 0$, equals 4.

Normal displacement resulting from normal compression or decompression obeys a hyperbolic law (43) or (49) with the initial stiffness k_n determined by the initial stress (44a).

Dilatancy does not occur when the shear deformation exceeds the residual value u_r. (61) and (62).

---

* Numbers in parentheses refer to equations of Chapter 8.

TABLE A-2

Structure of the Computer Program

```
MAIN PROGRAM (JETTY)

    - Read data
    - Preprocess data; e.g. the global stress strain matrix is
      assembled from the material properties.
    - Assemble the stiffness matrix:  CALL STIFF

          SUBROUTINE STIFF

          - For each element, compute the element stiffness matrix
            and the initial loads.  For solids, CALL TRIA; for joints,
            CALL JSTIF
          - Assemble the structural stiffness matrix and net initial
            load vector {R}
          - Introduce constraints
          Return

    - Solve for displacements:  call a suitable equation solver to
      invert the stiffness matrix.  Return the inverse in (KS).
    - Initialize displacements {u}.

    - ENTER ITERATIVE LOOP
    - Compute incremental displacements and add to {u}:  {u} =
      (KS){R} + {u}
    - Compute solid stresses from known displacements:  CALL STRESS

          SUBROUTINE STRESS

          For each solid element in turn:
          - Assemble total nodal point displacements
          - Form the element stress-displacement matrix
          - Compute the element stress change and add to the element
            initial stresses.
          Return

    - Compute joint stresses and define new initial load increments
      {R}  :  CALL JSTR

          SUBROUTINE JSTR

          For each joint element in turn:
          - Assemble the total nodal point displacements
          - Compute joint deformations in local coordinates
          - Compare with the theoretical model for joint closing,
            opening, shearing, and dilatancy.
          - Compute corrective force increments and store for the
            next iteration as equivalent initial stresses
          - Transform force increments to global coordinates and
            store in {R}.
          Return

    - Write output displacements and stresses
    - Leave iterative loop if the last iteration
    - Write restart information on tape
    End
```

INPUT INSTRUCTIONS FOR JETTY

The following cards constitute the data deck:

1. TITLE CARD

   Any title message desired; it will head the output.

2. FIRST CONTROL CARD (7I5)

   Field:
   (1) number of nodal points
   (2) number of elements
   (3) number of nodal points in the list of fixed points
   (4) number of solid material types
   (5) number of joint material types
   (6) number of the first iteration of this run (previous + 1)
   (7) number of iterations to be computed this run.

3. SECOND CONTROL CARD (5F10.0)

   Field:
   (1) 1.0 if gravity forces are to be computed; otherwise blank
   (2) the acceleration of gravity
   (3) pseudo-static acceleration coefficient k (acceleration = kg)
   (4) direction of the inertia force of the applied acceleration -- an angle in degrees counter-clockwise from x.
   (5) the mass density of water (this establishes the system of dimensions)

4. Data sets follow in any order. A heading card precedes each set and identifies it.

   HEADING CARDS FOR DATA SETS; begin in column 1; 9-72 are optional.

   NODAL POINT DATA
   ELEMENT DATA
   HYDRAULIC DATA
   RESIDUAL STRESS DATA
   BOUNDARY CONSTRAINTS
   SOLID ELEMENT PROPERTIES
   JOINT ELEMENT PROPERTIES
   INTERMEDIATE PRINTOUT DESIRED (no data set follows this card)

   a) NODAL POINT DATA (I5,4F10.0)

   Field:
   (1) nodal point number
   (2) x coordinate
   (3) y coordinate
   (4) applied external force in x direction
   (5) applied external force in y direction

   Omitted nodal points are linearly interpolated without external forces. The last nodal point (highest numbered) must be input.

   b) ELEMENT DATA (6I5)

Field:
(1) element number
(2) material number (joints are assigned numbers higher than solids).
(3) nodal point I
(4) nodal point J
(5) nodal point K
(6) nodal point L (blank for solid elements)

Nodal points are designated in counter-clockwise sequence. In joints, the long sides must be IJ and KL (see figure 8-20). Elements may be omitted; the numbers of omitted elements are assigned by incrementing the corner numbers of the previous element and the previous material number is assigned. The highest numbered element must be input.

c) HYDRAULIC DATA (I5,3F10.0)

Field:
(1) element number
(2) "Head" - Head is 0 for a solid element above the water table and 1.0 for a solid element below the water table.
(3) $\partial h/\partial x$ (negative for flow in the direction of positive x)
(4) $\partial h/\partial y$ (negative for flow in the direction of positive y)

If elements are omitted from the list, the gradients will be assigned equal to those on the preceding card. The highest numbered element to receive hydraulic data must be input.

d) RESIDUAL STRESS DATA (I5,3F10.0)

Field:
(1) element number
(2) $\sigma_{x,o}$
(3) $\sigma_{y,o}$
(4) $\tau_{xy,o}$

Elements may be omitted, in which case the residual stresses will be assigned equal to those on the preceding card. The highest numbered element to receive residual stress must be input.

e) BOUNDARY CONSTRAINT DATA (8(2I5))

(1) nodal point
(2) ICODE.....
if ICODE = 0, no constraint is introduced (this is a convenience when running multiple problems with changing boundary conditions)
if ICODE = 1, x displacement is zero ("rollers" parallel to y)
if ICODE = 2, y displacement is zero ("rollers" parallel to x)
if ICODE = 3, x and y displacements are both zero (fixed)
(3) next nodal point
(4) ICODE for next nodal point, etc. up to 8 nodal points per card.

At least one node must be constrained

f) SOLID ELEMENT PROPERTIES (I5,7F10.0)

Field:
(1) material number; materials must be numbered in sequence starting from 1.
(2) mass density
(3) E_s - modulus of elasticity in direction parallel to s
(4) E_n - modulus of elasticity in direction parallel to n
(5) G_{sn} - shear modulus in the sn plane
(6) ν_{sn} - Poisson's ratio giving strain in the n direction due stress applied parallel to the s direction
(7) ν_{st} - Poisson's ratio giving strain in the t direction due to stress applied parallel to the s direction.
(8) α --direction of the s axis, measured counter-clockwise from the x axis (see figure 8-4)

g) JOINT ELEMENT PROPERTIES (I5,7F10.0,F5.0)

Field:
(1) material number
(2) q_u - the unconfined compressive strength of the asperities (negative)
(3) T_o/q_u - the ratio of tensile to compressive strength of wall rock
(4) k_s - joint shear stiffness (dimensions force/length^3)
(5) B_o - the ratio of residual to peak strength at low normal stress
(6) V_{mc} - the maximum amount a joint can close from an initial seating load (positive)
(7) ξ - the seating load for measuring V_{mc} (negative)
(8) ϕ_μ - the friction angle for a smooth joint
(9) i_o - the dilatancy angle at zero normal pressure

5. FINAL CONTROL CARDS (beginning in column 1)

First Card: (one of the following)

START - will execute program, forming the structural stiffness matrix from the input information
STOP - will stop before executing program; no further card is needed.
RESTART - will continue a previous problem without accelerating the solution, reading required continuation data from tape 7. In this case, the data cards can be omitted, i.e. the entire deck consists of the title card (1) and the first and second control cards (2) and (3), followed by RESTART and the last control card. The structural stiffness matrix will not be computed, but will be read from tape 7.

Last card (one of the following):

STOP - continuation information will not be written on tape at the end of the problem.
SAVE STIFFNESS - continuation information will be written on

tape at the end of the problem.

ACCELERATE - will write new stiffnesses, corresponding to a variable stiffness iteration, on tape at the end of the problem. On restarting the problem, the full data deck must be used, and the first final control card must be START

Examples of restart information:

a) A run followed by a normal restart. (It will be necessary to use the appropriate control cards to get and attach tape 7).

First run:
Standard data set
START
SAVE STIFFNESS

Second run:
Title card + first and second control cards
RESTART
STOP

b) A run followed by an accelerated restart (control cards must be used to get and attach tape 7)

First run:
Standard data set
START
ACCELERATE

Second Run:
Standard data set
START
STOP

EXAMPLES OF INPUT AND OUTPUT

Example 3

Data coding form 1 shows the input information for example 3 of Chapter 8 (figure 8-10). In order to show the complete stiffness matrix in the output, no nodal points were fixed; however since at least one nodal point must be assigned a boundary constraint, node 1 was assigned ICODE equal to zero (leaving it free). The data deck includes the card INTERMEDIATE which causes all print statements to be executed. Output number 1 shows the formation of element stiffness matrices and the assembly of the structural stiffness matrix and load vector.

Example 4

The input information for example 4 of Chapter 8 (figure 8-19) is listed on coding form 2. Note the use of heading cards and the intentional omission of some residual stress cards which will be generated inside the program. Computer output number 2 shows the results of the first, second, fifth, and tenth iterations. The output gives the total displacements and solid element stresses in global coordinates (x,y) and joint deformations and stresses in local coordinates (s,n). Remember that tension is positive; the minor principal stress σ_3 will be the largest compression. The "orientation" given is the direction of σ_1 measured counter-clockwise from x. Output number 3 is a complete printout with intermediate results for computation of the first iteration.

DATA CODING FORM 1

```
EXAMPLE NØ 3 (WITH COMPLETE PRINTØUT) NØ NØDAL PØINTS FIXED
    6    3    0    1    1    1    1
         blank card
INTERMEDIATE
NØDAL PØINT DATA
    1    0.        0.
    2    0.        0.
    3    0.        1.
    4    2.        0.
    5    2.        0.
    6    2.        1.
ELEMENT DATA
    1    2    1    4    5    2
    2    1    2    5    3
    3    1    6    3    5
SØLID ELEMENT PRØPERTIES
    1    0.        1000.     1000.     417.      0.2       0.2       0.
JØINT ELEMENT PRØPERTIES
    2   -1000.    +0.1       500.      0.6       0.005    -1.0       30.      5.0
BØUNDARY CØNDITIØNS
    1    0
RESIDUAL STRESS DATA
    1   -10.0     -5.0      -1.0
    3   -10.0     -5.0      -1.0
START
STOP
```

DATA CODING FORM 2

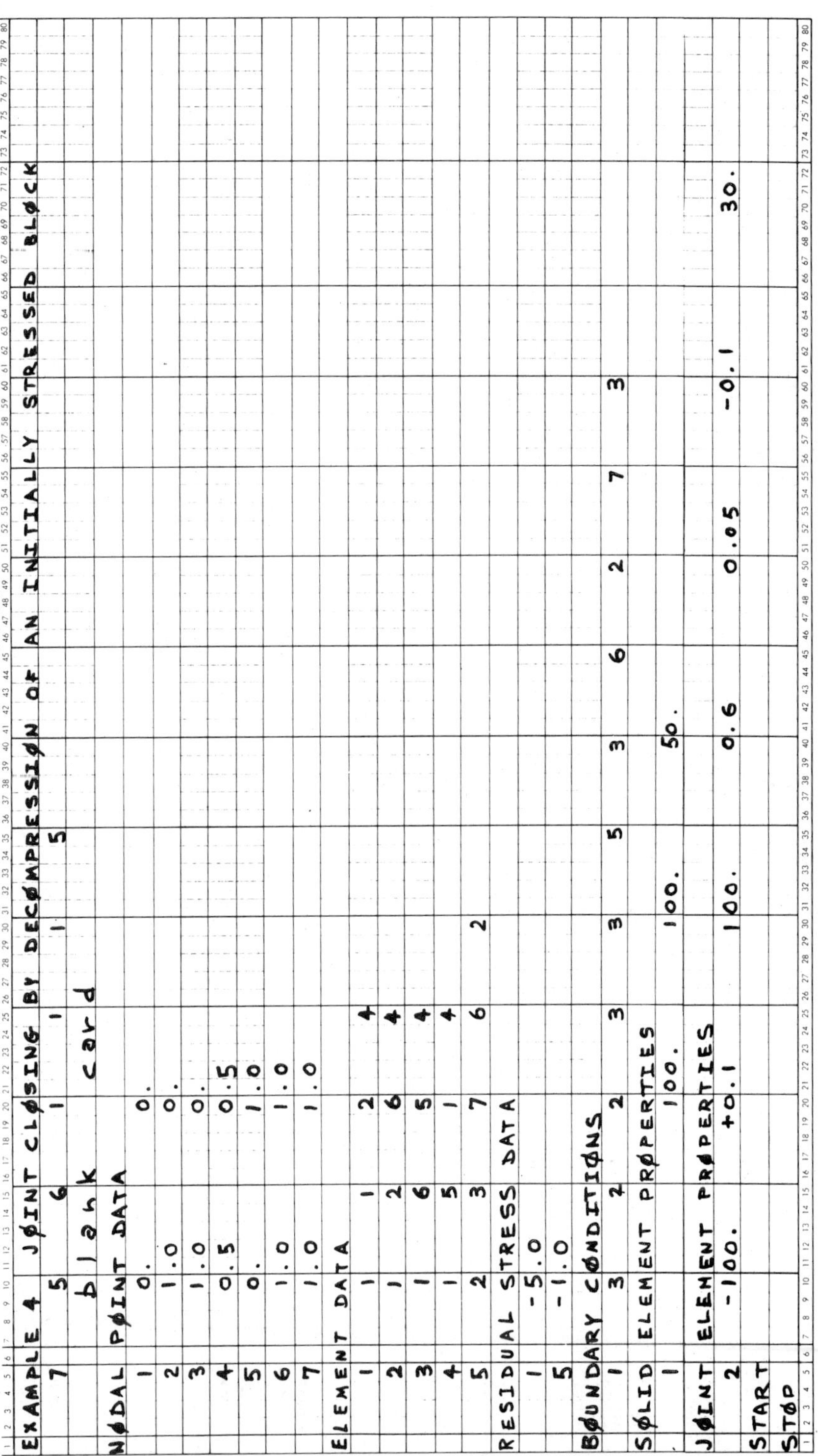

```
EXAMPLE 4 JØINT CLØSING BY DECØMPRESSIØN ØF AN INITIALLY STRESSED BLØCK
    7    5    6    1    1    1    5
         blank card
NØDAL PØINT DATA
    1    0.        0.
    2    1.0       0.
    3    1.0       0.
    4    0.5       0.5
    5    0.        1.0
    6    1.0       1.0
    7    1.0       1.0
ELEMENT DATA
    1    1    1    2    4
    2    1    2    6    4
    3    1    6    5    4
    4    1    5    1    4
    5    2    3    7    6    2
RESIDUAL STRESS DATA
    1   -5.0
    5   -1.0
BØUNDARY CØNDITIØNS
    1    3    2    2    3    3    5    3    6    2    7    3
SØLID ELEMENT PRØPERTIES
    1             100.      100.      50.
JØINT ELEMENT PRØPERTIES
    2   -100.     +0.1      100.      0.6       0.05     -0.1       30.
START
STØP
```

COMPUTER OUTPUT 1

```
        EXAMPLE NUMBER 3 WITH COMPLETE PRINTOUT OF INTERMEDIATE RESULTS

 NUMBER OF NODAL POINTS      =  6
 NUMBER OF ELEMENTS          =  3
 NUMBER OF CONSTRAINED NODES=  1
 NUMBER OF SOLID MATERIALS   =  1
 NUMBER OF JOINT TYPES       =  1
 NUMBER OF FIRST ITERATION   =  1
 NUMBER OF ITERATIONS        =  1
 GRAVITY LOADING - ON/OFF    =  0.
 ACCELERATION DUE TO GRAVITY=  0.
 PSEUDO STATIC ACCELERATION =  0.
 DIRECTION OF ACCELERATION   = -0.
 MASS DENSITY OF WATER       = -0.

  NODE      XORD        YORD        XLOAD       YLOAD       NODE      XORD        YORD        XLOAD       YLOAD
   1        0.          0.     0.          0.                2        0.          0.     0.          0.
   3        0.          1.00   0.          0.                4        2.00        0.     0.          0.
   5        2.00        0.     0.          0.                6        2.00        1.00   0.          0.

    ELEMENT    MAT TYPE        NODE LIST                ELEMENT    MAT TYPE        NODE LIST
       1          2          1    4    5    2              2          1          2    5    3    0
       3          1          6    3    5    0

  SOLID MATERIAL PROPERTIES
 MATERIAL TYPE                                   =  1
 MASS DENSITY                                    =  0.
 YOUNGS MODULUS - PARALLEL TO ANISOTROPY=    .1000E+04
 YOUNGS MODULUS - NORMAL TO ANISOTROPY   =    .1000E+04
 SHEAR MODULUS - GSN                             =    .4170E+03
 POISSONS RATIO  - SN                            =    .2000
 POISSONS RATIO  - ST                            =    .2000
 ORIENTATION OF LAYERS (ANGLE S-X)               =  0.

 STRESS-STRAIN MATRIX IN LOCAL CO-ORDINATES
        .11111E+04      .27778E+03     0.
        .27778E+03      .11111E+04     0.
       0.              0.               .41700E+03

 TRANSFORMATION MATRIX
        .10000E+01     0.              0.
       0.               .10000E+01    -0.
       0.             -0.               .10000E+01

 STRESS-STRAIN MATRIX IN GLOBAL COORDINATES
        .11111E+04      .27778E+03     0.
        .27778E+03      .11111E+04     0.
       0.              0.               .41700E+03

  JOINT PROPERTIES
 JOINT TYPE NUMBER                               =  2
 WALL ROCK COMPRESSIVE STRENGTH                  =  -.1000E+04
 RATIO OF TENSILE TO COMPRESSIVE STRENGTH=    .1000
 SHEAR STIFFNESS                                 =    .5000E+03
 RATIO OF RESIDUAL TO PEAK SHEAR STRENGTH=    .6000E+00
 MAXIMUM NORMAL CLOSURE                          =    .5000E-02
 SEATING LOAD                                    =  -.1000E+01
 FRICTION ANGLE OF A SMOOTH JOINT                =    .3000E+02
 DILATANCY ANGLE                                 =    .5000E+01
```

```
ELEMENT        RESIDUAL  STRESSES    XX,YY,XY           ELEMENT        RESIDUAL  STRESSES    XX,YY,XY
   1     -.1000E+02    -.5000E+01    -.1000E+01            2     -.1000E+02    -.5000E+01    -.1000E+01
   3     -.1000E+02    -.5000E+01    -.1000E+01

  NODE  DISPLACEMENT CODE
   1               -0

JOINT RESIDUAL SHEAR STRESS =  -.1000E+01
JOINT RESIDUAL NORMAL STRESS=  -.5000E+01

JOINT STIFFNESS - LOCAL COORDINATES
  .2500E+03  0.          .2500E+03  0.          -.2500E+03  0.          -.2500E+03  0.
 0.          .5000E+04  0.          0.          0.          0.          0.          -.5000E+04
  .2500E+03  0.          .2500E+03  0.          -.2500E+03  0.          -.2500E+03  0.
 0.          0.          0.          .5000E+04  0.          -.5000E+04  0.          0.
 -.2500E+03  0.          -.2500E+03  0.          .2500E+03  0.          .2500E+03  0.
 0.          0.          0.          -.5000E+04  0.          .5000E+04  0.          0.
 -.2500E+03  0.          -.2500E+03  0.          .2500E+03  0.          .2500E+03  0.
 0.          -.5000E+04  0.          0.          0.          0.          0.          .5000E+04

JOINT STIFFNESS AND FORCE VECTOR - GLOBAL COORDS. - ELEMENT  1
  .2500E+03  0.          .2500E+03  0.          -.2500E+03  0.          -.2500E+03  0.          -.1000E+01
 0.          .5000E+04  0.          0.          0.          0.          0.          -.5000E+04  -.5000E+01
  .2500E+03  0.          .2500E+03  0.          -.2500E+03  0.          -.2500E+03  0.          -.1000E+01
 0.          0.          0.          .5000E+04  0.          -.5000E+04  0.          0.          -.5000E+01
 -.2500E+03  0.          -.2500E+03  0.          .2500E+03  0.          .2500E+03  0.           .1000E+01
 0.          0.          0.          -.5000E+04  0.          .5000E+04  0.          0.           .5000E+01
 -.2500E+03  0.          -.2500E+03  0.          .2500E+03  0.          .2500E+03  0.           .1000E+01
 0.          -.5000E+04  0.          0.          0.          0.          0.          .5000E+04   .5000E+01

STRAIN DISPLACEMENT MATRIX
     -.5000E+00  0.          .5000E+00  0.          -0.         0.
     0.          -.1000E+01  0.         -0.          0.          .1000E+01
     -.1000E+01  -.5000E+00 -0.          .5000E+00   .1000E+01 -0.

STRESS-DISPLACEMENT TRANSFORMATION MATRIX
     -.5556E+03  -.2778E+03  .5556E+03  0.          0.          .2778E+03
     -.1389E+03  -.1111E+04  .1389E+03  0.          0.          .1111E+04
     -.4170E+03  -.2085E+03  0.          .2085E+03   .4170E+03  0.

STIFFNESS MATRIX - ELEMENT  2                                                                    LOAD VECTOR
  .6948E+03   .3474E+03  -.2778E+03  -.2085E+03  -.4170E+03  -.1389E+03                         -.6000E+01
  .3474E+03   .1215E+04  -.1389E+03  -.1042E+03  -.2085E+03  -.1111E+04                         -.5500E+01
 -.2778E+03  -.1389E+03   .2778E+03  0.          0.           .1389E+03                          .5000E+01
 -.2085E+03  -.1042E+03  0.           .1042E+03   .2085E+03  0.                                   .5000E+00
 -.4170E+03  -.2085E+03  0.           .2085E+03   .4170E+03  0.                                   .1000E+01
 -.1389E+03  -.1111E+04   .1389E+03  0.          0.           .1111E+04                           .5000E+01

STRAIN DISPLACEMENT MATRIX
      .5000E+00  0.          -.5000E+00  0.          -0.         0.
     0.           .1000E+01  0.          -0.          0.          -.1000E+01
      .1000E+01   .5000E+00 -0.          -.5000E+00  -.1000E+01 -0.

STRESS-DISPLACEMENT TRANSFORMATION MATRIX
      .5556E+03   .2778E+03  -.5556E+03  0.          0.          -.2778E+03
      .1389E+03   .1111E+04  -.1389E+03  0.          0.          -.1111E+04
      .4170E+03   .2085E+03  0.          -.2085E+03  -.4170E+03  0.
```

```
STIFFNESS MATRIX - ELEMENT  3
   .6948E+03   .3474E+03  -.2778E+03  -.2085E+03  -.4170E+03  -.1389E+03
   .3474E+03   .1215E+04  -.1389E+03  -.1042E+03  -.2085E+03  -.1111E+04
  -.2778E+03  -.1389E+03   .2778E+03  0.          0.           .1389E+03
  -.2085E+03  -.1042E+03  0.           .1042E+03   .2085E+03  0.
  -.4170E+03  -.2085E+03  0.           .2085E+03   .4170E+03  0.
  -.1389E+03  -.1111E+04   .1389E+03  0.          0.           .1111E+04

                                                                                        LOAD VECTOR
                                                                                         .6000E+01
                                                                                         .5500E+01
                                                                                        -.5000E+01
                                                                                        -.5000E+00
                                                                                        -.1000E+01
                                                                                        -.5000E+01

LOAD VECTOR AND STIFFNESS MATRIX
   .2500E+03  0.          -.2500E+03  0.          0.          0.           .2500E+03  0.          -.2500E+03  0.
  0.          0.
  0.           .5000E+04  0.          -.5000E+04  0.          0.          0.          0.          0.          0.
  0.          0.
  -.2500E+03  0.           .9448E+03   .3474E+03  -.4170E+03  -.1389E+03  -.2500E+03  0.          -.2778E+02  -.2085E+03
  0.          0.
  0.          -.5000E+04   .3474E+03   .6215E+04  -.2085E+03  -.1111E+04  0.          0.          -.1389E+03  -.1042E+03
  0.          0.
  0.          0.          -.4170E+03  -.2085E+03   .6948E+03  0.          0.          0.          0.           .3474E+03
  -.2778E+03  -.1389E+03
  0.          0.          -.1389E+03  -.1111E+04  0.           .1215E+04  0.          0.           .3474E+03  0.
  -.2085E+03  -.1042E+03
   .2500E+03  0.          -.2500E+03  0.          0.          0.           .2500E+03  0.          -.2500E+03  0.
  0.          0.
  0.          0.          0.          0.          0.          0.          0.           .5000E+04  0.          -.5000E+04
  0.          0.
  -.2500E+03  0.          -.2778E+02  -.1389E+03  0.           .3474E+03  -.2500E+03  0.           .9448E+03  0.
  -.4170E+03  -.2085E+03
  0.          0.          -.2085E+03  -.1042E+03   .3474E+03  0.          0.          -.5000E+04  0.           .6215E+04
  -.1389E+03  -.1111E+04
  0.          0.          0.          0.          -.2778E+03  -.2085E+03  0.          0.          -.4170E+03  -.1389E+03
   .6948E+03   .3474E+03
  0.          0.          0.          0.          -.1389E+03  -.1042E+03  0.          0.          -.2085E+03  -.1111E+04
   .3474E+03   .1215E+04
  -.1000E+01  -.5000E+01  -.5000E+01  -.5000E+00  -.4000E+01   .4500E+01  -.1000E+01  -.5000E+01   .5000E+01   .5000E+00
   .6000E+01   .5500E+01
```

COMPUTER OUTPUT 2

```
        EXAMPLE 4 JOINT CLOSING BY DECOMPRESSION OF AN INITIALLY STRESSED BLOCK

NUMBER OF NODAL POINTS      =  7
NUMBER OF ELEMENTS          =  5
NUMBER OF CONSTRAINED NODES=   6
NUMBER OF SOLID MATERIALS   =  1
NUMBER OF JOINT TYPES       =  1
NUMBER OF FIRST ITERATION   =  1
NUMBER OF ITERATIONS        = 10
GRAVITY LOADING - ON/OFF    = -0.
ACCELERATION DUE TO GRAVITY= -0.
PSEUDO STATIC ACCELERATION = -0.
DIRECTION OF ACCELERATION   = -0.
MASS DENSITY OF WATER       = -0.

  NODE     XORD      YORD      XLOAD     YLOAD      NODE     XORD      YORD      XLOAD     YLOAD
   1       -0.       -0.   -0.        -0.            2       1.00      -0.   -0.        -0.
   3        1.00     -0.   -0.        -0.            4        .50       .50 -0.        -0.
   5       -0.       1.00 -0.         -0.            6       1.00      1.00 -0.         -0.
   7        1.00     1.00 -0.         -0.

   ELEMENT    MAT TYPE     NODE LIST              ELEMENT    MAT TYPE     NODE LIST
     1          1        1   2   4   -0             2          1        2   6   4   -0
     3          1        6   5   4   -0             4          1        5   1   4   -0
     5          2        3   7   6    2

  ELEMENT        RESIDUAL  STRESSES     XX,YY,XY       ELEMENT       RESIDUAL  STRESSES     XX,YY,XY
     1       -.5000E+01    -0.           -0.              2       -.5000E+01    -0.           -0.
     3       -.5000E+01    -0.           -0.              4       -.5000E+01    -0.           -0.
     5       -.1000E+01    -0.           -0.

   NODE  DISPLACEMENT CODE
    1            3
    2            2
    3            3
    5            3
    6            2
    7            3

  SOLID MATERIAL PROPERTIES
 MATERIAL TYPE                                =  1
 MASS DENSITY                                 = -0.
 YOUNGS MODULUS - PARALLEL TO ANISOTROPY=     .1000E+03
 YOUNGS MODULUS - NORMAL TO ANISOTROPY   =    .1000E+03
 SHEAR MODULUS - GSN                          =    .5000E+02
 POISSONS RATIO   - SN                        = -0.
 POISSONS RATIO   - ST                        = -0.
 ORIENTATION OF LAYERS (ANGLE S-X)            = -0.

  JOINT PROPERTIES
JOINT TYPE NUMBER                             =  2
WALL ROCK COMPRESSIVE STRENGTH                =  -.1000E+03
RATIO OF TENSILE TO COMPRESSIVE STRENGTH=     .1000
SHEAR STIFFNESS                               =   .1000E+03
RATIO OF RESIDUAL TO PEAK SHEAR STRENGTH=     .6000E+00
MAXIMUM NORMAL CLOSURE                        =   .5000E-01
SEATING LOAD                                  =  -.1000E+00
FRICTION ANGLE OF A SMOOTH JOINT              =   .3000E+02
DILATANCY ANGLE                               = -0.
```

ITERATION NUMBER 1

NODE	X-DISPLACEMENT	Y-DISPLACEMENT
1	-.000000	.000000
2	.013333	-.000000
3	.000000	-.000000
4	.006667	.000000
5	-.000000	.000000
6	.013333	-.000000
7	.000000	-.000000

SOLID ELEMENT STRESSES

ELEMENT	XX-STRESS	YY-STRESS	XY-STRESS	MAJOR PRINCIPAL	MINOR PRINCIPAL	ORIENTATION
1	-.3667E+01	.9715E-14	.2842E-13	.7105E-14	-.3667E+01	90.16
2	-.3667E+01	.4815E-32	.2356E-13	0.	-.3667E+01	90.16
3	-.3667E+01	-.9715E-14	0.	-.7105E-14	-.3667E+01	90.16
4	-.3667E+01	-.6019E-32	.4857E-14	0.	-.3667E+01	90.16

	JOINT DISPLACEMENTS			JOINT STRESSES		
ELEMENT	SHEAR	NORMAL	ROTATION	SHEAR	NORMAL	MOMENT
5	-.2407E-34	-.1333E-01	-.6106E-15	-.2407E-32	-.3667E+01	-.3053E-13

ITERATION NUMBER 2

NODE	X-DISPLACEMENT	Y-DISPLACEMENT
1	-.000000	.000000
2	.006869	.000000
3	.000000	.000000
4	.003434	.000000
5	-.000000	-.000000
6	.006869	.000000
7	.000000	.000000

SOLID ELEMENT STRESSES

ELEMENT	XX-STRESS	YY-STRESS	XY-STRESS	MAJOR PRINCIPAL	MINOR PRINCIPAL	ORIENTATION
1	-.4313E+01	.1021E-18	.8882E-14	.1421E-13	-.4313E+01	90.16
2	-.4313E+01	.2407E-32	.5329E-14	0.	-.4313E+01	90.16
3	-.4313E+01	-.1021E-18	-.3553E-14	-.1421E-13	-.4313E+01	-90.16
4	-.4313E+01	-.3009E-32	.5104E-19	0.	-.4313E+01	90.16

	JOINT DISPLACEMENTS			JOINT STRESSES		
ELEMENT	SHEAR	NORMAL	ROTATION	SHEAR	NORMAL	MOMENT
5	.3009E-34	-.6869E-02	-.8327E-16	.3009E-32	-.4313E+01	-.3081E-13

```
ITERATION NUMBER     5


NODE       X-DISPLACEMENT        Y-DISPLACEMENT
  1              -.000000              -.000000
  2               .004009               .000000
  3               .000000               .000000
  4               .002005              -.000000
  5              -.000000              -.000000
  6               .004009               .000000
  7               .000000               .000000

SOLID ELEMENT STRESSES

         ELEMENT        XX-STRESS        YY-STRESS        XY-STRESS      MAJOR PRINCIPAL  MINOR PRINCIPAL  ORIENTATION
               1       -.4599E+01       -.1583E-14        .3553E-14      -.1421E-13       -.4599E+01            90.16
               2       -.4599E+01        .1505E-32       -.9671E-16      0.               -.4599E+01           -90.16
               3       -.4599E+01        .1583E-14       -.4441E-14       .1421E-13       -.4599E+01           -90.16
               4       -.4599E+01       -.1749E-32       -.7915E-15      0.               -.4599E+01           -90.16

                            JOINT DISPLACEMENTS                        JOINT STRESSES
      ELEMENT         SHEAR         NORMAL      ROTATION        SHEAR       NORMAL       MOMENT
        5        .3385E-34    -.4009E-02     .2776E-16    .3385E-32   -.4599E+01   -.2365E-13


ITERATION NUMBER    10


NODE       X-DISPLACEMENT        Y-DISPLACEMENT
  1              -.000000              -.000000
  2               .003915               .000000
  3               .000000               .000000
  4               .001958              -.000000
  5              -.000000              -.000000
  6               .003915               .000000
  7               .000000               .000000

SOLID ELEMENT STRESSES

         ELEMENT        XX-STRESS        YY-STRESS        XY-STRESS      MAJOR PRINCIPAL  MINOR PRINCIPAL  ORIENTATION
               1       -.4608E+01       -.7697E-15        .8882E-15      -.1421E-13       -.4608E+01            90.16
               2       -.4608E+01        .1471E-32        .1273E-14      0.               -.4608E+01            90.16
               3       -.4608E+01        .7697E-15       0.               .1421E-13       -.4608E+01            90.16
               4       -.4608E+01       -.1703E-32       -.3849E-15      0.               -.4608E+01           -90.16

                            JOINT DISPLACEMENTS                        JOINT STRESSES
      ELEMENT         SHEAR         NORMAL      ROTATION        SHEAR       NORMAL       MOMENT
        5        .2749E-34    -.3915E-02    -.2776E-16    .2749E-32   -.4608E+01   -.2226E-13
```

COMPUTER OUTPUT 3

```
        EXAMPLE 4   JOINT CLOSING  COMPLETE PRINTOUT OF FIRST ITERATION

NUMBER OF NODAL POINTS      =  7
NUMBER OF ELEMENTS          =  5
NUMBER OF CONSTRAINED NODES=  6
NUMBER OF SOLID MATERIALS   =  1
NUMBER OF JOINT TYPES       =  1
NUMBER OF FIRST ITERATION   =  1
NUMBER OF ITERATIONS        =  1
GRAVITY LOADING - ON/OFF    = -0.
ACCELERATION DUE TO GRAVITY= -0.
PSEUDO STATIC ACCELERATION = -0.
DIRECTION OF ACCELERATION  = -0.
MASS DENSITY OF WATER       = -0.

  NODE      XORD        YORD        XLOAD       YLOAD      NODE      XORD        YORD        XLOAD       YLOAD
   1        -0.         -0.    -0.          -0.             2        1.00        -0.    -0.          -0.
   3         1.00       -0.    -0.          -0.             4         .50         .50 -0.          -0.
   5        -0.         1.00 -0.            -0.             6        1.00        1.00 -0.          -0.
   7         1.00       1.00 -0.            -0.

  ELEMENT    MAT TYPE        NODE LIST              ELEMENT    MAT TYPE        NODE LIST
     1          1         1    2    4    -0            2          1         2    6    4    -0
     3          1         6    5    4    -0            4          1         5    1    4    -0
     5          2         3    7    6     2

 ELEMENT        RESIDUAL  STRESSES      XX,YY,XY         ELEMENT        RESIDUAL  STRESSES      XX,YY,XY
    1      -.5000E+01    -0.             -0.               2      -.5000E+01    -0.             -0.
    3      -.5000E+01    -0.             -0.               4      -.5000E+01    -0.             -0.
    5      -.1000E+01    -0.             -0.

  NODE  DISPLACEMENT CODE
   1              3
   2              2
   3              3
   5              3
   6              2
   7              3

 SOLID MATERIAL PROPERTIES
MATERIAL TYPE                           =  1
MASS DENSITY                            = -0.
YOUNGS MODULUS - PARALLEL TO ANISOTROPY=    .1000E+03
YOUNGS MODULUS - NORMAL TO ANISOTROPY   =    .1000E+03
SHEAR MODULUS - GSN                     =    .5000E+02
POISSONS RATIO  - SN                    = -0.
POISSONS RATIO  - ST                    = -0.
ORIENTATION OF LAYERS (ANGLE S-X)       = -0.

STRESS-STRAIN MATRIX IN LOCAL CO-ORDINATES
       .10000E+03    0.              0.
      0.              .10000E+03     0.
      0.             0.               .50000E+02

TRANSFORMATION MATRIX
       .10000E+01    0.              0.
      0.              .10000E+01    -0.
      0.            -0.               .10000E+01
```

```
STRESS-STRAIN MATRIX IN GLOBAL COORDINATES
       .10000E+03     0.              0.
     0.                .10000E+03     0.
     0.               0.               .50000E+02

 JOINT PROPERTIES
JOINT TYPE NUMBER                             =  2
WALL ROCK COMPRESSIVE STRENGTH                = -.1000E+03
RATIO OF TENSILE TO COMPRESSIVE STRENGTH=      .1000
SHEAR STIFFNESS                               =  .1000E+03
RATIO OF RESIDUAL TO PEAK SHEAR STRENGTH=      .6000E+00
MAXIMUM NORMAL CLOSURE                        =  .5000E-01
SEATING LOAD                                  = -.1000E+00
FRICTION ANGLE OF A SMOOTH JOINT              =  .3000E+02
DILATANCY ANGLE                               = -0.


 STRAIN DISPLACEMENT MATRIX
     -.1000E+01  0.          .1000E+01  0.          -0.          0.
     0.          -.1000E+01  0.          -.1000E+01  0.           .2000E+01
     -.1000E+01  -.1000E+01  -.1000E+01  .1000E+01   .2000E+01  -0.

 STRESS-DISPLACEMENT TRANSFORMATION MATRIX
     -.1000E+03  0.          .1000E+03  0.          0.           0.
     0.          -.1000E+03  0.          -.1000E+03  0.           .2000E+03
     -.5000E+02  -.5000E+02  -.5000E+02  .5000E+02   .1000E+03   0.

 STIFFNESS MATRIX - ELEMENT  1
   .3750E+02   .1250E+02  -.1250E+02  -.1250E+02  -.2500E+02  0.
   .1250E+02   .3750E+02   .1250E+02   .1250E+02  -.2500E+02  -.5000E+02
  -.1250E+02   .1250E+02   .3750E+02  -.1250E+02  -.2500E+02  0.
  -.1250E+02   .1250E+02  -.1250E+02   .3750E+02   .2500E+02  -.5000E+02
  -.2500E+02  -.2500E+02  -.2500E+02   .2500E+02   .5000E+02  0.
  0.          -.5000E+02  0.          -.5000E+02  0.           .1000E+03

LOAD VECTOR
-.1250E+01
0.
 .1250E+01
0.
0.
0.

 STRAIN DISPLACEMENT MATRIX
      .1000E+01  0.          .1000E+01  0.          -.2000E+01  0.
     0.          -.1000E+01  0.          .1000E+01  0.           0.
     -.1000E+01  .1000E+01   .1000E+01  .1000E+01   0.          -.2000E+01

 STRESS-DISPLACEMENT TRANSFORMATION MATRIX
      .1000E+03  0.          .1000E+03  0.          -.2000E+03  0.
     0.          -.1000E+03  0.          .1000E+03  0.           0.
     -.5000E+02  .5000E+02   .5000E+02  .5000E+02   0.          -.1000E+03

 STIFFNESS MATRIX - ELEMENT  2
   .3750E+02  -.1250E+02   .1250E+02  -.1250E+02  -.5000E+02   .2500E+02
  -.1250E+02   .3750E+02   .1250E+02  -.1250E+02  0.          -.2500E+02
   .1250E+02   .1250E+02   .3750E+02   .1250E+02  -.5000E+02  -.2500E+02
  -.1250E+02  -.1250E+02   .1250E+02   .3750E+02  0.          -.2500E+02
  -.5000E+02  0.          -.5000E+02  0.           .1000E+03  0.
   .2500E+02  -.2500E+02  -.2500E+02  -.2500E+02  0.           .5000E+02

LOAD VECTOR
 .1250E+01
0.
 .1250E+01
0.
-.2500E+01
0.

STRAIN DISPLACEMENT MATRIX
      .1000E+01  0.          -.1000E+01  0.          -0.          0.
     0.          .1000E+01   0.          .1000E+01   0.          -.2000E+01
      .1000E+01  .1000E+01   .1000E+01  -.1000E+01  -.2000E+01  -0.
```

```
STRESS-DISPLACEMENT TRANSFORMATION MATRIX
     .1000E+03  0.          -.1000E+03  0.          0.          0.
    0.          .1000E+03  0.           .1000E+03  0.          -.2000E+03
     .5000E+02   .5000E+02   .5000E+02  -.5000E+02  -.1000E+03  0.

STIFFNESS MATRIX - ELEMENT  3                                                      LOAD VECTOR
   .3750E+02   .1250E+02  -.1250E+02  -.1250E+02  -.2500E+02  0.                   .1250E+01
   .1250E+02   .3750E+02   .1250E+02   .1250E+02  -.2500E+02  -.5000E+02           0.
  -.1250E+02   .1250E+02   .3750E+02  -.1250E+02  -.2500E+02  0.                   -.1250E+01
  -.1250E+02   .1250E+02  -.1250E+02   .3750E+02   .2500E+02  -.5000E+02           0.
  -.2500E+02  -.2500E+02  -.2500E+02   .2500E+02   .5000E+02  0.                   0.
  0.          -.5000E+02  0.          -.5000E+02  0.           .1000E+03           0.

STRAIN DISPLACEMENT MATRIX
    -.1000E+01  0.          -.1000E+01  0.           .2000E+01  0.
    0.           .1000E+01  0.          -.1000E+01  0.          0.
     .1000E+01  -.1000E+01  -.1000E+01  -.1000E+01  0.           .2000E+01

STRESS-DISPLACEMENT TRANSFORMATION MATRIX
    -.1000E+03  0.          -.1000E+03  0.           .2000E+03  0.
    0.           .1000E+03  0.          -.1000E+03  0.          0.
     .5000E+02  -.5000E+02  -.5000E+02  -.5000E+02  0.           .1000E+03

STIFFNESS MATRIX - ELEMENT  4                                                      LOAD VECTOR
   .3750E+02  -.1250E+02   .1250E+02  -.1250E+02  -.5000E+02   .2500E+02           -.1250E+01
  -.1250E+02   .3750E+02   .1250E+02  -.1250E+02  0.          -.2500E+02           0.
   .1250E+02   .1250E+02   .3750E+02   .1250E+02  -.5000E+02  -.2500E+02           -.1250E+01
  -.1250E+02  -.1250E+02   .1250E+02   .3750E+02  0.          -.2500E+02           0.
  -.5000E+02  0.          -.5000E+02  0.           .1000E+03  0.                    .2500E+01
   .2500E+02  -.2500E+02  -.2500E+02  -.2500E+02  0.           .5000E+02           0.

JOINT RESIDUAL SHEAR STRESS =  0.
JOINT RESIDUAL NORMAL STRESS=  -.1000E+01

JOINT STIFFNESS - LOCAL COORDINATES
   .2500E+02  0.           .2500E+02  0.          -.2500E+02  0.          -.2500E+02  0.
  0.           .1000E+03  0.          0.          0.          0.          0.          -.1000E+03
   .2500E+02  0.           .2500E+02  0.          -.2500E+02  0.          -.2500E+02  0.
  0.          0.          0.           .1000E+03  0.          -.1000E+03  0.          0.
  -.2500E+02  0.          -.2500E+02  0.           .2500E+02  0.           .2500E+02  0.
  0.          0.          0.          -.1000E+03  0.           .1000E+03  0.          0.
  -.2500E+02  0.          -.2500E+02  0.           .2500E+02  0.           .2500E+02  0.
  0.          -.1000E+03  0.          0.          0.          0.          0.           .1000E+03

JOINT STIFFNESS AND FORCE VECTOR - GLOBAL COORDS. - ELEMENT  5
   .1000E+03  0.          0.          0.          0.          0.          -.1000E+03  0.           .5000E+00
  0.           .2500E+02  0.           .2500E+02  0.          -.2500E+02  0.          -.2500E+02  -0.
  0.          0.           .1000E+03  0.          -.1000E+03  0.          0.          0.           .5000E+00
  0.           .2500E+02  0.           .2500E+02  0.          -.2500E+02  0.          -.2500E+02  -0.
  0.          0.          -.1000E+03  0.           .1000E+03  0.          0.          0.          -.5000E+00
  0.          -.2500E+02  0.          -.2500E+02  0.           .2500E+02  0.           .2500E+02  0.
  -.1000E+03  0.          0.          0.          0.          0.           .1000E+03  0.          -.5000E+00
  0.          -.2500E+02  0.          -.2500E+02  0.           .2500E+02  0.           .2500E+02  0.

LOAD VECTOR AND STIFFNESS MATRIX
   .1000E+21   .2500E+02  -.1250E+02  -.1250E+02  0.          0.          -.7500E+02  -.2500E+02   .1250E+02   .1250E+02
  0.          0.          0.          0.
   .2500E+02   .1000E+21   .1250E+02   .1250E+02  0.          0.          -.2500E+02  -.7500E+02  -.1250E+02  -.1250E+02
  0.          0.          0.          0.
  -.1250E+02   .1250E+02   .1750E+03  -.2500E+02  -.1000E+03  0.          -.7500E+02   .2500E+02  0.          0.
```

```
  .1250E+02  -.1250E+02   0.          0.
 -.1250E+02   .1250E+02  -.2500E+02   .1000E+21   0.         -.2500E+02   .2500E+02  -.7500E+02   0.          0.
  .1250E+02   .1250E+02   0.         -.2500E+02
 0.          0.          -.1000E+03   0.          .1000E+21   0.          0.          0.          0.          0.
 0.          0.           0.          0.
 0.          0.           0.         -.2500E+02   0.          .1000E+21   0.          0.          0.          0.
 0.         -.2500E+02    0.          .2500E+02
 -.7500E+02  -.2500E+02  -.7500E+02   .2500E+02   0.          0.          .3000E+03   0.         -.7500E+02   .2500E+02
 -.7500E+02  -.2500E+02   0.          0.
 -.2500E+02  -.7500E+02   .2500E+02  -.7500E+02   0.          0.          0.          .3000E+03   .2500E+02  -.7500E+02
 -.2500E+02  -.7500E+02   0.          0.
  .1250E+02  -.1250E+02   0.          0.          0.          0.         -.7500E+02   .2500E+02   .1000E+21  -.2500E+02
 -.1250E+02   .1250E+02   0.          0.
  .1250E+02  -.1250E+02   0.          0.          0.          0.          .2500E+02  -.7500E+02  -.2500E+02   .1000E+21
 -.1250E+02   .1250E+02   0.          0.
 0.          0.           .1250E+02   .1250E+02   0.          0.         -.7500E+02  -.2500E+02  -.1250E+02  -.1250E+02
  .1750E+03   .2500E+02  -.1000E+03   0.
 0.          0.          -.1250E+02   .1250E+02   0.         -.2500E+02  -.2500E+02  -.7500E+02   .1250E+02   .1250E+02
  .2500E+02   .1000E+21   0.         -.2500E+02
 0.          0.           0.          0.          0.          0.          0.          0.          0.          0.
 -.1000E+03  0.           .1000E+21   0.
 0.          0.           0.         -.2500E+02   0.          .2500E+02   0.          0.          0.          0.
 0.         -.2500E+02    0.          .1000E+21
 -.2500E+01  0.           .2000E+01   0.          .5000E+00   0.          0.          0.         -.2500E+01   0.
  .2000E+01  0.           .5000E+00   0.

INVERSE OF STIFFNESS MATRIX
  .1000E-19  -.1316E-38   .1930E-20   .1316E-38   .1930E-38   .6579E-57   .3333E-20   .7895E-21   .1228E-38  -.1316E-38
  .1404E-20   .1316E-38   .1404E-38   .6579E-57
 -.1316E-38   .1000E-19  -.7895E-21   .2193E-39  -.7895E-39   .5263E-57   .8333E-21   .2632E-20   .1316E-38   .3114E-38
  .7895E-21   .1886E-38   .7895E-39   .5263E-57
  .1930E-20  -.7895E-21   .6491E-02   .7895E-21   .6491E-20   .3947E-39   .1667E-02  -.5263E-03   .1404E-20  -.7895E-21
  .1754E-03   .7895E-21   .1754E-21   .3947E-39
  .1316E-38   .2193E-39   .7895E-21   .1000E-19   .7895E-39   .2500E-38  -.8333E-21   .2368E-20  -.1316E-38   .1886E-38
 -.7895E-21   .6140E-39  -.7895E-39   .2500E-38
  .1930E-38  -.7895E-39   .6491E-20   .7895E-39   .1000E-19   .3947E-57   .1667E-20  -.5263E-21   .1404E-38  -.7895E-39
  .1754E-21   .7895E-39   .1754E-39   .3947E-57
  .6579E-57   .5263E-57   .3947E-39   .2500E-38   .3947E-57   .1000E-19   .4815E-53   .1184E-38  -.6579E-57   .5263E-57
 -.3947E-39   .2500E-38  -.3947E-57  -.2500E-38
  .3333E-20   .8333E-21   .1667E-02  -.8333E-21   .1667E-20   .6525E-53   .4167E-02   .3316E-17   .3333E-20  -.8333E-21
  .1667E-02   .8333E-21   .1667E-20   .4366E-53
  .7895E-21   .2632E-20  -.5263E-03   .2368E-20  -.5263E-21   .1184E-38   .2602E-17   .3421E-02  -.7895E-21   .2632E-20
  .5263E-03   .2368E-20   .5263E-21   .1184E-38
  .1228E-38   .1316E-38   .1404E-20  -.1316E-38   .1404E-38  -.6579E-57   .3333E-20  -.7895E-21   .1000E-19   .1316E-38
  .1930E-20  -.1316E-38   .1930E-38  -.6579E-57
 -.1316E-38   .3114E-38  -.7895E-21   .1886E-38  -.7895E-39   .5263E-57  -.8333E-21   .2632E-20   .1316E-38   .1000E-19
  .7895E-21   .2193E-39   .7895E-39   .5263E-57
  .1404E-20   .7895E-21   .1754E-03  -.7895E-21   .1754E-21  -.3947E-39   .1667E-02   .5263E-03   .1930E-20   .7895E-21
  .6491E-02  -.7895E-21   .6491E-20  -.3947E-39
  .1316E-38   .1886E-38   .7895E-21   .6140E-39   .7895E-39   .2500E-38   .8333E-21   .2368E-20  -.1316E-38   .2193E-39
 -.7895E-21   .1000E-19  -.7895E-39   .2500E-38
  .1404E-38   .7895E-39   .1754E-21  -.7895E-39   .1754E-39  -.3947E-57   .1667E-20   .5263E-21   .1930E-38   .7895E-39
  .6491E-20  -.7895E-39   .1000E-19  -.3947E-57
  .6579E-57   .5263E-57   .3947E-39   .2500E-38   .3947E-57  -.2500E-38   .3263E-53   .1184E-38  -.6579E-57   .5263E-57
 -.3947E-39   .2500E-38  -.3947E-57   .1000E-19
```

```
ITERATION NUMBER    1


NODE     X-DISPLACEMENT     Y-DISPLACEMENT
   1            -.000000            .000000
   2             .013333           -.000000
   3             .000000           -.000000
   4             .006667            .000000
   5            -.000000            .000000
   6             .013333           -.000000
   7             .000000           -.000000

SOLID ELEMENT STRESSES

        ELEMENT     XX-STRESS        YY-STRESS        XY-STRESS      MAJOR PRINCIPAL MINOR PRINCIPAL ORIENTATION

STRAIN DISPLACEMENT MATRIX
     -.1000E+01  0.           .1000E+01  0.           -0.          0.
     0.          -.1000E+01  0.          -.1000E+01  0.           .2000E+01
     -.1000E+01  -.1000E+01  -.1000E+01   .1000E+01   .2000E+01 -0.

STRESS-DISPLACEMENT TRANSFORMATION MATRIX
     -.1000E+03  0.           .1000E+03  0.           0.           0.
     0.          -.1000E+03  0.          -.1000E+03  0.           .2000E+03
     -.5000E+02  -.5000E+02  -.5000E+02   .5000E+02   .1000E+03  0.
             1      -.3667E+01       .9715E-14        .2842E-13        .7105E-14      -.3667E+01         90.16

STRAIN DISPLACEMENT MATRIX
      .1000E+01  0.           .1000E+01  0.           -.2000E+01  0.
     0.          -.1000E+01  0.           .1000E+01  0.           0.
     -.1000E+01   .1000E+01   .1000E+01   .1000E+01  0.           -.2000E+01

STRESS-DISPLACEMENT TRANSFORMATION MATRIX
      .1000E+03  0.           .1000E+03  0.           -.2000E+03  0.
     0.          -.1000E+03  0.           .1000E+03  0.           0.
     -.5000E+02   .5000E+02   .5000E+02   .5000E+02  0.           -.1000E+03
             2      -.3667E+01       .4815E-32        .2356E-13      0.               -.3667E+01         90.16

STRAIN DISPLACEMENT MATRIX
      .1000E+01  0.           -.1000E+01  0.           -0.          0.
     0.           .1000E+01  0.           .1000E+01  0.           -.2000E+01
      .1000E+01   .1000E+01   .1000E+01  -.1000E+01  -.2000E+01 -0.

STRESS-DISPLACEMENT TRANSFORMATION MATRIX
      .1000E+03  0.           -.1000E+03  0.           0.           0.
     0.           .1000E+03  0.           .1000E+03  0.           -.2000E+03
      .5000E+02   .5000E+02   .5000E+02  -.5000E+02  -.1000E+03  0.
             3      -.3667E+01      -.9715E-14      0.               -.7105E-14      -.3667E+01         90.16

STRAIN DISPLACEMENT MATRIX
     -.1000E+01  0.           -.1000E+01  0.           .2000E+01  0.
     0.           .1000E+01  0.           -.1000E+01  0.           0.
      .1000E+01  -.1000E+01  -.1000E+01  -.1000E+01  0.            .2000E+01

STRESS-DISPLACEMENT TRANSFORMATION MATRIX
     -.1000E+03  0.           -.1000E+03  0.           .2000E+03  0.
     0.           .1000E+03  0.           -.1000E+03  0.           0.
      .5000E+02  -.5000E+02  -.5000E+02  -.5000E+02  0.            .1000E+03
             4      -.3667E+01      -.6019E-32        .4857E-14      0.               -.3667E+01         90.16
```

```
                                JOINT DISPLACEMENTS                        JOINT STRESSES
     ELEMENT       SHEAR           NORMAL        ROTATION       SHEAR          NORMAL          MOMENT

JOINT DISPLACEMENT VECTOR - LOCAL COORDINATES
  -.1566E-52  -.1833E-19  -.5220E-53  -.1833E-19  -.3947E-39  -.1333E-01  -.4815E-34  -.1333E-01
       5    -.2407E-34  -.1333E-01  -.6106E-15  -.2407E-32  -.3667E+01  -.3053E-13

UP=  -.4666E-01  UR=  -.1866E+00  TORP=   .4666E+01  TORR=   .2868E+01  TOR=  -.2407E-32
DILAT=  0.             DELV=  -.1333E-01

UP=  -.4666E-01  UR=  -.1866E+00  TORP=   .4666E+01  TORR=   .2868E+01  TOR=  -.2407E-32
DILAT=  0.             DELV=  -.1333E-01

INCREMENTAL FORCES FOR JOINT
 -0.           .9697E+00  -0.              .9697E+00  0.           -.9697E+00  0.          -.9697E+00

JOINT STIFFNESSES
  0.          0.          0.          0.          0.          0.          0.          0.          .1000E+03   .2000E+03

INITIAL STRESSES AT END OF INCREMENT
     SHEAR          NORMAL          MOMENT                     SHEAR         NORMAL         MOMENT
  -.5000E+01  -0.              -0.                         -.5000E+01  -0.              -0.
  -.5000E+01  -0.              -0.                         -.5000E+01  -0.              -0.
  0.             -.2939E+01   -.2665E-13

INCREMENTAL LOAD VECTOR
  0.          0.          -.9697E+00  0.          .9697E+00  0.          0.          0.          0.          0.
  -.9697E+00  0.           .9697E+00  0.
```

LISTING OF FINITE ELEMENT PROGRAM - - "JETTY"

```
      PROGRAM JETTY(INPUT,OUTPUT,TAPE5=INPUT,TAPE6=OUTPUT,TAPE7)
C
C PROGRAM FOR ANALYSIS OF JOINTED ROCK STRUCTURES USING THE FINITE
C       ELEMENT METHOD WRITTEN BY DR. C. ST. JOHN WITH THE COLLABORATION
C       OF DR R E GOODMAN, USING THE THEORETICAL DISCUSSION OF CHAPTER 8
C       IT IS A SMALL PROGRAM WITH LIMITED HORIZONS.  ITS PRINCIPAL
C       PURPOSE IS EDUCATIONAL, I.E. TO DEMONSTRATE HOW  THE NUMERICAL
C       MODEL CAN BE CODED IN PRACTISE.   THE EQUATIONS REFERRED TO IN
C       THE NUMEROUS COMMENT CARDS HEREIN ARE EQUATIONS OF CHAPTER 8
C
C  LIST OF MAIN ARRAYS AND THEIR USES
C
C  MTYPE= ELEMENT MATERIAL OR JOINT TYPE
C  KS   = GLOBAL STIFFNESS MATRIX
C  R    = VECTOR OF NODAL FORCES
C  RXY  = 2*N MATRIX EQUIVALENT TO R - USED FOR INPUT AND OUTPUT
C  U    = VECTOR OF NODAL DISPLACEMENTS - USED INITIALY FOR WATER FORCE
C  UXY  = 2*N MATRIX EQUIVALENT TO U - USED FOR INPUT AND OUTPUT
C  X    = VECTOR OF X-ORDINATES OF NODES
C  Y    = VECTOR OF Y-ORDINATES OF NODES
C  ICODE= VECTOR OF DISPLACEMENT CONDITIONS
C       = 1   - FIXED IN X DIRECTION
C       = 2   - FIXED IN Y DIRECTION
C       = 3   - FIXED IN X AND Y DIRECTION
C  NOD  = N*4 MATRIX OF NODES DEFINING ELEMENTS
C  RESID= N*3 MATRIX OF RESIDUAL STRESSES
C  SIGNO= N*3 MATRIX OF INITIAL STRESSES
C  C    = GLOBAL STRESS STRAIN MATRIX FOR SOLID ELEMENT PROPERTIES
C         ALSO USED TO STORE JOINT ELEMENT PROPERTIES
C  RO   = VECTOR OF MASS DENSITY OF DIFFERENT MATERIAL TYPES
C  G    = SET TO ZERO IF GRAVITY NOT ACTIVE
C       = SET TO ONE  IF GRAVITY ACTIVE
C  KJS  = VECTOR OF JOINT SHEAR STIFFNESSES
C  KJN  = VECTOR OF JOINT NORMAL STIFFNESSES
C  KJSNU= NEW JOINT SHEAR STIFFNESSES
C  KJNNU= NEW JOINT NORMAL STIFFNESSES
C  ITER1= NUMBER OF FIRST ITERATION THIS RUN
C  ITERN= NUMBER OF LAST  ITERATION THIS RUN
C  NODES= NUMBER OF NODAL POINTS
C  NELEM= NUMBER OF ELEMENTS
C  NBOUN= NUMBER OF NODES WITH DISPLACEMENT CONSTRAINTS
C  NSOL = NUMBER OF SOLID ELEMENT TYPES
C  JOINT= NUMBER OF JOINT ELEMENT TYPES
C  ITER = NUMBER OF ITERATIONS THIS RUN
C  GRAV = ACCELERATION DUE TO GRAVITY
C  ACCEL= ACCELERATION NOT DUE TO GRAVITY - IN  G
C  DIR  = DIRECTION OF ACCEL - + ANTI CLOCKWISE FROM HORIZONTAL
C  LBAD = SET TO ZERO. BECOMES 1 IF AN ERROR IS ENCOUNTERED
C  NUM  = VECTOR OF NODES WITH DISPLACEMENT CONSTRAINTS
C  HEAD = 1.0 FOR SOLID ELEMENT ABOVE WATER TABLE
C       = 0.0 FOR SOLID ELEMENT BELOW WATER TABLE
C
      COMMON /MESH/ X(25),Y(25),R(50),U(50),KS(50,50),
     1 NOD(23,4),RESID(23,3),SIGNO(23,3),MTYPE(23),
     2              NODES,NELEM,NSOL,NFREE
      REAL          KS

      COMMON /STFF/ KJS(23),KJN(23),KJSNU(23),KJNNU(23),ITER1,ITERN,IT
      REAL          KJS,KJN,KJSNU,KJNNU
      COMMON /BOUN/ NUM(20),ICODE(20),NBOUN
```

```
      COMMON /CONS/ RO(10),ACELX,ACELY,NRES,INTER,C(3,3,10)
      DIMENSION  IDAT(13),ICARD(9),E(3,3),T(3,3),H(3,6),
     1           RXY(2,25),UXY(2,25),CJ(9,10)
      EQUIVALENCE (R(1),RXY(1,1)),(U(1),UXY(1,1)),(CJ(1,1),C(1,1,1))

      DATA IDAT /8HNODAL PO,8HELEMENT ,8HHYDRAULI,8HRESIDUAL,8HBOUNDARY,
     18HSOLID EL,8HJOINT EL,8HSTART   ,8HSTOP    ,8HRESTART ,8HINTERMED,
     28HACCELERA,8HSAVE STI/
C
      DIMENSION  B(50), SCRTCH(150)
C     THESE ARE NEEDED FORTHE MATRIX INVERSION SUBROUTINE
C
C READ THE TITLE CARD AND THEN WRITE IT
C
      READ (5,1000) ICARD
      WRITE(6,2000) ICARD
C
C READ ESSENTIAL CONTROL DATA AND WRITE IT OUT IMMEDIATELY
C
      READ(5,1001) NODES,NELEM,NBOUN,NSOL,JOINT,ITER1,ITER,G,GRAV,ACCEL,
     1 DIR,RW
      IF (ITER1 .EQ. 0)  ITER1 = 1
      WRITE(6,2001)NODES,NELEM,NBOUN,NSOL,JOINT,ITER1,ITER,G,GRAV,ACCEL,
     1 DIR,RW
      ITERN = ITER1+ITER-1
C
C SET SOME INITIAL VALUES
C
      NTOT= NSOL+JOINT
      NRES=0
      LBAD=0
      INTER = 1
      PI=ATAN(1.0)*4.0
C
C SET SOME OF THE ARRAYS TO ZERO BEFORE STARTING
C
      NFREE=NODES+NODES
C
      DO 1  I=1,50
      R(I)=0.0
      U(I)=0.0
    1 CONTINUE
      DO 2 I=1,NELEM
      KJS(I)=0.0
      KJN(I)=0.0
      KJNNU(I)=0.0
      KJSNU(I)=0.0
      DO 2 J=1,3
      RESID(I,J)=0.0
      SIGNO(I,J)=0.0
    2 CONTINUE
C
C NOW CALCULATE TOTAL ACCELERATION COMPONENTS - EQUATION 66A IN TEXT
C
      DIR=DIR*PI/180.
      ACELX=ACCEL*COS(DIR)*GRAV
C
C G CAN TAKE VALUE 1.0 OR 0.0 CORRESPONDING TO GRAVITY ON OR OFF
C
      ACELY=(-G+ACCEL*SIN(DIR))*GRAV
C
C READ NEXT DATA CARD -THIS MUST BE A HEADER CARD FOR A DATA SET
C OR A START / STOP / RESTART / INTER
```

```
C
  100 READ (5,1000) ICARD
C
C DETERMINE WHICH DATA SET FOLLOWS
C
      J=ICARD(1)
      DO 130  I=1,13
      IF(J.EQ.IDAT(I)) GO TO 150
  130 CONTINUE

      WRITE (6,2002) ICARD
      LBAD=LBAD+1
      GO TO 100

  150 GO TO (10,20,30,40,50,60,70,80, 700,95,5,600,650), I
C
C ALL INTERMEDIATE RESULTS DESIRED ON PRINTOUT
C
    5 INTER = 0
      GO TO 100
C
C READ NODAL POINT INFORMATION
C
   10 L=0
   11 READ(5,1004) N,X(N),Y(N),RXY(1,N),RXY(2,N)
C
C A SIMPLE NODAL POINT GENERATOR AND ERROR SEARCH
C
      IF(L.EQ.0) GO TO 12
      ZX=N-L
      DY=(Y(N)-Y(L))/ZX
      DX=(X(N)-X(L))/ZX
   12 L = L + 1
      IF (N - L)  15,14,13
   13 X(L)=X(L-1)+DX
      Y(L)=Y(L-1)+DY
      GO TO 12
   14 IF (NODES - N)  15,16,11
C
C NODAL POINT DATA ERROR       WRITE ERROR MESSAGE
C
   15 WRITE(6,2004) N
      LBAD=LBAD+1
      IF(N.LT.NODES) GO TO 11
C
C WRITE OUT NODE DATA
C
   16 WRITE(6,2003) (K,X(K),Y(K),RXY(1,K),RXY(2,K),K=1,NODES)
      GO TO 100
C
C ELEMENT DATA - MATERIAL TYPE AND NODAL POINTS
C
   20 N=0
   21 READ(5,1005)  K,MAT,(NOD(K,I),I=1,4)
C
C A SIMPLE ELEMENT GENERATOR AND ERROR SEARCH
C
      MTYPE(K)=MAT
      IF(MAT.GT.NTOT) GO TO 27
   22 N=N+1
      IF (K - N)  27,26,23
   23 N1=N-1
      MAT = MTYPE(N1)
```

```
      MTYPE(N)=MAT
      NOD(N,1)=NOD(N1,1)+1
      NOD(N,2)=NOD(N1,2)+1
      NOD(N,3)=NOD(N1,3)+1
      IF (MAT .GT. NSOL)  GO TO 24
      NOD(N,4)=0
      GO TO 25
   24 NOD(N,4)=NOD(N1,4)+1
   25 IF (K - N)  27,26,22
   26 IF (NELEM - N)  27,29,21
C
C ELEMENT DATA ERROR     - WRITE ERROR MESSAGE
C
   27 WRITE(6,2007) K
      LBAD=LBAD+1
      IF(K.LT.NELEM) GO TO 21
C
C WRITE OUT ELEMENT DATA
C
   29 WRITE(6,2006) (K,MTYPE(K),(NOD(K,I),I=1,4),K=1,NELEM)
      GO TO 100
C
C READ HYDRAULIC GRADIENT DATA FOR ELEMENTS
C
   30 L=0
   31 READ(5,1004) N,HEAD,DHDX,DHDY
C
C ZERO ENTRIES FOR JOINTS AS RESIDUAL STRESSES WILL BE EFFECTIVE
C FOR A SOLID ELEMENT HEAD =1.0 IF BELOW WATER TABLE
C                     HEAD = 0.0 IF ABOVE WATER TABLE
C
C
C CALCULATE AND THEN STORE TERMS REQUIRED FOR EQUATION 65
C PRESSURE GRADIENTS ARE STORED TO AVOID  HAVING TO STORE
C THE WATER TABLE SWITCH
C
      GW=GRAV*HEAD*RW
      UXY(1,N)=DHDX*GW
      UXY(2,N)=(DHDY-1.0)*GW
C
C GENERATE MISSING DATA AND CHECK FOR ERRORS
C
   32 L=L+1
      IF (N - L)  35,34,33
   33 L1=L-1
      UXY(1,L)=UXY(1,L1)
      UXY(2,L)=UXY(2,L1)
      GO TO 32
   34 IF (NELEM - N)  35,36,31
C
C HYDRAULIC GRADIENT CARD ERROR - WRITE MESSAGE
C
   35 WRITE(6,2010) N
      LBAD=LBAD+1
      IF(N.LT.NELEM) GO TO 31
C
C WRITE OUT PRESSURE GRADIENTS FOR ELEMENTS
C
   36 WRITE(6,2009) (K,UXY(1,K),UXY(2,K),K=1, NELEM)
      GO TO 100
C
C READ RESIDUAL STRESS CARDS
C NRES SET TO 1 TO INDICATE THAT RESIDUAL STRESSES HAVE BEEN READ
```

```
C
   40 NRES=1
      L=0
   41 READ(5,1004) N,RESID(N,1),RESID(N,2),RESID(N,3)
C
C GENERATE MISSING DATA AND THEN CHECK FOR ERRORS
C
   42 L=L+1
      IF (N - L)  45,44,43
   43 L1=L-1
      RESID(L,1)=RESID(L1,1)
      RESID(L,2)=RESID(L1,2)
      RESID(L,3)=RESID(L1,3)
      GO TO 42
   44 IF (NELEM - N)  45,46,41
C
C RESIDUAL STRESS CARD ERROR - WRITE ERROR MESSAGE
C
   45 WRITE(6,2013) N
      LBAD=LBAD+1
      IF(N.LT.NELEM) GO TO 41
C
C WRITE OUT RESIDUAL STRESS DATA
C
   46 WRITE(6,2012) (K,RESID(K,1),RESID(K,2),RESID(K,3),K=1,NELEM)
      GO TO 100
C
C READ LIST OF NODAL CONSTRAINTS AND THEN WRITE THEM OUT
C
   50 READ(5,1005) (NUM(K),ICODE(K),K=1,NBOUN)
      WRITE(6,2015) (NUM(K),ICODE(K),K=1,NBOUN)
      GO TO 100
C
C READ SOLID ELEMENT PROPERTIES AND THEN WRITE THEM
C
   60 WRITE(6,2017)
      DO 69 K=1,NSOL
      READ(5,1004) MAT,RO(MAT),ES,EN,GSN,PRSN,PRST,ALPHA
      WRITE(6,2018)MAT,RO(MAT),ES,EN,GSN,PRSN,PRST,ALPHA
C
C FORM STRESS-STRAIN MATRIX IN LOCAL CO-ORDINATES - EQUATION 20B
C
      RAT=EN/ES
      CONST=1.0/((1.0+PRST)*(1.0-PRST-2.0*RAT*PRSN*PRSN))
      E(1,1)=(ES-EN*PRSN*PRSN)*CONST
      E(2,1)=EN*PRSN*(1.0+PRST)*CONST
      E(3,1)= 0.0
      E(1,2)= E(2,1)
      E(2,2)=EN*(1.0-PRST*PRST)*CONST
      E(3,2)= 0.0
      E(1,3)= 0.0
      E(2,3)= 0.0
      E(3,3)=GSN
C
      IF (INTER .EQ. 0)  WRITE (6,999) ((E(J,L),L=1,3),J=1,3)
  999 FORMAT (44H0 STRESS-STRAIN MATRIX IN LOCAL CO-ORDINATES/
     1         (5X,3E15.5))
C
C NOW FORM THE TRANSFORMATION MATRIX  - EQUATION 23.
C
      ALPHA=ALPHA*PI/180.0
      COSA=COS(ALPHA)
      SINA=SIN(ALPHA)
```

```
      T(1,1)=COSA*COSA
      T(2,1)=SINA*SINA
      T(3,1)=-2.0*SINA*COSA
      T(1,2)= T(2,1)
      T(2,2)= T(1,1)
      T(3,2)=-T(3,1)
      T(1,3) = SINA*COSA
      T(2,3) = -T(1,3)
      T(3,3)= T(1,1)-T(2,1)
C
      IF (INTER .EQ. 0)  WRITE (6,998) ((T(J,L),L=1,3),J=1,3)
  998 FORMAT (23HC TRANSFORMATION MATRIX/(5X,3E15.5))
C
C NOW TRANSFORM TO GLOBAL CO-ORDINAL SCHEME - USE H AS TEMP. STORAGE
C
      DO 66 N=1,3
      DO 66 J=1,3
      H(J,N)=0.0
      DO 66 L=1,3
      H(J,N)=H(J,N)+E(J,L)*T(L,N)
   66 CONTINUE
      DO 67 N=1,3
      DO 67 J=1,3
      C(J,N,MAT)=0.0
      DO 67 L=1,3
      C(J,N,MAT)=C(J,N,MAT)+T(L,J)*H(L,N)
   67 CONTINUE
C
      IF (INTER .EQ. 0)  WRITE (6,997)  ((C(J,N,MAT),N=1,3),J=1,3)
  997 FORMAT (44H0 STRESS-STRAIN MATRIX IN GLOBAL COORDINATES/
     1         (5X,3E15.5))
C
C CHECK FOR ERRORS
C
      IF (NSOL .GE. MAT)  GO TO 69
      WRITE(6,2019)
      LBAD=LBAD+1
   69 CONTINUE
      GO TO 100
C
C READ JOINT PROPERTIES AND STORE IN C
C
   70 WRITE(6,2020)
      NTOT=NSOL+JOINT
      DO 75 J=1,JOINT
      READ(5,1004) MAT,(CJ(K,MAT),K=1,8)
C
C COMPRESSIVE STRENGTH NEGATIVE AND TENSILE TO COMPRESSIVE RATIO +VE
C MAXIMUM CLOSURE MUST BE POSITIVE
C SEATING LOAD MUST ALWAYS BE NEGATIVE
C
      CJ(1,MAT)=-ABS(CJ(1,MAT))
      CJ(2,MAT)=ABS(CJ(2,MAT))
      CJ(5,MAT)=ABS(CJ(5,MAT))
      CJ(6,MAT)=-ABS(CJ(6,MAT))
C
      WRITE(6,2021)MAT,(CJ(K,MAT),K=1,8)
C
C CHECK FOR ERRORS
C
      IF (NTOT .GE. MAT)  GO TO 75
      WRITE(6,2022)
      LBAD=LBAD+1
```

```
   75 CONTINUE
C
C IF THIS NOT THE FIRST RUN THEN JOINT STIFFNESS IS TO BE READ
C
      IF(ITER1.GT.1) READ(7) KJS,KJN
      GO TO 100
C
C DATA INPUT COMPLETE - PROCEED IF NO ERRORS
C
   80 IF (LBAD .EQ. 0)  GO TO 300
C
C ERRORS HAVE BEEN FOUND - WRITE MESSAGE
C
      WRITE(6,2023)
      STOP
C
C RESTART BY READING DATA FROM TAPE7
C
   95 READ(7)    C,X,Y,R,U,KS,NUM,NOD,RESID,SIGNO,MTYPE,ICODE,KJS,KJN
      GO TO 375
C
C CALL SUBROUTINE STIFF TO ASSEMBLE THE STIFFNESS MATRIX
C
  300 CALL STIFF
C
C CALL LIBRARY SUBROUTINE TO INVERT THE STIFFNESS MATRIX
C NFREE IS THE SIZE OF THE STIFFNESS MATRIX
C IN LARGER PROGRAMS, STIFFNESS MATRIX SHOULD BE NARROW BANDED
C
      IF (INTER .NE. 0)  GO TO 350
      WRITE(6,996)
  996 FORMAT (34H0 LOAD VECTOR AND STIFFNESS MATRIX)
      DO 995  I=1,NFREE
      WRITE(6,994)       (KS(I,J),J=1,NFREE)
  994 FORMAT(2X,10E12.4)
  995 CONTINUE
      WRITE (6,994) (R(I),I=1,NFREE)
C
  350 CALL LINV3F (KS,B,1,NFREE,50,-1.0,D2,SCRTCH,IER)
C
C  KS IS NOW THE INVERSE OF THE STIFFNESS MATRIX
C
C
      IF (INTER .NE. 0)  GO TO 359
      WRITE (6,993)
  993 FORMAT (29H0 INVERSE OF STIFFNESS MATRIX)
      DO 992  I=1,NFREE
  992 WRITE (6,994)  (KS(I,J),J=1,NFREE)
C
C SET INITIAL STRESSES TO RESIDUAL STRESSES BEFORE STARTING
C
  359 DO 360 M=1,NELEM
      SIGNO(M,1)=RESID(M,1)
      SIGNO(M,2)=RESID(M,2)
      SIGNO(M,3)=RESID(M,3)
  360 CONTINUE
C
C SET INITIAL DISPLACEMENTS TO ZERO
C
      DO 370 J=1,NFREE
      U(J)=0.0
  370 CONTINUE
C
```

```
C NOW ENTER MAIN ITERATIVE LOOP
C
  375 DO 500 IT=ITER1,ITERN
      WRITE(6,2026) IT
C
C DISPLACEMENTS ARE ACCUMULATED  - EACH ITERATION GIVES NEW INCREMENT
C DISPLACEMENT VECTOR (U)= U  + KS * LOAD VECTOR
C
      DO 400 J=1,NFREE
      DO 400 K=1,NFREE
      U(J)=U(J)+KS(J,K)*R(K)

  400 CONTINUE
C
C WRITE OUT NODAL DISPLACEMENTS
C
      WRITE(6,2024) (J,UXY(1,J),UXY(2,J),J=1,NODES)
C
C SET LOAD VECTOR TO ZERO FOR INCREMENTAL LOADS
C
      DO 450 J=1,NFREE
      R(J)=0.0
  450 CONTINUE
C
C CALL STRES  TO CALCULATE SOLID ELEMENT STRESSES
C
      CALL STRES
C
C CALL JSTR TO CALCULATE JOINT BEHAVIOUR AND NEW LOAD VECTOR
C
      IF(JOINT.GT.0) CALL JSTR
C
  500 CONTINUE
C
      GO TO 100
C
C ACCELERATE        WRITE NEW JOINT STIFFNESSES ONTO TAPE
C
  600 WRITE (7) KJSNU,KJNNU
      GO TO 700
C
C SAVE STIFFNESS   WRITE ALL DATA AND INVERTED STIFFNESS MATRIX ON TAPE
C
  650 WRITE (7) C,X,Y,R,U,KS,NUM,NOD,RESID,SIGNO,MTYPE,ICODE,KJS,KJN
C
  700 STOP
C
C INPUT FORMAT STATEMENTS
C
 1000 FORMAT(9A8)
 1001 FORMAT(7I5/5F10.0)
 1004 FORMAT(I5,7F10.0,F5.0)
 1005 FORMAT(16I5)
C
C OUTPUT FORMAT STATEMENTS
C
 2000 FORMAT(1H1,10X,9A8)
 2001 FORMAT(     30H0 NUMBER OF NODAL POINTS      =,I3 /
     1            30H  NUMBER OF ELEMENTS          =,I3 /
     2            30H  NUMBER OF CONSTRAINED NODES=,I3 /
     3            30H  NUMBER OF SOLID MATERIALS  =,I3 /
     4            30H  NUMBER OF JOINT TYPES       =,I3 /
     +            30H  NUMBER OF FIRST ITERATION  =,I3 /
```

```
    5             30H  NUMBER OF ITERATIONS         =,I3 /
    6             30H  GRAVITY LOADING - ON/OFF     =,F4.0 /
    7             30H  ACCELERATION DUE TO GRAVITY=,E12.4 /
    8             30H  PSEUDO STATIC ACCELERATION =,E12.4 /
    9             30H  DIRECTION OF ACCELERATION  =,E12.4 /
    A              30H  MASS DENSITY OF WATER       =,E12.4 )
 2002 FORMAT (30H0 FAULTY INPUT CARD CONTAINS  9A8)
 2003 FORMAT (1H02(3X,4HNODE6X,4HXORD8X,4HYORD7X,5HXLOAD7X,5HYLOAD5X)/
    1         (2(I7,2F12.2,2E12.4,3X)))
 2004 FORMAT (19H0 DATA ERROR - NODEI5)
 2006 FORMAT (1H02(5X,7HELEMENT4X,8HMAT TYPE7X,9HNODE LIST7X)/
    1         (2(I11,I10,5X,4I5)))
 2007 FORMAT (22H0 DATA ERROR - ELEMENTI5)
 2009 FORMAT (1H02(2X,7HELEMENT7X,5HDP/DX6X,10HDP/DY OR P)/
    1         (2(I9,2E14.4)))
 2010 FORMAT (47H0 DATA ERROR - HYDRAULIC GRADIENT CARD, ELEMENTI5)
 2012 FORMAT (1H02(3X,7HELEMENT9X,29HRESIDUAL STRESSES    XX,YY,XY7X)/
    1         (2(I10,3E15.4)))
 2013 FORMAT (44H0 DATA ERROR - RESIDUAL STRESS CARD, ELEMENTI5)
 2015 FORMAT (1H05X,23HNODE  DISPLACEMENT CODE/(I9,I15))
 2017 FORMAT(1H0,2X,25HSOLID MATERIAL PROPERTIES )
 2018 FORMAT(     2X,40HMATERIAL TYPE                              =,I3   /
    1             2X,40HMASS DENSITY                               =,E12.4 /
    2             2X,40HYOUNGS MODULUS - PARALLEL TO ANISOTROPY=,E12.4 /
    3             2X,40HYOUNGS MODULUS - NORMAL TO ANISOTROPY  =,E12.4 /
    +             2X,40HSHEAR MODULUS - GSN                        =,E12.4 /
    4             2X,40HPOISSONS RATIO  - SN                       =,F8.4  /
    5             2X,40HPOISSONS RATIO  - ST                       =,F8.4  /
    7             2X,40HORIENTATION OF LAYERS (ANGLE S-X)          =,F8.4  )
 2019 FORMAT (37H0 DATA ERROR - MATERIAL PROPERTY CARD)
 2020 FORMAT(1H0,20H  JOINT PROPERTIES  )
 2021 FORMAT(     1X,41HJOINT TYPE NUMBER                          =,I3    /
    1             1X,41HWALL ROCK COMPRESSIVE STRENGTH             =,E12.4 /
    2             1X,41HRATIO OF TENSILE TO COMPRESSIVE STRENGTH=,F8.4  /
    3             1X,41HSHEAR STIFFNESS                            =,E12.4 /
    4            1X,41HRATIO OF RESIDUAL TO PEAK SHEAR STRENGTH=,E12.4 /
    5             1X,41HMAXIMUM NORMAL CLOSURE                     =,E12.4 /
    V            1X,41HSEATING LOAD                                =,E12.4 /
    6             1X,41HFRICTION ANGLE OF A SMOOTH JOINT           =,E12.4 /
    7             1X,41HDILATANCY ANGLE                            =E12.4/)
 2022 FORMAT (34H0 DATA ERROR - JOINT PROPERTY CARD)
 2023 FORMAT (41H0 DATA ERRORS DETECTED - PROGRAMME HALTED)
 2024 FORMAT (44H0 NODE     X-DISPLACEMENT     Y-DISPLACEMENT/
    1         (I6,2F19.6))
 2026 FORMAT (18H1 ITERATION NUMBERI5//)
C
      END


      SUBROUTINE STIFF
C
C SUBROUTINE TO ASSEMBLE MODEL STIFFNESS MATRIX AND LOAD VECTOR
C AT END OF SUBROUTINE THE MATRIX IS READY FOR INVERSION
C
      COMMON /MESH/ X(25),Y(25),R(50),U(50),KS(50,50),
     1 NOD(23,4),RESID(23,3),SIGNO(23,3),MTYPE(23),
     2              NODES,NELEM,NSOL,NFREE
      REAL          KS
      COMMON /BOUN/ NUM(20),ICODE(20),NBOUN
      COMMON /FORC/ KSUB(8,8),F(8),LBAD,M,H(3,6)
      REAL          KSUB
```

```
      DIMENSION RES(3),XT(4),YT(4),LAB(4)
C
C SET STIFFNESS KS TO ZERO
C
      LBAD = 0
      DO 100 J=1,NFREE
      DO 100 I=1,NFREE
      KS(I,J)=0.0
  100 CONTINUE
C
C FOR EACH ELEMENT CALCULATE CONTRIBUTION TO GLOBAL STIFFNESS AND LOADS
C
      DO 450 M=1,NELEM
      MAT=MTYPE(M)
C
C IS IT A JOINT OR A TRIANGULAR ELEMENT
C
      NUMB = 3
      IF (MAT .GT. NSOL)  NUMB = 4
C
C EXTRACT NODAL COORDINATES FROM NODAL COORDINATE VECTORS
C
      DO 250 I=1,NUMB
      N=NOD(M,I)
      XT(I)=X(N)
      YT(I)=Y(N)
  250 CONTINUE
C
C SELECT JOINT OR TRIANGULAR STIFFNESS SUBROUTINE
C
      M2=M+M
      RES(1)=RESID(M,1)
      RES(2)=RESID(M,2)
      RES(3)=RESID(M,3)
      IF (MAT .GT. NSOL)  GO TO 270
C
      DPDX=U(M2-1)
      DPDY=U(M2)
      CALL TRIA(0,MAT,XT,YT,RES,DPDX,DPDY)
      IF (LBAD .NE. 0)  450, 280
C
  270 CALL JSTIF(MAT,XT,YT,RES)
      IF (LBAD .GT. 0)  GO TO 450
C
C RETAIN TRANSFORMED RESIDUAL STRESSES
C
      RESID(M,1)=RES(1)
      RESID(M,2)=RES(2)
      RESID(M,3)=RES(3)
C
C NOW ADD THE ELEMENT CONTRIBUTIONS KSUB AND F TO KS AND R
C ASSEMBLE VECTOR INDICATING LOCATION OF ELEMENT CONTRIBUTION
C
  280 DO 300 I=1,NUMB
      J=NOD(M,I)
      LAB(I)=J+J-1
  300 CONTINUE
C
      DO 400 I=1,NUMB
      II=LAB(I)
      IK=I+I-1
      I2=II+1
C
```

```
C LOAD CONTRIBUTION
C
      R(II)=R(II)+F(IK)
      R(I2)=R(I2)+F(IK+1)
      DO 400 J=1,NUMB
      JJ=LAB(J)
      JK=J+J-1
      J2=JJ+1
C
C STIFFNESS CONTRIBUTION
C
      KS(II,JJ)=KS(II,JJ)+KSUB(IK   ,JK   )
      KS(II,J2)=KS(II,J2)+KSUB(IK   ,JK+1)
      KS(I2,JJ)=KS(I2,JJ)+KSUB(IK+1,JK   )
      KS(I2,J2)=KS(I2,J2)+KSUB(IK+1,JK+1)
  400 CONTINUE
  450 CONTINUE
      IF (LBAD .GT. 0)  GO TO 600
C
C NOW INTRODUCE NODAL CONSTRAINTS
C
C ICODE = 1  - FIXED IN X DIRECTION
C ICODE = 2  - FIXED IN Y DIRECTION
C ICODE = 3  - FIXED IN X AND Y DIRECTION
C
C SET DIAGONAL TERM TO VERY HIGH STIFFNESS
C EFFECTIVELY FIXES NODE IN REQUIRED DIRECTION
C FOR GIVEN DISPLACEMENTS SET R(JJ)=DISPLACEMENT*1.E+20
C
      DO 500 I=1,NBOUN
      L=ICODE(I)
      IF(L.EQ.0) GO TO 500
      J=NUM(I)
      JJ = J + J
      IF (L .EQ. 2)  GO TO 470
      KS(JJ-1,JJ-1) = KS(JJ-1,JJ-1) + 1.0E20
      IF (L .EQ. 1)  GO TO 500
  470 KS(JJ,JJ)     = KS(JJ,JJ)     + 1.0E20
  500 CONTINUE
C
C
      RETURN
C
C ERRORS DETECTED
C
  600 WRITE(6,2000)
 2000 FORMAT (46HC PROGRAM STOPPED AS ERRORS HAVE BEEN DETECTED)
      STOP
C
      END



      SUBROUTINE TRIA(KUT,MAT,X,Y,RESID,DPDX,DPDY)
C
      COMMON /CONS/ RO(10),ACELX,ACELY,NRES,INTER,C(3,3,10)
      COMMON /FORC/ KSUB(8,8),F(8),LBAD,M,H(3,6)
      REAL          KSUB
      DIMENSION X(4),Y(4),RESID(3)
      REAL  LNORT(3,6)
C
```

```
C SUBROUTINE TO CALCULATE STIFFNESS MATRIX(KSUB) OF TRIANGULAR ELEMENT
C TOGETHER WITH CONTRIBUTION TO LOAD VECTOR FROM RESIDUAL STRESSES,
C GRAVITY LOADING, ACCELERATION LOADING AND WATER FORCES (F)
C
C IF KUT IS EQUAL TO 1 THIS ROUTINE WAS CALLED FROM SUBROUTINE STRESS
C TO CALCULATE THE STRESS-DISPLACEMENT TRANSFORMATION MATRIX (H) ONLY
C
C START BY SHIFTING ORIGIN SO THAT LOCAL COORDINATES FOR FIRST
C NODE ARE  0.0,0.0
C
      XJ=X(2)-X(1)
      YJ=Y(2)-Y(1)
      XK=X(3)-X(1)
      YK=Y(3)-Y(1)
C
C NOW CALCULATE TWICE THE AREA OF THE TRIANGLE
C
      AREA=XJ*YK-XK*YJ
C
C IF THIS AREA IS ZERO OR NEGATIVE THERE IS AN ERROR IN THE DATA
C RETURN IF AN ERROR HAS BEEN FOUND
C
      IF(AREA.GT.0.0) GO TO 10
      LBAD=LBAD+1
      WRITE(6,2000)  LBAD,M
 2000 FORMAT (7H0 ERRORI3,9H, ELEMENTI5,22H HAS NON-POSITIVE AREA)
      RETURN
C
C FORM STRAIN DISPLACEMENT MATRIX -LNORT -EQUATION 16 IN TEXT
C
C NOW CALCULATE THE NECESSARY TERMS - TOGETHER WITH THEIR SIGNS
C
   10 XK=-XK/AREA
      YK= YK/AREA
      XJ= XJ/AREA
      YJ=-YJ/AREA
      XKJ=-XK-XJ
      YJK=-YJ-YK
C
C SET LNORT AS EQUATION 16
C
      LNORT(1,1)= YJK
      LNORT(2,1)= 0.0
      LNORT(3,1)= XKJ
      LNORT(1,2)= 0.0
      LNORT(2,2)= XKJ
      LNORT(3,2)= YJK
      LNORT(1,3)= YK
      LNORT(2,3)= 0.0
      LNORT(3,3)= XK
      LNORT(1,4)= 0.0
      LNORT(2,4)= XK
      LNORT(3,4)= YK
      LNORT(1,5)= YJ
      LNORT(2,5)= 0.0
      LNORT(3,5)= XJ
      LNORT(1,6)= 0.0
      LNORT(2,6)= XJ
      LNORT(3,6)= YJ
C
      IF (INTER .EQ. 0)  WRITE (6,999)  ((LNORT(I,J),J=1,6),I=1,3)
  999 FORMAT (28H0 STRAIN DISPLACEMENT MATRIX/(5X,6E12.4))
C
```

```
C
C NOW CALCULATE STRESS-DISPLACEMENT TRANSFORMATION MATRIX ( H)
C GIVEN BY     C * LNORT   ,WHERE C IS THE STRESS-STRAIN MATRIX
C FOR THE RELEVANT MATERIAL TYPE ( MAT)
C
      DO 100  J=1,6
      DO 100  I=1,3
      H(I,J)= 0.0
      DO 100  K=1,3
      H(I,J)=H(I,J) + C(I,K,MAT)*LNORT(K,J)
  100 CONTINUE
C
      IF (INTER .EQ. 0)  WRITE (6,998)  ((H(I,J),J=1,6),I=1,3)
  998 FORMAT (43H0 STRESS-DISPLACEMENT TRANSFORMATION MATRIX/
     1         (5X,6E12.4))
C
C NOW RETURN IF THIS SUBROUTINE WAS CALLED FROM STRES
C
      IF(KUT.EQ.1) RETURN
C
C NOW COMPLETE THE FORMATION OF THE STIFFNESS MATRIX ( KSUB)
C FIRST MULTIPLY LNORT BY THE AREA OF THE ELEMENT
C
      AREA=AREA/2.0
      DO 200 J=1,6
      DO 200 I=1,3
       LNORT(I,J)=LNORT(I,J)*AREA
  200 CONTINUE
C
C CALCULATE (LNORT) TRANSPOSE * H - EQUATION 24A IN TEXT
C
      DO 300 J=1,6
      DO 300 I=1,6
      KSUB(I,J)=0.0
      DO 300 K=1,3
      KSUB(I,J)=KSUB(I,J)+LNORT(K,I)*H(K,J)
  300 CONTINUE
C
C CALCULATE CONTRIBUTIONS OF GRAVITY LOADING, ACCELERATION LOADING
C AND WATER FORCES TO LOAD VECTOR -EQUATIONS 65 AND 66A
C
      CONST=AREA/3.0
      XCOMP=(RO(MAT)*ACELX-DPDX)*CONST
      YCOMP=(RO(MAT)*ACELY-DPDY)*CONST
      F(1)=XCOMP
      F(2)=YCOMP
      F(3)=XCOMP
      F(4)=YCOMP
      F(5)=XCOMP
      F(6)=YCOMP
C
C NOW ADD THE CONTRIBUTION OF THE RESIDUAL STRESSES - - EQNS. 27 AND 27A
C
      IF(NRES.EQ.0) GO TO 500
      DO 400 J=1,6
      DO 400 I=1,3
      F(J)=F(J)-LNORT(I,J)*RESID(I)
  400 CONTINUE
C
  500 IF (INTER .EQ. 0)  WRITE (6,997) M,((KSUB(I,J),J=1,6),F(I),I=1,6)
  997 FORMAT (28H0 STIFFNESS MATRIX - ELEMENTI3,73X,11HLOAD VECTOR/
     1         (2X,6E12.4,28X,E12.4))
C
```

```
      RETURN
      END



      SUBROUTINE JSTIF(MATT,X,Y,RESID)
C
C SUBROUTINE TO CALCULATE STIFFNESS MATRIX AND LOAD VECTOR FOR JOINT
C THIS SUBROUTINE IS DELIBERATELY LONGWINDED SO THAT ALTERATIONS
C TO THE JOINT STIFFNESS MATRIX MAY BE MADE WITH EASE
C
      COMMON /STFF/ KJS(23),KJN(23),KJSNU(23),KJNNU(23),ITER1,ITERN,IT
      REAL          KJS,KJN,KJSNU,KJNNU
      COMMON /CONS/ RO(10),ACELX,ACELY,NRES,INTER,C(3,3,10)
      COMMON /FORC/ KSUB(8,8),F(8),LBAD,M,H(3,6)
      REAL          KSUB
      DIMENSION X(4),Y(4),RESID(3)
      REAL L,KSO,KNO
C
C START BY CHECKING THAT IT IS A VALID JOINT
C
      IF (X(1) .NE. X(4))  GO TO 10
      IF (X(2) .NE. X(3))  GO TO 10
      IF (Y(1) .NE. Y(4))  GO TO 10
      IF (Y(2) .NE. Y(3))  GO TO 10
      GO TO 50
C
C ELEMENT DATA ERROR
C RETURN IF PREVIOUS ERRORS FOUND
C
   10 WRITE(6,2000) M
 2000 FORMAT (32H0 JOINT ELEMENT DATA   -   ELEMENTI5)
      LBAD=LBAD+1
      RETURN
C
C SET MATT TO LOCAL VARIABLE MAT     (SPEEDS UP EXECUTION)
C
   50 MAT=MATT
C
C CALCULATE JOINT DIMENSIONS    -   L= JOINT LENGTH
C
      DX=X(2)-X(1)
      DY=Y(2)-Y(1)
      L=SQRT(DX*DX+DY*DY)
C
C CONSTANTS FOR FURTHER CALCULATION
C
      COSA=DX/L
      SINA=DY/L
      S2=COSA*SINA
      CC=COSA*COSA
      SS=SINA*SINA
      C2=CC-SS
C
C ARE THERE RESIDUAL STRESSES - IF SO , TRANSFORM THEM
C
      IF (NRES .GT. 0)  GO TO 80
C
C RESIDUAL SHEAR STRESS IS ZERO
C
      TORSN=0.0
      SIGMA = 0.
```

```
      GO TO 90
   80 TORSN=-S2*RESID(1)+S2*RESID(2)+C2*RESID(3)
      SIGMA= SS*RESID(1)+CC*RESID(2)-2.0*S2*RESID(3)
C
C
C RETAIN TRANSFORMED STRESSES FOR FUTURE USE
C
   90 RESID(1) = TORSN
      RESID(2)=SIGMA
      RESID(3)=0.0
C
      IF (INTER .EQ. 0)  WRITE (6,999)  TORSN,SIGMA
  999 FORMAT (31H0 JOINT RESIDUAL SHEAR STRESS =E12.4/31H  JOINT RESIDUA
     1L NORMAL STRESS=E12.4)
C
C JOINT STIFFNESS WILL HAVE BEEN READ FROM TAPE(7) IF THIS IS AN
C ACCELERATED RESTART . IF THE NORMAL RESTART WAS USED THIS SUBROUTINE
C WILL NOT HAVE BEEN ENTERED
C
      IF(ITER1.GT.1) GO TO 95
C
C CALCULATE JOINT NORMAL STIFFNESS - EQUATION 44A
C
      KJN(M)=-SIGMA*SIGMA/(C(2,2,MAT)*C(3,2,MAT))
C
C  NORMAL STIFFNESS MUST LIE WITHIN LIMITS 0.01 QU/VMC AND 100 QU/VMC
C
      SLOMIN=-0.01*C(1,1,MAT)/C(2,2,MAT)
      SLOMAX = 10000.*SLOMIN
      IF (KJN(M).GT.SLOMAX) KJN(M)=SLOMAX
      IF(KJN(M).LT.SLOMIN) KJN(M)= SLOMIN
C
C RECOVER JOINT SHEAR STIFFNESS
C
      KJS(M)=C(3,1,MAT)
C
C NOW SET UP JOINT STIFFNESS IN LOCAL COORDINATES   - EQUATION 35
C
   95 KSO = KJS(M)*L/4.0
      KNO = KJN(M)*L/2.0
      DO 100 J=1,8
      DO 100 I=1,8
      KSUB(I,J)=0.0
  100 CONTINUE
C
C NOTE THAT THE JOINT STIFFNESS MATRIX IS WRITTEN OUT IN FULL.
C THIS PERMITS READY ALTERATION TO A NON SYMMETRIC MATRIX .
C SINCE A STRAIGHT INVERSION OF THE GLOBAL STIFFNESS IS CARRIED OUT
C SUCH MODIFICATIONS WOULD CAUSE NO PROBLEMS
C
      KSUB(1,1)= KSO
      KSUB(3,1)= KSO
      KSUB(5,1)=-KSO
      KSUB(7,1)=-KSO
      KSUB(1,3)= KSO
      KSUB(3,3)= KSO
      KSUB(5,3)=-KSO
      KSUB(7,3)=-KSO
      KSUB(1,5)=-KSO
      KSUB(3,5)=-KSO
      KSUB(5,5)= KSO
      KSUB(7,5)= KSO
      KSUB(1,7)=-KSO
```

```
      KSUB(3,7)=-KSO
      KSUB(5,7)= KSO
      KSUB(7,7)= KSO
      KSUB(2,2)= KNO
      KSUB(2,8)=-KNO
      KSUB(4,4)= KNO
      KSUB(4,6)=-KNO
      KSUB(6,4)=-KNO
      KSUB(6,6)= KNO
      KSUB(8,2)=-KNO
      KSUB(8,8)= KNO
C
      IF (INTER .EQ. 0)  WRITE (6,998)  ((KSUB(I,J),J=1,8),I=1,8)
  998 FORMAT (37H0 JOINT STIFFNESS - LOCAL COORDINATES/(2X,8E12.4))
C
C
C PERFORM MULTIPLICATION TO TRANSFORM FROM LOCAL TO GLOBAL COORDS
C ECONOMIES ACHIEVED BY AVOIDING ZEROS  IN 37A
C USE OF SYMMETRY OF STIFFNESS MATRIX IS AVOIDED HERE
C TRANSFORM IN 2*2 BLOCKS AT A TIME
C FORM (K)*(T) IN EQUATION 38
C
      DO 200 I=1,4
      IK=I+I
      II=IK-1
      DO 200 J=1,4
      JK=J+J
      JJ=JK-1
      T11=KSUB(II,JJ)*COSA-KSUB(II,JK)*SINA
      T12=KSUB(II,JJ)*SINA+KSUB(II,JK)*COSA
      T21=KSUB(IK,JJ)*COSA-KSUB(IK,JK)*SINA
      T22=KSUB(IK,JJ)*SINA+KSUB(IK,JK)*COSA
C
C PREMULTIPLY BY (T) TRANSPOSE  EQUATION 38
C
      KSUB(II,JJ)=COSA*T11-SINA*T21
      KSUB(II,JK)=COSA*T12-SINA*T22
      KSUB(IK,JJ)=SINA*T11+COSA*T21
      KSUB(IK,JK)=SINA*T12+COSA*T22
  200 CONTINUE
C
C NOW CALCULATE LOADS CORRESPONDING TO RESIDUAL STRESSES
C CHANGE TO EFFECTIVE NORMAL STRESS BY ADDING WATER PRESSURE
C TRANSFORM TO GLOBAL COORDINATES IMMEDIATELY - EQUATION  40 COMPLETED
C
      SIGMA=RESID(2)
      L=L/2.0
      FXI =-L*TORSN*COSA + L*SIGMA*SINA
      FYI =-L*TORSN*SINA - L*SIGMA*COSA
      F(1)=-FXI
      F(2)=-FYI
      F(3)=-FXI
      F(4)=-FYI
      F(5)= FXI
      F(6)= FYI
      F(7)= FXI
      F(8)= FYI
C
      IF (INTER .EQ. 0) WRITE (6,997) M,((KSUB(I,J),J=1,8),F(I),I=1,8)
  997 FORMAT (61H0 JOINT STIFFNESS AND FORCE VECTOR - GLOBAL COORDS. - E
     1LEMENTI3/(2X,8E12.4,E14.4))
C
```

```
      RETURN
      END



      SUBROUTINE STRES
C
C SUBROUTINE TO CALCULATE SOLID ELEMENT STRESSES FROM DISPLACEMENTS
C
      COMMON /MESH/ X(25),Y(25),R(50),U(50),KS(50,50),
     1 NOD(23,4),RESID(23,3),SIGNO(23,3),MTYPE(23),
     2               NODES,NELEM,NSOL,NFREE
      REAL           KS
      COMMON /FORC/ KSUB(8,8),F(8),LBAD,M,H(3,6)
      REAL           KSUB
      DIMENSION DISP(6),DUMMY(3),XT(4),YT(4),SIG(6)
C
C WRITE HEADING FOR PRINTOUT OF STRESSES
C
      WRITE(6,2000)
C
C CALCULATE STRESS IN EACH SOLID ELEMENT IN TURN
C
      DO 400 M=1,NELEM
C
C IS IT A SOLID ELEMENT OR A JOINT
C
      MAT=MTYPE(M)
      IF (MAT .GT. NSOL)  GO TO 400
C
C ASSEMBLE VECTOR OF NODAL COORDINATES XT AND YT
C ALSO ASSEMBLE VECTOR (DISP)  OF ELEMENT NODAL DISPLACEMENTS
C
      DO 250 I=1,3
      II=I+I-1
      N=NOD(M,I)
      JJ=N+N-1
      XT(I)=X(N)
      YT(I)=Y(N)
      DISP(II)=U(JJ)
      DISP(II+1)=U(JJ+1)
  250 CONTINUE
C
C NOW CALL SUBROUTINE TRIA TO FORM ELEMENT STRESS-DISPLACEMENT
C TRANSFORMATION MATRIX -(H)
C DUMMY IS USED TO FILL UP UNUSED PART OF THE PARAMETER LIST
C
      CALL TRIA(1,MAT,XT,YT,DUMMY,DUM1,DUM2)
C
C NOW CALCULATE STRESSES - ADDING INCREMENT TO RESIDUAL STRESSES
C EQUATION 25 IN TEXT
C
      DO 300 I=1,3
      SIG(I)=RESID(M,I)
      DO 300 J=1,6
      SIG(I)=SIG(I)+H(I,J)*DISP(J)
  300 CONTINUE
C
C CALCULATE PRINCIPAL STRESSES FROM XX(SIG(1)),YY(SIG(2)),XY(SIG(3))
C
      CC=(SIG(1)+SIG(2))/2.0
      BB=(SIG(1)-SIG(2))/2.0
```

```
      CR=SQRT(BB*BB+SIG(3)*SIG(3))
      SIG(4)=CC+CR
      SIG(5)=CC-CR
      SIG(6)=28.698*ATAN2(SIG(3),BB)
C
C WRITE OUT ELEMENT STRESSES
C
      WRITE (6,2001)  M,SIG
  400 CONTINUE
C
      RETURN
C
 2000 FORMAT (24H0 SOLID ELEMENT STRESSES//10X,7HELEMENT6X,9HXX-STRESS7X
     1,9HYY-STRESS7X,9HXY-STRESS6X,43HMAJOR PRINCIPAL MINOR PRINCIPAL OR
     2IENTATION)
 2001 FORMAT (I17,5E16.4,F14.2)
C
      END
```

```
      SUBROUTINE JSTR
C
C SUBROUTINE  TO CALCULATE RESPONSE OF JOINTS TO NODAL DISPLACEMENTS.
C INCREMENTAL LOADS FOR NEXT ITERATION ARE AUTOMATICALLY CALCULATED.
C VECTOR R WILL FINALLY CONTAIN INCREMENTAL LOADS READY FOR
C CALCULATION OF NEW INCREMENTAL DISPLACEMENTS.
C
      COMMON /MESH/ X(25),Y(25),R(50),U(50),KS(50,50),
     1 NOD(23,4),RESID(23,3),SIGNO(23,3),MTYPE(23),
     2              NODES,NELEM,NSOL,NFREE
      REAL          KS
      COMMON /STFF/ KJS(23),KJN(23),KJSNU(23),KJNNU(23),ITER1,ITERN,IT
      REAL          KJS,KJN,KJSNU,KJNNU
      COMMON /CONS/ RO(10),ACELX,ACELY,NRES,INTER,C(3,3,10)
      DIMENSION LAB(4),USN(8),CJ(9,10),F(8)
      EQUIVALENCE  (CJ(1,1),C(1,1,1))
      REAL  L,LHLF,KSO,KNO,MO
C
      WRITE(6,2000)
C
C CONVERSION CONSTANT FOR DEGREES TO RADIANS
C
      CONV= ATAN(1.0)/45.0
C
C TAKE JOINTS ONE AT A TIME
C
      DO 800 M=1,NELEM
C
C SELECT JOINT TYPE
C
      MAT=MTYPE(M)
C
C IS IT A JOINT - IF NOT GO TO NEXT ELEMENT
C
      IF (MAT .LE. NSOL)  GO TO 800
C
C RECOVER BASIC JOINT INFORMATION READY FOR CALCULATION
C
      II=NOD(M,1)
      JJ=NOD(M,2)
```

```
      DX=X(JJ)-X(II)
      DY=Y(JJ)-Y(II)
C
C JOINT LENGTH AND ORIENTATION
C
      L=SQRT(DX*DX+DY*DY)
      COSA=DX/L
      SINA=DY/L
C
C POSITION VECTOR USED TO IN TRANFER OF DATA TO AND FROM GLOBAL VECTORS
C
      DO 200 I=1,4
      J=NOD(M,I)
      JJ=J+J-1
      LAB(I)=JJ
      II=I+I
C
C TRANSFORM DISPLACEMENTS TO LOCAL COORDINATES
C
      USN(II-1)= U(JJ)*COSA+U(JJ+1)*SINA
      USN(II)   =-U(JJ)*SINA+U(JJ+1)*COSA
  200 CONTINUE
C
      IF (INTER .EQ. 0)  WRITE (6,999)  USN
  999 FORMAT (47H0 JOINT DISPLACEMENT VECTOR - LOCAL COORDINATES/
     1          (2X,8E12.4))
C
C
C JOINT MAXIMUM CLOSURE     - EQUATION 44
C
      VM = -CJ(5,MAT)
      IF(RESID(M,2).LT.0.) VM = -CJ(5,MAT)*CJ(6,MAT)/RESID(M,2)
C
C JOINT SHEAR AND NORMAL STIFFNESSES
C
      KNO = KJN(M)
      KSO = KJS(M)
C
C RATIO OF RESIDUAL TO PEAK SHEAR STRNGTH
C
      BO = CJ(4,MAT)
C
C COMPRESSIVE STRENGTH OF WALL ROCK
C
      QU = CJ(1,MAT)
C
C  RATIO OF COMPRESSIVE TO TENSILE STRENGTH  LADANYI AND ARCH TERM N
C
      RAT = 1/CJ(2,MAT)
C
C TANGENT OF ANGLE OF FRICTION OF A SMOOTH JOINT
C
      TANTH = TAN(CJ(7,MAT)*CONV)
C
C TANGENT OF DILATATION ANGLE - ZERO CONFINING PRESSURE
C
      TANI = TAN(CJ(8,MAT)*CONV)
C
C CALCULATE THE JOINT DEFORMATION - TOTAL DISPLACEMENTS - EQUATION 28,29
C
      UO = (-USN(1)-USN(3)+USN(5)+USN(7))/2.0
      VO = (-USN(2)-USN(4)+USN(6)+USN(8))/2.0
      W  = ( USN(2)-USN(4)+USN(6)-USN(8))/L
```

```
C
C CALCULATE TOTAL STRESS THIS ITERATION  - EQUATION 32A,25,33
C TNS=SHEAR STRESS, SN=NORMAL STRESS, MO=MOMENT
C
      TNS = KSO*UO            +SIGNO(M,1)
      SN  = KNO*VO            +SIGNO(M,2)
      MO  =  L*L*L*KNO*W/4.0+SIGNO(M,3)
C
      WRITE(6,2001)M,UO,VO,W,TNS,SN,MO
C
C NOW TAKE EACH END OF THE JOINT IN TURN AND CHECK THE SHEAR AND
C NORMAL FORCES FOR EACH NODAL PAIR . FIRST I,L NODES ,THEN J,K
C VARIABLE SIGNE IS USED TO CONTROL THE SIGN OF TERMS IN THE B MATRIX
C (EQUATION 31) THIS MATRIX IS NOT SET UP BUT THE RELATIONSHIPS IT
C DEFINES ARE USED
C FIRST SET L TO HALF LENGTH OF JOINT
C
      LHLF = L / 2.0
      SIGNE = 1.0
      TORM =0.0
      FNOM = 0.0
      FNIM = 0.0
C
      DO 550 I=1,2
      SIGNE = -SIGNE
C
C CALCULATE NODAL FORCES FOR FIRST,PREVIOUS AND CURRENT ITERATIONS
C (EQ 31 IN TEXT ) THESE ARE EXTERNAL FORCES AT L AND THEN AT K
C
C
C FNIL = NORMAL FORCE FOR LAST ITERATION
C FSO  = INITIAL SHEAR FORCE
C FNO  = INITIAL NORMAL FORCE
C FSI  = CURRENT SHEAR FORCE
C FNI  = CURRENT NORMAL FORCE
C
      FNIL = SIGNO(M,2)*LHLF + SIGNE*SIGNO(M,3)/L
      FNO  = RESID(M,2)*LHLF + SIGNE*RESID(M,3)/L
      FSO  = RESID(M,1)*LHLF
      FSI  = TNS * LHLF
      FNI  = SN*LHLF + SIGNE*MO/L
C
C NORMAL FORCE PER UNIT AREA FOR THIS HALF OF THE JOINT IS SIGMA
C
      SIGMA = FNI/LHLF
C
C SHEAR STRENGTH ETC. SET TO ZERO IF JOINT IN TENSION
C
      IF(SIGMA.LT.0.0) GO TO 305
      TORP=0.0
      TORR=0.0
      TOR =0.0
      UP  =0.0
      UR  =0.0
      DILAT=0.0
      GO TO 460
C
C CALCULATE PEAK SHEAR STRENGTH - EQUATIONS 14 TO 16 OF CHAPTER 5
C FIRST CHECK IF NORMAL LOADING IS ABOVE THE COMPRESSIVE STRENGTH
C
  305 IF (QU .LT. SIGMA)  GO TO 320
C
C TRANSITION STRESS EXCEEDED
```

```
C
      AS=1.0
      VDOT=0.0
C
C LIMIT SIGMA/QU RATIO TO 1.0 SO THAT PEAK SHEAR STRENGTH IS LIMITED
C
      RATIO = 1.0
      GO TO 330
C
C BELOW TRANSITION STRESS - EQUATIONS 16A AND 16B OF CHAPTER 5
C
  320 RATIO=SIGMA/QU
      AS=1.0-SQRT((1.0-RATIO)**3)
      VDOT=(1.0-RATIO)**4*TANI
C
C EVALUATE EQUATION 14 OF CHAPTER 5   TORP IS PEAK SHEAR STRENGTH
C
  330 TERM1 = SIGMA*(1.0-AS)*(VDOT+TANTH)
      TERM2 = AS*QU*(SQRT(RAT+1.0)-1.0)/RAT*SQRT(1.0+RAT*RATIO)
      TERM3 = 1.0-(1.0-AS)*VDOT*TANTH
      TORP=ABS(TERM1+TERM2)/TERM3
C
C IF TRANSITION STRESS EXCEEDED THEN NO PEAK-RESIDUAL BEHAVIOUR
C TORR = RESIDUAL SHEAR STRENGTH FROM EQUATION 60
C
      TORR = TORP
      IF (QU .LT. SIGMA)  TORR = TORP*(BO + (1.0-BO)*SIGMA/QU)
C
C SELECT PEAK AND RESIDUAL SHEAR DISPLACEMENT - EQUATION 54 AND 54 A
C SIGN OF DISPLACEMENT DETERMINES THE EQUATION- SET SEL TO THE SIGN
C
      AUO = ABS(UO)
      SEL = UO/AUO
C
C UP = PEAK SHEAR DISPLACEMENT
C UR = RESIDUAL SHEAR DISPLACEMENT
C
      UP = (SEL*TORP-RESID(M,1))/KSO
      UR = (SEL*4.0*TORP-RESID(M,1))/KSO
C
C SELECT RANGE TO WHICH SHEAR DISPLACEMENT BELONGS
C AND CALCULATE LIMITING SHEAR STRESS CORRESPONDING TO DISPLACEMENT EQ55
C
      IF (AUO .GT. ABS(UP))  GO TO 400
C
C IN LINEAR RANGE DEFINED-BY SHEAR STIFFNESS - RANGE III
C
      TOR = KSO*UO + RESID(M,1)
      GO TO 450
C
  400 IF (AUO .GE. ABS(UR))  GO TO 430
C
C IN FALLING PORTION  - RANGE II OR IV
C
      TOR = SEL*(TORP-(UO-UP)/(UR-UP)*(TORP-TORR))
      GO TO 450
C
C IN RESIDUAL PORTION  - RANGE I OR V
C
  430 TOR = SEL*TORR
      AUO = ABS(UR)
C
C NOW CALCULATE DILATION - FIRST SET AUO TO ABS(UR) IF SHEAR
```

```
C BEHAVIOUR NON-LINEAR   - SEE EQUATION 62  (WILL BE ABS(UO) IF NOT RESET
C
  450 DILAT = (FNIL/(LHLF*QU)-1.0)**4 *TANI*(AUO+ABS(FSO)/(KSO*LHLF))
C
C CALCULATE SHEAR FORCE TO BE APPLIED NEXT ITERATION -EQUATION 57
C
  460 DELS = TOR*LHLF - FSI
C
C CALCULATE WHETHER NODAL PAIR OPENING OR CLOSING
C SET DELV TO TOTAL NORMAL DISPLACEMENT   - EQUATION 47A
C
      II=I+I
      IJ= 10-II
      DELV= -USN(II)+USN(IJ)
      IF(DELV) 520,510,500
C
C JOINT OPENING - EQUATION 47A
C
  500 DELN = (DELV/(VM-DELV)+1.0)*FNO - FNI
      GO TO 530
C
C JOINT UNCHANGED
C
  510 DELN = 0.0
      GO TO 530
C
C JOINT CLOSING - EQUATION 50A
C
  520 DELN = (DELV-VM*(FNI-FNO)/FNI)*KNO*LHLF
C
C COMPUTE EXTERNAL FORCES ON JOINT EQUIVALENT TO INCREASE OF RESIDUAL
C NORMAL STRESS - CAUSED BY DILATION
C
  530 DILN = -DILAT*FNI/(VM)
C
C NOW TRANSFER NODAL FORCES TO JOINT FORCE VECTOR
C
      F(II) = -DELN-DILN
      F(II-1)=-DELS
      F(IJ) =  DELN+DILN
      F(IJ-1)= DELS
C
      IF (INTER .EQ. 0)  WRITE (6,998) UP,UR,TORP,TORR,TOR,DILAT,DELV
  998 FORMAT (5H0 UP=E12.4,4H UR=E12.4,6H TORP=E12.4,6H TORR=E12.4,
     1        5H TOR=E12.4/8H  DILAT=E12.4,6H DELV=E12.4)
C
C THE FOLLOWING ARE USED TO SAVE MEAN SHEAR STRESSES AND NORMAL FORCES
C FOR CALCULATION OF NEW JOINT STIFFNESSES
C
      TORM = TORM+ TOR/2.0
      FNIM = FNIM + FNI
      FNOM = FNOM + FNO
  550 CONTINUE
C
      IF (INTER .EQ. 0)  WRITE (6,997)  F
  997 FORMAT (30H0 INCREMENTAL FORCES FOR JOINT/2X,8E12.4)
C
C
C CALCULATE INITIAL STRESSES CORRESPONDING TO PREVIOUS PLUS
C INCREMENT JUST CALCULATED        ( EQUATIONS 30 )
C L IS BACK TO FULL JOINT LENGTH
C
      SIGNO(M,1) = SIGNO(M,1) +(F(5)+F(7))/L
```

```
      SIGNO(M,2) = SIGNO(M,2) +(F(6)+F(8))/L
      SIGNO(M,3) = SIGNO(M,3) +(F(6)-F(8))*LHLF
C
C ROTATE FORCES TO GLOBAL COORDINATE AND ADD TO GLOBAL FORCE VECTOR R
C - SEE EQUATION 37
C
      DO 600 I=1,4
      II=I+I-1
      JJ=LAB(I)
      R(JJ) = R(JJ) - F(II)*COSA + F(II+1)*SINA
      R(JJ+1)=R(JJ+1)-F(II)*SINA - F(II+1)*COSA
  600 CONTINUE
C
      IF(IT.LT.ITERN) GO TO 800
C
C SET NEW JOINT SHEAR STIFFNESS
C
      KJSNU(M) = (TORM-RESID(M,1))/UO
      IF(DELV) 700,800,750
C
C JOINT CLOSING - NEW JOINT NORMAL STIFFNESS
C
  700 KJNNU(M) = FNIM/(VM*L)
      GO TO 800
C
C JOINT OPENING - SET NEW JOINT NORMAL STIFFNESS
C
  750 KJNNU(M) = FNOM/((VM-DELV)*L)
C
  800 CONTINUE
C
      IF (INTER .NE. 0)  GO TO 900
      WRITE(6,994) (KJS(I),KJN(I),I=1,NELEM)
  994 FORMAT (19H0 JOINT STIFFNESSES/(2X,10E12.4))
C
      WRITE(6,996)  (SIGNO(I,1),SIGNO(I,2),SIGNO(I,3),I=1,NELEM)
  996 FORMAT (38H0 INITIAL STRESSES AT END OF INCREMENT/
     1         2(6X,5HSHEAR6X,6HNORMAL6X,6HMOMENT11X)/
     2         (2X,3E12.4,10X,3E12.4))
C
      WRITE (6,995)  (R(I), I=1,NFREE)
  995 FORMAT (25H0 INCREMENTAL LOAD VECTOR/(2X,10E12.4))
C
  900 RETURN
C
 2000 FORMAT (1H025X,19HJOINT DISPLACEMENTS16X,14HJOINT STRESSES/7X,
     17HELEMENT5X,5HSHEAR6X,6HNORMAL5X,8HROTATION6X,5HSHEAR6X,6HNORMAL6X
     2,6HMOMENT)
 2001 FORMAT (I10,2X,6E12.4)
C
      END
```

appendix TWO
conversion factors

Abbreviations are shown in parentheses; $M = 10^6$; $K = 10^3$

LENGTH

1 inch (in)	=	25.4 millimeters (mm)
1 foot (ft)	=	0.3048 meters (m)
1 mile	=	1.60934 kilometers (Km)

AREA

1 in^2	=	6.4516×10^{-4} m^2
1 ft^2	=	0.092903 m^2
1 acre	=	4,046.86 m^2

VOLUME

1 fluid ounce (U.S.)	=	2.95735×10^{-5} m^3
1 liter	=	1×10^{-3} m^3
1 U.S. gallon	=	3.7854×10^{-3} m^3
1 U.K. gallon	=	4.5461×10^{-3} m^3

FLOW RATE

1 ft^3/sec	=	0.028317 m^3/sec
1 U.S. gallon/min	=	6.30902×10^{-5} m^3/sec

FORCE

1 dyne	=	1×10^{-5} newtons (N)
1 ounce	=	0.278014 N
1 pound (lb)	=	4.448222 N
1 kilogram (Kg)	=	9.80665 N
1 U.S. ton	=	8.89644 KN

PRESSURE

1 pound/foot^2 (psf)	=	47.88026 N/m^2
1 foot of water (at 60 degrees F)	=	2.9861 KN/m^2
1 pound/inch^2 (psi)	=	6.89476 KN/m^2
1 Kg/cm^2	=	0.0980665 MN/m^2
1 bar	=	0.1 MN/m^2

Note: 1 MN/m^2 = 1 MPa (MegaPascal) = 145.037 psi

For rough calculations, 1 bar = 1 Kg/cm^2 = 1 ton/ft^2

references

A list of abbreviations used follows the list of references. The numbers to the left of each entry identify the chapters to which the article is most pertinent. General references are assigned number 1. Government documents (U.S.) are available from NTIS, Springfield, Va. 22151; the AD number identifies the document for NTIS.

2 Aastrop, A., and Sallstrom, S.(1964), "Further treatment of problematic rock foundation at Bergeforsen Dam" Proc. 8th Cong. on Large Dams, Edinburgh, p 627

8 Agarwal, R.K., and Boshkov, S.H. (1969), "Stresses and displacements around a circular tunnel in a three layer medium" IJRM&MS, V6 n6

3 Arnold, K.J. (1941), "On spherical probability distributions" PhD Thesis, M.I.T.

5 Ashby, J. (1971), "Sliding and toppling modes of failure in models and jointed rock slopes" MsC Thesis, Imperial College, London

3,4 Badgeley, P.C. (1959), "Structural problems for the exploration geologist" (Harper)

8 Baker, L.E., Sandhu, R.S., and Shieh, W.Y. (1969), "Application of elasto-plastic analysis in rock mechanics by the finite element method, Proc. 11th Symp. on Rock Mech., p237

6 Banks, D.C., and Strohm, W.E. (1974), "Calculations of rock slide velocities" Proc. 3rd Cong. ISRM, Denver, V2B, p839

4 Barr, D.J. (1969) "Use of side looking air-borne radar (SLAR) imagery for engineering soils studies", U.S. Army Eng. Topog. Lab., Ft. Belvoir, Va., Tech. Report 46-TR (AD 701-902)

3 Barton, C.M. (1974), "Simplified procedures for the vector summation and statistical analysis of spherically distributed point clusters" CSIRO (Australia) Div. of Appl. Geomechanics, Tech. Rep. n20

7 Barton, N.R. (1971a),"A model study of the behaviour of steep excavated rock slopes" PhD thesis, Imperial College, London

5 Barton, N.R. (1971b), "A relationship between joint roughness and joint shear strength" Proc. Int. Symp. on Rock Fracture, Nancy, (ISRM), paper 1-8

5 Barton, N.R. (1972), "A model study of rock-joint deformation" IJRM&MS V9 n5

5 Barton, N.R. (1974a), "A review of the shear strength of filled discontinuities in rock" Saertrykk, Fjellsprengningsteknikk Bergmekanikk, 1973,(TAPIR, Norway) - Broch, Heltzen, and Johannesen, ed. Chapter 19

5 Barton, N.R. (1974b), "Estimating the shear strength of rock joints" Proc. 3rd Cong. ISRM, Denver, V2A, p219

2 Barton, N., Lien, R., and Lunde, J. (1975), "Analysis of rock quality and support practise in tunneling and a guide for estimating support requirements" Rock Mech. (in press)

8 Baudendistel, M. (1972), "Interaction between the tunnel lining and the surrounding rock" (in German), Veroeffentlichungen Inst. Bodenmechanik und Felsmechanik, Univ. Fredericana, Karlsruhe

8 Baudendistel, M., Malina, H,, and Müller, L. (1970), "The effect of the geologic structure on the stability of an underground power house" (in German), Proc. 2nd Cong. ISRM, Belgrade, V2, paper 4-56

1 Bellier, J. (1967), "Le Barrage de Malpasset", Travaux, July

8 Benson, R.P., Sigvaldason, O.T., and Kierans, T.W. (1970), "In-situ and induced stresses at the Churchill Falls underground power house, Labrador" Proc. 2nd Cong. ISRM, Belgrade, V2, paper 4-60

4 Berents, H.P. (1961), "A retractable triple tube core barrel" Snowy Mtn Hydroelectric Authority, Cooma, Australia

2 Bergh-Christensen, J., and Selmer-Olsen, R. (1970), "On the resistance to blasting in tunnels" Proc. 2nd Cong. ISRM, Belgrade, V3, paper 5-7

1,2 Bernaix, J. (1966), "Contribution a l'étude de la stabilité des appuis de barrages, Etude géotechnique de la roche de Malpasset" PhD Thesis, Ecole Polytechnique, Paris (Dunod)

2 Bernaix, J. (1969), "New laboratory methods of studying the mechanical properties of rocks" IJRM&MS V6, p43

5 Bernaix, J. (1974), "Properties of rock and rock masses" Proc. 3rd Cong. ISRM, Denver, V1A, p9

8 Best, B.S. (1970), "An investigation into the use of finite element methods for analysing stress distributions in block jointed rock masses" PhD Thesis, James Cook Univ. of N. Queensland, Townsville, Australia

8 Bhattacharya, K., and Boshkov, S.H. (1970), "Determination of the stresses and displacements in slopes by the finite element method" Proc. 2nd Cong. ISRM, Belgrade, V3, paper 7-10

2 Bieniawski, Z.T. (1974a), "Geomechanics classification of rock masses and its application in tunneling" Proc. 3rd Cong. ISRM, Denver V2A, p27

2 Bieniawski, Z.T. (1974b), "Engineering classification of jointed rock masses, Author's reply to discussion", Trans. S.Afr. Inst. of Civil Eng., July

5 Bishop, A.W. (1966), "Soils and soft rocks as engineering materials" Innaugural Lecture, Imperial College, London

5 Bishop, A.W. (1973), " The influence of an undrained change in stress on the pore pressure in porous media of low compressibility" Geotechnique, V23, p435

2 Bjerrum, L. (1967) "Mechanism of progressive failure in slopes of over-consolidated plastic clay and clay shales" the Third Terzaghi Lecture, J. SM&FD, ASCE V93, p3

2 Bjerrum, L., Brekke, T.L., Moum, J., and Selmer-Olsen, R. (1963), "Some Norwegian studies and experiences with swelling materials in rock gouges" Rock Mech. and Eng. Geol. V1, p23

8 Blake, W. (1966), "Applic. of the finite element method of analysis in solving boundary value problems in rock mechanics" IJRM&MS, V3, p169

8 Blake, W. (1968), "Finite element model study of slope modification at the Kimbley pit", Trans. SME of AIME, B241, p525

8 Blake, W. (1971), "Rockburst research at the Galena mine, Wallace,Idaho U.S.B.Mines Tech. Prog. Rep. 39

4 Blyth, F.G.H. (1965), "Geologic maps and their interpretation" (Edward Arnold)

4 Bolstad, D.D. and Mahtab, M.A. (1974), "A Bureau of Mines direct reading azimuth protractor", U.S.B.Mines Inf. Circ. 8617

5 Bowden, F.P., and Tabor, D. (1965), "The friction and lubrication of solids" (Clarendon Press)

2 Bradley, D.E. (1954), "Replica techniques" Brit. J. of Appl. Physics, V5, p165

5 Brawner, C.O., Pentz, D.L., and Sharp, J.C. (1972), "Stability studies of a footwall slope in layered coal deposit" Proc. 13th Symp. on Rock Mech. (ASCE) p329

5 Bray, J.W. (1967), "A study of jointed and fractured rock" Rock Mech. and Eng. Geol., V5, n2 and n3

2,5 Brekke, T.L. and Howard, T.R. (1972), "Stability problems caused by seams and faults" Proc. 1st Rapid Excav. and Tunneling Conf. (AIME) V1, p25

2,5 Brekke, T.L., and Howard, T.R. (1973), "Functional classification of gouge material from seams and faults as related to stability problems in underground openings" Report from Univ. of Cal. to U.S. B. Mines, (AD-766-046)

2 Broch, E., and Franklin, J.A. (1972), "The point load strength test" IJRM&MS, V9, p669

5,7 Brown, E.T. (1968), "The influence of planar discontinuities on the shear strength of a rock-like material" PhD thesis, James Cook University of N. Queensland, Townsville, Australia

5,7 Brown, E.T. (1970a), "Modes of failure in jointed rock masses" Proc. 2nd Cong. ISRM, Belgrade, V2, paper 3-42

5 Brown, E.T. (1970b), "Strength of models of rock with intermittent joints" J.SM&FD, ASCE, V96, p1935

8 Brown, R.E. (1968), "A multi-layered finite element model for predicting mine subsidence" PhD thesis, Carnegie Mellon Univ., Pittsburgh, Pa.

2 Brune, G. (1965), "Anhydrite and gypsum problems in engineering geology" Engineering Geology, Bull. A.E.G., V2, n1

3 Bucher, W.H. (1944), "The stereographic projection, a handy tool for the practical geologist" Jour. Geol. V52, p191

8 Burman, B.C. (1971), "A numerical approach to the mechanics of discontinua", PhD Thesis, James Cook Univ. of N. Queensland, Townsville, Australia

2 Burton, A.N. (1965), "Discussion to a paper by D. Coates", IJRM&MS, V1, p105

5 Byerlee, J.D. (1967a), "Frictional characteristics of granite under high confining pressure", J. Geophysical Res. V 72, p3639

5 Byerlee, J.D. (1967b), "Theory of friction based on brittle fracture" J. Applied Physics, V38, p2928

8 Byrne, R.J. (1974), "Physical and numerical models in rock and soil slope stability" PhD Thesis, James Cook Univ. of N. Queensland, Townsville, Australia

8 Byskov, E. (1970), "The calculation of stress intensity factors using the finite element method- with cracked elements" Int. J. of Fracture Mech. V6, p159

8 Call, R.O. (1971), "Slope stability study of the Tazadit pit, Mauritania" Proc. 2nd Int. Conf. on Stability for Open Pit Mines, Vancouver (AIME)

8 Carnahan, B., Luther, H., and Wilkes, J. (1969) "Applied numerical methods" (John Wiley)

4 Cassines, R. (1972), "Remote sensing - - an evaluation of its impact on earth sciences" Geophysical Prospecting, V20, p142

2,4 Caterpillar Tractor Co. (1966), "Handbook of ripping, a guide to greater profits" Third edition (Peoria, Illinois))

8 Chang, C.Y. and Nair, K. (1974) "Development and application of a general computer program for evaluating stability of openings in rock" Proc. 3rd Cong. ISRM, Denver, V2B, p981

5,7 Chappel, B.A. (1972), "The mechanics of blocky material" PhD thesis, Australian National Univ., Canberra

7,8 Chappel, B.A. (1974), "Numerical and physical experiments with discontinua" Proc. 3rd Cong. ISRM, Denver, V2A, p118

5 Chenevert, M.E. and Gatlin, C. (1965), "Mechanical anisotropies of laminated sedimentary rocks" SPE Journal (AIME), V5, p67

8 Christiansen, L.M., Misterek, D.L., and Bowles, G.F. (1971), "Foundation analysis of Auburn dam site" Proc. Int. Symp. on Rock Fracture, Nancy (ISRM) paper 2-26

8 Clough, R.W. (1960), "The finite element method in plane stress analysis", Proc. 2nd ASCE Conf. on Electronic Computation, p345

8 Clough, R.W. (1965), "The finite element method in structural analysis", Chapter 7 in Stress Analysis, Zienkiewicz and Holister, eds.

4 Cluff, L.S., and Slemmons, D.B. (1971), "Wasatch fault zone features defined by low sun angle photography" in Environmental Geology of the Wasatch Front, Utah Geol. Assoc. Pub. n1 (Utah Geol. and Mineralogical Survey) pG1

2 Coates, D.F. (1964), "Classification of rocks for rock mechanics" IJRM&MS, V1, p421

1 Coates, D.F. (1967), "Rock mechanics principles" Canadian Department of Energy, Mines, and Resources, Monograph 874

2 Coates, D.F. and Parsons, R.C. (1966), "Experimental criteria for classification of rock substances" IJRM&MS V3, p181

4 Colvocoresses, A.P. (1970), "ERTS - a satellite imagery" Photogrammetric Engineering, V36, p555

2 Cording, E.J. and Mahar, J.W. (1974), "The effect of natural geological discontinuities on behavior of rock in tunnels" Proc. 2nd Rapid Excav. and Tunneling Conf. (AIME) V1, p107

4 Cornea, I, and Enescu, D. (1965), "Seismic prospecting of faults by means of apparatus located on a radial profile" (in French), Revue Roumaine Geol., Geog., and Geoph, V4, p191

5 Coulson, J.H. (1972), "Shear strength of flat surfaces in rock" Proc. 13th Symp. on Rock Mech. (ASCE) p77

4 Cumming, J.D. (1956), "Diamond drill handbook" (J.K. Smit, Toronto)

8 Cundall, P. (1971), "A computer model for simulating progressive large scale movements in blocky rock systems" Proc. Int. Symp. on Rock Fracture, Nancy (ISRM), paper 2-8

8 Cundall, P.A. (1974), "A computer model for rock mass behavior using interactive graphics", U.S. Army, Corps of Engineers, Technical Report MRD 2-74 (Missouri River Division)

7 Currie, J.B. (1966), "Experimental structural geology", Earth Science Reviews, V1, p51

8 Dahl, H.D. (1972), "Two and three dimensional elastic and elasto-plastic analyses" Proc. 5th Int. Strata Control Conf., London

8 Dahl, H.D., and Voight, B. (1969), "Isotropic and anisotropic plastic yield associated with cylindrical underground excavations", Proc. Int. Symp. on Large Permanent Underground Openings, Oslo (ISRM)

8 Dahl, H.D., and Parsons, R.C. (1972), "Ground control studies in the Humphrey no. 7 mine", Trans. Soc.Min. Eng of AIME, V252, p211

2,4 Dearman, W.R., and Fookes, P. (1972), "Engineering geological mapping in civil engineering practise" Qtly J. of Eng. Geol, V5, n4

4 De. Chambrier, P. (1953), "The microlog continuous dipmeter", Geophysics, V18, p929

2,4 Deere, D.U. (1963), Tech. description of rock cores for eng. purposes", Rock Mech. and Eng. Geol., V1, p18

2,4 Deere, D.U. (1968), "Geological considerations" Chapter 1 in Rock Mech. in Eng. Practise, Stagg and Zienkiewicz, eds. (John Wiley)

1 Deere, D.U., Hendron, A.J., Patton, F.D., and Cording, E.J., (1967), "Design of surface and near surface construction in rock", Proc. 8th Symp. on Rock Mech. (AIME), p237

2,5 Deere, D.U., and Miller, R.P. (1966), "Eng. classification and index properties for intact rock" Air Force Weapons Lab Tech Rep. AF WL-TR-65-116

2 Deere, D.U., and Patton, F.D. (1971), "Slope stability in residual soils", Proc. 4th Pan American Conf. on Soil Mechanics and Foundation Engineering, San Juan, p87

8 Dejean, M., Martin, F., and Raffoux, J.F. (1970), "Deformations of rock galleries during driving" (in French), Proc. 2nd Cong. ISRM, Belgrade, V2, paper 4-36

8 de la Cruz, R.V., and Goodman, R.E. (1969), "The borehole deepening method of stress measurement", Proc. Int. Symp. on Det. of Stresses in Rock Masses, Lisbon (ISRM) p.230

4 Denny, C.S., Warren, C.R., Dow, D.H., and Dale, W.J. (1968), "A descriptive catalog of selected aerial photographs of geologic features in the United States", U.S.G.S. Prof. Paper 590

8 deRouvray, A.L., and Goodman, R.E. (1972), "Finite element analysis of crack initiation in a block model experiment" Rock Mech., V4, p203

8 Desai, C.S., and Reese, L.C. (1970), "Stress-deformation and stability analyses of deep boreholes", Proc. 2nd Cong. ISRM,(Belgrade, V2, paper 4-13

8 Desai, C.S., and Abel, J.F. (1972), "Intro. to the finite element method" (Van Nostrand Reinhold)

4 Dobrin, M.B. (1960), "Intro. to geophysical prospecting, 2nd edition" (McGraw Hill)

8 Dolcetta, M. (1971), "Problems with large underground stations in Italy", Proc. Symp. on Underground Rock Chambers, (ASCE), p243

8 Dolcetta, M. (1972), "Rock load on the support structures of two large underground hydroelectric power plants", Proc. Int. Symp on Underground Openings, Lucerne (ISRM)

5 Donath, F.A. (1964), "Strength variation and deformational behavior in anisotropic rock" in State of Stress in the Earth's Crust, W.R. Judd, ed. (Elsevier), p281

5 Donath, F.A. (1968), "Experimental study of kink-band development in Martinsburg slate", Proc. Conf. on Research in Tectonics, Geol. Survey of Canada, Paper 68-52, p255

3,4 Donn, W.L., and Shimer, J.A. (1958), "Graphic methods in structural geology" (Appleton-Century-Crofts)

8 Dorn, W.S., and McCracken, D.D. (1972), "Numerical methods with Fortran IV case studies" (Wiley)

2 Douglas, P., and Voight, B. (1969), "Anisotropy of granites: a reflection of microscopic fabric", Geotechnique, V19, p376

8 Dubois, J.L. (1972), "The hyperplane perturbation method for the analysis of non-linearities", PhD Thesis, Univ. of California, Berk.

2 Duncan, N. (1969), "Engineering geology and rock mechanics", (Hill Books, Int. Textbook Co., London, S.W. 1) 2 volumes.

8 Duncan, J.M., and Chang, C.Y. (1970), "Non-linear analysis of stress and strain in soils", J.SM&FD,ASCE, V96, nSM5

4 Eaton, G.P., Martin, N.W., and Murphy, M.A. (1964), "Applic. of gravity measurements to some problems in engineering geology", Engin. Geology, Bull. A.E.G., V1, p6

2 Ege, J.R. (1968), "Stability index for underground structures in granitic rock", Geol. Soc. America Memoir 110

7 Einstein, H.H., Nelson, R.A., Bruhn, R.W., and Hirschfeld, R.C., (1970), :Model studies of jointed rock behavior", Proc. 11th Symp. on Rock Mech. (AIME), p83

5 Engelder, J.T. (1974), "Coefficients of friction for sandstone sliding on quartz gouge", Proc. 3rd Cong. ISRM, Denver, V2A, p499

8 Ergatoudis, I, Irons, B.M., and Zienkiewicz, O.C., (1968),"Curved isoparametric quadrilateral elements for finite element analysis", Int. J. of Solids and Structures, V4, n1, p31

7 Erguvanli, K.A., and Goodman, R.E. (1972), "Applic. of models to engineering geology for rock excavations", Bull. A.E.G., V9, p89

5 Eurenius, J., and Fagerstrom, H. (1969), "Sampling and testing of soft rock with weak layers", Geotechnique, V19, p133

5 Evdokimov, P.D., and Sapegin, D.D. (1967), "Stability, shear, and sliding resistance and deformation of rock foundations", (Israel Program for Scientific Translations, Jerusalem)

5 Evdokimov, P.D., and Sapegin, D.D. (1970), "Large scale field shear test on rock", Proc. 2nd Cong. ISRM, Belgrade, V2, paper 3-17

5,8 Fairhurst, C. (1964), "On the validity of the Brazilian test for brittle materials, IJRM&MS, V1, p535

4 Fanshawe, H.G., and Watkins, M.D. (1971), "A sparker survey used to select a marine dam site in Hong Kong", Qtly J. of Eng. Geol, V4,p25

2 Farran, J. (1950),"Etude pétrographique du comportement des roches dans leurs applications en Genie Civil" Bull. Société Hist. Nat. Toulouse, V85, p331

5 Fecker, E. (1970), "Geologische Kartierung des Gebietes nordwestlich von Neustadt/Weinstrosse sowie Bau und Anwendung eines Profilographen", Diplomarbeit Univ. Karlsruhe (Inst. fur Bodenmechanik und Felsmechanik)

2,5 Fecker, E., and Rengers, N. (1971), "Measurement of large scale roughness of rock planes by means of profilograph and geological compass", Proc. Int. Symp. on Rock Fracture, Nancy (ISRM) paper 1-18

2 Feld, J. (1966), "Rock as an engineering material" (Soil Test)

3 Fisher, R.A. (1953), "Dispersion on a sphere" Proc. Royal Soc. London, Ser. A, V217, p295

2 Fookes, P.G., Dearman, W.R., and Franklin, J.A. (1971), "Some engineering aspects of rock weathering with field examples from Dartmoor and elsewhere" Qtly J. of Eng. Geol., V4, n3

2 Fookes, P.G., and Horswill, P. (1970), "Discussion of engineering grade zones", Proc. Conf. on In-situ Testing of Soils and Rock, (Inst. of Civil Eng., London) p53

2 Fookes, P.G., and Parrish, D.G. (1969), "Observations on small scale structural discontinuities in the London Clay and their relationship to regional geology" Qtly J. of Eng. Geol., V1, p217

2 Franklin, J.A. (1970), "Observations and tests for engineering description and mapping of rocks", Proc. 2nd Cong ISRM, Belgrade, V1, paper 1-3

2 Franklin, J.A., Broch, E., and Walton, G. (1971), "Logging the mechanical character of rock", Trans. IMM, Sect. A, V81, p43

2 Franklin, J.A., and Chandra, R. (1972), "The slake durability test", IJRM&MS, V9, p325

7 Fumagalli, E. (1968), "Model simulation of rock mechanics problems" in Rock Mechanics in Eng. Practise, Zienkiewicz and Stagg, ed., (John Wiley)

7 Fumagalli, E. (1973), "Statical and geomechanical models" (Springer Verlag)

8 Gale, J. (1975), "A numerical, field, and laboratory study of flow in rocks with deformable fractures", PhD Thesis, Univ. of California, Berkeley

6 Gallico, A., and John, K.W. (1974), "Graphical stability analyses for rock abutments of an arch dam" Proc. 3rd Cong. ISRM, Denver, V2B, p884

2 Gamble, J. (1971), Durability - plasticity classification of shales and other argillaceous rocks" PhD Thesis, Univ. of Illinois

4 Gardner, M.E., and Johnson, C.G. (1971), "Engineering geologic maps for regional planning"in Environmental Planning and Geology (U.S.G.S. Off. of Res. and Tech., Washington, D.C.) p154

2,4 Geological Society of London Engineering Group Working Party (1970), "The logging of rock cores for engineering purposes", Qtly J. Eng. Geol. V3, p1

4 Geological Society of London Engineering Group Working Party (1972), "The preparation of maps and plans in terms of engineering geology" Qtly J. Eng. Geol, V5, n4 (W.R. Dearman, P.G. Fookes, E.G. Smith, and others)

4 Geyer, R.L., and Myung, J.I. (1971), "The 3-D velocity log: a tool for in-situ determination of the elastic moduli of rocks", Proc. 12th Symp. on Rock Mech. (AIME)

8 Ghaboussi, J., Wilson, E., and Isenberg, J. (1973), "Finite element analysis for rock joints and interfaces", J.SM&FD ASCE, V99, p833

8 Golden, A.L., and Troitsky, A.P. (1974), "Behaviour of the rock mass at the site of a rock fill dam", Proc. 3rd Cong. ISRM, Denver, V2B, p896

6 Goodman, L.E., and Warner, W.H. (1963), "Statics" (Wadsworth, Belmont, Cal.)

8 Goodman, R.E. (1966), "On the distribution of stresses around tunnels in non-homogeneous rocks", Proc. 1st Cong. ISRM, Lisbon V2, p249

5,8 Goodman, R.E. (1970), "The deformability of joints", in Determination of the In-situ Modulus of Deformation of Rock - A.S.T.M. STP 477, p174

5,7 Goodman, R.E. (1972), "Geological investigations to evaluate stability", Proc. 2nd Int. Conf. on Stability for Open Pit Mines, Vancouver (AIME)

5 Goodman, R.E. (1974), "The mechanical properties of joints", Proc. 3rd Cong. ISRM, Denver, V1A, p127

5,8 Goodman, R.E. and Dubois, J. (1972), "Duplication of dilatancy in analysis of jointed rocks", J.SM&FD, ASCE V98, p399

1 Goodman, R.E. and Duncan, J.M. (1971), "The role of structure and solid mechanics in the design of surface and underground excavations in rock", Proc. Conf. on Structure, Solid Mechanics, and Engineering Design, Part 2, paper 105, p1379 (John Wiley)

5 Goodman, R.E. and Ohnishi, Y. (1973), "Undrained shear testing of jointed rock", Rock Mechanics, V5, p129

8 Goodman, R.E. and St. John, C. "Static finite element analysis of jointed rock", in Numerical Methods in Geotech. Eng.,Christian and Desai, eds. (McGraw-Hill), in press

6 Goodman, R.E., and Seed, H.B. (1966), "Earthquake induced displacements in sand embankments", J.SM&FD ASCE, V92, p125

6,8 Goodman, R.E. and Taylor, R.L. (1967), "Methods of analysis for rock slopes and abutments: a review of recent developments" Proc. 8th Symp.on Rock Mechanics (AIME), p303

8 Goodman, R.E. Taylor, R.L., and Brekke, T.L. (1968), "A model for the mechanics of jointed rock", J.SMFD ASCE, V94, p637

4 Goodman, R.E., Van, T.K., and Heuze, F. (1972), The measurement of rock deformability in boreholes", Proc. 10th Symp. on Rock Mech., (AIME), p523

4 Grant, F.S., and West, G.F. (1965), "Interpretation theory in applied geophysics", (McGraw-Hill)

2 Griffith, J.H. (1937), "Physical properties of typical American rocks", Bull. Iowa Eng. Experiment Sta. n131, p43

4 Griffiths, D.H., and King, R.F. (1965), "Applied geophysics for engineers and geologists" (Pergamon Press)

5 Grishin, M.M., and Evdokimov, P.D. (1961), "Shear strength of structures built on rock", Proc. 5th Int. Conf. on Soil Mech. and Found. Eng., Paris, p649

8 Grob, H., Kovari, K., and Vannotti. (1970), "Practical applications of the finite element method to the stability of tunnels" in German, Proc. 2nd Cong. ISRM, Belgrade, V2, paper 4-69

4 Guyod, H., and Shane, L.E. (1969), "Geophysical well logging, Volume I" (Hubert Guyod, Houston, Texas)

2 Habib, P., and Bernaix, J. (1966), "La fissuration des roches", Proc. 1st Cong. ISRM, Lisbon, V1, p185

4 Hackman, R.J. (1956), "The stereoslope comparator as an instrument for measuring angles of slopes in stereoscopic models", Photogrammetric Eng., V22, n5

2 Hamrol, A. (1961), "A quantitative classification of the weathering and weatherability of rocks", Proc. 5th Int. Conf. on Soil Mech. and Foundation Eng, Paris, V2, p771

8 Han, Chin Yuen (1972), "The technique of obtaining equipotential

lines of groundwater flow in slopes using electrically conductive paper", MSc Thesis, Imperial College, London.

5 Handin, J., and Stearns, D.W. (1964), "Sliding friction of rock", Trans. Amer. Geophysical Union, V45, p103

4 Harbaugh, J.W., Bonham-Carter, G., Graeme, (1970), "Computer simulation in geology" (Wiley-Interscience)

5 Haverland, M.L., and Slebir, E.J., (1972), "Methods of performing and interpreting in-situ shear tests", Proc. 13th Symp. on Rock Mech., (ASCE), p107

2 Hawkes, I., and Mellor, M. (1970), "Uniaxial testing in rock mechanics laboratories", Engineering Geology, V.4, n3

5 Hayashi, M. (1966), "Strength and dilatancy of brittle, jointed (rock) masses - - the extreme value stochastics and anisotropic failure mechanism", Proc. 1st Cong. ISRM, Lisbon, V1, p295

5 Hayashi, M., and Fujiwara, Y. (1968), "A mechanism of anisotropic dilatancy and shear failure of laminately jointed rock masses", Tech. Rep. C67006, Central Res. Inst. of Electric Power Industry, Tokyo

8 Hayashi, M., and Hibino, S.(1968), "Progressive relaxation of rock masses during excavation of an underground cavity", Proc. Int. Symp. on Rock Mechanics, Madrid (ISRM)

5 Hayashi, M., and Kitihara, Y. (1970), "Anisotropic dilatancy and strength of jointed rock masses, and stress distribution in fissured rock masses", in Rock Mechanics in Japan, V1, p85 (Japan Soc. of Civil Engineers)

4 Heiland, C.A. (1946), "Geophysical exploration" (Prentice Hall)

6 Hendron, A.J. Jr., Cording, E.J., and Aiyer, A.K. (1971), "Analytical and graphical methods for the analysis of slopes in rock masses", U.S.Army Corps of Engineers, Nuclear Cratering Group Rep 36

8 Heuze, F.E., and Goodman, F.E. (1967), "Mechanical properties and in-situ behavior of the Chino limestone, Crestmore mine, Riverside Cal.", Proc. 9th Symp. on Rock Mechanics,(AIME)

8 Heuze, F.E., and Goodman, R.E. (1970), "The design of room and pillar structures in competent, jointed rock, example: the Crestmore mine, Cal.", Proc. 2nd Cong. ISRM, V2, paper 4-41

8 Heuze, F.E., Goodman, R.E., and Bornstein, A. (1971), "Numerical analyses of deformability tests in jointed rock", Rock Mech. V3, p13

6 Heuze, F.E., and Goodman, R.E. (1972),"Three dimensional approach for design of cuts in jointed rock", Proc. 13th Symp. on Rock Mechanics (ASCE), p347

3,4 Higgs, D.V., and Tunell, G. (1959), "Angular relations of lines and planes", (Wm C. Brown, Dubuque, Iowa)

2,5 Hodgson, R.A. (1961), "Classification of structures on joint surfaces", Amer. J. of Sci. V259, p493

6,8 Hoeg, K., and Muraka, R. (1974), "Probabilistic analysis and design of a retaining wall", J. Geotech. Eng. Div., Proc. ASCE, V100, p349

6 Hoek, E. (1970), "Estimating the stability of excavated slopes in open cast mines", Trans. I.M.M., Sect. A, V79, pA109

1 Hoek, E. (1971), "Rock engineering", Innaugural lecture, Imperial College, London

1 Hoek, E., and Bray, J. (1974), "Rock slope engineering" (IMM,Lon)

2 Hoek, E. and Franklin, J.A. (1968), "A simple triaxial cell for field or laboratory testing of rock", Trans. IMM, Sect.A, p22

6 Hoek, E., Bray, J., and Boyd, J. (1973), "The stability of a rock slope containing a wedge resting on two intersecting discontinuities", Qtly J. Eng. Geol., V6, p1

5 Hoek, E., and Pentz, D.L. (1968), "The stability of open pit mines", Rock Mech. Res. Rep. n5, Imperial College, London

8 Hofman, H. (1970), "The deformation process of a regularly jointed discontinuum during the excavation of a cut" (in German), Proc. 2nd Cong. ISRM, Belgrade, V3, paper 7-1

4 Hollingsworth, J. (1974), "Two dimensional structural defect analysis techniques", CSIRO (Australia, Div. of Appl. Geomech. Tech. Rep. 21

5 Horino, F.G., Hoskins, J.R., and Ellickson, M.L. (1968), "A method of measuring surface texture of rock", U.S.B.Mines R.I. 7095

5 Horn, H.M., and Deere, D.U. (1962), "Frictional characteristics of minerals", Geotechnique, V12, p319

5 Hoskins, E.R., Jaeger, J.C., and Rosengren, K.J. (1968), "A medium scale direct friction experiment" IJRM&MS, V5, p143

8 Hoyaux, B., and Ladanyi, B. (1970), "Gravitational sress field around a tunnel in soft ground", Can. Geotech. Jour., V7, p54

2 Iida, R., and others (1970), "Geological rock classification of dam foundations", in Rock Mech. in Japan, V.1, p171 (Japan Soc. Civ Eng)

5,8 Iida, R., and Kobayashi, S. (1974),"Theoretical study on the non-elastic behavior of jointed rock masses" Jour. of Research, Public

Works Res. Inst., Ministry of Construction, V16, p35 (in English)

5 Iida, R., Hojo, K, and Harada, J. (1974), "In-situ tests and theoretical studies on the relations between looseness and deformation characteristics of jointed rock masses" Proc. 3rd Cong. of ISRM, Denver, V2B, p719

2 Iliev, I.G. (1966), "An attempt to estimate the degree of weathering of intrusive rocks from their physico-mechanical properties", Proc. 1st Cong. ISRM, Lisbon, V1, p109

2 Ingram, R.L. (1954), "Terminology for the thickness of stratification and parting units in sedimentary rocks", Bull. G.S.A., V65, p937

3 Irving, E. (1964), "Paleomagnetism and its applications to geological and geophysical problems" (John Wiley)

8 Isenberg, J. (1972), "Analytic modelling of rock structure interaction", Report from Agbabian Assoc. to U.S.B.Mines, (AD 749-373)

4,7 Jack, H. (1969), "Tunnel mapping methods, a discussion of a paper by Cooper" Bull. A.E.G., V6, p151

2 Jacquet, P.A., and Mencarelli, E. (1959), "Technique non-destructive pour l'étude des surfaces de cassure au microscope électronique", Compte Rendu Acad. Sci. Paris, V248, p2477

1 Jaeger, C. (1972), "Rock mechanics and engineering" (Cambridge Univ. Press)

5 Jaeger, J.C. (1959), "The frictional properties of joints in rock", Geofisica Pura e Applicata, V 43, p148

5 Jaeger, J.C. (1960), "Shear fracture of anisotropic rocks", Geol. Magazine, V97, p65

5 Jaeger, J.C. (1971), "Friction of rocks and the stability of rock slopes - Rankine Lecture", Geotechnique, V21, p97

1 Jaeger, J.C., and Cook, N.G.W. (1969), "Fundamentals of rock mechanics", (Methuen)

5 Jaeger, J.C. and Rosengren, K.J. (1969), "Friction and sliding of joints" Proc. Australasian Inst. of Min. and Metallurgy, n229, p93

4 Jakosky, J.J. (1950), "Exploration geophysics" 2nd edition, (Trija)

4 Janke, N.C. (1972),"Field measurements with common equipment", Photogrammetric Eng., V38, p37

3 Jeran, P.W., and Mashey, J.R. (1970), "A computer program for the stereographic analysis of coal fractures and cleats", U.S.B.Mines Inf. Circ. 8454

2,6 John, K.W. (1962), "An approach to rock mechanics" J SM&FD ASCE V88, p1

6 John, K.W. (1968), "Graphical stability analysis of slopes in jointed rock", J.SMFD, ASCE, V94, nSM2

5,7 John, K.W. (1970), "Civil engineering approach to evaluate strength of a regularly jointed rock", Proc 11th Symp. on Rock Mech. (AIME), p69

6 John, K.W. (1970), "Three dimensional stability analysis of slopes in jointed rock" Proc. Johannesburg Symp on Rock Mech, South Africa

5 Jouanna, P. (1972), "Essais de percolation au laboratoire sur des échantillons de micaschiste soumis à des contraintes", Proc. Symp. on Percolation through Fissured Rock, Stuttgart, (ISRM) paper T2-F

5,7 Kandaurov, I.I., Ouvarov, L.A., and Karpov, N.M. (1974), "Contrainte dans les modèles des massifs rocheux fissurés sans poussée horizontale" Proc. 3rd Cong. ISRM, Denver, V2A, p157

8 Kawamoto, T. (1970), "On the states of stress and deformation of natural slopes", in Rock Mechanics in Japan, (Japan Soc. Civ. Eng), V1, p85

4 Keller, G.V., and Frischknecht, F.C. (1970), "Electrical methods in geophysical prospecting" (Pergamon)

2 Kiersch, G.A., and Treasher, R.C. (1955), "Investigations, aerial and engineering geology - Folsom Dam project, Central Cal" Econ. Geol. V50, p271

4 Kiraly, L. (1969), "Statistical analysis of fractures (orientation and density)" Sonderdruck aus der Geologischen Rundschau, Band 59, p125

3 Kirschke, D. (1970), "Die programmiersprache GELI 1 in Fortran IV Version für die Rechenlage IBM 360" Clausthaler Tektonische Hefte 10, p133

4 Kneuper, G. et al. (1967), "New results with a reflection seismic method to locate tectonic faults in coal mining" Bergbauwissenschaften V14, p428

2 Komarnitskii, N.N. (1968), "Zones and planes of weakness in rocks and slope stability" (Consultants Bureau, N.Y.)

7 Korbin, G.E. (1975), "A model study of spiling reinforcement in underground openings" PhD Thesis, Univ. of Cal. Berkeley

4 Krumbein, W.C., and Graybill, F.A. (1965), "An introduction to statistical models in geology" (McGraw-Hill)

5 Krsmanovic, D. (1967), "Initial and residual shear strength of hard rocks", Geotechnique, V17, p145

7 Krsmanovic, D. (1971), "On the results of measurement of stresses and strains in the rock mass of a geostatical model of the Grancarevo Dam" Publ. of the Inst. of Geotechnics and Found. Eng., Faculty of Civil Engineering, Sarajevo, Yugoslavia

5 Krsmanovic, D. and Langof, Z. (1964), "Large scale laboratory tests of the shear strength of rocky material" Rock Mech. and Eng. Geol. Supplement II, p20

7 Krsmanovic, D., and Langof, Z. (1971), "On the results of testing the Geostatical model of the Grancarevo Dam"(see Krsmanovic, D. (1971))

5 Krsmanovic, D., and Popovic, M. (1966), "Large scale tests of the shear strength of limestone", Proc. 1st Cong. ISRM, Lisbon, V1,p773

8 Kruse, G.H. (1971), "Power plant chamber under Oroville Dam", Proc. Symposium on Underground Rock Chambers (ASCE) p333

2 Kruse, G.H., Zerneke, K.L., Scott, J.B., Johnson, W.S., and Nelson, J.S. (1970), "Approach to classifying rock for tunnel liner design", Proc. 11th Symp. on Rock Mech. (AIME), p169

1 Krynine, D., and Judd, W. (1959), "Principles of engineering geology and geotechnics" (McGraw-Hill)

8 Kulhawy, F.H., and Duncan, J.M. (1972), "Stresses and movements in Oroville Dam", JSM&FD ASCE, V98, p653

5 Kutter, H.K. (1971), "Stress distribution in direct shear test samples", Proc. Int. Symp. on Rock Fracture, Nancy, (ISRM), paper 2-6

5 Ladanyi, B., and Archambault, G. (1970), "Simulation of shear behaviour of a jointed rock mass" Proc. 11th Symp. on Rock Mechanics, (AIME) p105

5 Ladanyi, B., and Archambault G. (1972), "Evaluation de la résistance au cisaillement d'un massif rocheux fragmenté" Proc. 24th Int. Geol. Cong. Montreal, Section 13, p249

7 Lajtai, E.Z. (1967), "The influence of interlocking rock discontinuities on compressive strength (model experiments)" Rock Mech. and Eng. Geol V5, n4 p217

5 Lajtai, E.Z. (1969a), "Shear strength of weakness planes in rock" IJRM&MS, V6 p499

5 Lajtai, E.Z. (1969b), "Mechanics of second order faults and tension gashes", Bull. G.S.A., V80, p2253

5 Lane, K.S., and Heck, W.J. (1964), "Triaxial testing for strength of rock joints" Proc. 6th Symp. on Rock Mech., Univ. of Missouri, p98

8 Lane, R.G. (1970), "An investigation into the deformation of a combined dam and powerhouse structure" Proc. 2nd Cong. ISRM, Belgrade

V3, paper 8-2

7 Langhaar, H.L. (1951), "Dimensional analysis and theory of models" (John Wiley)

4 Lattman, L.H., and Ray, R.G. (1965), "Aerial photographs in field geology" (Holt, Rinehart and Winston)

2 Lauffer, H. (1958), "Gebirgsklassifizierung für den Stollenbau" Geologie und Bauwesen, V24, p46

4 Lauterbach, R. (1953), "Mikromagnetik, ein Hilfsmittel geologischer Erkundung" Wiss. Z. Karl Marx Univ. Leipzig, V3, p223

8 Lee, K.L., and Shen, C.K. (1969), "Horizontal movements related to subsidence" J SMFD ASCE V95, p139

4,2 Leet, L.D., and Judson, S.(1971), "Physical geology, 4th edition" (Prentice-Hall)

4 Library of Congress Geography and Map Division (1973), "The bibliography of cartography" (G.K.Hall and Co.)

4 Linkwitz, K. (1963), "Terrestrisch photogrammetrische Kluftmessung" Rock Mech. and Eng. Geol. V1, p153

5 Locher, H.G. (1968), "Some results of direct shear tests on rock discontinuities", Proc. 2nd Int. Symp on Rock Mech., Madrid (ISRM) V171

3,5 Lokin, P. (1974), "Discontinuity anisotropy of rock masses", Proc. 3rd Cong. ISRM, Denver V2A, p174

5 Lombardi, G. and DalVesco, E. (1966), "Die experimentelle Bestimmung der Reibungskoeffizienten für die Felswiderlager der Staumauer Contra (Verzasca)" Proc. 1st Cong. ISRM Lisbon, V1, p571

6 Londe, P. (1965), "Une methode d'analyse à trois dimensions de la stabilité d'une rive rocheuse" Annales des Ponts et Chaussées, n1, p37

1 Londe, P. (1967), "La mécanique des roches - - discipline nouvelle au service de la construction des grands barrages" Science Progres - La Nature, n3384, April (Dunod, Paris)

6 Londe, P., Vigier, G., and Vormeringer, R. (1969), "Stability of rock slopes, a three dimensional study" J SM&FD ASCE, V95,nSM7

6 Londe, P., Vigier, G., and Vormeringer, R. (1970), "Stability of rock slopes, graphical methods" J SM&FD ASCE V96, nSM4

2 Loughnan, F.C. (1969), "Chemical weathering of the silicate minerals" (Elsevier)

4 Louis, C., and Maini, T. (1970), "The determination of hydraulic

parameters in jointed rock" Proc. 2nd Cong. ISRM, Belgrade, V1, paper 1-32

4 Lugeon, M. (1933), "Barrages et geologie", (Dunod, Paris)

2 Lumb, P. (1962), "The properties of decomposed granite", Geotechnique, V12, p226

3 Mahtab, M.A., Bolstad, D.D., Alldredge, J.R., and Shanley, R.J. (1972), "Analysis of fracture orientations for input to structural models of discontinuous rock", U.S.B.Mines R.I. 7669

8 Mahtab, M.A., and Goodman, R.E. (1970), "Three dimensional finite element analysis of jointed rock slopes", Proc. 2nd Cong. ISRM, Belgrade, V3, paper 7-12

4 Maini, Y.N.T. (1971), "In-situ hydraulic parameters in jointed rock, their measurement and interpretation", PhD Thesis, Imperial College, London

8 Malina, H. (1969), "The numerical determination of stresses and deformations in rock taking into account discontinuities" Proc. 19th Coll. on Geomech., Salzburg

8 Markland, J.T., (1972), "A useful technique for estimating the stability of rock slopes when the rigid wedge slide type of failure is expected" Imperial College Rock Mech. Res. Rep. n19

2,5 Martin, G.R., and Millar, P.J. (1974), "Joint strength characteristics of a weathered rock", Proc. 3rd Cong. ISRM, Denver, V2A, p263

2 Martna, J. (1970), "Engineering problems in rocks containing pyrrhotite", Proc. Int. Symp. on Large Permanent Underground Openings, Oslo, Brekke and Jorstad, eds. (ISRM), p87

5 Mathews, K.E. (1970), "Excavation design in hard and fractured rock at the Mount Isa Mine, Australia", MSc. Thesis, Univ. of Queensl.

4 McGregor, K. (1967), "The drilling of rock", (C.R.Books Ltd, London)

2 McKee, E.D., and Weir, G.W. (1953), "Terminology for stratification and cross stratification in sedimentary rocks" Bull. G.S.A., V64, p381

5 McLamore, R., and Gray, K.E. (1967), "The mechanical behavior of anisotropic sedimentary rocks", Trans. A.S.M.E., p62

3 McMahon, B.K. (1967), "Indices related to the mechanical properties of jointed rock", Proc. 9th Symp. on Rock Mech. (AIME), P117

4 Meissner, R. (1961), "Wave front diagrams from uphole shooting", Geophysical Prospecting, V9, p40

4 Mekel, J.F., Savage, J.F., and Zorn, H.C. (1964), "Slope measurements and estimations from aerial photos", Int. Training Centre for Aerial Survey (ITC), Delft, publ. Series B, n26

5 Mencl, V. (1965), "Dilatancy of rocks", Rock Mech. and Eng. Geol. V3, p58

4 Merifield, P.M., et. al. (1969), "Satellite imagery of the earth" Photogrammetric Eng. V35, p654

4 Moffit, F.H. (1959), "Photogrammetry", (International Textbook)

4 Mollard, J.D. (1962), "Photo analysis and intcrpretation in engineering geology investigations, a review" Reviews in Engineering Geology (G.S.A.), Fluhr and Leggett, eds. V1

8 Morgenstern, N.R., and Guther, H. (1972), "Seepage into an excavation in a medium possessing stress dependent permeability", Symp. on Percolation Through Fissured Rock, Stuttgart (ISRM), paper T2-C

5 Morland, L.W. (1974), "Continuum model of regularly jointed mediums", J. Geoph. Res., V79, p357

2 Morlier, P. (1968), "Relation entre la fissuration et la célerité des ondes dans les roches fissurées", Colloque sur la Fissuration des Roches - Revue de L'Industrie Mineral, num. sp. 15 May, p.16

2,4 Moye, D.G. (1955), "Engineering geology for the Snowy Mtn. scheme Australia" J. Inst. of Engineers, Australia, V27, p287

4 Moye, D.G. (1967), "Diamond drilling for foundation exploration", Trans. Inst. of Engineers, Australia, V9, p95

1 Müller, L. (1963), "Der felsbau", volume 1 (F.Enke, Stuttgart)

1 Müller, L. (1964), "The rock slide in Vajont Valley", Rock Mech. and Eng. Geol., V2, p148

7 Müller, L. (1966), "Der progressive Bruch in geklufteten Median" Proc. 1st Cong. ISRM, Lisbon, V1, p679

1 Müller, L. (1968), "New considerations on the Vajont slide", Rock Mech. and Eng. Geol., V6, p1

2 Müller, L., and Hofmann, H. (1970), "Selection, compilation, and assessment of geological data for the slope problem" Proc. Symp. on Planning Open Pit Mines, Johannesburg, S. Afr.

7 Müller, L., and Pacher, F. (1965), "Modellversuche fur Klarung der Bruchgefahr geklufteter Medien", Rock Mech. and Eng. Geol. Supp.II, p7

4 Musgrave, A.W. (1967), "Seismic refraction prospecting", (Soc.

Expl. Geoph., Tulsa)

4 Myung, J., and Baltosser, R.W. (1972), "Fracture evaluation by the borehole logging method", Proc. 13th Symp. on Rock Mech. (ASCE)

2 Nascimento, U. (1970), "O problema da alterabilidades das rochas em engenharia civil", Revista Fomento, V8, n2

5 Nascimento, U., and Teixeira, H. (1971), "Mechanisms of internal friction in soils and rocks", Proc. Symp. on Rock Fracture, Nancy (ISRM), paper 2-3

6 Newmark, N. (1965), "Effects of earthquakes on dams and embankments", Geotechnique, V15, n140

2 Nickelson, R.P., and Hough, V.N.D. (1967), "Jointing in the Appalachian plateau of Pennsylvania" Bull. G.S.A., V78, p609

8 Noorishad, J., Witherspoon, P.A., and Maini, Y.N.T. (1972), "The influence of fluid injection on the state of stress in the earth's crust", Proc. Symp. on Percolation Through Fissured Rocks, Stuttgart, (ISRM), paper T2-H

2 Nur, A. and Simons, G. (1970), "The origin of small cracks in igneous rocks", IJRM&MS, V7, p307

1 Obert, L., and Duvall, W. (1967), "Rock mechanics and the design of structures in rock" (John Wiley)

2 Ollier, C. (1969), "Weathering" (Elsevier)

5 Ohnishi, Y. (1973), "Lab. measurement of induced water pressures in jointed rocks", PhD Thesis, Univ. of California, Berkeley

2 Onodera, T.F. (1963), "Dynamic investigation of foundation rocks in-situ", Proc. 5th Symp. on Rock Mech. (Macmillan), p517

2 Onodera, T.F. (1970), "Activities in rock mechanics in the Japanese Society of Soil Mechanics and Foundation Eng." in Rock Mech. in Japan, V1 (Jap. Soc. Civ. Eng.)

8 Oudenhoven, T.F., Babcock, C.O., and Blake, W. (1972), "A method for the prediction of stresses in an isotropic inclusion or orebody of irregular shape", U.S.B.Mines, R.I. 7645

8 Pariseau, W.G., and Stout, K. (1972), "Open pit mine stability: the Berkeley pit" Proc. 13th Symp. on Rock Mech. (ASCE), p367

8 Pariseau, W.G., Voight, B., and Dahl, H.D. (1970), "Finite element analyses of elastic plastic problems in the mechanics of geologic media", Proc. 2nd Cong. ISRM, Belgrade, V2, paper 3-45

5 Paterson, M.S., and Weiss, L.E. (1966), "Experimental deformation and folding in phyllite", Bull. G.S.A., V77, p343

5 Patton, F.D. (1966), "Multiple modes of shear failure in rock", Proc. 1st Cong. ISRM, Lisbon, V1, p509

5 Paulding, B.W. (1970), "Coefficient of friction of natural rock surfaces", J.SM&FD ASCE, V96m nSM2

2 Pender, M.J. (1971), "Some properties of weathered graywacke" Proc. 1st Australia-New Zealand Conf. on Geomechanics, V1, p423

3 Phillips, F.C. (1946), "An introduction to crystallography", (Longmans, Green)

3 Phillips, F.C. (1971), "The use of stereographic projection in structural geology, 3rd edition" (Edward Arnold)

8 Phukan, T.A., Lo, K., and La Rochelle, P. (1969), "Stresses and deformations of vertical slopes in elasto-plastic rocks", Proc. 11th Symp. on Rock Mech. (AIME), p193

3 Pincus, H.J. (1953), "The analysis of aggregates of orientation data in the earth sciences", J. of Geol., V61, p482

4 Pincus, H.J. (1973), "A modified pocket transit for measuring strike on the underside of surfaces", IJRM&MS, V10, p83

2,5 Piteau, D.R. (1973), "Characterizing and extrapolating rock joint properties in engineering practice", Rock Mech. Supp. II, p5

4 Polak, E.J. (1963), "The measurements of, relation between, and factors affecting the properties of rocks", Proc. 4th Australia, New Zealand Conf. on Soil Mech. and Found. Eng., p220

4 Polak, E.J. (1971), "Seismic attenuation in engineering site investigations", Proc. 1st Australia, New Zealand Conf. on Geomechanics Melbourne, V1, p430

5 Pratt, H.R., Black, A.D., and Brace, W.F. (1974), "Friction and deformation of jointed quartz diorite", Proc. 3rd Cong. ISRM, Denver, V2A, p306

2,4 Price, N. (1966), "Fault and joint development in brittle and semi-brittle rock" (Pergamon Press)

4 Ragan, D.M. (1973), "Structural geology - - an introduction to geometrical techniques, 2nd edition", (John Wiley)

5 Raleigh, C.B., and Paterson, M.N. (1965), "Experimental deformation of serpentine and its tectonic implications", J. Geoph. Res. V70, p3965

4 Ramsay, J.G. (1967), "Folding and fracturing of rocks" (McGraw-Hill)

4 Ray, R.G. (1960), "Aerial photographs in geologic interpretation

and mapping", U.S.G.S. Prof. Paper 373

2 Reichmuth, D.R. (1963), "Correlation of force-displacement data with physical properties of rock for percussive drilling systems", Proc. 5th Symp. on Rock Mech. (Pergamon), p49

4 Rengers, N. (1967), "Terrestrial photogrammetry: a valuable tool for engineering geological purposes" Rock Mech. and Eng. Geol., V5, p150

2,5 Rengers, N. (1970), "The influence of surface roughness on the friction properties of rock planes", Proc. 2nd Cong. ISRM, Belgrade, V1, paper 1-31

5 Rengers, N. (1971), "Unebenheit und Reibungswiderstand von Gesteinstrennflächen", Veröffentlichungen des Inst. fur Bodenmechanik und Felsmechanik der Univ. Fredericanna in Karlsruhe, V47

8 Rescher, O.J. (1968), "Rock reinforcement of the cavern at Veytaux by rock anchors and pneumatically applied concrete"(in German) Rock Mech. and Eng. Geol., Supp. 4, p216

8 Reyes, S.F. and Deere, D.U. (1966), "Elastic plastic analysis of underground openings by the finite element method", Proc. 1st Cong. ISRM, Lisbon, V2, p477

2,5 Richards, L.R. (1972), "Classification and weathering of near surface, jointed rock" MsC Thesis, Imperial College

5 Richards, L.P. (1973), "Strength properties of Delabole slates", Imperial College Rock Mech. Res. Rep. n22

5 Ripley, C.F. and Lee, K.L. (1961), "Sliding friction tests on sedimentary rock specimens", Trans. 7th Int. Cong. on Large Dams, V4, p657

4 Robinson, J.B., Brown, N.P., and Benns, F.T. (1967), "Drilling equipment for difficult coring conditions; new type core lifter and triple tube core barrel", Proc. Austalasian Inst. of Min. and Metal., n224, p37

5 Robinson, L.H., and Holland, W.E. (1969), "Some interpretation of pore fluid effects in rock failure", Proc. 11th Symp. on Rock Mech. (AIME), p585

4,5 Rocha, M. (1971), "A new method of integral sampling of rock masses", Rock Mech. V3, p1

4 Rocha, M. (1974), "Recent possibilities of studying foundations of concrete dams", Proc. 3rd Cong. ISRM, Denver, V1A, p879

4,5 Rocha, M., and Barroso, M. (1971), "Some applications of the new integral sampling technique", Proc. Int. Symp. on Rock Fracture, Nancy (ISRM), paper 1-21

4 Rocha, M., DaSilveira, A., et. al. (1970), "Characterization of the deformability of rock masses by dilatometer tests", Proc. 2nd Cong. ISRM, Belgrade, V1, paper 2-32

8 Rodatz, W., and Wittke, W. (1972), "Interaction between deformation and percolation in fissured, anisotropic rock" (in German), Proc. Symp. on Percolation Through Fissured Rock, Stuttgart (ISRM) paper T2-I

3 Rodgers, J. (1952), "The use of equal area or other projections in the statistical treatment of joints", Bull. G.S.A., V.63, p427

5 Romero, S. (1968), "In-situ direct shear tests on irregular surface joints filled with clayey material", Proc. Int. Symp. on Rock Mech., Madrid (ISRM), V1, p189

7 Rosenblad, J.L. (1970), "Failure modes of models of jointed rock masses", Proc. 2nd Cong. ISRM, Belgrade, V2, paper 3-11

4,5 Rosengren, K.J. (1968), "Rock mechanics of the Black Star open cut, Mount Isa" PhD Thesis, Australian National Univ., Canberra

4 Rosengren, K. (1970), "Diamond drilling for structural purposes at Mt. Isa" Industrial Diamond Review, p388

4 Ross-Brown, D.M., and Atkinson, K.B. (1972), "Terrestrial photogrammetry in open pits; part 1 - description and use of the photo theodelite in mine surveying", Trans. IMM, Sect. A, V81, p205

4 Ross-Brown, D.M., Wickens, E.H., and Markland, J. (1973), "Terrestrial photogrammetry in open pits; part 2 - an aid to geol. mapping" Trans. IMM, Sect. A, V82, p115

4 Roxtrom, F.B. (1961), "New core orientation device" Econ. Geol. V56, p1310

4 Roxtrom, F.B. (1961), "Craelius automatic core orienter, inclined drill holes", Can. Min. J., V82, p60

5 Rowe, P.W. (1962), "The stress dilatancy relation for static equilibrium of an assembly of particles in contact", Proc. Royal Soc., London, A., V269, p500

5 Rowe, P.W., Barden, L., and Lee, I. K. (1964), "Energy components during the triaxial cell and direct shear test", Geotechnique, V14, p247

6 Ruiz, M.D. (1974), "Minimum anchoring cost for stabilization of rock slopes" Proc. 3rd Cong. ISRM, Denver, V2B, p813

5 Ruiz, M., and Camargo, F. (1966), "A large scale field test on rock", Proc. 1st Cong. ISRM, Lisbon, V1, p257

5 Ruiz, M.D., Camargo, F.P., Midea, N.F., and Nieble, C.M. (1968), "Some considerations regarding the shear strength of rock masses",

Proc. Int. Symp. on Rock Mech., Madrid (ISRM), paper 2-5, p159

2 Ruxton, B.P., and Berry, L. (1957), "Weathering of granite and associated erosional features in Hong Kong", Bull. G.S.A., V68, p1263

2 Rzhevsky, V., and Novik, G. (1971), "The physics of rocks", translated by A.K. Chaterjee (Mir publishers, Moscow)

4 Sabarly, F. (1965), "Mesures des permeabilities en place", a report prepared by Geoconeil, 31, Route de Versailles, 78170 La Celle, Saint Cloud, France

2 Sahores, J. (1962), "Contribution to the study of the mech. phen. accompanying hydration of anhydrite" (in French), Centre d'etudes et de Recherche de L'Industries des Liants Hydrauliques, Pub. Tech. n126

8 St. John, C.M. (1971), "Numerical and observational methods of determining the behaviour of rock slopes in open cast mines", Imperial College, PhD Thesis.

8 St. John, C. (1972), "Finite element analyses of two and three dimensional jointed structures - computer programmes" Imperial Coll. Rock Mech. Res. Rep. n13 and appendix

4 Savage, J.F. (1965), Terrestrial photogrammetry for geol. purposes", Int. Train. Centre for Aerial Survey (ITC), Delft, publ. series B, n33, p41

2 Schneider, B. (1967), "Moyens nouveaux de reconnaissance des massifs rocheux", Annales de l'Inst. Tech. du. Batiments et des Travaux Public, Supp, V20, p1055

5 Schneider, H.J. (1974), "Rock friction - a laboratory invest." Proc. 3rd Cong. ISRM, Denver, V2A, p311

2,5 Scott, R.F. (1963), "Prin. of soil mechanics" (Addison-Wesley)

2 Selmer-Olsen, R. and Blindheim, O.T. (1970), "On the drillability of rock by percussive drilling", Proc. 2nd Cong. ISRM, Belgrade, V2, paper 5-8

2 Serafim, J.L. (1964), "Rock mechanics considerations in the design of concrete dams", in State of Stress in the Earth's Crust, Judd, ed. (Elsevier), p611

5 Serafim, J.L. and del Campo, A. (1965), "Interstitial pressures on rock foundations of dams", J.SM&FD, ASCE, V91, p65

5 Serafim, J.L., and Lopes, J.B. (1961), "In-situ shear tests and triaxial tests on foundation rocks of concrete dams", Proc. 5th Int. Conf. on Soil. Mech. and Found. Eng., Paris, V1, p533

8 Serata, S. (1970), "Prerequisites for applic. of finite element method to solution cavities and conventional mines", Proc. 3rd Symp.

on Salt, Rau and Delwig, eds. (Northern Ohio Geol. Soc., Cleveland) V2, p249

3 Shanley, R.J., and Mahtab, M.A. (1974), "A computer program for clustering data points on the sphere", U.S.B.Mines Inf. Circ. 8624

3 Shanley, R.J., and Mahtab, M.A. (1975), "FRACTRAN: a computer code for analysis of clusters defined on the unit hemisphere" U.S.B. Mines Inf. Circ. 8671

5 Sharp, J.C., and Maini, Y.N.T. (1972), "Fund. considerations on the hyd. char. of joints in rock", Proc. Symp. on Perc. Through Fissured Rock, Stuttgart (ISRM)

2,4 Sherard, J.L., Cluff, L.S., and Allen, C.R. (1974), "Potentially active faults in dam foundations", Geotechnique, V24, p367

4 Simpson, B. (1968), "Geological Maps", (Pergamon)

5 Skempton, A.W. (1954), "The pore pressure coefficients A and B" Geotechnique, V4, p143

4 Slichter, L.B. (1955), "Geophysics applied to prospecting for ores" Economic Geology, 50th Anniversary Volume, Part II, p885

4,5 Snow, D.T. (1965), "A parallel plate model of permeable fractured media", PhD Thesis, Univ. of California, Berkeley

4,5 Snow, D.T. (1968), "Rock fracture spacings, openings, and porosities" J.SM&FD ASCE, V94, p73

4,5 Snow, D.T. (1968), "Fracture deformation and change of permeability and storage upon change of fluid pressure" Qtly, Colo. School of Mines, V63,n1

4,5 Snow, D.T. (1970), "The frequency and apertures of fractures in rock", IJRM&MS, V7,p23

2 Spears, D.D., and Taylor, R.K. (1972), "The influence of weathering on the composition and engineering properties of in-situ coal measures rocks", IJRM&MS, V9, p729

8 Stacey, T.R. (1969), "Application of the finite element method in the field of rock mechanics with particular reference to slope stability" S. Afr. Mech. Eng., n5, p131

8 Stacey, T.R. (1970), "The stresses surrounding open pit mine slopes", Proc. Symp. on Planning Open Pit Mines, Johannesburg

8 Stagg, K.G., and Treharne, G. (1970), "Some experiments on the ultimate bearing capacity of rock=type materials", Proc. 2nd Cong. ISRM Belgrade, V1, paper 2-33

4 Stahl, R.L. (1973), "Detection and delineation of faults by sur-

face resistivity measurements - Gas Hills region, Wyoming", U.S.B.Mines R.I. 7824

4 Stahl, R.L. (1974), "Detection and delineation of faults by surface resistivity measurements - Schwartzsalder Mine, Jefferson Co, Colo." U.S.Bureau of Mines R.I. 7975

3 Starkey, J. (1970), "A computer programme to construct spherical projections", Experimental and Natural Rock Deformation (Springer)

2 Steffen, O.K.H. and Jennings, J.E. (1972), "Definition of design joints for two dimensional rock slope analysis" Proc. 3rd Cong. ISRM, Denver, V2B, p827

8 Stephansson, O. (1971), "Stability of single openings in horizontally bedded rock", Engineering Geology, V5, p5

7 Stimpson, B. (1970), "Modelling materials for engineering rock mechanics", IJRM&MS, V7, p77

4,5 Stimpson, B., Metcalfe, R.G., and Walton, G. (1970), "A new technique for sealing and packing rock and soil samples" Qtly J. Eng Geol. V3, p127

1 Talobre, J.A. (1967), "La mécanique des roches et ses applications", second edition (Dunod) - first edition: 1957

8 Tardieu, B., and Pouyet, P. (1974), "Proposition d'un modèle de joint tridimensionnel courbe", Proc. 3rd Cong. ISRM, Denver, V2B, p833

2,5 Taylor, R.K. (1973), "Physical and physical-chemical changes in coal measure rocks", PhD Thesis, Univ. of Durham (U.K.)

2 Terzaghi, K. (1946), "Introduction to tunnel geology" in Rock Tunneling with Steel Supports, by Proctor and White (Commercial Shearing and Stamping Co., Youngstown, Ohio)

3 Terzaghi, R.D. (1965), "Sources of error in joint surveys" Geotechnique, V15, p287

8 Thompson, E.G., and Sayles, F.H. (1972), "In-situ creep analysis of room in frozen soil", J.SM&FD, ASCE V98, p899

8 Tincelin, E., Fine, J., and Vouilie, C.T. (1972), "A comparison of forecasts calculated by the finite element method and the actual behaviour of mine workings using the small pillar method", Proc. 5th Int. Strata Control Conf., London

2 Tourenq, C. (1967), "Mise en evidence de la microfissuration des roches", Compte Rendu des seances de la Societé Géol. de France, V22

2 Tourenq, C., and Denis, A. (1970), "La résistance à la traction des roches" Lab. des Ponts et Chaussées, Paris, Rapport de Recherche 4

2 Tourenq, C., and Fourmaintraux, D. (1974), "L'indice de qualité des roches, quelques applications" Proc. 2nd Cong IAEG, Sao Paulo, V1, paper 4-20

2 Tourenq, C., Fourmaintraux, D., and Denis, A. (1971), "Propagation des ondes et discontinuités des roches", Proc. Int. Symp. on Rock Fracture, Nancy (ISRM)

7 Tufo, M. (1971), "Preliminary investigation and construction of a three dimensional geostatical model of the Grancarevo Dam", Publ. of the Inst. of Geotechnics and Found. Eng., Faculty of Civil Engineering, Sarajevo, Yugoslavia

4 Trantina, J.A., and Cluff, L.S. (1963), "NX bore hole camera" ASTM STP 351

7 Trollope, D.H. (1966), "The stability of trapezoidal openings in rock masses" Rock Mech. and Eng. Geol. V4, p232

3,4 Turner, F.J., and Weiss, L.E.(1963), "Structural analysis of metamorphic tectonites - Chapter 3 'Graphic treatment of fabric data'" (McGraw-Hill)

2 Underwood, L.B. (1967), "Classification and identification of shales", J SM&FD, ASCE V93, p97

8 Valliappan, S. (1968), "Non-linear stress analysis of two dimensional problems with special reference to rock and soil mechanics", PhD Thesis, Univ. Wales, Swansea (Dept of Civ. Eng.)

4 Van Nostrand, R.G., and Cook, K.L. (1966), "Interpretation of resistivity data", U.S.G.S. Prof. Paper 499

2 Vargas, M. (1953), "Some engineering properties of residual clay soils occurring in Southern Brazil", Proc. 3rd Int. Conf. on Soil Mech. and Found. Eng., Zurich, V1, p67

2 Verdier, J. (1969a), "Examen ultramicroscopique de quelques roches métamorphiques de sites de barrages", 2eme Colloque Sur La Fissuration des Roches, Revue de L'Industrie Minerale, Numero Special 15 July

Verdier, J. (1969b), "Fine microscope and ultramicroscope examination of a few metamorphic rocks from dam sites" Eng. Geol., V3, p233

2 Verdier, J. and Deicha, G. (1971), "La répartition des inclusions fluides épigénetiques comme témoin de la microfissuration des roches" Proc. Int Symp on Rock Fracture, Nancy (ISRM) paper 1-10

4 Voloshin, V., Nixon, D.D., and Timberlake, L.L. (1968), "Oriented core, a new technique in engineering geology", Bull. A.E.G., V5, p37

2,5 Vutukuri, V.S., Lama, R.D., and Saluja, S.S. (1974), "Handbook on mechanical properties of rocks - - testing techniques and results" (TransTech Publications)

8 Waddell, G.G. (1971), "Technique of measuring initial deformation around an opening, analysis of two raise bore tests", U.S.B.Mines R.I. 7505

8 Waddell, G.G., Crocker, T.J., and Skinner, E.H. (1971), "Tunnel relaxation method for determining the initial and long term deformation around an underground opening", Proc. 12th Symp. on Rock Mechanics, (AIME), p903

1 Wahlstrom, E.E. (1973), "Tunneling in rock" (Elsevier)

7 Walker, P.E. (1972), "The shearing behaviour of a jointed rock model", Ground Eng., V5, p24

4 Wallace, R.E. (1950), "Determination of dip and strike by indirect observations in the field and from aerial photographs, J. Geol., V58, p269

5 Wallace, G.B., and Olsen, O.J. (1966), "Foundation testing techniques for arch dams and underground power plants" Testing Tech. for Rock Mech., ASTM STP 402, p272

2 Walsh, J. (1965a), "The effect of cracks on the compressibility of rock", J. Geoph. Res. V70, n2

2 Walsh, J. (1965b), "The effect of cracks on the uniaxial elastic compression of rock", J. Geoph. Res. V70, n2

2 Walsh, J.D. (1974), "Detection of rock defects by fluorescence photography", CSIRO, Australia, Div. of Applied Geomech., Tech.Memo.12

8 Wang, F.O., Panek, L.A., and Sun, M.C. (1971), "Stability analysis of underground openings using a Coulomb-Navier shear failure criterion" Trans. SME of AIME, V250, p317

8 Wang, F.D., Sun, M.C., and Ropchan, D.M. (1972), "Computer program for pit slope stability analysis by the finite element stress analysis and limiting equilibrium method" U.S.B.Mines R.I. 7685

4 Warren, C.R., Schmidt, D.L., Denny, C.S., and Dale, W.J. (1969), "A descriptive catalog of selected aerial photographs of geologic features in areas outside the U.S.", U.S.G.S. Prof. Paper 591

2 Watkins, M.D. (1970), "Terminology for describing the spacing of discontinuities of rock masses" Qtly J. Eng. Geol. V3, p193

3 Watson, G.S. (1966), "The statistics of orientation data", J. Geol., V74, p786

5 Wawersik, W.R. (1975), "Technique and apparatus for strain measurements on rock in constant confining pressure experiments", Rock

Mech. (in press)

5 Wawersik, W.R., and Brown, W.S. (1973), "Creep fracture of rock" Report from Univ. Utah to U.S.B.Mines,(AD-738-002)

2 Weinbrandt, R.M., and Fatt, I. (1970), "Scanning electron microscope study of the pore structure of sandstone", Proc. 11th Symp. on Rock Mech. (AIME), p629

2 Weinert, H.H. (1964), "Basic igneous rocks in road foundations" Bull. National Inst. of Road Res. V2,p1 (S.Afr.)

2 Weinert, H.H. (1965), "Climatic factors affecting the weathering of igneous rocks", Agr. Meter. V2, p27

8 West, L.J., and Perry, R.M. (1969), "Rock mechanics studies for a high cut at a nuclear power station in Penn." Proc. 11th Symp. on Rock Mech.

4,6 Wickens, E.H., and Barton, N.R. (1971), "Application of photogrammetry to the stability of excavated rock slopes", Photogrammetric Record, (Photogrammetry Soc. London) V6, n37

4 Wiebenga, W.A. and Polak, E.J. (1969), "Great Lake North engin. Geophysical surveys, Tasmania, 1951-1959", Australia Dept. of National Development, Bureau of Min. Resources, Geol, and Geophysics, Bull 101

2 Williams, H., Turner, F., and Gilbert, C. (1958), "Petrography", (Freeman)

4 Williams, J.C.C. (1969), "Simple photogrammetry" (Academic Press)

8 Wilson, E.L. (1972), "Solid SAP, a static analysis program for three dimensional solid structures", Univ. of Cal., Berkeley, Dept. of Civil Eng., Div. of S.E.S.M. Report

4 Wilson, G. (1967), "The geometry of cylindrical and conical folds" Proc. of the Geologists' Association, V78, p179

8 Winkel, B.V., Gerstle, K.H., and Ko, H.Y. (1972), "Analysis of time dependent deformations of openings in salt media", IJRM&MS, V9, p249

5 Withers, J.H. (1964), "Sliding resistance along discontinuities in rock masses", PhD Thesis, Univ. of Illinois

6 Wittke, W. (1965), "Methods to analyze the stability of rock slopes with and without additional loading" (in German), Rock Mech. and Eng. Geol., Supp. II, p52

6 Wittke, W. (1966), "Berechnungsmöglichkeiten den Standsicherheit von Boschungen in Fels",(Deutsche Gesellschaft fur Erd un Grundbau)

6 Wittke, W. (1967), "Influence of the shear strength of the joints

on the design of prestressed anchors to stabilize a rock slope", Proc. Geotech. Conf., Oslo, paper 4-11, p311

8 Wittke, W., Rissler, P., and Semprich, S. (1972), "Three dimensional and turbulent flow through fissured rock according to discontinuous and continuous models" (in German), Proc. Symp. on Percolation Through Fissured Rock, Stuttgart (ISRM)

8 Wittke, W., Rodatz, W., and Wallner, M. (1972), "Three dimensional calculation of the stability of caverns, slopes, and foundations in anisotropic, jointed rock, by means of the finite element method", Deutsche Geotechnik, V1, n1

8 Wright, F.D. (1972), "Arching action in cracked roof beams", Proc. 5th Int. Strata Control Conf., London

8 Wylie, C.R. Jr. (1960), "Advanced engineering mathematics", (McGraw-Hill)

8 Yoshida, M., and Yoshimura, K. (1970), "Deformation of rock mass and stress in concrete lining around the machine hall of Kisenyama underground power plant" Proc. 2nd Cong. ISRM, Belgrade, V2, paper 4-29

8 Zienkiewicz, O.C. (1967), "The finite element method in structural and continuum mechanics" (McGraw-Hill)

8 Zienkiewicz, O.C. (1970), "Three dimensional finite element analysis of the Tachien arch dam", Water Power, V22, p173

8 Zienkiewicz, O.C. (1971), "The finite element method in engineering science" (McGraw-Hill)

8 Zienkiewicz, O.C., Best, B., Dullage, C., and Stagg, K.G. (1970), "Analysis of non-linear problems in rock mechanics with particular reference to jointed rock systems", Proc. 2nd Cong. ISRM, Belgrade, V3, paper 8-14

8 Zienkiewicz, O.C., and Cheung, Y,K. (1966a), "Application of the finite element method to problems of rock mechanics", Proc. 1st Cong. ISRM, Lisbon, V1, p661

8 Zienkiewicz, O.C., and Cheung, Y.K. (1966b), "Stress in anisotropic media with particular reference to problems of rock mechanics - - finite element analysis" J. of Strain Analysis, V1, p172

5 Zienkiewicz, O.C., and Stagg, K.G. (1966), "The cable method of in-situ testing", Proc. 1st Cong. ISRM, Lisbon, V1, p667

8 Zienkiewicz, O.C., Watson, M., and King, I.P. (1968), "A numerical method of viscoelastic stress analysis", IJRM&MS, V10, p807

8 Zienkiewicz, O.C., Valliappan, S., and King, I.P. (1968), "Stress analysis of rock as a no tension material" Geotechnique, V18, p56

4 Zimmer, P.W. (1963), "Orientation of small diameter drill core", Economic Geology, V58, p1313

ABBREVIATIONS

A.E.G.	Association of Engineering Geologists
A.I.M.E.	American Institute of Mining and Metallurgy
A.S.C.E.	American Society of Civil Engineers
A.S.M.E.	American Society of Mechanical Engineers
A.S.T.M.	American Society for Testing and Materials
Can.	Canadian
C.S.I.R.O.	Central Scientific and Industrial Research Organization (Australia)
G.S.A.	Geological Society of America
I.A.E.G.	International Association of Engineering Geologists
IJRM&MS	International Journal of Rock Mechanics and Mining Science
I.M.M.	Institution of Mining and Metallurgy (London)
J.	Journal
R.I.	Report of Investigation
S.E.S.M.	Structural Engineering and Structural Mechanics (University of California, Dept. of C.E.)
S.M.E.	Society of Mining Engineers (AIME)
SM&FD	Soil Mechanics and Foundations Division (ASCE)
S.T.P.	Special Technical Publication
U.S.B.Mines	United States Bureau of Mines, U.S. Department of the Interior
U.S.G.S.	United States Geological Survey, U.S. Department of the Interior

*

subject index

author index